Progress in Nonlinear Differential Equations and Their Applications
Volume 15

Topological Nonlinear Analysis
Degree, Singularity, and Variations

Michele Matzeu
Alfonso Vignoli
Editors

Birkhäuser
Boston • Basel • Berlin

Editors:
Michele Matzeu
Department of Mathematics
University of Rome, Tor Vergata
00133 Rome, Italy

Alfonso Vignoli
Department of Mathematics
University of Rome, Tor Vergata
00133 Rome, Italy

Library of Congress Cataloging-in-Publication Data

Topological nonlinear analysis : degree, singularity, and variations /
    Michele Matzeu, Alfonso Vignoli, editors
        p.   cm. -- (Progress in nonlinear differential equations and
    their applications ; v. 15))
        Includes bibliographical references.

ISBN-13:978-1-4612-7584-8   e-ISBN-13:978-1-4612-2570-6
DOI: 10.1007/978-1-4612-2570-6

    1. Nonlinear functional analysis.   2. Topological algebras.
I. Matzeu, M. (Michele)   II. Vignoli, Alfonso, (1940-
III. Series.
QA321.5.T67 1994   94-5251
515.'355--dc20   CIP

ISBN-13:978-1-4612-7584-8

Reformatted from authors' disks.

9 8 7 6 5 4 3 2 1

Contents

Preface

Topological tools in Nonlinear Analysis had a tremendous development during the last few decades. The three main streams of research in this field, Topological Degree, Singularity Theory and Variational Methods, have lately become impetuous rivers of scientific investigation. The process is still going on and the achievements in this area are spectacular.

A most promising and rapidly developing field of research is the study of the role that symmetries play in nonlinear problems. Symmetries appear in a quite natural way in many problems in physics and in differential or symplectic geometry, such as closed orbits for autonomous Hamiltonian systems, configurations of symmetric elastic plates under pressure, Hopf Bifurcation, Taylor vortices, convective motions of fluids, oscillations of chemical reactions, etc... Some of these problems have been tackled recently by different techniques using equivariant versions of Degree, Singularity and Variations.

The main purpose of the present volume is to give a survey of some of the most significant achievements obtained by topological methods in Nonlinear Analysis during the last two-three decades. The survey articles presented here reflect the personal taste and points of view of the authors (all of them well-known and distinguished specialists in their own fields) on the subject matter. A common feature of these papers is that of starting with an historical introductory background of the different disciplines under consideration and climbing up to the heights of the most recent results. As a consequence, we obtain a very dynamic picture of the state of affairs. Actually, we hope to be able in the near future to involve other distinguished specialists to get their own versions on these topics. Most probably a never-ending fascinating tale!

Finally let us mention the fact that most of the material of this book was presented by the authors at the Topological Analysis Workshop on Degree, Singularity and Variations, held in May 1993 at Villa Campitelli, Frascati, near Rome.

Contributors

Antonio Ambrosetti, Scuola Normale Superiore, Piazza dei Cavalieri 7, 56100 Pisa, Italy

Vieri Benci, Istituto di Matematiche Applicate "U. Dini", Università di Pisa, 56100 Pisa, Italy

James Damon, Department of Mathematics, University of North Carolina, Chapel Hill, NC 27514, USA

E. N. Dancer, Department of Mathematics, Statistics and Computing Science, The University of New England, Armidale, NSW 2351, Australia

Jorge Ize, Departamento de Matemáticas y Mecánica, ITMAS-UNAM, Apartado Postal 20-726, D. F. Mexico 20

Michele Matzeu, Dipartimento di Matematica, Università degli Studi di Roma, "Tor Vergata", Via della Ricerca Scientifica, 00133 Rome, Italy

Paul H. Rabinowitz, Department of Mathematics, University of Wisconsin-Madison, Van Vleck Hall, 480 Lincoln Drive, Madison, WI 53706-1388

Alfonso Vignoli, Dipartimento di Matematica, Università degli Studi di Roma, "Tor Vergata", Via della Ricerca Scientifica, 00133 Rome, Italy

Claude Viterbo, Département de Mathématique, Bât. 425, Université Paris-Sud, F-91405 Orsay Cedex, France

Variational Methods and Nonlinear Problems: Classical Results and Recent Advances

Antonio Ambrosetti

Scuola Normale Superiore
56100 Pisa, Italy

1. Introduction

Around the end of the Twenties two memoires, a first one by Morse [63] and a second one by Lusternik and Schnirelman [59], marked the birth of those variational methods known under the name of *Calculus of Variation in the Large*. These tools are mainly concerned with the existence of critical points, distinct from minima, which give rise to solutions of nonlinear differential equations. The elegance of the abstract tools and the broad range of applications to problems that had been considered of formidable difficulty, such as the existence of closed geodesics on a compact anifold or the problem of minimal surfaces, have rapidly made the Calculus of Variation in the Large a very fruitful field of research.

The natural fields of application of these theories are all those nonlinear variational problems where the search of minima is not satisfactory, or useless, or else impossible, and thus solutions have to be found by means of min-max procedures.

In particular, the Lusternik–Schnirelman theory of critical points has been largely emploied in Nonlinear Analysis because it does not require any *a-priori* nondegeneracy assumption. An important progress was marked by the works of Browder [35], Krasnoselski [55], Palais [65], Schwartz [76] and Vainberg [87] who extended the theory to infinite dimensional manifolds. For example, a typical result is that a C^1 functional f has infinitely many critical points on any infinite dimensional Hilbert sphere S, provided: (i) f is bounded from below on S; (ii) f satisfies the Palais-Smale, (PS) for short, compactness condition; and (iii) f is even. These critical points give rise to solutions of nonlinear elliptic eigenvalue problems. A short review of these classical results is made in Sections 2 and 3.

A further step forward was taken by studying critical points of functionals which are possibly unbounded from below. The *Mountain-Pass* Theorem and the *Linking* Theorems establish that critical points can be found by means of appropriate min-max procedures, provided f satisfies the (PS) condition and some suitable geometric assumptions. An extension of the Lusternik–Schnirelman theory which covers a class of symmetric functionals, unbounded from below, has also been obtained. These abstract tools will be discussed in Section 4, while some applications to semilinear Elliptic Boundary value Problems are given in Section 5. Here we limit ourselves to an overview of some more classical results concerning the existence and multiplicity of solutions for sub-linear and super-linear Dirichlet boundary problems, as well as to give some indications on further topics of great interest, where the research is still very active.

Finally, in Section 6 we briefly discuss some recent advances concerning periodic motions of second order Hamiltonian systems with singular potentials, like those arising in Celestial Mechanics. This also proved to be a topic where the power of variational methods apply and, although several interesting results have been obtained in these last years, many important problems are still left open and will require a great deal of future researches.

This survey is necessarily incomplete. The interested reader can find other material on variational methods in the books [44], [62], [82] and in the survey papers [64], [6].

2. Lusternik–Schnirelman Theory

Let M be a Riemannian $C^{1,1}$ manifold modelled on a Hilbert space E and let $f \in C^1(M, R)$. We will use the following notation:

$$f^b = \{u \in M \ : \ f(u) \leq b\}$$
$$f_a = \{u \in M \ : \ f(u) \geq a\}$$
$$f_a^b = \{u \in M \ : \ a \leq f(u) \leq b\}$$

A critical points of f on M is a $u \in M$ such that $df(u) \neq 0$. We set $K = \{u \in M \ : \ f'(u) = 0\}$ and $K_c = \{u \in K \ : \ f(u) = c\}$. We say that c is a critical level of f on M whenever $K_c \neq \emptyset$.

The following examples show that critical points are naturally related with differential problems.

Example 2.1. Let $\mathcal{V}$ be a compact Riemannian manifold with a Riemannian structure $< \cdot, \cdot >$. The problem of finding closed geodesics on $\mathcal{V}$

amounts to look for critical points of the functional

$$f(u) = \int <\dot{u}, \dot{u}> \, ds$$

on the manifold $\Lambda^1(\mathcal{V}) = H^{1,2}(S^1, \mathcal{V})$. Let us point out that f achieves its minimum on the constant loops and thus non-constant closed geodesics correspond to min-max critical level of f on $\Lambda^1(\mathcal{V})$.

Example 2.2. Let $\Omega \subset R^N$ be a bounded domain with smooth boundary $\partial \Omega$ and let $H_0^1 = H_0^{1,2}(\Omega)$ denote the usual Sobolev space, endowed with the Dirichlet norm

$$\|u\|^2 = \int_{\Omega} |\nabla u|^2 \, dx.$$

Let $F : R \to R$ be a C^1 function such that

$$F(u) \in L^1(\Omega) \quad \forall u \in H_0^1, \tag{2.1}$$

and consider the functional $f : H_0^1 \to R$,

$$f(u) = \int_{\Omega} F(u(x)) dx.$$

If u is a critical point of f on the Hilbert sphere $M = \{u \in H_0^1 : \|u\|^2 = 1\}$, then according to the Lagrange multiplier rule, there exists $\lambda \in R$ such that

$$\int_{\Omega} F'(u)v \, dx = \lambda \int_{\Omega} \nabla u \cdot \nabla v \, dx, \quad \forall v \in H_0^1,$$

Hence u is a weak solution of the nonlinear eigenvalue problem

$$\begin{cases} -\lambda \Delta u = F'(x, u), & x \in \Omega, \\ u = 0 & x \in \partial \Omega \end{cases} \tag{2.2}$$

where Δ denotes the Laplace operator. Let us point out that when $F(u) = |u|^2$, (2.2) becomes a linear eigenvalue problem, that possesses infinitely many characteristic values $\mu_k > 0$ (or infinitely many eigenvalues $\lambda_k = 1/\mu_k$) and infinitely many corresponding eigenvectors e_k, which are nothing but the critical points of the quadratic functional $\int_{\Omega} |u|^2 \, dx$ on M. We will see that such a multiplicity result holds true for a broad class of even functions F. In any case, it is clear that, apart from the minimum, other critical points have to be found by min-max procedure.

In order to find critical points of f on M one defines the *category*.

a) *The Lusternik–Schnirelman category.* Let M be a topological space and $X \subset M$. We say that X is contractible in M if there exists a homotopy $h \in C([0,1] \times X, M)$ such that

$$\begin{cases} h(0,u) = u, & \forall u \in X \ ; \\ h(1,u) = u_0 \in M, & \forall u \in X. \end{cases}$$

The (Lusternik–Schnirelman) category of X relative to M, $cat(X, M)$, is the least integer $n \geq 1$ such that there exist n subsets $X_1, \ldots, X_n$ of M such that for all $i = 1, \ldots, n$,

    1. X_i is closed in M;

    2. $X \subset X_1 \cup \ldots \cup X_n$;

    3. X_i is contractible in M.

We set $cat(\emptyset, M) = 0$ and define $cat(X, M) = +\infty$ if there there are no integers with the above properties. We also set $cat(M) = cat(M, M)$.

Example 2.3. Let $S^m = \{x \in R^{m+1} \ : \ |x| = 1\}$. Clearly, $cat(S^m) = 2$. On the contrary, if we consider the manifold $P^m = S^m/\mathbf{Z}_2$ obtained indentifying antipodal points on S^m then, using the Borsuk-Uhlam Antipodensatz, one can show that

$$cat(P^m) = m + 1.$$

Moreover, if E is an infinite dimensional, separable Hilbert space, and $S = \{u \in E \ : \ \|u\| = 1\}$ then $cat(S) = 1$ while, letting $P^\infty = S/\mathbf{Z}_2$, one finds that $cat(P^\infty) = +\infty$.

Example 2.4. Let $\Lambda = \{u \in H^{1,2}(S^1, R^n) \ : \ u(t) \neq 0\}$ denote the manifold of H^1 loops which do not cross the origin $x = 0$. If $n = 2$, Λ has infinitely many connected components Λ_k, each containing the loops whose degree (i.e. the Poincaré rotation number) is k. More in general, one can show that for all $n \geq 2$ ther results: $cat(\Lambda) = +\infty$, see [46].

Among the properties of the category let us recall the following ones:

    (i) $X \subset Y \subset M \Rightarrow cat(X, M) \leq cat(Y, M)$;

    (ii) $cat(X \cup Y, M) \leq cat(X, M) + cat(Y, M)$;

    (iii) If X is closed and φ is a deformation of X in M, i.e. $\varphi \in C(X, M)$ is homotopic to the inclusion map, then $cat(X, M) \leq cat(\varphi(X), M)$;

    (iv) If M is a manifold (or, more im general an ANR, see [65]) and $K \subset M$ is compact, then $cat(K, M) < \infty$ and there exists a neighbourhood U of K such that $cat(K, M) = cat(U, M)$.

b) *Min-max levels.* The category can be used to define critical levels of min-max type. Let us set

$$\mathcal{X}_k = \{X \subset M \ : \ X \text{ is closed and } cat(X, M) \geq k\}$$

and define

$$c_k = \inf_{X \in \mathcal{X}_k} \sup_{u \in X} f(u). \tag{2.3}$$

Plainly one has that $c_k \leq c_{k+1}$. We will show that, under suitable assumptions, the category of M provides a lower bound for the number of critical points of a functional f constrained on M.

In order to carry out this program, we need to introduce a compactness condition and to define deformations.

We say that the pair (f, M) satisfies the $(PS)_c$ condition (in some case we will simply say that $(PS)_c$ holds) if every sequence $u_n \in M$ such that $f(u_n) \to c$ and $df(u_n) \to 0$, has a convergent subsequence. When $(PS)_c$ holds for all $c \in R$, we say that (PS) holds.

As for deformations, they will roughly be nothing but the steepest descent flow of f. Actually, to deal with C^1 functionals, it is convenient to introduce the notion of *Pseudo-Gradient Vector Field*. Let M be a Finsler manifold of class $C^{1,1}$, $f \in C^1(M, R)$ and let $M^* = \{u \in M \ : \ df(u) \neq 0\}$.

Lemma 2.5. *There exists a Lipschitz vector field* $\chi : M^* \to TM$ *such that*

1. $\|\chi(u)\|_u < 2\|df(u)\|_u^*$;
2. $\langle df(u), \chi(u) \rangle > \|df(u)\|_u^{*2}$;

where $\|\cdot\|_u$ *and* $\|\cdot\|_u^*$ *denote the norm in* $T_u M$ *and in the cotangent bundle, respectively.*

The above χ is called a *pseudo-gradient vector field* for f.

Let us first suppose that $c \in R$ is not a critical level, namely that $K_c = \emptyset$. We set

$$A = f^{c-2\epsilon} \cup f_{c+2\epsilon}$$
$$B = f_{c-\epsilon}^{c+\epsilon}$$
$$h(u) = \frac{d(u, A)}{d(u, A) + d(u, B)}$$

and consider the solution $\sigma = \sigma(t, u)$ of the Cauchy problem

$$\begin{cases} \frac{d}{dt}\sigma(t, u) = -h(\sigma(t, u))\frac{\chi(\sigma(t,u))}{\|\chi(\sigma(t,u))\|} \\ \sigma(0, u) = u. \end{cases} \tag{2.4}$$

One readily verifies that (2.4) defines a flow $\sigma \in C([0, +\infty), M)$ and there results

$$\frac{d}{dt} f(\sigma(t, u)) = \langle df(\sigma(t, u)), \sigma'(t, u)\rangle$$

$$= -\frac{h(\sigma(t, u))}{\|\chi(\sigma(t, u))\|} \cdot \langle df(\sigma(t, u)), \chi(\sigma(t, u))\rangle \leq 0,$$

by the property (1) of the pseudogradient χ and since $h \geq 0$. Moreover, suppose that $(PS)_c$ holds. Then there exists $\epsilon > 0$ such that

$$\|df(u)\| \geq 4\epsilon, \qquad \forall u \in f_{c-\epsilon}^{c+\epsilon}. \tag{2.5}$$

Let $\sigma(t, u) \in f_{c-\epsilon}^{c+\epsilon}$. Using the properties of the pseudogradient, one finds that

$$f(\sigma(t, u)) = f(u) + \int_0^t \frac{d}{dt} f(\sigma(t, u)) dt$$

$$= f(u) - \int_0^t \frac{1}{\|\chi(\sigma(t, u))\|} \cdot \langle df(\sigma(t, u)), \chi(\sigma(t, u))\rangle$$

$$< f(u) - \frac{1}{2} \int_0^t \|df(\sigma(t, u))\| dt,$$

and (2.5) implies

$$f(\sigma(1, u)) \leq c - \epsilon \qquad \forall u \in f^{c+\epsilon}.$$

In other words, $\eta(u) = \sigma(1, u)$ defines a deformation in M such that $\eta(u) \in f^{c-\epsilon}$ whenever $u \in f^{c+\epsilon}$. This argument can be extended, with some minor changes, to cover the case in wich K_c is not empty, yielding

Lemma 2.6. (Deformation Lemma) *Assume that M is a $C^{1,1}$ manifold and that $f \in C^1(M, R)$. Let U denote any neighbourhood of K_c and suppose that $(PS)_c$ holds.*

Then there exist $\epsilon > 0$, small enough, and a deformation $\eta \in C(M, M)$ such that

1. $\eta(u) = u \qquad \forall u \in f^{c-2\epsilon} \cup f_{c+2\epsilon}$;
2. $f(\eta(u)) \leq f(u) \qquad \forall u \in M$;
3. $f(\eta(u)) \leq c - \epsilon \qquad \forall u \in f^{c+\epsilon} \setminus U$.

As a first consequence of the Deformation Lemma one can prove

Theorem 2.7. *Let M be a $C^{1,1}$ manifold and $f \in C^1(M, R)$. Suppose that (PS) holds and that f is bounded from below and from above on M. Then $cat(M) < +\infty$.*

Roughly, under the above assumptions, the critical set K is compact and all M can be deformed into any neighbourhood U of K. Then the Theorem follows from the properties of category.

c) *A Theorem of the LS type.* The preceding arguments allow us to estimate the number of critical points by means of the category.

Theorem 2.8. *Let M be a $C^{1,1}$ manifold, $f \in C^1(M, R)$ and let (PS) holds. Then*

1. *for all $1 \le k \le cat(M)$, any finite c_k is a critical level;*
2. *if $c_k = +\infty$ for some k, then $\sup\{f(u) : u \in K\} = +\infty$;*
3. *if $c_k = c_{k+1} = \ldots = c_{k+m}$ for some $m \ge 1$, then $cat(K_{c_k}, M) \ge m + 1$.*

In particular, if f is bounded from below on M, then f has at least $cat(M)$ critical points on M.

As a first application, let us take as M a unit sphere $M = \{u \in E : \|u\| = 1\}$, where E is a separable, infinite dimensional Hilbert space and let consider the corresponding infinite dimensional projective space $P^\infty = M/\mathbf{Z}_2$. Recall that, according to Example 2.3, one has $cat(M) = +\infty$. Suppose that $f \in C^1(E, R)$ is even, in such a way that it induces a functional, still denoted by f, on P^∞. Then Theorem 2.8 yields

Theorem 2.9. *Let E be a separable, infinite dimensional Hilbert space and let $f \in C^1(E, R)$ be even. Suppose that f is bounded below on $M = \{u \in E : \|u\| = 1\}$ and that (f, M) satisfies (PS). Then f has infinitely many (pairs) of critical points on M.*

d) *Critical points in presence of symmetry.* When one deals with problems which inherit a symmetry, one can make use of a more direct approach. Precisely, let us consider the case in which the manifold M and the functional f are invariant under the action of a compact topological group G. Let Σ_G denote the class of all closed subsets $A \subset M$ which are G-invariant. An *index* for (G, Σ_G) or, in short, a G-index is a map $\gamma : \Sigma_G \to \mathbf{N} \cup \{+\infty\}$ satisfying the following properties $(X, Y \in \Sigma_G)$:

1. $\gamma(X) \ge 0, \quad \gamma(X) = 0$ iff $X = \emptyset$;
2. $X \subset Y$ then $\gamma(X) \le \gamma(Y)$;

3. $\gamma(X \cup Y) \leq \gamma(X) + \gamma(Y)$;

4. if $h \in C(X, M)$ is G-equivariant, then $\gamma(X) \leq \gamma(h(X))$

5. if X is compact and $X \cap Fix(G) = \emptyset$ then $\gamma(X) < +\infty$, and there exists a G-invariant neighbourhood U of X such that $\gamma(\overline{U}) = \gamma(X)$;

6. if $u \notin Fix(G)$ then $\gamma(Gu) = 1$.

Above, Gu denotes the G-orbit of u. An example of index is the category corresponding to the trivial group $G = \{1\}$. Another example important for the applications is the case when $G = \mathbf{Z}_2$ with action $u \to -u$.

A $\mathbf{Z}_2$-index, usually called *genus*, is defined on the class $\Sigma = \Sigma_{\mathbf{Z}_2}$ of closed, symmetric subsets $A \subset E$, with $0 \notin A$, as the least integer n such that there exists an odd map $\phi \in C(A, R^n)$, with $\phi(u) \neq 0$, for all $u \in A$. For example, it turns out that $\gamma(S^m) = cat(P^m) = m + 1$. Problems which inherit a $\mathbf{Z}_2$ symmetry are nothing but *even* functionals.

Another important case is that of the S^1-index $\gamma(A)$ defined (see [23]) as the least $n \in \mathbf{N}$ such that there exists a map $\phi \in C(A, \mathbf{C}^n \setminus \{0\})$ an $k \in \mathbf{N}$, with $\phi \circ \theta = e^{ik\theta}\phi$, for all $\theta \in S^1$. We set $\gamma(A) = +\infty$ if there are no such integers and $\gamma(\emptyset) = 0$. For example, one can show that the set

$$A = \{(x \cos t + y \sin t, x \sin t - y \cos t) : x, y \in R^m, |x|^2 = |y|^2 = 1\}$$

has S^1-index equal to m. The search of multiple closed orbits on a Hamiltonian surface is a typical S^1-invariant problem where the S^1 index defined above can be used. See [44], see also the survey of Rabinowitz in this volume.

One can use the *index* to define critical levels for G-invariant functionals f on an G-invariant manifold M by setting

$$c_k = \inf_{\gamma(A) \geq k} \sup_{u \in A} f(u).$$

All the arguments sketched in points b) and c) above can modified when dealing with symmetric problems. For example, if f and M are G-invariant, the Pseudo-Gradient Vector Field defined in Lemma 2.5 turns out to be G-equivariant. As a consequence the deformation η of the Deformation Lemma is G-equivariant, too. This permits to carry out the preceding arguments leading to the following theorem

Theorem 2.10. *Let M be a $C^{1,1}$ manifold and let $f \in C^1(M, R)$. Suppose that M and f are invariant under the action of a compact group G, that $M \cap Fix(G) = \emptyset$ and let γ denote a G-index. Suppose further that f is bounded from below on M and that (PS) holds. Then f has at least $\gamma(M)$ critical orbits Gu on M.*

Clearly, Theorem 2.9 is nothing but a particular case of this latter result.

3. Applications to Nonlinear Eigenvalues

Theorem 2.9 applies to nonlinear variational eigenvalue problems such as

$$C(u) = \lambda u \qquad (3.1)$$

when the operator $C : E \rightarrow E$ (with E say, a Hilbert space) is variational, namely when there exists a C^1 functional $f : E \rightarrow R$ such that $f'(u) = C(u)$. Indeed, letting as before $M = \{u \in E : \|u\| = r\}$, critical points of f constrained on M give rise to solutions of (3.1). Suppose that C is an odd operator, so that f is even. Then we can apply Theorem 2.9 whenever (PS) holds and f is bounded below on M. One shows that is indeed the case provided f satisfies

(f1) $f(u) < 0 \quad \forall u \in E, u \neq 0, f(0) = 0$, and $f'(u) = 0$ iff $u = 0$;

(f2) f' is compact and f is weakly continuous.

Let us point out that under such conditions $(PS)_c$ holds for all $c < 0$, only. However, this suffices to prove

Theorem 3.1. *Let $f \in C^1(E, R)$ satisfy (f1), (f2) and be even. Then the nonlinear euigenvalue problem*

$$f'(u) = \lambda u, \qquad \|u\| = r,$$

has infinitely many (pairs of) solutions.

As an application, let us consider the elliptic eigenvalue problem (2.2). Keeping the notation introduced in Example 2.2, let us suppose that

(F1) $F(u) < 0, \quad \forall u \neq 0$;

(F2) $F \in C^1(\Omega, R)$ and there exist $a_1, a_2 > 0$ and $p < 2^* = \frac{2N}{N-2}$ such that

$$|F(u)| \leq a_1 + a_2|u|^p;$$

Hereafter we will suppose that $N > 2$: if $N = 2$, p can be arbitrary. Here we take $E = H_0^1(\Omega)$ and define $f : E \rightarrow R$ by setting

$$f(u) = \int_\Omega F(u(x))dx.$$

Since, by the Sobolev embedding theorem one has that $E \subset L^{2^*}(\Omega)$, then the well known results concerning Nemitski operators, see for example [12], yield

$$F(u) \in L^1(\Omega) \qquad \forall u \in H_0^1(\Omega).$$

and f is well defined on E. Plainly, f is of class C^1 and verifies (f1). Moreover, since p is strictly less than 2^*, then the embedding of $H_0^1(\Omega)$ into $L^p(\Omega)$ is compact and it follows that (f2) holds. Then one finds

Theorem 3.2. *Suppose that $F \in C^1(R)$ satisies (F1) and (F2). Then (2.2) has infinitely many (pairs) of solutions.*

Remark 3.3. Since $(PS)_c$ holds for all $c < 0$, then, according to Theorem 2.7, one infers that $cat(f^c, M) < +\infty$ for all $c < 0$. On the other side $cat(M) = +\infty$ and therefore the critical levels c_k satisfy

$$c_k \uparrow 0 \qquad \text{as } k \to \infty.$$

As a consequence, one deduces that (2.2) possesses a sequence of solutions (μ_k, u_k) such that $\mu_k \uparrow 0$ as $k \to \infty$.

When in (F2) $p = 2^* = 2N/(N-2)$, one still has that $F(u) \in L^1$ for all $u \in H_0^1$ and hence the functional f makes still sense, but (f2) does not hold any more. The following identity shows that in such a case (2.2) can have the trivial solution, only. Let ν_x denote the unit outward normal at $x \in \partial\Omega$ and u_ν the normal derivative of u along ν.

Lemma 3.4. [67] *Let u be a smooth solution of (2.2). Then there results*

$$N \int_\Omega F(u)dx + \frac{2-N}{2} \int_\Omega uF'(u)dx = \frac{1}{2} \cdot \lambda \int_{\partial\Omega} u_\nu^2 (x \cdot \nu_x)d\sigma.$$

As an immediate consequence one infers

Corollary 3.5. *If $x \cdot \nu_x \geq 0$ on $\partial\Omega$, namely if Ω is star-shaped with respect to 0, then the nonlinear eigenvalue problem*

$$-\lambda\Delta u = |u|^{p-2}u, \quad x \in \Omega; \quad u = 0, \quad x \in \partial\Omega$$

has only the trivial solution, whenever $p \geq \frac{2N}{N-2}$ $(N > 2)$.

4. Unbounded Functionals

When f is not bounded below (nor above) the preceding theory does not apply. We will see later on that there are several interesting problems where unbounded functionals arise. In this section we will survey some results which extend the Lusternik–Schnirelman theory to some classes of unbounded functionals.

A model case consists in functionals $f \in C^1(E, R)$ of the type

$$f(u) = \frac{1}{2}\|u\|^2 - \Phi(u) \tag{4.1}$$

where $\Phi(u) = o(\|u\|^2)$ near $u = 0$ and there exists $u_0 \in E$, $u_0 \neq 0$ such that

$$\lim_{t \to +\infty} t^{-2}\Phi(tu_0) = -\infty.$$

A functional of this form arises, for example, dealing with superlinear elliptic boundary value problems such as

$$-\Delta u = F'(u), \quad x \in \Omega; \quad u = 0, \quad x \in \partial\Omega$$

when $F(u)$ behaves like $|u|^p$, with $p > 2$, near $u = 0$ and as $|u| \to \infty$. This and other applications will be given in the next section. A functional which behaves like (4.1) has a strict, local minimum at $u = 0$, with $f(0) = 0$; moreover there exists $e \neq 0$ such that $f(e) < 0$. If this is the case, we will say that f *has the Mountain Pass geometry.* Let us define

$$\mathcal{B} = \{\beta \in C([0,1], E) : \beta(0) = 0, \ \beta(1) = e\}$$

and

$$b = \inf_{\beta \in \mathcal{B}} \max_{t \in [0,1]} f(\beta(t)). \tag{4.2}$$

Note that $b > 0$ because any $\beta \in \mathcal{B}$ crosses the boundary of a ball $\partial B(r)$, whenever $0 < r < \|e\|$, and $\min\{f(u) : \|u\| = r\} > 0$. Moreover, if f satisfies the $(PS)_b$ condition, then we can use the Deformation Lemma to infer that the sublevel $f^{b+\epsilon}$ can be deformed through η into $f^{b-\epsilon}$ whenever b is not a critical level for f. On the other side, by the definition of b there exists $\beta \in \mathcal{B}$ such that

$$\max_{t \in [0,1]} f(\beta(t)) \leq b + \epsilon.$$

Since $f(0) = 0$ and $f(e) \leq 0$, taking ϵ possibly smaller in such a way that $b - 2\epsilon > 0$, point (1) of the Deformation Lemma implies that the path $\eta \circ \beta$ satisfies

$$\eta(\beta(0)) = 0, \qquad \eta(\beta(1)) = e,$$

and hence $\eta \circ \beta \in \mathcal{B}$. From the above argument one also has

$$\max_{t \in [0,1]} f(\eta \circ \beta(t)) \leq b - \epsilon,$$

a contradiction with the definition (4.2) of b. In conclusion one can state

Theorem 4.1. (Mountain Pass) *Let $f \in C^1(E, R)$ satisfy:*

(f3) $f(0) = 0$ *and* $\exists a, r > 0$ *such that* $f(u) > 0 \quad \forall 0 < \|u\| \leq r$, *and* $f(u) \geq a \quad \forall u \in \partial B(r)$;

(f4) $\exists e \in E, e \neq 0$, *such that* $f(e) \leq 0$. *Let b be defined as in (4.2) and suppose that f satisfies $(PS)_b$. Then $b \geq a$ is a critical level of f.*

When f is even one can obtain a multiplicity result on the line of the Lusternik–Schnirelman type Theorem 2.9. Set $E_+ = \{u \in E : f(u) \geq 0\}$ and suppose that f is even, satisfies (f3) and

(f5) for all $E^k \subset E$, with $dim(E^k) = k$, one has that $E^k \cap E_+$ is bounded.

Generalizing the class $\mathcal{B}$, we define

$$\mathcal{S} = \{\sigma \in C(E, E) : \sigma \text{ is an odd homeomorphism and } \sigma(B(1)) \subset E_+\}$$

By (f3), letting $\sigma_r : u \to ru$, one has that $\sigma_r \in H$. We set

$$\mathcal{A}_k = \{A \subset E : A \in \Sigma, \gamma(A \cap \sigma(\partial B(1))) \geq k, \forall \sigma \in \mathcal{S}\}.$$

Here Σ, respectively γ, denotes the class of closed $\mathbf{Z}_2$-invariant subsets of $E \setminus \{0\}$, resp. the *genus*, introduced in subsection 2-c. Using (f3), (f5) and the properties of the genus, one shows that the classes $\mathcal{A}_k$ satisfy

1. $\mathcal{A}_k \neq \emptyset$ for all $k \in \mathbf{N}$;
2. $\mathcal{A}_k \subset \mathcal{A}_{k+1}$;
3. if $A \in \mathcal{A}_k$ and $\gamma(U) \leq k' < k$, then $\overline{A - U} \in \mathcal{A}_{k-k'}$;
4. let η be an odd homeomorphism such that $\eta^{-1}(E_+) \subset E_+$; then $A \in \mathcal{A}_k \Rightarrow \eta(A) \in \mathcal{A}_k$.

Let us explicitly point out that the deformation η of Lemma 2.6 satisfies the condition $\eta^{-1}(E_+) \subset E_+$ required for the validity of point (4) before.

The classes $\mathcal{A}_k$ can be used to define min-max levels by setting

$$b_k = \inf_{A \in \mathcal{A}_k} \max\{f(u) : u \in A\}.$$

The following theorem holds:

Theorem 4.2. *Let $f \in C^1(E, \mathbf{R})$ be even and satisfy (f3), (f5) and $(PS)_c$, for all $c > 0$. Then*

1. $b_{k+1} \geq b_k \geq a > 0$
2. $K_{b_k} \neq \emptyset$
3. *if $b_k = b_{k+1} = \cdots = b_{k+m}$, for some $m \geq 1$ then $\gamma(K_{b_k}) \geq m + 1$ (and hence K_{b_k} contains infinitely many critical points).*

In particular, f possesses infinitely many critical points on E.

For other multiplicty results we refer to [13].

Theorem 4.1 can be extended to handle the case in which $u = 0$ is not a local minimum but a saddle point.

Suppose $E = X \oplus Y$, with $\dim X < +\infty$ and set $X_\rho = \{x \in X : \|x\| \leq \rho\}$. Given $e \in Y$ with $\|e\| = 1$, we define

$$Z_\rho(e) = X_\rho \oplus [0, \rho e],$$

and denote by

$$\partial Z_\rho(e)$$

its boundary relative to $X \oplus Re$. Let

$$\mathcal{F} = \{\varphi \in C(Z_\rho(e), E) : \varphi(u) = u \, \forall u \in \partial Z_\rho(e)\}$$

and

$$b = b(\rho, e) = \inf_{\varphi \in \mathcal{F}} \max_{u \in Z_\rho(e)} f(\varphi(u)). \tag{4.3}$$

Theorem 4.3. (Linking Theorem) *Let $f \in C^1(E, R)$ and suppose*

(f6) $f(0) = 0$ and $\exists a, r > 0$ such that $f(y) \geq a \, \forall y \in Y$, $\|y\| = r$;

(f7) $\exists e \in Y \setminus \{0\}$ and $\rho > r$ such that $f(u) < 0$, $\forall u \in \partial Z_\rho(e)$.

Assume also that f satisfies $(PS)_b$ with b given by (4.3). Then $b \geq a$ is a critical level for f.

The proof relies in showing that for all $\varphi \in \mathcal{F}$ the manifolds $\partial Z_\rho(e)$ and $\partial X_r = \{x \in X : \|x\| = r\}$ link, in the sense that

$$\varphi(Z_\rho(e)) \cap \partial X_r \neq \emptyset, \quad \forall \varphi \in \mathcal{F}.$$

This implies that $b \geq a > 0$ and an argument similar to that sketched for the Mountain Pass Theorem shows that b is indeed a critical level.

It is worth pointing out that the condition $\dim X < +\infty$ can be eliminated provided one deals with a more specific class of functionals. This improvement is particularly relevant when one is concerned with Hamiltonian systems. We will not discuss here this kind of result, but refer to [29].

For other results related to the Mountain-Pass theorem, we refer, amnong other papers, to [50], [70].

We end this section with a result concerning the Morse index of a Mountain-Pass or Linking critical point. A first result in this direction was obtained in [5] for a nondegenerate Mountain-Pass critical point. Further improvements dealing with possibly degenerate Mountain-Pass and Linking critical points have been found in [53] and [20], [27], [56], respectively; see also [88].

Let $f \in C^2(E, R)$ and let u be a critical point of f. The Morse index $m(u)$ of u is the dimension of the subspace $\{v \in E : f''(u)[v, v] < 0\}$.

Theorem 4.4. *Let E be a Hilbert space and let $f \in C^2(E, R)$ satisfy the assumptions of the Linking Theorem 4.3. Moreover, suppose that $f''(u)$ has a discrete spectrum for all $u \in E$. Then f has a critical point u, with $f(u) \geq a$ and such that $m(u) \leq \dim(X) + 1$.*

5. Elliptic Dirichlet Problems

As anticipated before, the Mountain Pass and the Linking Theorem can be used to obtain existence and multiplicty results for many nonlinear problems. Referring to the article of Paul Rabinowitz in this volume, for applications to the existence of periodic solutions of nonlinear Hamiltonian systems, that marked an important progress in the research in such a field, we will survey here elliptic, Dirichlet, semilinear problems such as

$$\begin{cases} -\Delta u = F'(u), & x \in \Omega, \\ u = 0 & x \in \partial\Omega \end{cases} \tag{5.1}$$

Setting for $u \in E = H_0^1(\Omega)$,

$$f(u) = \frac{1}{2}\|u\|^2 - \int_\Omega F(u(x))dx,$$

solutions of (5.1) correspond to critical points of f on E. Depending on the behaviour of F near $u = 0$ and at infinity, critical points of f can be found by means of the preceding abstract Theorems. In order to highlight the kind of results one can obtain, we will limit our discussion to some model nonlinearities. The interested reader can find more general problems and more details in the original papers. See also the book [82] and [6] and references therein.

Hereafter λ_k denotes the eigenvalues of the Laplace operator $-\Delta$ with zero Dirichlet boundary conditions.

1) We first consider the case when

$$F(u) \simeq \frac{1}{2}\lambda \cdot |u|^2 - |u|^p, \text{ with } p > 2. \tag{5.2}$$

Note that, unlike in Section 3 here and in the sequel λ is a *given* real parameter.

The functional corresponding to the above F has the form

$$f(u) \simeq \frac{1}{2}\|u\|^2 - \frac{1}{2}\lambda \cdot \int_\Omega |u|^2 dx + \int_\Omega |u|^p dx$$

and is bounded below on E and corcive. Thus f achieves its minimum on all E. Such a minimum is different from 0 provided $\lambda > \lambda_1$ and gives rise to a nontrivial solution of (5.1). More precisely, working with the positive (resp. negative) part of F' and using the the maximum principle, one can show that (5.1) has a positve (resp. negative) solution. Taking these two solutions as base points, the Mountain Pass Theorem yields a third critical point v. Since the trivial solution $u = 0$ has Morse index ≥ 2 whenever $\lambda > \lambda_2$, Theorem 4.4 implies that the Mountain-Pass level carries a critical point v, with $v \neq 0$ (see Theorem 6.1 of [6]). Finally, if F is an even function, one can use the Lusternik–Schnirelman theory to obtain multiple solutions. Precisely, whenever $\lambda > \lambda_k$ the inf-max level c_k, defined in subsection d) of Section 2, are negative and hence f has at least k nontrivial critical points. We can sumarize these results as follows. Let F be of the form (5.2)

1. if $\lambda > \lambda_1$ then (5.1) has a positive and a negative solution;
2. if $\lambda > \lambda_2$ then (5.1) has a third nontrivial solution;
3. if F is even and $\lambda > \lambda_k$ then (5.1) has at least k (pairs of) non-trivial solutions.

Let us point out that here no restrictions on p like in (F2) are needed, because one can substitute F with a truncated nonlinearity and use the maximum principle.

2) Let us consider a nonlinearity of the form

$$F(u) \simeq \frac{1}{2}\lambda \cdot |u|^2 + |u|^p, \text{ with } 2 < p < 2^* = \frac{2N}{N-2}. \tag{5.3}$$

If $\lambda < \lambda_1$ the Poincaré inequality implies that the quadratic part

$$\frac{1}{2}\|u\|^2 - \frac{1}{2}\lambda \cdot \int_\Omega |u|^2 dx$$

is equivalent to the usual norm $\|u\|^2_{H^{1,2}}$. Then the corresponding functional

$$f(u) \simeq \frac{1}{2}\|u\|^2 - \frac{1}{2}\lambda \cdot \int_\Omega |u|^2 dx - \int_\Omega |u|^p dx$$

is of the form (4.1) and has therefore the Mountain-Pass geometry, namely satisfies $(f3)$ and $(f4)$. In addition one shows that (PS) holds and thus an application of Theorem 4.1 yields the existence of a nontrivial solution for (5.1). Here it is important to assume $2 < p < 2^*$. To highlight the role of such a condition, let us indicate the arguments we use to prove the (PS) condition in the simpler case when $F(u) = |u|^p$. Let $u_n \in E$ be such that $f(u_n) \to b$ and $f'(u_n) \to 0$. As a first step, one derives a bound on u_n as follows. From $|(f'(u_n)|u_n)| \leq \epsilon\|u_n\|$ one infers that

$$p \int_\Omega |u|^p dx \leq \|u_n\|^2 + \epsilon\|u_n\|,$$

while $f(u_n) \leq a_1$ yields

$$\|u_n\|^2 \leq a_2 + 2 \int_\Omega |u|^p dx,$$

and therefore

$$\|u_n\|^2 \geq a_2 + \frac{2}{p}\|u_n\|^2 + \frac{2}{p}\epsilon\|u_n\|.$$

Since $p > 2$ it follows that $\|u_n\| \leq a_3$. Then, up a subsequence $u_n \to u$ weakly in E. Now, since $p < 2^*$, E is compactly embedded in L^p and $u_n \to u$ strongly in L^p. Finally, from $f'(u_n) \to 0$ one readily infers that u_n strongly converges in E. It is worth mentioning that all the preceding arguments can be carried over whenever F satisfies $(F2)$ and

(F3) $F(0) = 0$ and $\exists\, \theta \in]0, \frac{1}{2}[$ and $r > 0$ such that $F(u) \leq \theta u F'(u)$ for all $|u| \geq r$.

Note that condition (F3) is in fact a superlinearity condition and implies, in particular, that

$$\lim_{|u|\to\infty} F(u) = +\infty.$$

Moreover, by taking the positive part of F' and using the maximum principle, one actually finds a positive solution of (5.1). The existence of an additional solution that changes sign has been proved in [89].

When $\lambda \geq \lambda_1$, f satisfies $(f6)$ and $(f7)$ and one applies Theorem 4.3 to find a solution (not positive, in general). Finally, if F is even then $(f5)$ holds and one uses Theorem 4.2 to show that (5.1) has infinitely many solutions.

3) The question of finding multiple solutions of (5.1) when F has the form (5.2) but is not even, has also been addressed. In [4] a perturbed problem of the form

$$\begin{cases} -\Delta u = |u|^{p-2}u + \epsilon h(x,u), & x \in \Omega, \\ u = 0 & x \in \partial\Omega \end{cases} \qquad (5.4)$$

$(2 < p < 2^*)$ has been studied, proving that for all integer k, (5.4) has at least k solutions, provided ϵ is small enough. Under stronger restrictions on p the preceding multiplicity result can be greatly improved, showing that there exists $p^* < 2^*$ such that the problem

$$\begin{cases} -\Delta u = |u|^{p-2}u + h(x,u), & x \in \Omega, \\ u = 0 & x \in \partial\Omega \end{cases}$$

has infinitely many solutions whenever $2 < p < p^*$, see [18], [72] and [80]. A further improvement has been obtained in [20], by means of Morse theoretical arguments. Roughly, it has been shown that the problem

$$\begin{cases} -\Delta u = |u|^{p-2}u + h(x), & x \in \Omega, \\ u = 0 & x \in \partial\Omega \end{cases} \qquad (5.5)$$

has infinitely many solutions provided $2 < p < (2N-2)/(N-2)$.

Let us point out that to prove these multiplicity results one takes advantage of the fact that the unperturbed problems has a sequence of solutions such that $f(u_n) \to +\infty$. It is worth recalling that this is not true in the case 1), namely when $F(u) = \frac{1}{2}\lambda|u|^2 - |u|^p$ and a general multiplicity result for (5.1) like the previous one, has not been yet established when $F(u) \simeq \frac{1}{2}\lambda|u|^2 - |u|^p + \epsilon h(x,u)$.

Let us also mention that Bahri has proved in [17] that (5.5) has infinitely many solutions for h in a dense subset of L^2, for all $2 < p < 2^*$.

4) Another kind of multiplicity results is that one related to the effect of the domain topology. For example, it has been shown, see [26], that problem

$$\begin{cases} -\Delta u = |u|^{p-2}u, & x \in \Omega, \\ u = 0 & x \in \partial\Omega \end{cases}$$

has at least $cat(\Omega) + 1$ positive solutions, provided ε is positive and small enough.

5) When

$$F(u) \simeq q(x)u^p, \quad 2 < p < 2^*,$$

and $q(x)$ changes sign in Ω, the existence of positive solutions for problems like (5.1) has been recently studied in [3]. As before, a positive solution of (5.1) exists for all $\lambda < \lambda_1$. In addition, letting $e_1 > 0$ denote the first eigenfunction of $-\Delta$ on Ω with zero Dirichlet boundary conditions, and assuming that

$$\int_\Omega q(x)e_1(x)\,dx < 0,$$

it is shown, roughly, that there exists $\lambda^* > \lambda_1$ such that (5.1) has two positive solutions for all $\lambda \in (\lambda_1, \lambda^*)$, and one positive solution for $\lambda = \{\lambda_1, \lambda^*\}$.

6) The interplay between a sublinear and a superlinear term has been studied in [7], see also [32]. Consider the problem

$$\begin{cases} -\Delta u = \lambda |u|^{q-2}u + |u|^{p-2}u, & x \in \Omega, \\ \quad u = 0 & x \in \partial\Omega \end{cases} \tag{5.6}$$

with $0 < q < 2 < p$. One proves

1. there exists $\Lambda > 0$ such that (5.6) has a minimal positive solution u_λ for all $\lambda \in (0, \Lambda)$, a positive weak solution for $\lambda = \Lambda$ and no positive solutions for $\lambda > \Lambda$;
2. if, in addition, $p < 2^*$, then (5.6) has a second positive solution $v_\lambda > u_\lambda$, for all $\lambda \in (0, \Lambda)$; moreover there exists $\lambda^* > 0$ such that (5.6) has infinitely many solutions with negative energy and infinitely many solutions with positive energy for all $\lambda \in (0, \lambda^*)$.

Roughly, assuming $p < 2^*$, the corresponding functional

$$f(u) = \frac{1}{2}\|u\|^2 - \lambda \int_\Omega |u|^q dx - \int_\Omega |u|^p dx$$

has, for $\lambda > 0$ and small, a negative local minimum and a mountain pass solution.

Actually, a first positive solution of (5.6) can be found using the method of the sub and super-solutions. This topological approach has the advantage of leading to the maximal interval of existence of positive solutions and to require no restriction on p. It is also possible to show that a positive solution exists such that it is a local minimum for f. The existence of a second positive solution at positive level is then obtained by means of the Mountain Pass procedure. Finally, for $\lambda > 0$ and sufficiently small, one can prove the stronger multiplicity result stated in point 2 above, using

Theorems 2.9 (as far as the solutions at negative level are concerned) and 4.2 for those at positive level.

7) Let us consider the case in which

$$F(u) = \frac{1}{2^*}|u|^{2^*}. \tag{5.7}$$

As in Lemma 3.4 and Corollary 3.5, it is easy to see that the corresponding Dirichlet problem

$$\begin{cases} -\Delta u = \lambda u + u^{2^*-1}, & x \in \Omega, \\ \qquad u = 0 & x \in \partial\Omega \end{cases} \tag{5.8}$$

only has the trivial solution, whenever $\lambda \leq 0$ and Ω is star-shaped with respect to the origin. The corresponding functional f does not satisfy the (PS) condition because the embedding of $H^{1,2}$ into L^{2^*} is no longer compact. On the other hand, a direct bifurcation analysis indicates that nontrivial solutions of (5.8) branch off from each λ_k to the left and hence, at least locally, (5.8) has nontrivial solutions in any left neighbourhood of λ_k. The first global result for this problem has been found in a celebrated paper [34] by Brezis and Nirenberg. Using variational methods they have established that (5.8) has a positive solution for all $0 < \lambda < \lambda_1$ and $N > 3$ (if $N = 3$, positive solutions exist provided $\lambda_* < \lambda < \lambda_1$, for some $\lambda_* > 0$). Roughly, the idea is that, although the embedding of $H^{1,2}$ into L^{2^*} is not compact, $(PS)_c$ still holds whenever $c < S^{N/2}/N$, where S denotes the best Sobolev constant for R^N.

This paper has stimulated a great deal of work on critical Sobolev exponent problems and the topic is one of considerable current research interest. We limit ourselves here to mention a few references, dealing with: (i) existence of (possibly non-positive) solutions for $\lambda > \lambda_1$ [36], [14]; (ii) multiplicity results [38]; (iii) uniqueness of positive solutions [79], [90]; (iv) p-Laplace operator [48]; (v) existence results for $\lambda = 0$ when Ω is star-shaped [19], [39], [43], [68]; (vi) the Dirichlet problem for the equation of constant mean curvature [33], [81]. Elliptic problems with critical Sobolev exponent and satisfying Neumann boundary conditions have also been studied, [1], [2].

8) When (PS) breaks down the concentration-compactness method developed by P.L. Lions [57], [58], also proves to be very useful. Applications include the existence of ground states of field equations [30], [31]; and of positive solutions in exterior domains [25], to cite only a few.

When dealing with problems on unbounded domains, another strategy is to consider approximated equations in an increasing sequence of bounded domains and then pass to the limit. This approach has been followed, for example, dealing with the existence of vortex rings in an axisymmetric ideal fluid filling all of R^3, [47] and [15]. It is worth mentioning that in some cases (see [15]) it is just the variational characterization of the approximated solution that allows us to find the appropriate estimates and to perform the limiting process.

Lack of compactness may arise for several other reasons. For example, the problem can be invariant under the action of a noncompact group. This is the case of homoclinic orbits of Hamiltonian systems, discussed in the survey of Rabinowitz in this volume.

Another class of problems where (PS) breaks down is that of second order Hamiltonian systems with singular potentials, which is studied in the next section.

6. Singular Potentials

Another class of problems where the lack of compactness arise is the search of periodic solutions to second order Hamiltonian systems with singular potentials. This is the case, for example, of perturbed Keplerian systems

$$\ddot{q} + \frac{q}{|q|^3} + W'(q) = 0 \tag{6.1}$$

whith $W \in C^1(R^N, R)$. These problems are variational in nature and it turns out that critical point theory, like that one discussed above, can be adapted to find periodic solutions of equations such as (6.1).

In the last years there has been a considerable amount of work on this subject and we will survey below some of the results obtained so far. For a more complete discussion we refer to Sections 11, 12 and 13 of [6] or to the forthcoming monograph [11].

a) *Existence of T-periodic solutions.* Suppose $V \in C^1(R \times R^N \setminus \{0\}, R)$ is such that

(V1) $V(t + T, x) = V(t, x)$ for all $t \in R$ and $x \in R^N \setminus \{0\}$;

(V2) $V(t, x) \to -\infty$ as $x \to 0$ (uniformly in t).

and consider the second order Hamiltonian system

$$\ddot{q} + V'(t, q) = 0, \tag{6.2}$$

where $V' = \partial V/\partial q$. Smooth T-periodic solutions of (6.2), such that $q(t) \neq 0$ for all t, in the sequel referred to as *noncollision* orbits, are critical points of the Lagrangian functional

$$f(u) = \int_0^T \left[\frac{1}{2}|\dot{u}|^2 - V(t, u) \right] dt, \tag{6.3}$$

defined on the loop space

$$\Lambda = \{u \in H^1(S_T, R^N) : u(t) \neq 0 \ \forall t\}\,.$$

Hereafter $S_T = [0, T]/\{0, T\}$. To see the kind of problems that one has to face, let us consider the case of the Kepler potential, when $V(x) = -1/|x|$ and f becomes

$$f(u) = \int_0^T \left[\frac{1}{2}|\dot{u}|^2 + \frac{1}{|u|}\right] dt\,.$$

First of all, inf $f = 0$ but the minimum is not achieved. Indeed, for any sequence $x_n \in R^N$ such that $|x_n| \to +\infty$ one has $f(x_n) \to 0$ and $f'(x_n) = 0$. In other words, f does not satisfy the (PS) condition at level $c = 0$. This and the fact that Λ has infinite Lusternik–Schnirelman category (see Example 2.4) suggests to use minmax methods to find critical points of f. However, to do that, one has to overcome a second problem concerning the behaviour of f near $\partial \Lambda$. For example, again in the case of the Kepler problem, let $q_n = q_{e_n}$ denote a family of T-periodic elliptic orbits of the Kepler equation

$$\ddot{q} + \frac{q}{|q|^3} = 0$$

with eccentricity equal to $e_n > 0$, $e_n \to 0$. Such a q_n is plainly a (PS) sequence, because $f'(q_n) \equiv 0$ and $f(q_n) \equiv$ constant, but q_n converges to the collision orbit corresponding to q_e, with eccentricity $e = 0$, which does not belong to Λ.

Coming back to the general case, the problem concerning the behaviour of f at the boundary $\partial \Lambda$ can be overcome whenever V satisfies the so called *strong force condition* introduced in [49]:

(SF) $V(t, x) \leq -\frac{1}{|x|^2}$ near $x = 0$.

Actually if (SF) holds then one proves that

$$\int_0^T V(t, u_n(t))\, dt \to -\infty \tag{6.4}$$

whenever $u_n \in \Lambda$, $u_n \to u \in \partial \Lambda$, weakly in H^1 and uniformly in $[0, T]$.

It is clear that, whenever (6.4) holds, one can use the Lusternik–Schnirelman theory to find critical point s of f on Λ. The following lemma shows how to handle the lack of (PS) at level $c = 0$.

Lemma 6.1. *Suppose that V satisfies $(V1)$, (SF) and that, uniformly in t, there results*

(V3) $\lim_{|x|\to+\infty} V(t, x) = 0$;

(**V4**) $\lim_{|x| \to +\infty} V'(t, x) = 0$;

(**V5**) $V(t, x) < 0, \forall |x| \neq 0$.

Then

      1. *f satisfies the $(PS)_c$ condition at any level $c > 0$;*

      2. *for $\epsilon > 0$ small enough one has $\text{cat}(f^\epsilon, \Lambda) \leq 2$.*

From point 2) above, one deduces that the minmax levels c_k defined in (2.3) are positive for all $k \geq 3$. Hence, according to point 1) of the preceding lemma, such c_k are critical levels for f on Λ and one finds

Theorem 6.2. *Suppose that $V \in C^1(R \times R^N \setminus \{0\}, R)$ satisfies $(V1)$, $(V2)$, (SF) and $(V3 - 4 - 5)$. Then (6.2) has infinitely many T-periodic (noncollision) solutions.*

Theorem 6.2 is a particular case of more general results contained in [8], see also [51]. Subsequently, a great deal of work dealing with this kind of problems have appeared, see for example [60], [75] and the references in [11]. Theorem 6.2 can also be used as a basic tool when dealing with potentials that do not satisfy (SF) but merely $(V2)$. Indeed, one can consider the approximated problems

$$\ddot{q} + V'(t, q) + \delta \frac{q}{|q|^4} = 0, \tag{6.5}$$

corresponding to the perturbed potential

$$V_\delta(t, x) = V(t, x) - \frac{\delta}{|x|^2},$$

which satisfy the (SF) condition. Hence one can apply Theorem 6.2 to find, for all $\delta > 0$, a noncollision solution q_δ of (6.5). Moreover, one can choose q_δ in such a way that $\|q_\delta\| \leq a_1$, with a_1 independent of δ. It follows that, up to a subsequence, q_δ converges weakly in H^1 and uniformly in $[0, T]$ to some $q \in H^1(S_T, R^N)$ that has the following properties:

      1. $\text{meas}(\Gamma[q]) = 0$, where $\Gamma[q] = \{t \in [0, T] : q(t) = 0\}$;

      2. $q \in C^2([0, T] \setminus \Gamma[q], R^N)$ and solves (6.2).

Such a q is called a *weak solution* of (6.2). The existence of a weak solution has been established in [21]. In certain situations, it is also possible to estimate by Morse theoretical arguments, the number of collisions or even to show that the weak solution is a non-collision orbit, namely a solution q such that $\Gamma[q] = \emptyset$, see [41] and [84].

The preceding results have been extended to n-body type problems, namely second order systems with potential $V : R^{kn} \to R$ of the form

$$V(x) = \frac{1}{2} \sum_{i \neq j} V_{ij}(x_i - x_j),$$

where $x_i \in R^k$, $x = (x_1, \ldots, x_n) \in R^{kn}$ and V_{ij} satisfies (V1–2). Clearly, when $k = 3$ and

$$V_{ij}(\xi) = -\frac{m_i m_j}{|\xi|},$$

the potential V is nothing but the Keplerian interaction among n bodies in R^3 with mass m_i and position x_i.

The main new difficulty when dealing with n-body type problem is concerned with the (PS) condition. For example, dealing with the 3-body problem, the (PS) breaks down not only at level $c = 0$ but also, roughly, at the critical levels of the two-body problems corresponding to any pair of particles. Without entering into detail, let us recall that the 3-body problem has been studied in [22] by means of the Morse theory, while minmax arguments have been used in [61] to handle general n-body type problems.

Let us point out that, when $V_{ij}(\xi) = V_{ji}(\xi)$, the problem becomes considerably simpler and has been studied by more direct arguments in [40].

For other results on this important subject, including the regularity of the weak solutions in the case that V_{ij} do not satisfy (SF), see [78] and the Thesis [77], [86].

b) *Closed orbits with prescribed energy.* Dealing with autonomous potentials, one can seek for solutions of

$$\ddot{q} + V'(q) = 0 \tag{6.6}$$

$$\frac{1}{2}|\dot{q}|^2 + V(q) = h, \tag{6.7}$$

where h is a prescribed number. To find solutions of (6.6)-(6.7) one can use the following variant of the classical Maupertuis principle. From now on we denote by S^1 the unit circle and set

$$\Lambda = \{u \in H^1(S^1, R^N) : u(t) \neq 0, \ \forall t\}$$

Lemma 6.3. *Let $V \in C^1(R^N \setminus \{0\}, R)$ and let $g : \Lambda \to R$ denote the functional*

$$g(u) = \frac{1}{2} \int_0^1 |\dot{u}|^2 \, dt \cdot \int_0^1 (h - V(u)) \, dt.$$

Suppose that u is a critical point of g on Λ such that $g(u) > 0$ and set

$$\omega^2 = \frac{\int_0^1 (h - V(u))\, dt}{\frac{1}{2} \int_0^1 |\dot{u}|^2\, dt}.$$

Then $q(t) := u(\omega t)$ is a solution of (6.6)-(6.7).

Repeating the procedure used before, one considers the perturbed potential $V_\delta(x) = V(x) - \delta |x|^{-2}$ and the corresponding perturbed functional

$$g_\delta(u) = \frac{1}{2} \int_0^1 |\dot{u}|^2 dt \cdot \int_0^1 (h - V_\delta(u))\, dt.$$

If u_δ is a critical point of g_δ on Λ with $g(u_\delta) > 0$, then appropriate estimates will allow us to show that u_δ, respectively ω_δ, converges weakly to some $u \in E$, respectively $\omega_\delta \to \omega \neq 0$, and $q(t) := u(\omega t)$ is a weak solution of (6.6)–(6.7). Looking for critical points of g_δ at positive level one encounters an additional problem due to the specific feature of the functional, in the sense that, even if V_δ satisfies (SF), there could exist (PS) sequences $u_n \to 0$. For example, when $V_\delta(x) = -|x|^{-\alpha} - a|x|^{-2}$ with $0 < \alpha < 1$, a direct calculation shows that the sequences

$$u_n = r_n e^{i2\pi t},$$

with $r_n \to 0$, satisfy

$$g_\delta(u_n) = 2\pi^2 r_n^2 (h + r_n^{-\alpha} + \delta_n^{-2}) \to 2\pi^2 \delta.$$

and

$$g'_\delta(u_n) \to 0.$$

Hence they are (PS) sequences which converge to $u = 0$. In order to overcome this problem one remarks that any critical point u of g_δ satisfy the relation $(g'_\delta(u) \mid u) = 0$ which implies

$$\int_0^1 [V(u) + \frac{1}{2} V'(u) \cdot u] dt = h. \tag{6.8}$$

If $V \in C^2$ and

$$3V'(x) \cdot x + V''(x) x \cdot x \neq 0, \qquad \forall x \neq 0,. \tag{6.9}$$

then (6.8) defines a C^1 manifold M_h diffeomorphic to Λ and critical points of g_δ constrained on M_h verify $g'_\delta(u) = 0$. In addition, if V satisfies

$$V'(x) \cdot x > 0, \qquad \forall x \neq 0, \tag{6.10}$$

then

$$g_\delta(u) = 0, \qquad (u \in M_n) \quad \text{iff} \quad u(t) \equiv \text{const.} \qquad (6.11)$$

Let us also point out that $M_h \neq \emptyset$ whenever $h < 0$ and

$$\liminf_{|x| \to +\infty} [V(x) + \frac{1}{2} V'(x) \cdot x] \geq 0 \qquad (6.12)$$

and

$$\lim_{|x| \to 0} [V(x) + \frac{1}{2} V'(x) \cdot x] = -\infty. \qquad (6.13)$$

Moreover one has that

$$\text{cat}(M_h) = \text{cat}(\Lambda) = +\infty. \qquad (6.14)$$

The advantage of working such a manifold relies on the fact that g_δ satisfies the (PS) condition on M_h. Precisely the following lemma holds

Lemma 6.4. *Suppose that V satisfies (6.9), (6.10), (6.12) and*

(V6) *$\exists \alpha \in (0, 2)$ such that $V'(x) \cdot x \geq -\alpha V(x)$, $\forall x \neq 0$;*

(V7) *$\exists r > 0$, $\beta \in (0, 2)$ such that $V'(x) \cdot x \leq -\beta V(x)$, $\forall 0 < |x| < r$.*

Then g_δ satisfies (PS) on M_h. Moreover, for all $a > 0$ the sublevels g_δ^a are closed.

Let us point out that (V6–7) imply (6.13). The preceding lemma and (6.14) allow us to use the Lusternik–Schnirelman theory. Letting $X = \{u \in M_h : u(t) \equiv \xi\}$, one plainly finds that $\text{cat}(X, M_h) = 2$ and thus, using (6.11), it follows that the critical levels c_k are positive whenever $k \geq 3$. From the preceding discussion one infers there exists a critical point u_δ of g_δ, such that $g_\delta(u_\delta) > 0$. Let ω_δ denote the corresponding value indicated in Lemma 6.3. One shows that there exists $u \in E$ and $\omega \neq 0$ such that:

1. $u_\delta \rightharpoonup u$, weakly;

2. $\omega_\delta \to \omega$.

In conclusion one has

Theorem 6.5. [9] *Suppose that V satisfies (6.9), (6.10), (6.12), (V6) and (V7). Then for all $h < 0$ (6.6)–(6.7) has a periodic weak solution.*

Improvements of the preceding result can be found in [10], [73] and [85]. The first paper deals with a class of symmetric n-body problems; in the second one the existence of a *collision* solution for potentials which behave like $-|x|^{-\alpha}$ $(0 < \alpha < 2)$ near $x = 0$, is established under mild assumptions; the last paper is concerned with the existence of weak solutions without assuming (6.9). Under the further hypothesis that $V(x) = -|x|^{-\alpha}+$ higher order terms, the regularity of such a solution is investigated by means of Morse theory. For example, one can show the existence of a noncollision solution, provided $1 < \alpha < 2$.

We end this survey by a short discussion of the recent paper [16]. For the sake of brevity we will be sketchy and refer to the quoted paper for more details. Given $h < 0$, the region of admissible motions of solutions of (6.6)–(6.7) is the bounded set

$$\Omega_h = \{x \in R^N \backslash \{0\} : V(x) \leq h\}.$$

Then is makes sense to search solutions of (6.6)–(6.7) under assumptions on V in Ω_h only. Actually, if V is regular, namely if $V \in C^1(R^N, R)$, solutions exist whenever the set $\{x \in R^N : V(x) \leq h\}$ is compact and $V'(x) \neq 0$ on $\partial\Omega_h := \{x \in R^N : V(x) = h\}$, see for example [24], [28], [52], [54]. When V is singular, the admissible set Ω_h is not compact any more and assumptions on V at the singularity $x = 0$ is in order. Actually, in the model case $V(x) = -|x|^{-\alpha}$, problem (6.6)–(6.7), with $h < 0$, has no solution if $\alpha \geq 2$, although Ω_h is bounded and $V'(x) \cdot x > 0$ on $\partial\Omega_h$. In contrast, when $V(x) \simeq -|x|^{-\alpha}$ with $0 < \alpha < 2$ solutions exist, as the following theorem shows.

Theorem 6.6. *Let $V \in C^1(R^N \backslash \{0\}, R)$ and suppose Ω_h is bounded and $V'(x) \neq 0$ on $\partial\Omega_h$. Moreover, let us assume there exist $a, r > 0$ and $\alpha \in (0, 2)$ such that*

  (V8) $V(x) \geq -\frac{\alpha}{|x|^2}$ $\forall 0 < |x| \leq r$;

  (V9) $V'(x) \cdot x \geq 0$ $\forall 0 < |x| \leq r$;

  (V10) $\lim_{|x| \to 0}[V(x) + \frac{1}{2}V'(x) \cdot x] = -\infty$.

Then (6.6)–(6.7) has a weak periodic solution.

Moreover, under a further assumption on the behaviour of V near $x = 0$ one can be more precise about the regularity of the solutions.

When V is even the preceding result can be greatly sharpened.

Theorem 6.7. *Let $V \in C^2(R^N \backslash \{0\}, R)$ satisfy the preceding assumptions. Moreover suppose V is even and there exists $\rho > 0$ such that*

  (V11) $V(x) = -|x|^{-\alpha} + \psi(|x|)$, $\forall 0 < |x| \leq \rho$, *where ψ satisfies*

$$s^{\alpha}\psi(s), s^{\alpha+1}\psi'(s), s^{\alpha+2}\psi''(s) \to 0 \quad as \quad s \to 0.$$

Then (6.6)–(6.7) has a non-sollision solution, provided that either

$$N \geq 4 \text{ and } 0 < \alpha < 2, \text{ or}$$
$$N = 3 \text{ and } 1 < \alpha < 2.$$

These results have been proved by using an approach which is quite different from that one employed for proving Theorem 6.5. Let $\chi_n \in C^2(R)$ denote functions such that

$$\begin{cases} \chi_n(s) = -n & \text{for } s \leq -n - 1/n \\ \chi_n(s) \leq s + 1/n & \text{for } s \in [-n - 1/n, -n/2] \\ \chi_n(s) = s & \text{for } s \geq -n/2 \end{cases}$$

we define a sequence of smooth potentials by setting

$$V_n(x) = \begin{cases} \chi_n(V(x)) & \text{for } x \neq 0 \\ -n & \text{for } x = 0. \end{cases}$$

Plainly, for n large one has that $\{x \in R^N : V_n(x) \leq h\} = \Omega_h \cup \{0\}$. Then, according to the results cited before, the perturbed problem

$$\ddot{q} + V_n(q) = 0 \tag{6.15}$$

has a periodic solution q_n with energy h:

$$\frac{1}{2}|\dot{q}_n|^2 + V(q_n) = h. \tag{6.16}$$

Roughly, if u_n is a critical point of $g_n(u) = \frac{1}{2}\int_0^1 |\dot{u}|^2 \cdot \int_0^1 (h - V_n(u))$ at positive level and ω_n is the corresponding ratio

$$\omega_n^2 = \frac{\int_0^1 (h - V_n(u_n))}{\frac{1}{2}\int_0^1 |\dot{u}_n|^2} > 0,$$

then, similarly as in Lemma 6.3, $q_n(t) = u_n(\omega_n t)$ is a solution of (6.15), satisfying (6.16). Here the Linking Theorem applies to g_n[1] yielding a critical point u_n such that $g_n(u_n) > 0$.

---

[1] Actually, in order to show that g_n satisfy the (PS) condition, V_n is replaced by a suitable $\tilde{V}_n$ which is $= V_n$ in $\Omega_h \cup \{0\}$ and is such that $\tilde{V}_n(x) \to +\infty$ as $x \to \bar{x} \in \partial\Omega_{h+\epsilon}$ (see [28] for more details). However, to keep notation simpler, we write hereafter V_n instead of $\tilde{V}_n$.

The subsequent step is to show that

$$u_n \to u \neq 0, \qquad \omega_n \to \omega \neq 0, \tag{6.17}$$

and that $q(t) := u(\omega t)$ is a week solution of (6.6)–(6.7). For this, one clearly needs to find appropriate a-priori bounds. When V is even this is rather easy. Indeed, one can introduce the subspace

$$E_0 = \{u \in H^1(S^1, R^N) : u(t + \tfrac{1}{2}) = -u(t)\}$$

and work on $\Lambda_0 = \Lambda \cap E_0$. Since V is even it follows that critical points of g_n on Λ_0 coincide with critical points of g_n on Λ. Moreover,

$$\|\dot{u}\|^2 := \int_0^1 |\dot{u}|^2 dt$$

is a norm in E_0 equivalent to the usual one. Then the functional g_n has the Mountain Pass geometry because

$$g_n''(0)[w, w] = (n + h)\|\dot{w}\|^2$$

and, letting $z(t) = \xi e^{2\pi i t}$, $\xi \in R^N$, $\|\dot{z}\|^2 = 1$,

$$g_n(sz) = \frac{1}{2}s^2 \cdot \int_0^1 [h - V(sz)]dt \to -\infty \quad \text{as } s \to +\infty.$$

Let b_n, u_n, denote the Mountain Pass critical level, the Mountain Pass critical point respectively, of g_n found by using Theorem 4.1. From (V8) it follows

$$g_n(sz) = \frac{1}{2}s^2 \cdot \int_0^1 [h - V(sz)]dt \le \frac{1}{2}hs^2 + a_1 s^{2-\alpha}, \tag{6.18}$$

for $s > 0$ small enough. This readily implies there exists $\bar{b}$ such that $b_n \le \bar{b}$. Moreover, since $V_n(x) \le V_1(x)$ one infers that there exists $\underline{b} > 0$ such that $\underline{b} \le b_n$. These uniform bounds allow us to show that (6.17) holds and hence $q(t) = u(\omega t)$ is a weak solution of (6.6)–(6.7).

When V is not even, the question is more subtle and critical points u_n of g_n such that $\underline{b} \le g_n(u_n) \le \bar{b}$ have to be found by a suitable, unusual linking procedure.

Finally, Morse theoretical arguments allow us to estimate the number of sollisions of q, yielding, for example, the existence of a noncollision solution in the case covered by Theorem 6.7.

To give an idea of how the number of collisions of the solution q is related to the Morse index $m(u_n)$ of the approximated critical points u_n

(n large), let us sketch below the argument. First of all, recall that, by a straight calculation

$$g_n''(u_n)[w, w] = D_n \cdot \int_0^1 [|\dot{w}|^2 - \frac{1}{\omega_n^2} V_n''(u_n)w \cdot w]dt$$

$$- 2 \int_0^1 \dot{u}_n \cdot \dot{w} dt \cdot \int_0^1 V_n'(u_n) \cdot w dt, \qquad (6.19)$$

where

$$D_n = \int_0^1 [h - V_n(u_n)]dt > 0.$$

Suppose now that $\tau \in \Gamma[q]$. From $u(\tau) = 0$ on infers that there exists a neighbourhood J of τ such that $|u_n(t)| \leq \rho$ for all $t \in J$. Then from assumption (V11) it follows that u_n is planar for $t \in J$:

$$u_n(t) = r_n(t)e^{i\theta_n(t)}, \qquad t \in J.$$

If $w(t)$ is chosen in such a way that its support supp$[w]$ is contained in J and $\dot{w}(t) \cdot \dot{u}_n(t) = 0$ for all $t \in$ supp$[w]$, (6.19) becomes

$$g_n''(u_n)[w, w] = D_n \cdot \int_J [|\dot{w}|^2 - \frac{1}{\omega_n^2} V_n''(u_n)w \cdot w]dt. \qquad (6.20)$$

Since

$$V_n''(u_n)w \cdot w = \chi_n''(V(u_n))(V'(u_n) \cdot w)^2 + \chi_n'(V(u_n))V''(u_n)w \cdot w$$
$$\geq \chi_n'(V(u_n))V''(u_n)w \cdot w, \qquad (6.21)$$

and, according to (V11),

$$V''(u_n)w \cdot w \simeq \alpha \frac{w^2}{|u_n|^{\alpha+2}} - \alpha(\alpha + 2)\frac{u_n \cdot w}{|u_n|^{\alpha+4}},$$

up to higher order terms. If, in addition, one takes w in such a way that $w(t) \cdot u_n(t) \equiv 0$ on supp$[w] \subset J$ then (6.21) yields

$$\int_J V_n''(u_n)w \cdot w dt \geq \int_J \chi_n'(V(u_n))V''(u_n)w \cdot w \, dt$$

$$\simeq \alpha \int_J \chi_n'(V(u_n))\frac{w^2}{|u_n|^{\alpha+2}}dt \qquad (6.22)$$

Next, note that the angular momentum M_n is conserved and $M_n \to 0$, because u has a collision. First, let us suppose that $M_n = 0$ for n large. In such a case u_n lies on straight lines:

$$u_n(t) = \xi_n r_n(t), \qquad \xi_n \in R^N.$$

Plainly, $\omega_n \to \omega$, $u_n \to u$, in the C^2 topology away from $\Gamma[q]$. Moreover, as a consequence of (6.16), one has that $u(t) = \xi r(t)$ satisfies

$$\frac{1}{2}\omega^2 \dot{r}^2 + V(r) = h.$$

From this it follows that

$$r(t) \simeq (t - \tau)^{2/(\alpha+2)}$$

near $t = \tau$ and hence

$$J_n := \alpha \int_J \chi_n'(V(u_n))\frac{w^2}{|u_n|^{\alpha+2}} dt \to +\infty,$$

whenever $w(0) \neq 0$. Let us fix any index n such that

$$J_n > \omega_n^2 \qquad (6.23)$$

(recall that ω_n are bounded), and let $e_1, \ldots, e_{N-1}$ be an orthonormal basis in R^{N-1} such that $e_j \cdot \xi_n = 0$ for all $j = 1, \ldots, N-1$. Moreover, let $\varphi(t)$ a smooth real valued function such that: (i) $\varphi(0) \neq 0$; (ii) $\int \dot{\varphi}^2 = 1$; and (iii) $\text{supp}[\varphi] \subset J$ and set $w_j(t) = e_j \varphi(t)$. Then $w_j(t) \cdot u_n(t) = \dot{w}_j(t) \cdot \dot{u}_n(t) = 0$ for all $t \in J$ and (6.20) applies. Using (6.22) and (6.23), (6.19) yields

$$g_n''(u_n)[w, w] = D_n \cdot \int_J [\dot{\varphi}^2 - \frac{1}{\omega_n^2}V_n''(u_n)w \cdot w]dt$$

$$\leq D_n \cdot (1 - \int_J \chi_n'(V(u_n))V''(u_n)w \cdot w\, dt)$$

$$\simeq D_n \cdot (1 - \frac{J_n}{\omega_n^2}) < 0$$

for all such w. Since this can be repeated for each $\tau \in \Gamma[q]$ one deduces that the number of possible collisions $\nu := \#(\Gamma[q])$ satisfies

$$(N - 1)\nu \leq m(u_n),$$

for n sufficiently large. The arguments can also be carried out when $M_n \neq 0$, possibly, and yield to an estimate such as

$$i(\alpha)(N - 2)\nu \leq m(u_n),$$

(remark that u_n is now merely planar near $\tau = 0$ and hence the one can find $N - 2$ test functions w orthogonal to u_n) where

$$i(\alpha) = \max\{\kappa \in N : \kappa < \frac{2}{2 - \alpha}\}.$$

This, jointly with Theorem 4.4, leads to bound the number of collisions.

Finally, when V is even, the result becomes rather sharp, because u_n is a Mountain Pass critical point and therefore $m(u_n) \leq 1$, proving, as stated in Theorem 6.7, that $\nu = 0$, namely that (6.6)–(6.7) has a non-collision solution.

There are still several questions concerning singular potentials that should be studied. For example, to cite the most important one, let us recall the existence of periodic motions of the n-body type problem, having prescribed energy $h < 0$. The only known result on such a topic is the paper [10] dealing with symmetric n-body type potentials.

References

[1] Adimurthi and Mancini G., *Geometry and topology of the boundary in the critical Neumann problem*, to appear.

[2] Adimurthi, Pacella, F., and Yadava S.L., *Interaction between the geometry of the boundary and positive solutions of a semilinear Neumann problem with critical nonlinearity*, to appear.

[3] Alama, S. and Tarantello, G., *On semilinear elliptic equations with indefinite nonlinearities*, Calculus of Variat. (to appear)

[4] Ambrosetti, A., *A perturbation theorem for superlinear boundary value problems*, M.R.C. Tech. Summ. Rep. (1974).

[5] Ambrosetti, A., *Elliptic equations with jumping nonlinearities*, J. Math. Phys. Sci. **18-1** (1984), 1-12.

[6] Ambrosetti, A., *Critical points and nonlinear variational problems*, Suppl. Bulletin de la Soc. Math. de France, Mem. N.49, Tome **120**, 1992.

[7] Ambrosetti, A., Brezis, H., and Cerami, G., *Combined effects of concave and convex nonlineatities in some elliptic problems*, Jour. Funct. Anal. (to appear)

[8] Ambrosetti, A. and Coti Zelati, V., *Critical point with lack of compactness and singular dynamical systems*, Ann. Mat. Pura Appl. **149** (1987), 237-259.

[9] Ambrosetti A. and Coti Zelati, V., *Closed orbits with fixed energy for singular Hamiltonian Systems*, Archive Rat. Mech. Analysis, **112** (1990), 339-362.

[10] Ambrosetti, A. and Coti Zelati, V., *Closed orbits with fixed energy for a class of N-body problems*, Ann. Inst. Poincaré Anal. Nonlinéaire. **9** (1992), 187-200, and 337-338.

[11] Ambrosetti, A. and Coti Zelati, V., *Periodic Solutions of Singular Lagrangian Systems*, Birkhäuser, 1993.

[12] Ambrosetti, A. and Prodi, G., *A Primer of Nonlinear Analysis*, Cambridge Univ. Press, 1993.

[13] Ambrosetti, A. and Rabinowitz, P.H., *Dual variational methods in critical points theory and applications*, J. Funct. Anal. **14** (1973), 349-381.

[14] Ambrosetti, A. and Struwe, M., *A note on the problem* $-\Delta u = \lambda u + u |u|^{2^*-2}, u \in H_0^1$, $\lambda > 0$, Manus. Math. **54** (1986), 373-379.

[15] Ambrosetti, A. and Struwe, M., *Existence of steady vortex rings in an ideal fluid*, Arch. Rat. Mech. Anal. **108**, 2 (1989), 97-109.

[16] Ambrosetti, A. and Struwe, M., *Periodic motions of conservatiove systems with singular potentials*, Nonlin. Diff. Equat. Appl. (to appear)

[17] Bahri, A., *Topological results on a certain class of functionals and applications*, J. Funct. Anal. **41** (1981), 397-427.

[18] Bahri, A. and Berestycki, H., *A perturbation method in critical point theory and applications*, Trans. Amer. Math. Soc. **267** (1981), 1-32.

[19] Bahri, A. and Coron, J.M., *On a nonlinear elliptic equation involving the critical Sobolev exponent: the effect of the topology of the domain*, Comm. Pure Appl. Math. **41** (1988), 253-294.

[20] Bahri, A. and Lions, P.L., *Morse index of some min-max critical points I. Application to multiplicity results*, Comm. Pure Appl. Math. **41** (1988), 1027-1037.

[21] Bahri, A. and Rabinowitz, P.H., *A minimax method for a class of Hamiltonian systems with singular potentials*, J. Funct. Anal. **82** (1989), 412-428.

[22] Bahri, A. and Rabinowitz, P.H., *Solutions of the three-body problem via critical points of infinity*, Ann. Inst. Poincaré Anal. Nonlinéaire **8** (1991), 561-649.

[23] Benci, V., *A geometrical index for the group S^1 and some applications to the research of periodic solutions of O.D.E.'s*, Comm. Pure Appl. Math. **34** (1981), 393-432.

[24] Benci, V., *Closed geodesics fior the Jacobi metric and periodic solutions of prescribed energy of natural Hamiltonian systems*, Ann. Inst. Poincaré Anal. Nonlinéaire **1** (1984), 401-412.

[25] Benci, V. and Cerami, G., *Positive solutions of some nonlinear elliptic problems in exterior domains*, Arch. Rat. Mech. Anal. **99** (1987), 283-300.

[26] Benci, V. and Cerami, G., *The effect of the domain topology on the number of positive solutions of nonlinear elliptic problems*, Arch. Rat. Mech. Anal. **114** (1991), 79-93.

[27] Benci, V. and Fortunato, D., *Subharmonic solutions of prescribed minimal period for non-autonomous differential equations*, in: Recent Advances in Hamiltonian systems, World Scientific, 1987.

[28] Benci, V. and Giannoni, F., *A new proof of the existence of a brake orbit*, in: Advanced topics in the theory of dynamical systems, Academic Press, 1990.

[29] Benci, V. and Rabinowitz, P.H., *Critical point theorems for indefinite functionals*, Invent. Math. **52** (1979), 241-273.

[30] Berestycki, H. and Lions, P.L., *Nonlinear scalar field equations*, Arch. Rat. Mech. Anal. **82** (1983), 313-376.

[31] Berestycki, H. and Lions, P.L., *Existence of stationary states in nonlinear scalar field equations*, in: Bifurcation Phenomena in Mathematical Physics and related topics, Ridel Publ. Co., 1980.

[32] Boccardo, L., Escobedo, M., and Peral, I., *A Dirichlet problem involving critical exponent*, to appear.

[33] Brezis, H. and Coron, J.M., *Multiple solutions of H-systems and the Rellich conjecture*, Comm. Pure Appl. Math **37** (1984), 149-187.

[34] Brezis, H. and Nirenberg, L., *Positive solutions of nonlinear elliptic equations involving critical Sobolev exponents*, Comm. Pure Appl. Math. **36** (1983), 437-477.

[35] Browder, F., *Infinite dimensional manifolds and nonlinear eigenvalue problems*, Ann. of Math. **82** (1965), 459-477.

[36] Capozzi, A., Fortunato, D., and Palmieri, G., *An existence result for nonlinear elliptic problems involving critical Sobolev exponent*, Ann. Inst. Poincaré Anal. Nonlinéaire **2** (1985), 463-470.

[37] Capozzi, A., Greco, C., and Salvatore, A., *Lagrangian systems in presence of singularities*, Proc. Amer. Math. Soc. **102** (1988), 125-130.

[38] Cerami, G., Solimini, S., and Struwe, M., *Some existence results for superlinear elliptic boundary value problems involving critical Sobolev exponent*, J. Funct. Anal. **69** (1986), 209-212.

[39] Coron, J.M., *Topologie et cas limite des injections de Sobolev*, C. R. Acad. Sci. Paris **299** (1984), 209-212.

[40] Coti Zelati, V., *Periodic solutions for N-body type problems*, Ann. Inst. Poincaré Anal. Nonlinéaire **7** (1990), 477-492.

[41] Coti Zelati, V. and Serra, E., *Collision and noncollision solutions for a class of Keplerian-like dynamical systems*, to appear.

[42] Dancer, E.N., *The effect of the domain shape on the number of positive solutions of certain nonlinear equations*, J. Diff. Equat. **74** (1988), 120-156.

[43] Ding, W.Y., *Positive solutions of $-\Delta u + u^{(N+2)/(N-2)} = 0$ on contractible domains*, to appear.

[44] Ekeland, I., *Convexity Methods in Hamiltonian Mechanics*, Springer, 1990.

[45] Ekeland, I., *On the variational principle*, J. Math. Anal. Appl. **47** (1974), 324-353.

[46] Fadell, E. and Husseini, S., *A note on the category of free loop space*, Proc. Amer. Math. Soc. **107** (1989), 527-536.

[47] Fraenkel, L.E. and Berger, M.S., *A global theory of steady vortex rings in an ideal fluid*, Acta Math. **132** (1974), 13-51.

[48] Garcia, J. and Peral, I., *Multiplicity of solutions for elliptic problems with critical exponent or with a nonsymmetric term*, Trans. Amer. Math. Soc. **323** (1991), 877-895.

[49] Gordon, W.B., *Conservative dynamical systems involving strong forces*, Trans. Amer. Math. Soc. **204** (1975), 113-135.

[50] Ghoussoub, N. and Preiss, D., *A general mountain pass principle for locating and classifying critcal points*, Ann. Inst. Poincaré Anal. Non-linéaire, **6** (1989), 321-330.

[51] Greco, C., *Periodic solutions of a class of singular Hamiltonian systems*, J. Nonlinear Analysis T.M.A. **12** (1988), 259-270.

[52] Hayashi, K., *Periodic solutions of classical Hamiltonian systems*, Tokio J. Math. **6** (1983), 473-486.

[53] Hofer, H., *A note on the topological degree at a critical point of mountain-pass type*, Proc. Amer. Math. Soc. **90** (1984), 309-315.

[54] Hofer, H. and Zehnder, E., *Periodic solutions on hypersurfaces and a result by C. Viterbo*, Inv. Math. **90** (1987), 1-9.

[55] Krasnoselskii, M.A., *Topological Methods in the theory of nonlinear integral equations*, Pergamon, Oxford, 1965.

[56] Lazer, A.C. and Solimini, S., *Nontrivial solutions of operator equations and Morse indices of critical points of min-max type*, J. Nonlin. Anal. T.M.A. **12** (1988), 761-775.

[57] Lions, P.L., *The concentration-compactness principle in the Calculus of Variaitons. The locally compact case (part 1 and 2)*, Ann. Inst. Poincaré Anal. Nonlinéaire (1984), 109-145 and 223-283.

[58] Lions, P.L., *The concentration-compactness principle in the Calculus of Variaitons. The limit case (part 1 and 2)*, Riv. Iberoamer. **1** (1985), 45-121 and 145-201.

[59] Lusternik, L. and Schnirelman, L., *Méthode topologique dans les problémes varationelles*, Hermann, Paris, 1934.

[60] Majer, P., *Ljusternik-Schnirelman theory without Palais-Smale condition and singular dynamical systems*, Ann. Inst. Poincaré Anal. Nonlinéaire **8** (1991), 459-476.

[61] Majer, P. and Terracini, S., *Periodic solutions to some N-body type problems*, to appear.

[62] Mawhin, J. and Willem, M., *Critical Point Theory and Hamiltonian Systems*, Springer, 1989.

[63] Morse, M., *Relations between the critical points of a real function of n independent variables*, Trans. Amer. Math. Soc. **27** (1925), 345-396.

[64] Nirenberg, L., *Variational and topological methods in nonlinear problems*, Bull. Amer. Math. Soc. 4-3 (1981), 267-302.

[65] Palais, R., *Lusternik–Schnirelman theory on Banach manifolds*, Topology **5** (1966), 115-132.

[66] Palais, R. and Smale, S., *A generalized Morse theory*, Bull. Amer. Math. Soc. **70** (1964), 165-171.

[67] Pohozaev, S.I., *Eigenfunctions of the equations* $\Delta u + \lambda f(u) = 0$, Soviet Math. **5** (1965), 1408-1411.

[68] Passaseo, D., *Multiplicity of positive solutions of nonlinear elliptic equations with critical Sobolev exponent in some contractible domains*, Manus. Math. **65** (1989), 147-166.

[69] Pucci, P. and Serrin, J., *Extension of the mountain pass theorem*, J. Funct. Anal. **59** (1984), 185-210.

[70] Pucci, P. and Serrin, J., *A mountain pass theorem*, J. Diff. Equat. **60** (1985), 142-149.

[71] Rabinowitz, P.H., *Some minmax theorems and applications to nonlinear partial differential equations*, in: Nonlinear Analysis, Academic Press, 1978.

[72] Rabinowitz, P.H., *Multiple critical points for perturbed symmetric functionals*, Trans. Amer. Math. Soc. **272** (1982), 753-770.

[73] Rabinowitz, P.H., *A note on periodic solutions of prescribed energy for singular Hamiltonian systems*, to appear.

[74] Rabinowitz, P.H. and Tanaka, K., *Some results on connecting orbits for a class of Hamiltonian systems*, Math. Zeit.**206** (1991), 473-499.

[75] Salvatore, A., *Multiple periodic solutions for Hamiltonian systems with singular potentials*, Rend. Mat. Acc. Lincei, Ser. IX, **3** (1992), 111-120.

[76] Schwartz, J.T., *Generalizing the Lusternik–Schnirelman theory of critical points*, Comm. Pure Appl. Math. **17** (1964), 307-315.

[77] Serra, E., *Dynamical systems with singular potentials: existence and qualitative properties of periodic motions*, Ph. D. Thesis, SISSA, Trieste, 1991.

[78] Serra, E. and Terracini,S., *Collisionless periodic solutions to some three-body like problems*, Arch. Rat. Mech Anal. (to appear).

[79] Srikanth, P.N., *Uniqueness of solutions of nonlinear Dirichlet problems*, Diff. Int. Equations, **6** (1993), 663-670.

[80] Struwe, M., *Infinitely many critical points for functionals which are not even and applications to superlinear boundary value problems*, Manus. Math. **32** (1980), 335-364.

[81] Struwe, M., *Large H-surfaces via the mountain-pass- lemma*, Math. Annalen **270** (1985), 441-459.

[82] Struwe, M., *Variational Methods*, Springer, 1990.

[83] Szulkin, A., *Ljusternik-Schnirelmann theory on C^1 manifolds*, Ann. Inst. Poincaré Anal. Nonlinéaire **5** (1988), 119-139.

[84] Tanaka, K., *Prescribed energy problem for a singular Hamiltonian system with a weak force*, to appear.

[85] Tanaka, K., *Non-collision solutions for a second order singular Hamiltonian system with weak forces*, J. Funct. Anal. (to appear).

[86] Terracini, S., *Periodic solutions for dynamical systems with Keplerian type potentials*, Ph. D. Thesis, SISSA, Trieste, 1990.

[87] Vainberg, M.M., *Variational Methods for the Study of Nonlinear Operators*, Hoden-Day, San Francisco, 1964.

[88] Viterbo, C., *Indice de Morse des points critiques obtenus par minmax*, Ann. Inst. Poincaré Anal. Nonlinéaire **5** (1988), 221-225.

[89] Wang, Z.Q., *On a superlinear elliptic equation*, Ann. Inst. Poincaré Anal. Nonlinéaire **8** (1991), 43-58.

[90] Wang, Z.Q., *Uniqueness of positive solutions of $\Delta u + u + u^p = 0$ in a ball*, Comm. P.D.E. **17** (1992), 1141-1164.

Introduction to Morse Theory
A New Approach

Vieri Benci

Istituto di Matematiche Applicate "U. Dini"
Università di Pisa
56100 Pisa, Italy

To the Memory of Charles Conley

INTRODUCTION

This paper present a new approach to Morse theory with the aim to give to the unexperienced reader an extra tool for working in the critical point theory. Of course this presentation depends on the taste of the writer and the applications are chosen among the ones more familiar to him.

The first Chapter presents a new construction of the abstract theory based on the Conley Index [Co,78]. Actually the Conley Index Theory for isolated invariant sets is more general and abstract than it is needed for the applications which we have in mind; moreover it is based more on homotopy than on homology. However, the leading ideas of this theory can be used to construct the Morse theory in a more flexible way.

The second Chapter applies the Morse theory to the study of geodesics in Riemannian geometry. This is the first and one of the most interesting applications of infinite dimensional Morse theory and it deserves a special attention in any introduction to it.

The third Chapter is devoted to the Morse theory for geodesics in Lorentzian manifolds; this is a natural extension of the Riemannian theory, but it presents a lot of difficult problems which only in recent years begin to be faced. This topic has been chosen because of the interest of the author in it.

Finally, we have a chapter devoted to elliptic partial differential equations. In this field, the Morse theory has been applied much less than it deserves. For example, the ideas contained in the pioneering paper of Marino and Prodi on this subject, [Ma,Pr,75], did not have the attention which should be expected. Probably the main reasons are the following ones: (a) the functional involved are usually of class C^1 and not C^2; (b) the Morse theory is simpler for the Morse function i.e. for functions whose critical points are not degenerate, and in P.D.E. this generic assumption is not natural; (c) sometimes, people working in P.D.E.'s feel uneasy in using algebraic topology.

We hope that our approach can overcome these difficulties and we hope that the Morse theory will get a bigger attention from people working in this field.

We did not have enough time to write an other chapter on the applications of Morse theory to the study of periodic solutions of Hamiltonian systems. This work, started by Conley, Zahnder [Co,Ze,84] and, with a different approach by Ekeland [Ek,84] (cf. also the book [Ek,90]), is beautiful and full of interesting results. We just mention a recent paper of Dell'Antonio, D'Onofrio and Ekeland,

[De,DO,Ek,ta], in which the Morse theory has been exploited to prove that the Hamiltonian systems, convex and even, are not ergodic in general. We refer to [De,ta] for a survey paper on this subject. We refer also to [Be,91] for a treatment of the Morse theory for Hamiltonian systems more in the spirit of the approach given here.

Regarding the references, I did not attempt to include all significant papers in Morse theory and the in other topics treated here and I apologize to those people whose fine contributions to these subjects have not been cited. The inclusion of many of my papers must be interpreted as a consequence of the author's knowledge and the choice of the topics and not as any measure of their relative importance to the field.

CONTENTS

Chapter I   - The abstract theory.

Point Theorem, the Equivariant Linking theorem, the Box theorem.

Chapter II - An applications to Riemannian geometry.

II.1 - Riemannian manifolds.
Covariant derivative, affine connections, curvature.

II.2 - Geodesics.
Definition of geodesics, the energy functional, the geodesics equation, the exponential map.

II.3 - The Morse theory for geodesics.
The space $\Omega^1(\mathfrak{M};x_0,x_1)$, the Morse relations.

II.4 - The index theorem.
The second variation of E, the Jacobi fields and the geometrical index, the index theorem

Chapter III - An applications to the space-time geometry.

III.1 - Introduction
Lorentian geometry and General Relativity, geodesics in Lorentian manifolds, causal type of the geodesics

III.2 - Some example of Lorentian manifolds.
The Minkowsky space-time, the static space-time, space-time of split-type.

III.3 - Morse theory for Lorentian manifolds.
Morse theory for space-times of split-type, Morse theory for static space-times.

III.4 - Preliminary lemmas.
The "J" functional, the first order variational principle, the second order variational principle.

III.5 - Proof of the Morse relations for static space

times

The "J" functional for static space-times, the index theorem for static space-times, proof of the Morse relations for static space-time.

Chapter IV - Some applications to a semilinear elliptic equation

IV.1 - Introduction

The model problem, the functional setting and differentiability, Morse index and nodal regions.

IV.2 - The sublinear case.

The Box theorem for elliptic equations, the three solutions theorem.

IV.3 - The superlinear case.

Existence of infinitely many solutions.

IV.4 - Morse relation for positive solutions

The model problem, the main result, a particular case.

IV.5 - The functional setting.

The manifold M_ε, the constrained critical points.

IV.6 - Some hard analysis.

the number $m(\varepsilon,\rho)$, the barycentre, the number $m^*(\varepsilon,\rho,\gamma)$, estimates on $m(\varepsilon,\rho)$ and $m^*(\varepsilon,\gamma)$.

IV.7 - The photography method.

Estimates on the range of β, the map ϕ_ε.

IV.8 - The topology of the strips.

Computation of $\mathcal{P}_\lambda(E_\varepsilon^{m(\varepsilon,r)}, E_\varepsilon^\delta)$, computation of $\mathcal{P}_\lambda(W_0^1(\Omega), E_\varepsilon^{m(\varepsilon,r)})$, proof of the main results.

C H A P T E R   I
T H E   A B S T R A C T   T H E O R Y

The main point of Morse theory consists in the "Morse relations", i. e. in the relations between two polynomials, the Morse polynomial and the Poincaré (or Betti) polynomial. The Morse polynomial depends on the critical points of a function and it will be discussed in this section. The Poincaré polynomial is a topological invariant and it will be introduced in section I.1.2.

I.1 - THE MORSE INDEX.

Throughout this section Λ will be a Hilbert manifold of class C^1, Ω an open subset of Λ and f a functional of class C^1 in a neighborhood of $\overline{\Omega}$.

Critical points.

Definition I.1.1. *A point $x \in \Lambda$ is called a critical point of f if*

$$df(x) = 0$$

If x is not a critical point is called regular point. The set of critical points of f will be denoted by K_f.

$a \in \mathbb{R}$, is called critical value if there exists $x \in K_f$ such that $f(x) = a$; $a \in \mathbb{R}$ is called regular value for f if it is not a critical value.

We spend few words to recall the definition of the differential of a function. First of all we recall that given a tangent vector v at Λ in the point x there exists a smooth curve

$$\gamma_v : (-\varepsilon, \varepsilon) \longrightarrow \Lambda$$

such that

$$\gamma_v(o) = x$$
$$\dot{\gamma}_v(o) = v$$

Actually, some times, the class of equivalence of such curves is taken as definition of tangent vector. The set of all tangent vectors at Λ in the point x has a natural structure of vector space and it is called tangent space at Λ in the point x an denoted by $T_x\Lambda$. The set

$$T\Lambda = \bigcup_{x \in \Lambda} T_x\Lambda$$

is called tangent bundle.

Remark I.1.2. By a well known theorem of Whitney, every differentiable manifold Λ is diffeomorphic to a submanifold of $\mathbb{R}^N$ (for a suitable N); having this picture in mind, it is easy to visualize the meaning of tangent vectors.

If f is a C^1-function, the differential of f at x,

$$df(x): T_x\Lambda \longrightarrow \mathbb{R}$$

can be defined as follows:

$$df(x)[v] = \frac{d}{dt} f(\gamma_v(t))\Big|_{t=0}$$

Now, let

$$\gamma_{v,w} : \mathcal{U} \longrightarrow \Lambda$$

(where $\mathcal{U}$ is a neighborhood of 0 in $\mathbb{R}^2$) be a function of two variables t and s such that:

$$\gamma_{v,w}(o) = x$$
$$\partial_t \gamma_{v,w}(o) = v$$
$$\partial_s \gamma_{v,w}(o) = w$$

If x is a critical point of f, and f is of class C^2 in a neighborhood of x we define the Hessian form of f at x

$$H^f(x): T_x\Lambda \times T_x\Lambda \longrightarrow \mathbb{R},$$

as follows

$$H^f(x)[v,w] = \frac{\partial^2}{\partial t \partial s} \, f(\gamma_{v,w}(t,s)) \Big|_{t=0,s=0}.$$

Since x is a critical point of f, it is not difficult to see that the bilinear form $H^f(x)$ is well defined (i.e. it depends only on v and w, but not on $\gamma_{v,w}$).

Remark I.1.3 Notice that $H^f(x)[v,w]$ can be defined only on the critical points. In $\mathbb{R}^N$, by virtue of its affine structure, it can be defined everywhere, and it corresponds to the second differential; but it depends on the affine structure and not only on its differential structure.

The Morse index of a critical point.

Now we can define the Morse index of a critical point.

Definition I.1.4. *Let $x \in K_f$ be a critical point such that $H^f(x)$ is defined. The (restricted) Morse index of x is the maximal dimension of a subspace of $T_x\Lambda$ on which $H^f(x)$ is negative definite and it is denoted by $m(x,f)$ (or simply by $m(x)$).*

The nullity of x is the dimension of the kernel of $H^f(x)$ (i.e. the subspace consisting of all v such that $H^f(\dot{x})[v,w] = 0$ for all $w \in T_x\Lambda$.

The large Morse index is the sum of the restricted

Morse index and the nullity and it will be denoted by $m^*(x,f)$.

Notice that if f is defined in an infinite dimensional manifold it is possible that $m(x,f) = \infty$.

Nondegenerate critical points.

If Λ is a finite dimensional manifold, a critical point is called nondegenerate if its nullity is O. In infinite dimension this definition is more delicate.

Definition I.1.5. *Let* $x_0 \in K_f$ *be a critical point such that* $H^f(x_0)$ *is defined.* x_0 *is called nondegenerate if there exist a splitting* $H^+ \oplus H^-$ *of* $T_{x_0}\Lambda$ *and a constant* $\nu > 0$ *such that*

(i) $H^f(x_0)[v,v] \geq \nu \|v\|^2$ $\forall v \in H^+$

(ii) $H^f(x_0)[v,v] \leq -\nu \|v\|^2$ $\forall v \in H^-$

Then, if L is a selfadjoint operator such that
$$H^f(x_0)[v,v] = \langle Lv,v \rangle,$$
x_0 is nondegenerate if and only if L is invertible.

The Morse index for functionals on manifolds of class C^1.

In applications, particularly in applications to partial differential equations, sometimes, it is necessary to consider functionals of class $\mathscr{C}^1$ or/and functionals defined on manifolds of class $\mathscr{C}^1$. In this section we show that sometimes it is possible to define the Morse index also in this case. In the next chapter the definition of

Morse index will be extended even to more general situations.

First we need a lemma.

Lemma I.1.6. *Let Λ be of class C^1, f and Ω as above and $x \in K_f(\bar{\Omega})$. Assume that there exists a C^1-chart (U, ϕ) (with $x \in U$), such that $\phi(x)$ is a nondegenerate critical point of $f \circ \phi^{-1}$ according to definition I.1.5*

Then, for any C^1-chart (V, ψ) (with $x \in V$), $f \circ \psi^{-1}$ has the second derivative in $\psi(x)$, $\psi(x)$ is a nondegenerate critical point of $f \circ \psi^{-1}$ and

$$m(\psi(x), f \circ \psi^{-1}) = m(\phi(x), f \circ \phi^{-1}).$$

Proof. Let $\tilde{U}, \tilde{V}$ be open subsets of the Hilbert space where Λ is modeled, such that

$\phi : U \cap V \longrightarrow \tilde{U}$ and $\psi : U \cap V \longrightarrow \tilde{V}$ are homeomorphisms

Then, there exists a diffeomorphism of class C^1,

$$\theta : \tilde{V} \longrightarrow \tilde{U}.$$

Let $\alpha = \phi(x)$ and $\beta = \psi(x)$. Denoting by L the Hessian at $\phi^{-1}(x) = \alpha$ for $f \circ \phi^{-1}$, we have, for any $y \in \tilde{U}$,

$$f \circ \phi^{-1}(y) = f \circ \phi^{-1}(\alpha) + \frac{1}{2}<L(y-\alpha), y-\alpha> + o(|y-\alpha|^2).$$

Therefore, since $\phi^{-1} \circ \theta = \psi^{-1}$ and $\theta(\beta) = \alpha$, for $z = \theta^{-1}(y) \in \tilde{V}$,

$$f \circ \psi^{-1}(z) = f \circ \phi^{-1}(\theta(z)) =$$
$$= f \circ \phi^{-1}(\alpha) + \frac{1}{2}<L(\theta(z)-\theta(\beta)), \theta(z)-\theta(\beta)> + o(|\theta(z)-\theta(\beta)|^2).$$

Since θ is differentiable at β,

$$\theta(z)-\theta(\beta) = d\theta(\beta)(z-\beta) + o(|z-\beta|),$$

then, since $d\theta(\beta)$ is a linear isomorphism,

$$f \circ \psi^{-1}(z) = f \circ \psi^{-1}(\beta) =$$
$$= \frac{1}{2}< L \circ d\theta(\beta)[z-\beta], d\theta(\beta)[z-\beta] > + o(|z-\beta|^2).$$

Therefore $f \circ \psi^{-1}$ is twice differentiable at β and its Hessian is given by $[d\theta(\beta)]^* \circ L \circ d\theta(\beta)$. Since $d\theta(\beta)$ is a linear isomorphism the proof of Lemma I.1.4 follows immediately.

□

By virtue of the above lemma, we can give the following definition:

Definition I.1.7. *Let Λ be of class $\mathscr{C}^1$, f and Ω as above and $x \in K_f(\overline{\Omega})$. We say that x is a nondegenerate critical point if there exists a C^1-chart (U, ϕ) (with $x \in U$) such that $\phi(x)$ is a nondegenerate critical point of $f \circ \phi^{-1}$ according to definition I.1.5. . In this case, we set*

$$m(x, f) = m(\phi(x), f \circ \phi^{-1}).$$

The Morse polynomial.

A function $f \in C^2$ is called a Morse function if all its critical point are non-degenerate. Notice that the critical points of a Morse function are isolated; hence if Λ is compact, they are a finite number.

Definition I.1.8. *Let f be a Morse function. The Morse polynomial of a set $K \subset K_f$ is defined as follows:*

$$m_\lambda(K) = \sum_{x \in K} \lambda^{m(x)},$$

with the convention that $\lambda^\infty = 0$.

Thus $m_\lambda(K)$ is a polynomial

$$\sum_k a_k \, \lambda^k$$

whose coefficients a_k are integer numbers representing the number of critical points in K_f having Morse index k.

Remark I.1.9. By virtue of the lemma I.1.6, the Morse polynomial makes sense also for functions of class C^1 whose critical points are not degenerate in the sense of definition I.1.7. If Λ is not compact, it is possible that K is an infinite set; in this case, the Morse polynomial becomes a formal series.

The algebraic and topologic structure of S.

The above remark makes useful to define the family S of the formal series in one variable λ with coefficients in $\mathbb{N} \cup \{+\infty\}$. On S the sum and the product are defined in the usual way:

$$\sum_k a_k \, \lambda^k + \sum_k b_k \, \lambda^k = \sum_k (a_k + b_k) \, \lambda^k$$

and

$$\sum_k a_k \, \lambda^k \cdot \sum_k b_k \, \lambda^k = \sum_k \left(\sum_{j=0}^{k} a_{k-j} b_k \right) \lambda^k;$$

(and we set, as usual, $0 \cdot \infty = 0$).

For the next development of the theory, it is necessary to impose to S an order structure. We define a relation of total order as follows:

(1-1) $\sum a_k \lambda^k < \sum b_k \lambda^k$ $\Leftrightarrow$ $\exists \ n \in \mathbb{N}: a_k = b_k$ *for any* $k \leq$

n-1 and $a_n < b_n$.

If $\mathcal{P} \in \mathbb{S}$ we set

$$c_k(\mathcal{P}) = a_k \Leftrightarrow \mathcal{P}(\lambda) = \sum_k a_k \cdot \lambda^k$$

We define the notion of limit in $\mathbb{S}$ in the following way:

(1-2) $\mathcal{R} = \lim_{n \to +\infty} \mathcal{P}_n \Leftrightarrow c_k(\mathcal{P}_n) \xrightarrow[n]{} c_k(\mathcal{R})$ *for any* $k \in \mathbb{N}$.

If we identify the formal series $\sum a_k \lambda^k$ with the sequence $\{a_k\}$, then the topology introduced by (1-2) is equivalent to the product topology on $\prod_{i \in 0}^{\infty} X_i$, where $X_i = \mathbb{N} \cup \{+\infty\}$; hence by Tychonoff's theorem, $\mathbb{S}$ is compact. If $\mathbb{A} \subseteq \mathbb{S}$, we denote with $\overline{\mathbb{A}}$ the closure of $\mathbb{A}$, i.e.

$$\overline{\mathbb{A}} = \left\{ \mathcal{P} \in \mathbb{S} \mid \exists \ \{\mathcal{P}_n\} \subseteq \mathbb{S}: \mathcal{P} = \lim_{n \to +\infty} \mathcal{P}_n \right\}$$

Now, it makes sense to define the infimum and the supremum as follows:

DEFINITION I.1.10 - *If* $\mathbb{A} \subseteq \mathbb{S}$, *we put*

$\mathcal{R} = \inf \mathbb{A}$ *if* $\mathcal{R} = \min \overline{\mathbb{A}}$

and

$\mathcal{R} = \sup \mathbb{A}$ *if* $\mathcal{R} = \max \overline{\mathbb{A}}$

We have the following result:

THEOREM I.1.11 – *For any set* $\mathbb{A} \subseteq \mathbb{S}$ *inf* $\mathbb{A}$ *and sup* $\mathbb{A}$ *exist and are unique.*

Proof. The uniqueness is trivial.

Existence of the infimum. We set

$$b_0 = min \left\{ c_0(\mathcal{P}) \mid \mathcal{P} \in \overline{\mathbb{A}} \right\}$$

$$\mathbb{B}_0 = \left\{ \mathcal{P} \in \overline{\mathbb{A}} \mid c_0(\mathcal{P}) = b_0 \right\}$$

$$b_n = min \left\{ c_n(\mathcal{P}) \mid \mathcal{P} \in \mathbb{B}_{n-1} \right\}$$

$$\mathbb{B}_n = \left\{ \mathcal{P} \in \mathbb{B}_{n-1} \mid c_n(\mathcal{P}) = b_n \right\}$$

Since the $\mathbb{B}_n$'s are compact and $\mathbb{B}_{n-1} \subseteq \mathbb{B}_n$ for every n, their intersection is not empty; $\mathcal{R} \in \bigcap_{n=0}^{\infty} \mathbb{B}_n$ is *min* $\overline{\mathbb{A}}$.

Existence of the supremum: we argue in the same way.

$\square$

Remark I.1.12 – Notice that the topology induced by this notion of convergence, is not the topology induced by the order relation. Then, we might have that

$$sup \; \mathbb{A} \neq inf \left\{ \mathcal{P} \in \mathbb{S} \mid \forall \; Q \in \mathbb{S}: \mathcal{P} \geq Q \right\}.$$

For example, take

$$\mathbb{A} = \left\{ n\lambda^0 \in \mathbb{S} \mid n \in \mathbb{N} \right\} \cup \left\{ n\lambda^0 + \lambda^1 \in \mathbb{S} \mid n \in \mathbb{N} \right\}.$$

In this case, we have $sup \; \mathbb{A} = \infty\lambda^0 + \lambda^1$, but

$$inf \left\{ \mathcal{P} \in \mathbb{S} \mid \forall \; Q \in \mathbb{S}: \mathcal{P} \geq Q \right\} = \infty\lambda^0.$$

Remark I.1.13 - If $\{P_k\}$ is a non-decreasing sequence and $\mathcal{R}$ = sup $\{P_k\}$ is a polynomial, then, for k large, $\{P_k\}$ is constantly equal to $\mathcal{R}$. In fact, by the definition (1-2), the values $c_n(\mathcal{R})$ are achieved by $c_n(P_k)$ for k sufficiently large.

I.2. - THE POINCARE' POLYNOMIAL

Definition of the Poincaré polynomial

The Poincaré polynomial of a topological couple is a topological invariant which carries the information on the homology of a topological couple (X,A).

Definition I.2.1. *Given an homology theory* $H_*(\cdot,\cdot,\mathbb{Z}_2)$ *and a topological couple* (X,A), *we set*

$$\mathcal{P}_\lambda(X,A) = \sum_{q \,\in\, \mathbb{N}} dim\left[H_q(X,A,\mathbb{Z}_2)\right] \cdot \lambda^q$$

Moreover we set $\mathcal{P}_\lambda(X) = \mathcal{P}_\lambda(X,\emptyset)$.

The natural numbers $dim\left[H_q(X,A,\mathbb{Z}_2)\right]$ are called Betti numbers (in fact, sometimes, the Poincaré polynomial is called the Betti polynomial). Notice that in general the Poincarè polynomial is not a "polynomial" but a formal series in $\mathfrak{S}$; however, it is a "real" polynomial in the more elementary situation, for example if X is a compact manifold and A is its boundary.

We have chosen the field of coefficients to be $\mathbb{Z}_2$ in order to avoid orientation problem. However the theory which we will developed holds also with any other field $\mathbb{K}$ of coefficients. The only difference arises in the actual computation of $\mathcal{P}_\lambda(X,A)$; for example, if we take the real projective space $P\mathbb{R}^2$ we have that

$$\mathcal{P}_\lambda(P\mathbb{R}^2) = \sum_{q=0}^{2} dim\left[H_q(P\mathbb{R}^2,\mathbb{Z}_2)\right] \cdot \lambda^q = 1 + \lambda + \lambda^2$$

but

$$\sum_{q=0}^{2} dim\left[H_q(P\mathbb{R}^2,\mathbb{Q})\right] \cdot \lambda^q = 1$$

In the following, we will remark the situations in which the choice of coefficients makes a difference.

Properties of the Poincaré polynomial.

Many properties of the homology theory can be transferred to the Poincaré polynomial and the operations and relations in S have a topological interpretation.

I.2.2. Theorem. *Let (X,A) and (Y,B) be a couple of topological spaces.*

(i) if (X,A) and (Y,B) are homotopically equivalent, then $\mathcal{P}_\lambda(X,A) = \mathcal{P}_\lambda(Y,B)$

(ii) if $X \cap Y = \varnothing$, then $\mathcal{P}_\lambda(X \cup Y, A \cup B) = \mathcal{P}_\lambda(X,A) + \mathcal{P}_\lambda(Y,B)$

(iii) $\mathcal{P}_\lambda(X \times Y) = \mathcal{P}_\lambda(X) \cdot \mathcal{P}_\lambda(Y)$

(iv) if (X,A,B) is a topological triple and B is a weak deformation retract of A

$$\mathcal{P}_\lambda(X,A) = \mathcal{P}_\lambda(X,B);$$

if A is a weak deformation retract of X,

$$\mathcal{P}_\lambda(X,B) = \mathcal{P}_\lambda(A,B);$$

(v) if (X,A,B) is a topological triple then there exists a $Q_\lambda = Q_\lambda(X,A,B) \in \mathbb{S}$, *such that*

$$\mathcal{P}_\lambda(X,A) + \mathcal{P}_\lambda(A,B) = \mathcal{P}_\lambda(X,B) + Q_\lambda;$$

(vi) let (X,A) be a topological pair, then if $\overline{C} \subset$ int A,

$$\mathcal{P}_\lambda(X,A) = \mathcal{P}_\lambda(X \backslash C, A \backslash C).$$

(vii) let $\varphi_1 : (X,A) \longrightarrow (Y,B)$, $\varphi_2 : (Y,B) \longrightarrow (Z,C)$ be two maps such that $(\varphi_2 \circ \varphi_1)_*$ is an isomorphism; then, $\exists \; Z_\lambda \in$ $\mathbb{S}$:

$$\mathcal{P}_\lambda(Y,B) = \mathcal{P}_\lambda(X,A) + Z_\lambda$$

(in particular, this happens e.g. if $(X,A) = (Z,C)$ and $\varphi_2 \circ \varphi_1$ is homotopically equivalent to the identity).

(viii) let M be a manifold and let $N \subset M$ be a closed (in M) submanifold of codimension n. If W is a subset of N closed in N, then

$$\mathcal{P}_\lambda(M, M \backslash W) = \lambda^n \, \mathcal{P}_\lambda(N, N \backslash W).$$

(in this case, if the coefficients field $\mathbb{K}$ is not $\mathbb{Z}_2$, we need to assume N and M to be orientable).

(ix) if x_0 is a single point, then $\mathcal{P}_\lambda(\{x_0\}) = 1$.

(x) if B_n is a n-dimensional ball, then

$$\mathcal{P}_\lambda(B_n , \partial B_n) = \lambda^n$$

Proof. (i)-(iv), (vi)-(vii) and (ix)-(x) are standard results in algebraic topology (cf. e.g. [Gr,Ha,81]. A proof of (v) can be found e.g. in [Be,91]. The proof of (viii) follows by Corollary 8.11.20 of [Do,72] (cf. also the remark below the Corollary) because the manifold N is an A.N.R (cf. [Pa,66,a]. Moreover notice that Thom Theorem, in this form, holds even if the dimension of the manifold M is infinite.   □

The Morse relations

For any $a < b \in \mathbb{R}$ set

$$f^b = \left\{ x \in \Lambda \mid f(x) \leq b \right\}$$

$$f^b_a = \left\{ x \in \Lambda \mid a \leq f(x) \leq b \right\}$$

Now we can state the Morse relations in their simplest form:

Theorem I.2.3 - *Let Λ be a smooth manifold, let $f: C^2(\Lambda)$ be a Morse function and let $a < b$ be two regular vales for f. Then if f^b_a is compact, we have*

$$\sum_{x \in K(f_a^b)} \lambda^{m(x)} = \mathcal{P}_\lambda(f^b, f^a) + (1 + \lambda) \, Q_\lambda$$

where $K(f_a^b) = K_f \cap f_a^b$ and $Q(t)$ *is a polynomial with integer nonnegative coefficients.*

In particular, if $\mathfrak{M}$ is a compact manifold, we have that

$$\sum_{x \in K(f_a^b)} \lambda^{m(x)} = \mathcal{P}_\lambda(\mathfrak{M}) + (1 + \lambda) \, Q_\lambda.$$

Notice that $\mathcal{P}_\lambda(\mathfrak{M})$ does not depend on f; then if we choose f very simple, it is possible to try to compute the Morse polynomial directly. In fact, if for some reason $Q_\lambda = 0$, then we have computed $\mathcal{P}_\lambda(\mathfrak{M})$ without using the tools of Algebraic topology. There is a very simple criterion which can be employed to see whether $Q_\lambda = 0$: the Lacunary Principle. This Principle simply states that if

$$m_\lambda(X) = \sum_{x \in K(f_a^b)} \lambda^{m(x)} = \sum_{x \in K(f_a^b)} a_k \lambda^k$$

is such that $a_k \neq 0 \Rightarrow a_{k+1} = 0$, then $Q_\lambda = 0$. (The proof is immediate). Using this method, for example, it is immediate to check that, for $n \geq 2$, $\mathcal{P}_\lambda(S^n) = 1 + \lambda^n$. (Of course this is true also for $n = 1$, but in this case it is necessary to use also (ix) and (vii) of theorem I.2.2).

If f_a^b is not compact, Th. I.2.3 is not longer valid as showed by simple examples. However, there are assumptions on the couple (Λ, f) which guarantee the validity of the

Morse relation. The most famous (but not the most general) condition is the condition of Palais and Smale:

Definition I.2.4. *We say that* f *satisfies the Palais–Smale condition in a set* $\Omega \subseteq \Lambda$ *(P.S. in* Ω*) if any sequence* $\{x_n\}_{n \in \mathbb{N}} \subset \Lambda$ *such that*

$$f(x_n) \xrightarrow[n]{} c \in \mathbb{R} \text{ and } f'(x_n) \xrightarrow[n]{} 0$$

has a subsequence which converges to $x \in \Lambda$.

Clearly, if f_a^b is compact and a, b are regular values, f satisfies P.S. in f_a^b; however it is possible for f to satisfy P.S. even if f_a^b is not locally compact.

Moreover, as we have seen in remark I.1.6, the Morse polynomial makes sense also for C^1-functions.

These remarks make natural the following definition:

Definition I.2.5. *Let* Ω *be an open set in* Λ*. A function* $f \in C^1(\bar{\Omega})$ *is called generalized Morse function if*

(i) the critical points of f *are not degenerate in the sense of definition I.1.5;*

(ii) f *satisfies P.S. in* Ω*;*

(iii) f *can be extended to a function of class* C^1 *in a neighborhood of* $\bar{\Omega}$*.*

The set of generalized Morse function will be denoted by $\mathcal{M}(\bar{\Omega})$*.*

Lemma I.2.6. *If* $f \in \mathcal{M}(\bar{\Omega})$ *and it is bounded in* Ω*, then it has a finite number of critical points.*

Proof. By (i) of Definition I.2.5, the critical points of f in Ω are isolated as we can easily see using the Taylor expansion of f on its critical points. Then, since f satisfies (ii) of Definition I.2.5, the number of critical points of f must be finite.

□

Now we can generalize the theorem I.2.3:

Theorem I.2.7. *Let Λ be a complete manifold of class C^1 and let $f \in M(f_a^b)$; then*

$$\sum_{x \in K(f_a^b)} \lambda^{m(x)} = \mathcal{P}_\lambda(f^b, f^a) + (1 + \lambda) \, Q_\lambda$$

where $K(f_a^b) = K_f \cap f_a^b$ and $Q(t)$ is polynomial with integer nonnegative coefficients.

Clearly Th. I.2.3 is an immediate consequence of Th. I.2.7. The proof of Th. I.2.7 is quite involved. We will prove it in section I.1.4, using the notion of "Conley block" which will be introduced in the next section.

I.3 - THE CONLEY BLOCKS

In this section we will introduce the notion of index pair and Conley block which will allow to prove the Morse relations in a very general context.

Preliminary notions and notations.

Let Λ be a Hilbert manifold and $T\Lambda$ its tangent bundle.

Consider a vector field $F: \Lambda \longrightarrow T\Lambda$ and denote by $\eta(t,x)$ the solution of the Cauchy problem

$$
\begin{cases}
\dfrac{d\eta}{dt} = F(\eta) \\[2ex]
\eta(0,x) = x
\end{cases}
\tag{3.1}
$$

Assume that F is such that

(3.1) *is well posed and*

for any $x \in \Lambda$, $\eta(t,x)$ is defined for any $t \in \mathbb{R}$. (3.2)

For any set $A \subset \Lambda$ put

$$
W_+(A) \equiv W_+(A,F) = \bigcap_{t \le 0} \eta(t,A) =
$$
$$
= \left\{ x \in A : \eta(t,x) \in A \text{ for any } t \ge 0 \right\}.
$$

$$
W_-(A) \equiv W_-(A,F) = \bigcap_{t \ge 0} \eta(t,A) =
$$
$$
= \left\{ x \in A : \eta(t,x) \in A \text{ for any } t \le 0 \right\},
$$

$$
G(A) \equiv G(A,F) = W_+(A,F) \cap W_-(A,F) =
$$
$$
= \left\{ x \in A : \eta(t,x) \in A \text{ for any } t \in \mathbb{R} \right\},
$$

The set $W_+(A)$ is usually called the positively maximal invariant set relatively to A (with respect to the flow η) and $W_-(A)$ the negatively maximal invariant set relatively to A. $G(A)$ is called the maximal invariant set in A. (We recall that a set $E \subseteq A$, is called *positively invariant* relatively to A if

$$
x \in E \text{ and } \eta([0,t],x) \cap \Lambda\backslash E \ne \varnothing \Rightarrow
$$

$$\exists t_* \in [0,t]: \eta(t_*,x) \notin A).$$

Moreover for any $A \subset \Lambda$ and $T \geq 0$ put

$$W_+^T(A) \equiv W_+^T(A,F) = \bigcap_{t\in[-T,0]} \eta(t,A),$$
$$W_-^T(A) \equiv W_-^T(A,F) = \bigcap_{t\in[0,T]} \eta(t,A),$$
$$G^T(A) \equiv G^T(A,F) = W_+^T(A) \cap W_-^T(A).$$

Remark I.3.1. *It is easy to verify that:*

(i) if A is closed, then $W_+(A)$, $W_-(A)$, $G(A)$, $W_+^T(A)$, $W_-^T(A)$ and $G^T(A)$ *are closed;*

(ii) $W_+(W_+^T(A)) = W_+(A)$, *for any $T \geq 0$ and for any A.*

(iii) $G(W_+^T(A)) = G(A)$, *for any $T \geq 0$ and for any A.*

(iv) $\eta_T(W_+^T(A)) = W_-^T(A)$

(v) $\eta_T(W_+^{2T}(A)) = G^T(A)$

(vi) $G^T(A) \backslash W_+(G^T(A)) = G^T(A) \backslash W_+(A).$

(vii) $G^{T_1}(G^{T_2}(A) = G^{T_1+T_2}(A)$

Conley blocks and their index.

For any closed set $A \subset \Lambda$, define the following set which we will call the exit set (relative to A):

$$\Gamma(A) = \left\{ x \in \partial A \mid \forall \; \varepsilon_0 \geq 0, \; \exists \; \varepsilon \in (0,\varepsilon_0): \eta(\varepsilon,x) \notin A \right\};$$

the points in $\Gamma(A)$ are called exit points with respect to A.

Notice that, even if A is closed, $\Gamma(A)$ need not to be closed. Then, it makes sense to define

$$\Sigma = \Sigma_F = \left\{ N \subseteq \Lambda : \Gamma(N) \text{ is closed} \right\}.$$

A set $A \in \Sigma$ is called Conley block.

Notice that

$$\Gamma(\eta(t,N)) = \eta(t,\Gamma(N)) \text{ for any } t \in \mathbb{R}.$$

Definition I.3.2. *If* $N \in \Sigma$, *we define the index of* N *as follows:*

$$I_\lambda(N) = I_\lambda(N,F) = \mathcal{P}_\lambda(N,\Gamma(N))$$

The notion of index pair.

Definition I.3.3. *A couple* (N,E) *of* <u>*closed*</u> *subsets of* Λ *(with* $E \subset N$*) is called index pair (with respect to* F*) if*

(i) *E is positively invariant relatively to N,i.e.*

$x \in E$ *and* $x \cdot [0,t] \cap \Lambda \backslash E \neq \emptyset$ $\Rightarrow$ $\exists t_* \in [0,t]$: $x \cdot t_* \notin N$.

(ii) *E is an exit set for N,i.e.*

$x \in N$ *and* $x \cdot [0,t] \cap \Lambda \backslash N \neq \emptyset$ $\Rightarrow$ $\exists t_* \in [0,t]$: $x \cdot t_* \in E$.

Clearly, if N ia a Conley block, (i.e. $\Gamma(N)$ is closed), the topological couple $(N,\Gamma(N))$ is an index pair of a particular type.

Construction of Conley blocks.

The next theorem gives a simple but very useful method to construct Conley blocks.

Theorem I.3.4. *Let* g_1 *and* g_2 *be two differentiable functions defined on* Λ. *Set*

$$N = \left\{ x \in \Lambda \mid g_j(x) \leq o, j = 1, 2 \right\} ; \Gamma =$$

$$\left\{ x \in \partial N \mid g_1(x) = o \right\}$$

and suppose that

$$x \in \partial N \text{ and } g_1(x) = 0 \Rightarrow \langle \nabla g_1(x), F(x) \rangle > 0$$
$$x \in \partial N \text{ and } g_2(x) = 0 \Rightarrow \langle \nabla g_2(x), F(x) \rangle < 0$$

Then N is a Conley block.

Proof. It is immediate to verify that $\Gamma(N) = \Gamma$ and that Γ is closed.

$\square$

Remark I.3.5. The condition on g_2 can be weakened as follows:

$$x \in \partial N \text{ and } g_2(x) = 0 \quad \Rightarrow \quad \frac{d}{dt}(g_2 \circ \eta(t,x))|_{t=0} \leq 0.$$

Theorem I.3.4 can be easily generalized:

Theorem I.3.6. *Let* g_i *(i = 1,...,k) be functions of class* C^1 *on* Λ. *Set*

$$N = \bigcap_{i=1}^{k} \{x \in \Lambda: g_i(x) \leq 0\}.$$

Suppose that for any $x \in \partial A$, *there exists* g_i *satisfying:*

$$g_i(x) = 0 \Rightarrow \langle \nabla g_i(x), F(x)\rangle \neq 0.$$

Then $N \in \Sigma_F$ *and*

$$I_\lambda(N) = \mathcal{P}_\lambda(N, \Gamma(N)).$$

The box.

An interesting application of Theorem I.3.4 is the following:

Theorem I.3.7. *Let* Λ *be an Hilbert space with scalar product* $\langle \cdot, \cdot \rangle$ *and let L be a continuous linear operator satisfying the following assumptions:*

L_1) *there exists* H^+ *and* H^- *closed subspaces of* Λ *such that*

$$L(H^\pm) \subset H^\pm \text{ and } H^- \oplus H^+ = \Lambda;$$

L_2) *there exists* $\nu > 0$ *such that*

$$\langle Lx,x \rangle \leq -\nu\|x\|^2 \text{ for any } x \in H^-, \langle Ly,y \rangle \geq \nu\|y\|^2 \text{ for any}$$
$$y \in H^+.$$

where $\|\cdot\|$ *is the norm induced by the scalar product* $\langle \cdot,\cdot \rangle$.

Let

$$F(z) = -Lz + \mathfrak{K}(z),$$

where $z = x + y$, $x \in H^-$, $y \in H^+$ *and* $\mathfrak{K} : \Lambda \longrightarrow \Lambda$ *is a map of class* $C^{0,1}_{loc}$ *such that there exists* $\rho > 0$ *satisfying:*

$$\|\mathfrak{K}(x,y)\| \leq \frac{\nu}{2}\cdot\|x\| \text{ if } \|x\| = \rho \text{ and } \|y\| \leq \rho, \qquad (3.3)$$

$$\|\mathfrak{K}(x,y)\| \leq \frac{\nu}{2}\cdot\|y\| \text{ if } \|y\| = \rho \text{ and } \|x\| \leq \rho. \qquad (3.4)$$

Then, setting

$$Q_\rho = \left\{ x+y \in \Lambda \ \middle| \ \|x\|, \|y\| \leq \rho \right\}$$

we have that $Q_\rho \in \Sigma_F$ *and*

$$I_\lambda(Q_\rho) = \lambda^m, \text{ where } m = \dim H^-.$$

Proof. Set

$$g_1(x+y) = \frac{1}{2}\|x\|^2 - \frac{1}{2}\rho^2 \text{ and } g_2(x+y) = \frac{1}{2}\|y\|^2 - \frac{1}{2}\rho^2$$

Then by (3.3) and (3.4), we have

$$x \in \partial Q_\rho \text{ and } g_1(x) = 0 \Rightarrow$$
$$\langle F(x,y),\nabla g_1(x) \rangle = \langle -Lx,x \rangle + \langle \mathfrak{K}(x,y),x \rangle \geq \frac{\nu}{2}\cdot\|x\|^2 > 0$$

and

$$x \in \partial Q_\rho \text{ and } g_1(x) = 0 \Rightarrow$$
$$\langle F(x,y),\nabla g_2(x) \rangle = \langle -Ly,y \rangle + \langle \mathfrak{K}(x,y),y \rangle \leq -\frac{\nu}{2}\cdot\|y\|^2 < 0$$

Then, setting

$$\Gamma = \left\{ x+y \in Q_\rho \ \middle| \ \|x\| = \rho \right\}$$

by Theorem I.3.4, Q_ρ is a Conley block and

$$I_\lambda(Q_\rho) = \mathcal{P}_\lambda(Q_\rho,\Gamma) = \mathcal{P}_\lambda(B_\rho \cap H^-,\partial B_\rho \cap H^-) = \lambda^m,$$

(cf. Theorem I.2.8(x)).

□

Properties of the index I_λ.

The next theorem describes the main properties of the index.

Theorem I.3.8. *Let $N \in \Sigma_F$. Then*

(i) $I_\lambda(N) = \mathcal{P}_\lambda(N, N \setminus W_+(N))$

(ii) $\eta(t,N) \in \Sigma_F$ *and* $I_\lambda(\eta(t,N)) = I_\lambda(N)$

(iii) $W_+^T(N) \in \Sigma_F$ *and* $I_\lambda(W_+^T(N)) = I_\lambda(N)$

(iv) $W_-^T(N) \in \Sigma_F$ *and* $I_\lambda(W_-^T(N)) = I_\lambda(N)$

(v) $G^T(N) \in \Sigma_F$ *and* $I_\lambda(G^T(N)) = I_\lambda(N)$

(vi) *if* N_1, $N_2 \in \Sigma_F$, *and for some* $T > 0$, $G^T(N_1) \subseteq G^T(N_2)$ *and* $G^T(N_2) \subseteq G^T(N_1)$ *then*

$$I_\lambda(N_1) = I_\lambda(N_2)$$

(vii) *if* N_1, $N_2 \in \Sigma_F$ *are disjoint sets, then* $N_1 \cup N_2 \in \Sigma_F$ *and*

$$I_\lambda(N_1 \cup N_2) = I_\lambda(N_2) + I_\lambda(N_2)$$

(viii) *let* F_1 *and* F_2 *be two vector fields which satisfy* (3.1) *and* (3.2); *if* $N \in \Sigma_{F_1}$ *and* $F_2 = F_1$ *on a neighborhood of* ∂N, *then* $N \in \Sigma_{F_2}$ *and*

$$I_\lambda(N,F_1) = I_\lambda(N,F_2).$$

(ix) *if for some* $T > 0$, $G^T(N) \subset int\ N$, *then*

$$I_\lambda(N) = \mathcal{P}_\lambda(int\ N, (int\ N) \setminus W_+(N))$$

where int N denotes the interior of N whenever N is the closure of an open subset of Λ, while if $N = cl\ M$ and M is a submanifold of Λ, int N denotes the manifold M itself.

(x) *let* $M \in \Sigma$ *and* $N \subset M$, *be two manifolds in Λ such that*

(1) $\overline{N}$ is positively invariant relatively to $\overline{M}$

(2) $G^T(\overline{M}) \subset M$ for some $T > 0$

(3) $\partial N \equiv \overline{N} \setminus N \subset \partial M \equiv \overline{M} \setminus M$

(4) $W_+(\overline{M}) \subset \overline{N}$

then $\overline{N} \in \Sigma$. Moreover, if N has codimension d in M,

$$I_\lambda(\overline{M}) = \lambda^d \, I_\lambda(\overline{N})$$

(in this case, if the coefficients field $\mathbb{K}$ is not $\mathbb{Z}_2$, we need to assume N and M to be orientable).

Proof. (i). By the definition of $\Gamma(N)$ it is evident that $\Gamma(N) \subset N \setminus W_+(N)$. Then, by lemma I.2.2.(iv), applied to the topological triple $(N, \ N \setminus W_+(N), \ \Gamma(N))$, it is sufficient to prove that $\Gamma(N)$ is a deformation retract of $N \setminus W_+(N)$.

In order to prove this, consider $x \in N \setminus W_+(N)$. By the definition of $W_+(N)$ there exists $t > 0$ such that $\eta(t,x) \notin N$. Let

$$\tau(x) = inf \left\{ t \in \mathbb{R}^+ : \eta(t,x) \notin N \right\}$$

We want to prove that τ is a continuous function. Let $x \in N \setminus W_+(N)$ and let x_k be a sequence in $N \setminus W_+(N)$ such that $x_k \longrightarrow x$. We will prove that $\tau(x_k) \longrightarrow \tau(x)$.

Choose any $\varepsilon_0 > 0$; then, there exists $\varepsilon \in (0, \varepsilon_0)$ such that

$$\eta(\tau(x)+\varepsilon, \ x) \notin N.$$

Then by the continuity of η, for k large enough, we have that

$$\eta(\tau(x)+\varepsilon, x_k) \notin N$$

and hence $\tau(x_k) \leq \tau(x) + \varepsilon$. Since ε can be chosen arbitrarily small, this proves that

$$\tau(x) \geq \underset{k \longrightarrow \infty}{lim\text{-}sup} \ \tau(x_k)$$

Now let $\tau = lim\text{-}inf\limits_{k \to \infty} \tau(x_k)$. If $\tau = +\infty$, the proof is finished. Then, suppose $\tau < +\infty$ and take a subsequence $\tau(x_{k_m})$ such that

$$\tau = lim_{m \to +\infty} \tau(x_{k_m}).$$

Since $\eta(\tau(x_{k_m}),x_{k_m}) \in \Gamma$ and Γ is closed, $\eta(\tau,x) \in \Gamma$, therefore, by the definitions of Γ and $\tau(x)$,

$$\tau(x) \leq \tau = lim\text{-}inf\limits_{k \to \infty} \tau(x_k).$$

Thus $\tau(x)$ is continuous, and the map

$$x \longmapsto \eta(\tau(x),x)$$

is a strong retraction of $N \setminus W_+(N)$ on $\Gamma(N)$.

(ii) since $\eta(t,\cdot)$ is an homeomorphism,

$$I_\lambda(N) = \mathcal{P}_\lambda(N, \Gamma(N)) = \mathcal{P}_\lambda(\eta(t,N), \eta(t,\Gamma(N))) =$$
$$= \mathcal{P}_\lambda(\eta(t,N), \Gamma(\eta(t,N))) = I_\lambda(\eta(t,N)).$$

(iii)   Put $C = N \setminus W_+^T(N)$. Then

$$(W_+^T(N), W_+^T(N) \setminus W_+(N)) = (N \setminus C, (N \setminus W_+(N)) \setminus C) \qquad (3.5).$$

Since

$$cl_N\, C \subset int_N\, N \setminus W_+(N)$$

by Lemma I.2.4(vi),

$$\mathcal{P}_\lambda(N, N \setminus W_+(N)) = \mathcal{P}_\lambda(N \setminus C, (N \setminus W_+(N)) \setminus C)$$

and, by the (3.5),

$$\mathcal{P}_\lambda(N, N \setminus W_+(N)) = \mathcal{P}_\lambda(W_+^T(N), W_+^T(N) \setminus W_+(N)) \qquad (3.6).$$

It is not difficult to prove that $W_+^T(N) \in \Sigma_F$. Then using the (3.6), Theorem I.3.8(i) and Remark I.3.1(ii),

$$I_\lambda(N) = \mathcal{P}_\lambda(N, N \setminus W_+(N)) =$$
$$= \mathcal{P}_\lambda(W_+^T(N), W_+^T(N) \setminus W_+(N)) = I_\lambda(W_+^T(N)).$$

(iv) By (iii), (ii) and Remark I.3.1(iv), we have

$$I_\lambda(N) = I_\lambda(W_+^T(N)) = I_\lambda(\eta(T, W_+^T(N))) = I_\lambda(W_-^T(N))$$

(v) By (iii), (ii) and remark I.3.1(v), we have

$$I_\lambda(N) = I_\lambda(W_+^{2T}(N)) = I_\lambda(\,\eta(T,W_+^{2T}(N))\,) = I_\lambda(G^T(N)).$$

Notice that, in order to prove the above inequality, we have used only the homeomorphism $\eta(-T,\cdot)$ and the excision map. Then, since $G^T(N)\backslash W_+(G^T(N))) = G^T(N)\backslash W_+(N)$, denoting by

$$i: W_+^{2T}(N) \longrightarrow N$$

the inclusion map, we have that the map

$$(i\circ\eta(-T,\cdot))_* : H_*(G^T(N),G^T(N)\backslash W_+(N)) \longrightarrow H_*(N,N\backslash W_+(N))$$

is an isomorphism.

Moreover, using the homotopy $H(\sigma,x) = \eta(-\sigma T,x)$ shows that the inclusion

$$j: (G^T(N),G^T(N)\backslash W_+(N)) \longrightarrow (N,N\backslash W_+(N))$$

is homotopically equivalent to the map $i\circ\eta(-T,\cdot)$, therefore

$$j_*: H_*(G^T(N),G^T(N)\backslash W_+(N)) \longrightarrow H_*(N,N\backslash W_+(N))$$

is an isomorphism $\hspace{2cm}$ (3.7)

(vi) by our assumptions,

$$G^{2T}(N_2) \subset G^T(N_1) \subset N_2,$$

then, we have the embedding

$$i_1: (G^{2T}(N_2), G^{2T}(N_2)\backslash W_+(N_2)) \longrightarrow (G^T(N_1), G^T(N_1)\backslash W_+(N_2))$$

and

$$i_2 : (G^T(N_1), G^T(N_1)\backslash W_+(N_2)\,) \longrightarrow (N_2, N_2\backslash W_+(N_2)\,).$$

By (3.7), $(i_2\circ i_1)_*$ is a isomorphism, then, by Theorem 2.2.(vii), there exists $Z_\lambda \in S$ such that

$$\mathcal{P}_\lambda(G^T(N_1), G^T(N_1)\backslash W_+(N_2)\,) = \mathcal{P}_\lambda(N_2, N_2\backslash W_+(N_2)) + Z_\lambda.$$

Since $G^T(N_1)\backslash W_+(N_2) = G^T(N_1)\backslash W_+(G^T(N_1))$, using (i) and (v), we have that

$$I_\lambda(N_1) = I_\lambda(G^T(N_1)) = \mathcal{P}_\lambda(G^T(N_1), G^T(N_1)\backslash W_+(G^T(N_1))) =$$
$$= \mathcal{P}_\lambda(N_2, N_2\backslash W_+(N_2)) + Z_\lambda = I_\lambda(N_2) + Z_\lambda$$

Arguing in the same way, we have also the existence of $\tilde{Z}_\lambda \in S$ such that $I_\lambda(N_1) + \tilde{Z}_\lambda = I_\lambda(N_2)$; hence $I_\lambda(N_1) = I_\lambda(N_2)$.

(vii) and (viii) are immediate consequence of the definition of I_λ.

(ix) we define the homotopy

$$H(t,x) = \begin{cases} \eta(t,x) & \text{if } t \leq \tau(x) \\ \eta(\tau(x),x) & \text{if } t \geq \tau(x) \end{cases}$$

where $\tau(x)$ is defined in the proof of (i). If $H(2T,x) \notin \Gamma(N)$, then $H(T,x) \in G^T(N) \subset \text{int } N$, hence $H(2T,x) \in \text{int } N$. Therefore $K(\sigma,x) = H(2\sigma T,x)$ is a weak deformation of N on $\text{int } N \cup \Gamma(N)$ and also a weak deformation retract of $N\backslash W_+(N)$ on $(\text{int } N \backslash W_+(N)) \cup \Gamma(N)$. Then, by Lemma I.2.4(iv) we have

$$I_\lambda(N) = \mathcal{P}_\lambda(N, N \backslash W_+(N)) =$$
$$= \mathcal{P}_\lambda(\text{int } N \cup \Gamma(N), (\text{int } N \backslash W_+(N)) \cup \Gamma(N))$$

The conclusion follows by the excision property (lemma 1.4.(vi)) taking $C = \Gamma(N)$.

(x) we have first of all to prove that $\Gamma(\overline{N})$ is closed. Let x_k be a converging sequence of exit points with respect to $\overline{N}$; by assumption (1) they are also exit points with respect to $\overline{M}$; thus they converge to a point x_0 which is an exit point with respect to $\overline{M}$ and which belongs to $\overline{N}$ (by (3)). Thus $x_0 \in \Gamma(\overline{N})$. Now, by (3), $\overline{N} \subset \overline{M}$, therefore $G^T(\overline{N}) \subset G^T(\overline{M})$ and, by (2) and (3), $G^T(\overline{N}) \subset N$. Then by (ix),

$$I_\lambda(\overline{M}) = \mathcal{P}_\lambda(M, M \backslash W_+(\overline{M}))$$
$$I_\lambda(\overline{N}) = \mathcal{P}_\lambda(N, N \backslash W_+(\overline{N}))$$

Moreover, by (4), $W_+(N) = W_+(M)$ and hence

$$I_\lambda(\overline{N}) = \mathcal{P}_\lambda(N, N \backslash W_+(\overline{M}))$$

Now, by (3), N is closed in M, and also $W_+(\overline{M}) \cap N$ is closed in N; then we can apply Theorem I.2.2 (viii) with $X = W_+(\overline{M})$ and we have that

$$\mathcal{P}_\lambda(M, M \backslash W_+(\overline{M})) = \lambda^d \mathcal{P}_\lambda(N, N \backslash W_+(\overline{M}))$$

Therefore the conclusion follows.

□

Remark I.3.9. *Let* $S = G(A)$ *with* $A \in \Sigma$. *Then* S *is an invariant maximal set in* A. *If the following condition* (*introduced in* [Be,91])

(C)*for every neighborhood* B *of* S, $G^T(A) \subset B$ *for some* $T > 0$,

is satisfied, then, by Theorem I.3.8.(vi), $I_\lambda(B)$ *(whenever* $B \in \Sigma$) *is independent of* B.

In particular, if Λ *is compact, we have*

$$I_\lambda(B) = \Sigma_{k \geq 0} \; dim \; H_k(Con \; S,\mathbb{K}) \cdot \lambda^k,$$

where Con S is the Conley index of S *(cf.* [Co,78]).

I.4 – THE MORSE RELATIONS.

The pseudogradient vector field.

The theory developed in the previous section concerns general flows. Now we can apply it to the study of the critical point of a C^1-functional f and to the Morse theory. To do this it is necessary to construct a vector field F, such that (3.1) and (3.2) holds and which is related to f by the following relation:

$$\forall \; x \in \Lambda \backslash K_f \qquad df(x)[F(x)] < 0$$

This relation implies that f is decreasing along the orbits flow η, i.e.

$$\frac{d}{dt}(f \circ \eta(t,x)) < 0 \; for \; any \; x \notin K_f \; and \; t \in \mathbb{R}.$$

If Λ has a Riemannian structure, $f \in C^{1,1}_{loc}$ and ∇f is bounded, then the vector field F can be obtained taking

$$F = -\nabla f$$

If ∇f is not bounded, then F can be obtained as follows:

$$F = \frac{-\nabla f}{1 + \|\nabla f\|}$$

However, in the applications, particularly in the applications to P.D.E. we are interested in functionals of class C^1. In this case the construction of F is more delicate.

As essentially proved in [Pa,66,a] the following Lemma holds:

Lemma I.4.1. *Given $f \in C^1(\Lambda)$, there exists a vector field $-F$ (called pseudo gradient vector field for f) such that*

(i) $F \in C^{0,1}_{loc}(\Lambda \backslash K_f)$

(ii) $\|F(x)\| \le M \cdot d(x, K_f)$

where M is a suitable constant

(iii) $\langle \nabla f(x), F(x) \rangle < 0$ *for any* $x \notin K_f$

Moreover, if f satisfies P. S. and it is bounded on an open set Ω, (iii) can be reinforced as follows

(iv) *for any neighborhood U of $K_f(\Omega)$, there exists $\nu = \nu(\Omega, U) > o$, such that*

$$\forall x \in \Omega \backslash U, \qquad \langle \nabla f(x), F(x) \rangle \le -\nu$$

Moreover, we can construct F near the critical point by the following lemma:

Lemma I.4.2. *Let $f \in C^1(\Lambda)$ and let x_0 be a nondegenerate critical point of f (in the sense of definition 1.3). Then, there exists a neighborhood U of x_0 and a vector field F*

$\in C^{0,1}(U)$ such that
$$\forall \, x \in U, \qquad \langle \nabla f(x), F(x) \rangle \leq -v \, \|\nabla f(x)\|^2$$
where $v > 0$ is a suitable constant.

Proof. let (U, ϕ) be a chart as in definition 1.3 and let L be the Hessian of $f \circ \phi^{-1}$ at the point $\phi(x_0)$ as in the proof of lemma 1.4. Then we can define F by the following formula:
$$F(x) = (d\phi^{-1}(x))^* \circ L \, [\phi(x) - \phi(x_0)]$$
i.e. F is L pulled back by ϕ^{-1}. It is not difficult to check that F satisfies the required properties.

□

The deformation theorem.

As in section 2 set
$$f_a^b = \{x \in \Lambda : a \leq f(x) \leq b\},$$
and, for $c \in \mathbb{R}$,
$$f^c = \{x \in \Lambda : f(x) \leq c\},$$

Theorem I.4.3. *(deformation theorem) Let $f \in \mathcal{M}(\Lambda)$. There exists a pseudo-gradient vector field $-F$ such that the Cauchy problem*
$$\begin{cases} \dfrac{d\eta}{dt} = F(\eta) \\[2mm] \eta(o,x) = x \end{cases}$$
is well posed, $\eta(t,x)$ is defined for any $x \in \Lambda$ and $t \in \mathbb{R}$ and

(i) $\qquad \dfrac{d}{dt}(f \circ \eta(t,x)) < 0$ *for any $x \notin K_f$ and $t \in \mathbb{R}$*

Moreover, for any neighborhood U of K_f,

(ii)   *if f is bounded on an open set* Ω, *there exists* $v =$ $v(U,\Omega) > o$, *such that*

$$\forall \ x \in \Omega \setminus U, \qquad \frac{d}{dt}(f \circ \eta(0,x)) = \langle \nabla f(x), \ F(x) \rangle \leq -v$$

(iii)   *if c is the only critical value of f in (a,b)*, $\exists \ T$ $= T(U,a,b) > o$ *such that*

$$G^T(f_a^b) \subseteq U$$

Proof. Let F_1 be the vector field given by Lemma I.4.1 and let F_2 be a vector field which coincides with the vector field given by lemma I.4.2 in an open neighborhood U of the critical points (which, by our assumptions are isolated). Then we can assume that U is given by the union of local charts. Now let ϕ be a Lipschitz continuous function which is 0 in $\Lambda \setminus U$ and which is 1 in an open neighborhood V $(\overline{V} \subset U)$ of the critical points.

Then, the vector field F such that, on every local chart is given by $F(x) = \phi(x) \ F_1(x) + (1 - \phi(x)) \ F_2(x)$ satisfies the properties (i) and (ii). Let us prove (iii).

Let U_0 be a neighborhood of the critical points such that its closure is included in U. Using standard argument (cf. e.g.[Ra,86]) we see that

$$\exists \ T_1 > 0 : x \in U_0 \Rightarrow \eta_t(x) \in U \ \forall t \in [-T_1,T_1],$$
$$\exists \ \varepsilon > 0 : G^{T_1}(f_{c-\varepsilon}^{c+\varepsilon}) \subset U,$$

and

$$\exists \ T_2 > 0 : G^{T_2}(f_a^b) \subset f_{c-\varepsilon}^{c+\varepsilon}.$$

Then the proof follows by Remark I.3.1(vii).

□

Remark I.4.4. If $f \in \mathcal{M}(\bar{\Omega})$ where Ω is an open set in Λ, then the same result of Th. I.4.3 holds, except that $\eta(t,x)$ is defined only for $x \in \bar{\Omega}$ and $t \leq \tau(x)$, the exit time from $\bar{\Omega}$.

The strip.

In the variational case, the strip f_a^b is the simplest set at which we can apply the theory of section 1.3

Theorem I.4.5. *Let f and F as above, let $a,b \in \mathbb{R}$ with $a < b$ and suppose that a is a regular value of f. Then $f_a^b \in \Sigma_F$ and*

$$I_\lambda(f_a^b, F) = \mathcal{P}_\lambda(f^b, f^a),$$

Proof. Take
$$g_1(x) = -f(x) + a \quad and \quad g_2(x) = f(x) - b$$
Then the conclusion follows by theorem I.3.4 and Remark I.3.5.

□

Notice that in this theorem is not necessary that F satisfies P.S. Moreover, in this case $I_\lambda(f_a^b, F)$ does not depend on F, but only on f: any vector field F which satisfies (i) of Th. I.4.3, gives the same index.

Blocks around nondegenerate critical points

The next theorem relates the Morse index of a nondegenerate critical point to the index I_λ:

Theorem I.4.6. *Let $f \in C^1(\Lambda)$, let x_0 be a nondegenerate critical point of f (in the sense of definition 1.3) and let F be as in lemma 4.2. Then there exists a neighborhood*

U_0 of x_0 such that $U_0 \in \Sigma$ and

$$I_\lambda(U_0) = \lambda^{m(x_0)}$$

Proof. Choosing a local chart V for x_0 we can assume to be in an Hilbert space. Let $Q \subset V$ be the set of Theorem 3.7 "centered" at x_0. If we take ρ sufficiently small, the assumptions of Theorem 3.7 are satisfied. Then the proof of theorem 4.6. follows.

□

Sections of Conley blocks.

In the variational case, we can produce sets in Σ by a section of sets in Σ.

Lemma I.4.7 *Let f and F be as in Theorem I.4.3, let $N \in \Sigma$ and let b be a regular value of f. Then, $N \cap f^b$ and $N \cap f_b$ $\in \Sigma$ and there exists $Q_\lambda = Q_\lambda(F) \in S$ such that*

$$I_\lambda(N \cap f^b) + I_\lambda(N \cap f_b) = I_\lambda(N) + (1+\lambda) \, Q_\lambda(F)$$

Proof. Set

$$A = N \cap f_b \quad and \quad B = N \cap f^b$$

Consider the triple $(\Gamma(N), B \cup \Gamma(N), N)$. By Theorem I.2.2.(v), there exists $Q \in S$ such that

$$\mathcal{P}_\lambda(N, B \cup \Gamma(N)) + \mathcal{P}_\lambda(B \cup \Gamma(N), \Gamma(N)) = \mathcal{P}_\lambda(N, \Gamma(N)) + (1+\lambda)Q_\lambda.$$

Moreover, since b is a regular value for f, it is not difficult to check that

$$\Gamma(B) = \Gamma(N) \cap B,$$

and

$$\Gamma(A) = (\Gamma(N) \cap A) \cup f^{-1}(b),$$

Then $\Gamma(B)$ and $\Gamma(A)$ are closed.

Now set

$$C_\varepsilon = \{\eta_t(x): t \in [0,\varepsilon],\ x \in f^{-1}(b)\}.$$

Since b is a regular value, if ε is sufficiently small, by Theorem I.2.2.(iv), we have

$$P_\lambda(A,\Gamma(A)) = P_\lambda(A \cup C_\varepsilon, \Gamma(A) \cup C_\varepsilon).$$

Moreover, following the proof of Theorem I.3.8.(i) and using Theorem I.2.2.(iv) and I.2.2.(vi) shows (if ε is sufficiently small)

$$P_\lambda(N, B \cup \Gamma(N)) = P_\lambda(A \cup C_\varepsilon, \Gamma(A) \cup C_\varepsilon),$$

therefore,

$$\mathcal{P}_\lambda(N, B \cup \Gamma(N)) = \mathcal{P}_\lambda(A, \Gamma(A)) = I_\lambda(A, F).$$

Moreover

$$B \cup \Gamma(N) \setminus (\Gamma(N) \setminus \Gamma(B)) = B$$

and, since b is a regular value,

$$cl_{B \cup \Gamma(N)}(\Gamma(N) \setminus \Gamma(B)) \subset int_{B \cup \Gamma(N)}(\Gamma(N)).$$

Therefore, by excision property,

$$P_\lambda(B \cup \Gamma(N), \Gamma(N)) = P_\lambda(B, \Gamma(B)) = I_\lambda(B, F).$$

Then, since $I_\lambda(N) = \mathcal{P}_\lambda(N, \Gamma(N))$, we get the conclusion.

$\square$

Corollary I.4.8. *Let f and F be as in Theorem I.4.3, let $N \in \Sigma$ and let b_k, $(k=1,...,n)$ be a sequence of regular values of f; moreover set $b_0 = -\infty$ and $b_{n+1} = +\infty$. Then, $N \cap f_{b_k}^{b_{k+1}} \in \Sigma$ and there exists $Q_\lambda = Q_\lambda(F) \in S$ such that*

$$\sum_{k=0}^{n} I_\lambda(N \cap f_{b_k}^{b_{k+1}}) = I_\lambda(N) + (1+\lambda)\, Q_\lambda(F)$$

Proof. It is an easy consequence of Lemma I.4.7.

$\square$

Now we are ready to prove Theorem I.2.7 in a more general form:

Theorem I.4.9. *Let Ω be an open set in Λ and let $f \in M(\overline{\Omega})$ with $\overline{\Omega} \in \Sigma$. If f bounded, then*

$$\sum_{x \in K(\Omega)} \lambda^{m(x)} = I_\lambda(\overline{\Omega}, F) + (1 + \lambda) Q_\lambda$$

where $K(\Omega) = K_f \cap \Omega$, $Q(t)$ is a polynomial with integer nonnegative coefficients and F is given by Theorem I.4.3.

(Note that, if $f \in M(\overline{\Omega})$, $K_f(\overline{\Omega}) = K_f(\Omega)$).

Theorem I.2.7 follows from the above theorem and Theorem I.4.5 if we take $\overline{\Omega} = f_a^b$.

First we need the following lemma.

Lemma I.4.10. *Let $\overline{\Omega} \in \Sigma$ and let $f \in M(\overline{\Omega})$. If f it bounded and it has only the critical value $c \in \mathbb{R}$, then*

$$I_\lambda(N) = \lambda^{m(x_1)} + \ldots + \lambda^{m(x_k)},$$

where $\{x_1, \ldots, x_k\}$ are the critical points of f and $m(x_j)$ is the Morse index of x_j (notice that we have only a finite number of critical points by virtue of Remark I.2.6).

Proof. Let $U_1, \ldots, U_k$ be disjoint neighborhoods of our critical points taken according to Theorem I.4.6, and let

$$U = \bigcup_{h=1}^{k} U_h$$

By theorem I.3.8.(vii) and theorem I.4.6, we have that

$$I_\lambda(U) = \sum_h I_\lambda(U_h) = \lambda^{m(x_1)} + \ldots + \lambda^{m(x_k)}.$$

Moreover, by Theorem I.4.3.(iii), there exists T large enough that $G^T(\overline{\Omega}) \subseteq U$. Then, by Theorem I.3.8. (vi), $I_\lambda(\overline{\Omega}) = I_\lambda(U)$ from which the conclusion.

$\square$

Proof of theorem I.4.9. Since $f \in \overline{M}(\Omega)$ and it is bounded, by Remark I.2.6, it has only a finite number of critical values $\{c_1,...,c_n\}$. Now let $\{b_0,...,b_{n+1}\}$ be a sequence of numbers such that

$$-\infty = b_0 < c_1 < b_1 < ... < b_n < c_n < b_{h+1} = +\infty$$

Then, by Corollary I.4.8,

$$\sum_{k=0}^{n} I_\lambda(\overline{\Omega} \cap f_{b_k}^{b_{k+1}}) = I_\lambda(\overline{\Omega}) + (1+\lambda) \, Q_\lambda$$

and the conclusion follows from lemma 4.10 with $\overline{\Omega}$ replaced by $\overline{\Omega} \cap f_{b_k}^{b_{k+1}}$. $\square$

The Morse relations when f is unbounded.

In theorem I.4.9, we have assumed that f is bounded, but this assumption can be partially removed if we allows both the Morse and the Poincaré polynomial to be formal series (in the space S).

Theorem I.4.11. *Let Ω be an open set in Λ and let $f \in \overline{M}(\Omega)$ with $\overline{\Omega} \in \Sigma$; if f it bounded from below on Ω, then*

$$\sum_{x \in K(\Omega)} \lambda^{m(x)} = I_\lambda(\overline{\Omega},F) + (1 + \lambda) \, Q_\lambda \qquad (4.1)$$

where $K(\Omega) = K_f \cap \Omega$ is a countable set, $Q(t)$ is a formal series in S and F is given by Theorem I.4.3.

Proof. Since every critical point of f is nondegenerate, f satisfies P.S. on Ω, and it is bounded from below, there exists $\{b_h\}_{h\in\mathbb{N}} \subset \mathbb{R}$ and $\{c_h\}_{h\in\mathbb{Z}} \subset \mathbb{R}$ such that

every b_h is a regular value for f,

$$b_0 < \inf_{\overline{\Omega}} f < b_1 < ... < b_h < b_{h+1} < ...,$$

$$\lim_{h\to+\infty} b_h = +\infty,$$

for any $h \in \mathbb{N}$, $f_{b_h}^{b_{h+1}} \cap K_f(\Omega) = f^{-1}(c_h) \cap K_f(\Omega)$,

for any $h \in \mathbb{N}$, $f^{-1}(c_h) \cap K_f(\Omega)$ is finite.

Then, by Lemma I.4.7 and Corollary I.4.8, arguing by induction on h gives, for any $h \in \mathbb{N}$, the existence of $Q_h \in S$, such that

$$(4.2) \quad m_\lambda\left(f^{b_h} \cap K_f(\overline{\Omega}),f\right) + I_\lambda(f_{b_h},F) = I_\lambda(\overline{\Omega},F) + (1+\lambda)Q_h(\lambda),$$

where $f^b = \{x \in \overline{\Omega}: f(x) \leq b\}$ and $f_b = \{x \in \overline{\Omega}: f(x) \geq b\}$.

Fix $k \in \mathbb{N}$. Our goal is to prove (4.1), using (4.2) and arguing on the coefficients relatively to any degree of the formal series in (4.1).

If the set M_k of points of $K_f(\Omega)$ having Morse index k is infinite, taking the limit in (4.2) when h goes to $+\infty$, gives immediately the prove of (4.1) relatively to the degree k, since the coefficient of degree k of $m_\lambda\left(f^{b_h} \cap K_f(\overline{\Omega}),f\right)$ is non decreasing with respect to h and tends to $+\infty$.

Now suppose that M_k is finite and let b a regular value such that

$$(4.3) \qquad\qquad b > \max_{M_k} f.$$

By (4.2), to prove (4.1) (relatively to the degree k) it is sufficient to prove that

(4.4)   *the coefficient of degree k of $I_\lambda(f_b, F)$ is zero.*

Now let $c > b$ be a regular value for f. Then, using the flow η and the excision property as in proving Lemma I.4.7, shows that

(4-5) $P_\lambda(\Gamma(\bar\Omega) \cup f^c, \ \Gamma(\bar\Omega) \cup f^b) = P_\lambda\left(f_b^c, (\Gamma(\bar\Omega) \cap f_b^c) \cup f^{-1}(b)\right).$

Since $(\Gamma(\bar\Omega) \cap f_b^c) \cup f^{-1}(b) = \Gamma(f_b^c)$, by (4.5) we have

(4.6) $P_\lambda(\Gamma(\bar\Omega) \cup f^c, \Gamma(\bar\Omega) \cup f^b) = P_\lambda(f_b^c, \Gamma(f_b^c)).$

Then, by Corollary I.4.8 with $N = f_b^c$, combining (4.3) and (4.6) gives

(4.7) $H_k(\Gamma(\bar\Omega) \cup f^c, \Gamma(\bar\Omega) \cup f^b)) = 0.$

Since c is a regular value using the flow and the excision property we also get

(4.8) $I_\lambda(f_b, F) = P_\lambda\left[f_b, (\Gamma(\bar\Omega) \cap f_b) \cup f^{-1}(b)\right] =$
$= P_\lambda\left[\bar\Omega, \Gamma(\bar\Omega) \cup f^b\right].$

Now consider the exact homology sequence

(4.9) $\longrightarrow H_k(\Gamma(\bar\Omega) \cup f^c, \Gamma(\bar\Omega) \cup f^b)) \xrightarrow{\ i_k^*\ } H_k(\bar\Omega, \Gamma(\bar\Omega) \cup f^b)$

$\xrightarrow{\ j_k^*\ } H_k(\bar\Omega, \Gamma(\bar\Omega) \cup f^c) \longrightarrow$

where i_k^* and j_k^* are the map induced by the inclusion maps.

If, by contradiction, (4.4) does not hold, by (4.8) there exists

$$\alpha \in H_k(\bar\Omega, \Gamma(\bar\Omega) \cup f^c), \quad \alpha \neq 0.$$

Now, denoting by Δ the support of α and choosing

$$c > \max_\Delta f,$$

gives, $j_k^*(\alpha) = 0$. Then, by the exactness of the homology sequence (4.9), there exists $\beta \in H_k(\Gamma(\bar\Omega) \cup f^c, \Gamma(\bar\Omega) \cup f^b))$ such that $i_k^*(\beta) = \alpha$, in contradiction with (4.7). Then (4.4) is proved and the proof of Theorem I.4.11 is complete.

□

I.5 - MORSE THEORY FOR DEGENERATE CRITICAL POINTS.

In this section we introduced a index i_λ generalizing the Morse polynomial m_λ when *the critical points of f are degenerate*. More in general, we define a Morse polynomial also for an isolated set K of critical points of f.

The class $\mathcal{F}(\overline{\Omega})$.

Now we shall describe a class $\mathcal{F}(\overline{\Omega})$ of C^1-functionals where the generalized Morse index will be defined.

For any $\varepsilon > 0$ and $A \subset \Lambda$, we shall put
$$N_\varepsilon(A) = \{x \in \Lambda: d(x,A) < \varepsilon\},$$
where d is the distance induced by the Hilbert structure of Λ. Now, consider the class
$$M_f^\varepsilon(\overline{\Omega}) = \left\{ g \in M(\overline{\Omega}): g(x) = f(x) \ for \ x \notin N_\varepsilon(K_f(\overline{\Omega})) \right\}.$$
where $M(\overline{\Omega})$ is defined at Definition 2.5.

Notice that
$$\varepsilon_1 < \varepsilon_2 \Rightarrow M_f^{\varepsilon_1}(\overline{\Omega}) \subset M_f^{\varepsilon_2}(\overline{\Omega}).$$
We define
$$\mathcal{F}(\overline{\Omega}) = \left\{ f \in C^1(\overline{\Omega}) \mid M_f^\varepsilon(\overline{\Omega}) \neq \emptyset \ for \ any \ \varepsilon > 0 \right\}.$$

Clearly, if $\Omega \subseteq \Lambda$ is bounded and Λ is a finite dimensional manifold, then $\mathcal{F}(\overline{\Omega}) = C^1(\overline{\Omega})$. In general we do not know how general this class is; however we can prove easily that many interesting functionals belong to this class. In fact, we have the following results:

Example I.5.1. *Let f of class C^2 in a neighborhood of $\bar{\Omega}$ and satisfying P.S. in $\bar{\Omega}$. Suppose that, for any degenerate critical point $x \in K_f(\bar{\Omega})$, the linear operator associated to $H^f(x)$ is a Fredholm operator (of index 0). Then $f \in \mathcal{F}(\bar{\Omega})$ (cf. [Ma,Pr,75]).*

Theorem I.5.2. *Let Λ be a separable Hilbert space and let f_0 satisfy the assumptions of Example 5.1 (for instance $f_0(x) = \frac{1}{2}\langle Lx, x \rangle$ where $L:\Lambda \longrightarrow \Lambda$ is a bounded selfadjoint invertible operator). Let $\psi \in C^1(\Lambda)$ be a functional whose gradient is completely continuos, i.e.*

$$\text{if } x_k \text{ weakly converges to } x,$$
$$\text{then } \psi'(x_k) \text{ strongly converges to } \psi'(x). \tag{5.1}$$

Assume that $f(x) = f_0(x) + \psi(x)$ is bounded in $\bar{\Omega}$ and it satisfies P.S. in Ω. Then $f \in \mathcal{F}(\bar{\Omega})$.

Proof. Since Λ is separable, there exists a sequence of linear subspaces of Λ, $\{E_n\}_{n \in \mathbb{N}}$, such that

$$E_n = span \{e_1, \ldots, e_n\}$$

where $\{e_i\}_{i \in \mathbb{N}}$ is a complete orthonormal system for Λ. Consider the orthogonal projection P_n of Λ on E_n and put

$$\psi_n(x) = \psi(P_n(x)).$$

The definition of P_n and assumptions (5.1) easily gives

$$\sup_{x \in B} \|\psi'_n(x) - \psi'(x)\| \xrightarrow[n]{} 0, \text{ for any bounded subset } B,$$

which implies that

$$\sup_{x \in B} |\psi_n(x) - \psi(x)| \xrightarrow[n]{} 0, \text{ for any bounded subset } B.$$

Since $\psi_{n}|_{E_n} : E_n \longrightarrow \mathbb{R}$, and E_n is finite dimensional, there exists a sequence of C^∞-maps, $\{\varphi_n\}_{n \in \mathbb{N}}$, defined on E_n such that

$$\left\| \psi_{n}|_{E_n} - \varphi_n \right\|_{C^1(E_n)} \leq 1/n.$$

Now, for any $x \in \Lambda$, set

$$\tilde{\psi}_n(x) = \varphi_n(P_n(x)).$$

Therefore for any bounded subset B,

(5-2) $\tilde{\psi}_n(x) \xrightarrow[n]{} \psi(x)$ in $C^1(B)$,

Now, set $K = K_f(\Omega)$, and for any $\varepsilon > 0$, let $\phi_\varepsilon \in C^2(\Lambda,[0,1])$ be a function such that

$$\phi_\varepsilon = 0 \ on \ \Lambda \setminus N_\varepsilon(K),$$
$$\phi_\varepsilon = 1 \ on \ N_{\varepsilon/2}(K),$$
$$\|\phi'_\varepsilon\| \ is \ bounded \ in \ \Lambda.$$

Now define

(5-3) $g_{\varepsilon,n}(x) := f_0(x) + \psi(x) + \phi_\varepsilon(x)(\psi_n(x) - \psi(x)) =$
$$= f(x) + \phi_\varepsilon(x)(\psi_n(x) - \psi(x)).$$

Since f satisfies P.S. and it is bounded on Ω, there exists $\nu(\varepsilon) > 0$ such that

$$\|f'(x)\| \geq \nu(\varepsilon) \ for \ any \ x \ in \ \Omega \setminus N_{\varepsilon/2}(K),$$

so

$$\|g'_{\varepsilon,n}(x)\| \geq$$
$$\|f'(x)\| - \|\phi'_\varepsilon(x)\| |\psi_n(x) - \psi(x)| - |\phi_\varepsilon(x)| \|\psi'_n(x) - \psi'(x)\|.$$

Since f satisfies P.S. and is bounded in Ω, K is compact and hence $N_\varepsilon(K)$ is bounded. Therefore, by (5-2), if n is sufficiently large, there are not critical points of $g_{\varepsilon,n}$ in $N_\varepsilon(K) \setminus N_{\varepsilon/2}(K)$ and hence in $\Omega \setminus N_{\varepsilon/2}(K)$. Moreover on $N_{\varepsilon/2}(K)$,

$$g_{\varepsilon,n} = f_0 + \tilde{\psi}_n.$$

Since $g_{\varepsilon,n}$ satisfies the assumptions of example 5.1 on

$N_{\varepsilon/2}(K)$ and it does not have critical points on $\partial N_{\varepsilon/2}(K_f)$, using the perturbation methods of Marino and Prodi (cf. [9]), $g_{\varepsilon,n}$ can be modified into a function $\tilde{g}_{\varepsilon,n}$ such that

- $\tilde{g}_{\varepsilon,n}\Big|_{N_{\varepsilon/2}(K)}$ *is a Morse function,*

- $\tilde{g}_{\varepsilon,n}$ *does not have critical ponts in* $N_{\varepsilon}(K)\backslash N_{\varepsilon/2}(K)$,

- $\tilde{g}_{\varepsilon,n} = f$ *on* $\Omega\backslash N_{\varepsilon}(K)$.

    Then, for n large, $\tilde{g}_{\varepsilon,n} \in M_f^{\varepsilon}(\overline{\Omega})$ and hence $f \in \mathcal{F}(\overline{\Omega})$.

                                                  □

Remark I.5.3. *Note that* $f \in \mathcal{F}(\overline{\Omega})$ *might not satisfy P.S. on* $\overline{\Omega}$. *This happens if* $K_f(\overline{\Omega})$ *is not compact. For example let* Λ *be an infinite dimensional Hilbert space and*

$$f(x) = \varphi(\|x\|^2)$$

where $\varphi : \mathbb{R}^+ \longrightarrow \mathbb{R}^+$ *and* $\varphi(s) = \begin{cases} 0 & \text{if } s \leq 1 \\ (s-1)^2 & \text{if } s \geq 1 \end{cases}$.

Let $\varphi_{\varepsilon} \in C^2(\mathbb{R}^+,\mathbb{R}^+)$ *be a function such that* $\varphi_{\varepsilon}''(0) \neq 0$ *and* $\varphi_{\varepsilon}'(s) = 0$ *if and only if* $s = 0$ *and* $\varphi_{\varepsilon}(s) = \varphi(s)$ *for any* $s \geq 1+\varepsilon$, *and set* $g_{\varepsilon}(x) = \varphi_{\varepsilon}(\|x\|^2)$; *then* $g_{\varepsilon} \in M_f^{\varepsilon}(\overline{\Omega}) \neq \varnothing$. *Another example is the map* $\sin(\cdot) \in \mathcal{F}(\mathbb{R})$.

Isolated critical sets and their Morse Index.

    Let $f \in C^1(\Omega)$; a compact set $K \subset K_f$ is called isolated critical set if there exists an open set ω such that $K = K_f \cap \omega$. The open set ω will be said isolating set (for K).

    Finally we define a Morse index for an isolated critical set $K \subset K_f$ which is the main point of this section. Let ω an isolating set for K.

Definition I.5.4. *We set*

$$(5.4) \qquad i_\lambda(K,f) \equiv i_\lambda(K,f,\omega) = \sup_{\varepsilon > 0} \left[\inf_{g \in \mathcal{M}_f^\varepsilon(\overline{\omega})} m_\lambda(K_g(\overline{\omega}),g) \right]$$

where m_λ, *inf and sup are given by Definitions 1.8, and 1.10 respectively.*

   The formal series $i_\lambda(K,f)$ *is called the (generalized) Morse Index of* K.

Remark I.5.5. *It is easy to see that the index* (5.4) *of an isolated critical set does not depend on the isolating set* ω.

Definition I.5.6. *If* $x \in K_f(\overline{\Omega})$ *is an isolated critical point, we call multiplicity of* x *the integer number* $i_1(\{x\},f)$. *Analogously it is possible to define the multiplicity of an isolated critical set.*

   Note that a critical set K contains at least $i_1(K,f)$ critical points if counted with their multiplicity.

Remark I.5.7. *If a critical point* x *has multiplicity* $h \in \mathbb{N}$, *then every perturbation of* f *producing nondegenerate critical points has at least* h *critical points nearby* x. *In fact, by the Remark I.I.13, for* ε *sufficiently small,*

$$\inf_{g \in \mathcal{M}_f^\varepsilon(\overline{\omega})} m_\lambda(K_g(\overline{\omega}),g)$$

is constant.

Some properties of i_λ.

The next theorem describes the first properties of the index i_λ and shows that it is a generalization of the Morse polynomial.

Theorem I.5.8. *Let* $f \in f \in \mathcal{F}(\bar{\Omega})$, *then*

(i) *if* x_0 *is a nondegenerate critical point of* f,
$$i_\lambda(\{x\}) = \lambda^{m(x,f)}.$$

(ii) *if* $K_1, K_2 \subset K$ *are isolated compact sets, and* $K_1 \cap K_2 = \emptyset$, *then,*
$$i_\lambda(K_1 \cup K_2, f) = i_\lambda(K_1, f) + i_\lambda(K_2, f),$$

(iii) *if* K *is a discrete set, then*

$$i_\lambda(K, f) = \sum_{x \in K} i_\lambda(x, f).$$

(iv) *if* $f \in M(\bar{\Omega})$, *then*

$$i_\lambda(K_f, f) = m_\lambda(K_f, f) = \sum_{x \in K_f} \lambda^{m(x,f)}.$$

Proof. Let (U_x, ϕ_x) a local chart for the nondegenerate critical point x and $\alpha_x = \phi_x(x)$. Putting

$$L = d\phi_x^{-1}(\alpha_x) \circ \left[H^{f \circ \phi_x^{-1}}(\alpha_x) \right]$$

in Theorem I.3.7 gives the existence of $\rho > 0$ such that $Q_\rho \subset U_x$, $Q_\rho \in \Sigma_F$ and

(5.5) $$I_\lambda(Q_\rho, F) = \lambda^{m(x)}.$$

Moreover ρ can be chosen so small that f satisfies P.S. on Q_ρ, and

(5.6) $$K_f(Q_\rho) = \{x\}.$$

Then, for any $\varepsilon > 0$, $f \in M_f^\varepsilon(Q_\rho)$ and, by Definition I.5.4,

(5-7) $$i_\lambda(\{x\}, f) \leq \lambda^{m(x)}.$$

Moreover, by Theorem I.4.9 and (5.5), for any $g \in M_f^\varepsilon(Q_\rho)$,

there exists $Q_g \in S$ such that
$$m_\lambda(\{x\},g) = \lambda^{m(x)} + (1+\lambda)Q_g(\lambda).$$
Therefore, by Definition I.5.4,

(5.8) $i_\lambda(\{x\},f) \geq \lambda^{m(x)},$

and combining (5.7) and (5.8) gives the proof of (i).

(ii) and (iii) are simple consequence of Definition I.5.4; (iv) follows by (iii) and (i).

□

By Theorem I.5.8 it turns out that that the multiplicity of a nondegenerate critical point is one.

Notice that it is possible, for a critical point to have multiplicity zero (consider for instance $f : \mathbb{R} \longrightarrow \mathbb{R}$, $f(x) = x^3$).

The generalized Morse relations.

Of course the index i_λ has been constructed in such a way that the Morse relations are still valid. In fact we have the following Theorem

Theorem I.5.9. *Assume* f *bounded from below on* Ω *and* $f \in M(\overline{\Omega})$. *If* F *is a vector field as in Theorem 4.3, and* $\overline{\Omega} \in \Sigma$, *then*

$$i_\lambda(K_f, f) = I_\lambda(\overline{\Omega}, F) + (1+\lambda)Q_\lambda,$$

where $Q_\lambda \in S$.

Proof. By the definitions 5.4 and 1.10, there are sequences $\varepsilon_k \to 0^+$, and g_k such that,

$$g_k \in M_f^{\varepsilon_k}(\overline{\omega}), \qquad \text{and} \qquad \lim_{k \to \infty} m_\lambda(K_{g_k}, g_k) = i_\lambda(K_f, f)$$

By Theorem 4.11, for any $k \in \mathbb{N}$, there exists $Q_\lambda^k \in \mathbb{S}$ such that

(5-9) $$m_\lambda(K_{g_k}, g_k) = I_\lambda(\bar{\Omega}, F) + (1+\lambda)Q_\lambda^k,$$

Taking the limit of the coefficients of the above equation, we conclude the proof.

□

If some of the coefficients of $i_\lambda(K_f, f)$ are $+\infty$, then, taking the limit (5-9), some of the coefficients of Q_λ might be not uniquely determined ; nevertheless the Morse relations holds applying the usual algebra to $+\infty$.

The meaning of the generalized Morse index for some C^1-functional.

In this section we will state a theorem which will be usefull when we apply The Morse theory to P.D.E.

Theorem I.5.10. *Let f be a function as in example 5.2, and let x_0 be an isolated critical point of f.*
Let

$$E_n = span \{e_1, ..., e_n\}$$

be a sequence of linear subspaces of Λ, where $\{e_i\}_{i \in \mathbb{N}}$ is a complete orthonormal system for Λ and set

$$V_n = x_0 + E_n.$$

Suppose that

(i) $f\big|_{V_n}$ is twice differentiable in x_0

(ii) for n sufficiently large, x_0 is a nondegenrate critical point of $f\big|_{V_n}$ and

$$m(x_0, f\big|_{V_n}) = k \ .$$

Then

$$i(x_0) = \lambda^k.$$

Proof. It is not restrictive to suppose that $x_0 = 0$. Let N be a Conley block such that $K_f \cap N = \{x_0\}$. Such a block can be constructed in the following way; take a neighborood V of x_0 such that $K_f \cap V = \{x_0\}$; if $\varepsilon > 0$ is sufficiently small and $c = f(x_0)$, take

$$N = V \cap G^1(f_{c-\varepsilon}^{c+\varepsilon})$$

($G^T(A)$ is defined in section 3). Now, let $g_{\varepsilon,n}$ be the function defined by (5-3) in the proof of Example 5.2 with ε small enough such that $N_\varepsilon(x_0) \subseteq N$ and n large enough such that (ii) holds. Then, by the construction of $g_{\varepsilon,n}$, we have that x_0 is the only critical point of $g_{\varepsilon,n}$ in N and

$$m(x_0, g_{\varepsilon,n}) = k.$$

Since N is a Conley block for $g_{\varepsilon,n}$, by the Morse relations for $g_{\varepsilon,n}$, we get:

$$\lambda^k = I_\lambda(N);$$

then, by the Morse relations for f,

$$i(x_0) = I_\lambda(N) = \lambda^k.$$

$\square$

The meaning of the generalized Morse index for degenerate critical points.

The next theorem states some information given by i_λ for the degenerate points of C^2-functionals.

Theorem I.5.11. *Let f satisfying the assumptions of Example I.5.1 and suppose that $K_f(\overline{\Omega})$ is compact. Then*

(i) if $i_\lambda(K,f) = \sum_{k\geq 0} a_k \lambda^k$ with $a_k \neq 0$, then there exists $x \in K$ such that

$$m(x) \leq k \leq m^*(x).$$

(ii) if we set $m(K) = \inf_{x \in K} m(x)$ and $m^*(K) = \sup_{x \in K} m^*(x)$, then

$$i_\lambda(K,f) = \sum_{k=m(K)}^{m^*(K)} a_k \lambda^k,$$

(iii) if x_0 is an isolated degenerate critical point and $i_\lambda(x_0,f) \neq 0$, then

$$i_\lambda(x_0,f) = \sum_{k=m(x_0)+1}^{k=m^*(x_0)-1} a_k \lambda^k,$$

Proof. (i) follows by the perturbation methods of Marino and Prodi (cf.[Ma,Pr,75]), and (since the Hessian is a Fredholm map) by the lower semicontinuity of the strict Morse index and the upper semicontinuity of the large Morse index. (ii) is a simple consequence of (i), while (iii) can be get following the proof of Theorem 6.13 of [Be,91].

□

Theorem I.5.12. *Let f be a function as in example 5.2, and let x_0 be a critical point of f.*

Let

$$E_n = span \{e_1,...,e_n\}$$

be a sequence of linear subspaces of Λ, where $\{e_i\}_{i \in \mathbb{N}}$ is a complete orthonormal system for Λ and set

$$V_n = x_0 + E_n.$$

Suppose that

(i) $f\big|_{V_n}$ is twice differentiable in x_0

(ii) for n sufficiently large, $m(x_0, f\big|_{V_n})$ and $m^*(x_0, f\big|_{V_n})$ do not depend on n. Then the same conclusion of Th. I.5.11 holds.

Proof. The proof can be obtained combining the arguments of Th. I.5.10 and Th. I.5.11.

□

Remark I.5.13. *Let Λ be an Hilbert space. Suppose that $f \in \mathcal{F}(\overline{\Omega})$, f is bounded on $\overline{\Omega}$, F is as in Theorem 4.3 and*

$$F = id + \psi$$

where id is the identity on Λ and $\psi : \overline{\Omega} \longrightarrow \Lambda$ is a compact operator.

Assume $\overline{\Omega} \in \Sigma$ and $F(x) \neq 0$ for any $x \in \partial\Omega$. Then by Theorem 4.9 and the definitions of i_λ, I_λ and the topological degree we see that

$$I_{-1}(\overline{\Omega}, F) = \deg(\overline{\Omega}, -F, 0), \tag{5.10}$$

and it is also possible to show that (5.10) can be generalized to situations where F is not a pseudo gradient vector field (nonvariational case) by the approach to Morse inequalities for the nonvariational case.

I.6- SOME EXISTENCE THEOREM:

In this section we will apply the results of the previous sections to obtain existence results of critical points with information to their Morse index.

The Mountain pass theorem

We start with the so called Mountain Pass Theorem which is very simple but it has proved to be very flexible and useful in many situations. This theorem is due to Ambrosetti and Rabinowitz [Am,Ra,73]. Then Hofer [Ho,85] was able to give a geometric description of the neighborhood of a critical point of "Mountain Pass" type and to relate it to the Morse index. Here we give a proof due to the author [Be,91].

Theorem I.6.1. *(Mountain-Pass Theorem). Let* $f \in \mathcal{F}(\mathfrak{M})$ *and let* $S \subset \mathfrak{M}$ *be a set which splits* $\mathfrak{M} \backslash S$ *in two connected components. Moreover assume that there exists a constant a such that*

i) $\inf_{x \in S} f > a,$

ii) $f(x_i) < a,$ *i=1,2. where* x_1 *and* x_2 *are two points belonging to different connected components of* $\mathfrak{M} \backslash S$.

Then

$$i_\lambda(K_f \cap f_a^b) = \lambda + \mathcal{Z}(\lambda)$$

where $b = \sup_{x \in \gamma} f$ *and* γ *is a continuous curve joining* x_1 *and* x_2 *and* $\mathcal{Z}(\lambda)$ *is a (possibly null) term which cannot be estimated with the available information. (Then, in particular we have that* $K_f \cap f_a^b \neq \emptyset$).

Moreover, if f *satisfies the assumptions of the Example I.5.1, there exists a critical point* $\bar{x} \in f_a^b$ *such that*

$$m(\bar{x}) \leq 1 \leq m^*(\bar{x}).$$

Proof. f^a has at least two connected components and hence $H_0(f^a)$ has at least two generators $[x_1]$ and $[x_2]$. Now consider the map

$$i_0 \colon H_0(f^a) \longrightarrow H_0(f^b)$$

induced by the natural embedding. Since x_1 and x_2 belong to the same connected component of f^b, then $i_0([x_1] - [x_2]) = 0$. Then by the exactness of the sequence,

$$\longrightarrow H_1(f^b, f^a) \xrightarrow{\ \partial_1\ } H_0(f^a) \xrightarrow{\ i_0\ } H_0(f^b) \longrightarrow$$

it follows that, $[x_1] - [x_2] \in Im\ \partial_1$. So $H_1(f^b, f^a,) \neq 0$. Then,

$$I_\lambda(f_a^b) = \lambda + \mathcal{Z}_1(\lambda).$$

Therefore, by theorem I.5.9,

$$i_\lambda(K_f \cap f_a^b) = = \lambda + \mathcal{Z}_1(\lambda) + other\ terms.$$

The second statement follows from Theorem I.5.10.(i).

□

The linking theorem.

The next theorem which we will present is the so called Linking Theorem which is a generalization of the Mountain Pass Theorem. This theorem is due to Rabinowitz [Ra,78,b]; cf also [Be,Ra,79], [Ba,Be,Fo,83] and the book of Rabinowitz [Ra,86] for some generalizations. The determination of the Morse index of the critical points of linking type and the proof of this theorem via the Morse Theory is due to the author, cf.[Be,87], [Be,91] and [Be,Fo,86]. Here we will give a proof taken from [Be,91].

Definition I.6.2 - *Let Q and S be two disjoint subsets of a topological space M (which in the application will be a*

Riemannian manifold). We suppose that Q is diffeomorphic to the (n+1)-dimensional ball. We say that ∂Q and S link if the embedding

$$i_n : H_n(\partial Q) \longrightarrow H_n(M \backslash S)$$

is not trivial.

Roughly speaking this means that ∂Q is not the boundary (in the sense of the homology) of any set contained in $M \backslash S$. For example, in Th. I.6.1, the set $\partial \gamma = \{x_1, x_2\}$ and S link.

Theorem I.6.3. *(Linking Theorem)* - *Let* $f \in \mathcal{F}(\mathfrak{M})$ *and let* $Q, S \subset \mathfrak{M}$ *be two sets such that* ∂Q *and* S *link and let* $\dim Q = n+1$. *Moreover assume that there exists a constant* $a \in \mathbb{R}$ *such that*

i) $\inf\limits_{x \in S} f > a$,

ii) $\sup\limits_{x \in \partial Q} f(x) < a$,

     Then

$$i_\lambda(K_f \cap f_a^b) = \lambda^{n+1} + \mathcal{Z}(\lambda)$$

where $b = \sup\limits_{x \in Q} f$ *and* $\mathcal{Z}(\lambda)$ *is a (possibly null) term which cannot be estimated with the available information. (Then, in particular we have that* $K_f \cap f_a^b \neq \varnothing$).

Moreover, if f satisfies the assumptions of Example I.5.1, there exists a critical point $\overline{x} \in f_a^b$ such that

$$m(\overline{x}) \leq n + 1 \leq m^*(\overline{x}).$$

Proof. Since the embedding

$$H_n(\partial Q) \xrightarrow{\ i_{1,n}\ } H_n(f^a) \xrightarrow{\ i_{2,n}\ } H_n(M \backslash S)$$

not trivial $[\partial Q]$ is a generator of a nonzero homology class in $H_n(f^a)$. Now consider the embedding

$$H_n(f^a) \xrightarrow{\ i_n\ } H_n(f^b);$$

since $i_n([\partial Q]) = [\partial Q] = \partial[Q] = 0$, by the exactness of the sequence

$$\dots..H_{n+1}(f^b,\ f^a) \xrightarrow{\ \partial_n\ } H_n(f^a) \xrightarrow{\ i_n\ } H_n(f^b) \xrightarrow{\ j_n\ } \dots.$$

it follows that, $[\partial Q] \in Im\ \partial_{n+1}$. Therefore $H_{n+1}(f^b,f^a,) \neq 0$. Then, $I_\lambda(f^b_a) = \lambda^{n+1} + Z(\lambda)$ ($Z(\lambda) \in S$ cannot be determined using only the information of the assumption of the theorem).

Then, by theorem I.5.9, $i_\lambda(K_f \cap f^b_a) = = \lambda^{n+1} + Z(\lambda) \neq 0$.

The second statement follows from Theorem I.5.10.(i).

$\square$

As a consequence of Th. I.6.3, we shall prove the Linking Theorem in its "classical" form.

Theorem I.6.4 *("Classical" Linking Theorem).* Let $\mathfrak{M}$ be an Hilbert space and $f \in \mathcal{F}(\mathfrak{M})$. Let $V_n \subset \mathfrak{M}$ be a n-dimensional subspace and suppose that there exists constants a, ρ, R_1 and R_2 (with $0 < \rho < R_1$) such that

i) $\inf\limits_{x\,\in\,S} f > a$, where $S = V_n^\perp \cap \partial B(0,\rho)$

ii) $\sup\limits_{x\,\in\,\partial Q} f(x) < a$, where

$$Q = \{y + se: y \in V_n \cap \overline{B(0,\rho)}, 0 \leq s \leq R\},$$
$$e \in V_n^\perp \text{ with } \|e\| = 1.$$

Then the same conclusion of Th. I.6.3 holds.

Proof. We have to prove that S and ∂Q link and apply Th.

I.6.3. It is not difficult to show that ∂Q is a generator of a nontrivial homology class in $H_n(\mathfrak{M}\backslash S)$. For the formal details we refer to [Be,91].

□

The saddle point theorem.

An other consequence of Th. I.6.3, it the Saddle Point Theorem. This theorem has been proved by Rabinowitz using minimax methods [Ra,78,a] (cf. also [Be,Ra,79], [Ba,Be,Fo,83] and [Ra,86] for various generalizations). Here, we will give a prove via Morse Theory (cf. [Be,87], [Be,91]) which allows to give an estimate on the Morse index of a critical point of saddle type.

Theorem I.6.5. *(Saddle Point Theorem). Let* $\mathfrak{M}$ *be an Hilbert space and* $f \in \mathcal{F}(\mathfrak{M})$. *Let* $V_{n+1} \subset \mathfrak{M}$ *be a* $(n+1)$-*dimensional subspace and suppose that there exists constants a and* ρ *(with* $\rho > 0$*) such that*

i) $\inf\limits_{x \,\in\, S} f > a$, *where* $S = (V_{n+1})^{\perp}$

ii) $\sup\limits_{x \,\in\, \partial Q} f(x) < a$, *where* $Q = V_{n+1} \cap \overline{B(0,\rho)}$

Then the same conclusion of Th. I.6.3 holds.

Proof. We have to prove that S and ∂Q link and apply Th. I.6.2. It is not difficult to show that ∂Q is a strong deformation retract of $\mathfrak{M}\backslash S = \mathfrak{M}\backslash(V_{n+1})^{\perp}$. Hence, the map $i_{n+1}\colon H_n(\partial Q) \longrightarrow H_n(\mathfrak{M}\backslash S)$ is an isomorphism, and since $H_n(\partial Q) \neq 0$, i_n is not trivial.

□

The equivariant linking theorem.

If the functional is even (i.e. equivariant with respect to a $\mathbb{Z}_2$ action), in some situations in which it is possible to apply the Mountain pass or the Linking theorem, it is possible to prove the existence of infinitely many solutions (cf.[Am,Ra,73]). Here we give a proof of this fact using the Morse theory. We refer to [Be,91] for functionals equivariant with respect to a general group.

Theorem I.6.6. *Let $\mathfrak{M}$ be a Riemannian manifold and let $f \in \mathcal{F}(\mathfrak{M})$ satisfy the following assumption:*

(6-1)   there exists $c < o$ such that $\mathcal{P}_\lambda(\mathfrak{M}, f^c) = 0$
  Then,
(a) if, 0 is a nondegenerate critical point, f has at least a critical point different from 0.
(b) if f is even, f has infinitely many critical points and K_f is unbounded.

Proof. (a) We have that
$$i_\lambda(K) = i_\lambda(0) + i_\lambda(K_0) = \lambda^{m(0)} + i_\lambda(K_0)$$
where $K_0 = K \backslash \{0\}$ and $m(0)$ is the Morse index of 0. Then by theorem I.5.9, and $(6-1)$,
$$\lambda^{m(0)} + i_\lambda(K_0) = (1+\lambda) \, Q_\lambda.$$
This equation forces $i_\lambda(K_0)$ to be different from 0.
(b) First, let us suppose that all the critical points of f are non degenerate; then
$$i_\lambda(K) = i_\lambda(0) + i_\lambda(K_0) = \lambda^{m(0)} + i_\lambda(K_0)$$
We argue indirectly, and suppose that K_0 is finite, then it contains an even number of critical points, in fact, if x

is a critical point, also $-x$ is a critical point. Hence

$$i_\lambda(K_0) = 2 \sum_k a_k \lambda^k.$$

Then by theorem I.5.9, and (ii),

$$\lambda^{m(0)} + 2 \sum_k a_k \lambda^k = (1+\lambda)\, Q_\lambda;$$

taking $\lambda = 1$ we get a contradiction, since the left-hand side of this equation is an odd number and the right-hand side is an even number. The fact that K_f is unbounded follows from Lemma I.2.6.

The general case, when f is not a Morse function, can be obtained using the results of section I.5.

□

Notice that (a) of the above theorem does not depend on the eveness of f; it can be considered as a variant of the linking theorem.

The above theorem can ba applied to functionals of the form

(6-2) $f(x) = \frac{1}{2}\langle Lx, x\rangle + \psi(x),$

where L is a bounded selfadjoint invertible operator and ψ' is compact. In fact we have the following result.

Theorem I.6.7. *Let $\mathfrak{M}$ be an Hilbert space and let f be the functional (6-2). Suppose that*

(i) $\lim_{\rho \to +\infty} f(\rho x) = -\infty$ *for any* $x \in \mathfrak{M} \setminus \{0\}$;

(ii) there exists $p > 2$ and $M \geq 0$ such that

$$0 \leq p\psi(x) \leq M + \langle \psi'(x), x\rangle \quad \text{for any } x \in \mathfrak{M},$$

Then $f \in \mathcal{F}(\mathfrak{M})$ and there exists $c < o$ such that $\mathcal{P}_\lambda(\mathfrak{M}, f^c)$

= 0. *Hence the conclusions of theorem I.6.6.*

Proof. $f \in \mathcal{F}(\mathfrak{M})$ by the example I.5.2. The proof of the second statement, can be found in [Be,91].

□

The Box theorem.

We conclude this section with a theorem which is very simple, but nevertheless has interesting applications (cf. section IV.5)

Theorem I.6.8. *(Box theorem) Let $\mathfrak{M}$ be an Hilbert space and let f be a functional of the form (6-2) which satisfies (P.S.) in bounded sets. Suppose that L satisfy the assumptions of Theorem 1.3.7 and that*

(6-3) $\|\psi'(x)\| \le M_1 \|x\|^\alpha + M_2$

for suitable constants M_1 and $M_2 \in \mathbb{R}$ and $\alpha \in (0,1)$. Then

$$i_\lambda(K_f) = \lambda^k + (1+\lambda) Q_\lambda,$$

where $k = dim \ H^-$.

Proof. If $\psi'(x)$ is Lipschitz continuous, the conclusion is an immediate consequence of Th. I.3.7 (with $\mathfrak{K} = \psi'$) and Th.I.5.9 (observe that, since f has the form (6-2) and satisfies (P.S.) in bounded sets, it belongs to $\mathcal{F}(Q_\rho)$). If ψ' is only continuous, arguing as in Lemma 3.2 of [Be,82] (cf. also Proposition A.23 of [Ra,84]), for every $\varepsilon > 0$, we can construct a compact Lipschitz continuous operator

$$\mathfrak{K}: W_0^1(\Omega) \longrightarrow W_0^1(\Omega),$$

such that

$$\left\| \mathfrak{K}(x) - \psi'(x) \right\| \leq \varepsilon$$

By the above formula and (2-4), we have that

$$\left\| \mathfrak{K}(u) \right\| \leq M_3 \; \|x\|^{\alpha_1} + M_3$$

for a suitable constants $M_3 \in \mathbb{R}$ and $\alpha_1 \in (0,1)$. From this formula it is not difficult to get the I.3.3. and apply Th. I.3.7. and Th.I.5.9.

□

CHAPTER   II

AN APPLICATIONS TO RIEMANNIAN GEOMETRY

One of the most interesting applications of Morse Theory lies in the study of geodesics in Riemannian geometry. In this Chapter, we will present an elementary approach to the basic notions of Riemannian geometry and the relation of Morse theory to the study of geodesics. We will omit most of the proofs referring to the literature.

II.1 – RIEMANNIAN MANIFOLDS.

Let $\mathfrak{M}$ be a smooth manifold embedded in the Euclidean space $\mathbb{R}^N$ with the usual scalar product $<\cdot,\cdot>$. We can think of a tangent vector in $T\mathfrak{M}$ as a vector in $\mathbb{R}^N$; in this way it is defined the scalar product of two vectors in $T_x\mathfrak{M}$, $x \in \mathfrak{M}$. The structure $(\mathfrak{M},<\cdot,\cdot>)$ which we have obtained is called Riemannian manifold.

There is also an intrinsic way to define a Riemannian manifold, but, by virtue of the well known Nash theorem, they are equivalent, and the one we have presented here is more intuitive.

Covariant derivative.

If $\gamma(t)$ is a differentiable curve in $\mathfrak{M}$, its derivative $\dot{\gamma}(t) = \partial_t \gamma(t)$ is a tangent vector in $T_{\gamma(t)}\mathfrak{M}$. The second derivative $\partial_t^2 \gamma(t) = \partial_t \dot{\gamma}(t)$ is a vector in $\mathbb{R}^N$ but, in general it does not belong to the tangent space $T_{\gamma(t)}\mathfrak{M}$; hence it is not intrinsic to $\mathfrak{M}$. However, since $\mathbb{R}^N$ has a scalar product, we can define the orthogonal projection P_x on $T_x\mathfrak{M}$. Thus it is a natural operation to project $\partial_t \dot{\gamma}(t)$ on $T_{\gamma(t)}\mathfrak{M}$. This operation is called covariant derivative (with respect to t). In symbols we have:

$$D_t \dot{\gamma}(t) = P_{\gamma(t)} \partial_t \dot{\gamma}(t)$$

Now let f be a function and v a tangent vector field defined on a manifold $\mathfrak{M}$. We recall that the Lie derivative of f in the direction of v is given by

$$L_v f(x) = \frac{d}{dt} f(\gamma_v(t))\bigg|_{t=0}$$

where γ_v is any curve such that

$$\gamma_v(o) = x$$
$$\dot{\gamma}_v(o) = v$$

(notice that $L_v f(x) = df(x)[v]$).

In a linear space we can also define the differential of a vector field w,

$$dw(x)[v] = \frac{d}{dt} w(\gamma_v(t))\bigg|_{t=0}$$

In the geometry literature $L_v f$ and $dw(x)[v]$ are also written as vf and $v[w]$, i.e. the vector fields are identified with the associated first order differential operators.

Using the Euclidean structure, can define this operation also on a Riemannian manifold:

$$D_v w(x) = P_x \, dw(x)[v].$$

The vector field $D_v w$ is called covariant derivative of w in the direction v; the operation which assigns to a couple of vector fields (v,w) its covariant derivative is called Levi-Civita connection. By its definition we have that

$$D_t \, \dot{\gamma}(t) = D_{\dot{\gamma}(t)} \, \dot{\gamma}(t)$$

moreover, if $\gamma(t)$ is any curve, we set $w(t) = w(\gamma(t))$ and

$$D_t \, w(t) = D_{\dot{\gamma}(t)} \, w(\gamma(t)).$$

Affine connections.

It is not difficult to check that the covariant derivative satisfies the following properties:

(i) $D_v w$ is bilinear as function of v and w;

(ii) $D_{fv} w = f \, D_v w$

(iii) $D_v(fw) = L_v f \, w + f \, D_v w$

Any operator on vector fields which satisfies (i),(ii) and (iii) is called affine connection, or simply connection. Moreover the Levi-Civita connection (i.e. the connection related to the covariant derivative), satisfies also the following property:

(iv) $\partial_t \langle w(t), z(t) \rangle = \langle D_v w(t), z(t) \rangle + \langle w(t), D_v z(t) \rangle$

The Levi-Civita connection is the only connection compatible with the Riemannian structure; i.e. the only one which satisfies (iv).

Now let us choose a local frame i.e. a basis $\left\{ e_1(x), \dots e_n(x) \right\}$ for $T_x \mathfrak{M}$ for x living in some open set which depends smoothly on x. Then we set

$$\Gamma^k_{i,j}(x) = \langle D_{e_k(x)} \, e_i(x), \, e_j(x) \rangle$$

The numbers $\Gamma^k_{i,j}(x)$ are called Christoffel symbols, and, by virtue of (i) they determine the connection and allow to make computations in local coordinates.

Curvature

Now, let

$$\sigma: \mathcal{U} \longrightarrow \mathfrak{M}$$

(where $\mathcal{U}$ is a neighborhood of 0 in $\mathbb{R}^2$) be a function of two variables t and s such that:

(1-1)
$$\begin{cases} \sigma(0,0) = x \\ \partial_t \, \sigma(t,0) = v(t) \\ \partial_s \, \sigma(0,s) = w(s) \end{cases}$$

if z is a vector field, we set

(1-2) $$R(v,w)z = -D_t D_s z + D_s D_t z$$

where $z = z(\sigma(t,s))$.

$R(v,w)z$ measures the extent to which the covariant derivative do not commute; it can be proved that it is a tensor. If we are in a Euclidean space of course $R(v,w)z$ vanishes, but it is different from 0 when our manifold (in some sense) is not flat; this imply that $R(v,w)z$ is a "measure" of the curvature: for this reason it is called (Riemann) curvature tensor.

Remark II.1.1. Given two vector fields v and w it is not always true that there exists a function σ which satisfies (1-1). If such a σ exists, then

$$L_v L_w - L_w L_v = \partial_t \partial_s - \partial_s \partial_t = 0$$

i.e. the operators L_v and L_w commute. But in general these differential operators do not commute; for example, in $\mathbb{R}^2$, take $v = y\ \mathbf{i}$ and $w = \mathbf{j}$; then $L_v = y\ \partial_x$ and $L_w = \partial_y$, so

$$L_v L_w - L_w L_v = y\ \partial_x \partial_y - \partial_y\ (y\ \partial_x) = \partial_x$$

The operator $[L_v, L_w] = L_v L_w - L_w L_v$ is called commutator of L_v and L_w and it is a first order differential operator. Thus, all the first order differential operator equipped with the anti-symmetric product $[\cdot,\cdot]$ form an Algebra which is called Lie algebra.

The tangent vector field associated to $[L_v, L_w]$ is denoted by $[v,w]$, so that we have

$$L_v L_w - L_w L_v = L_{[v,w]}$$

Notice that, if we operate on a manifold $\mathfrak{M}$, $[v,w]$ is a tangent vector field and its definition depends only on the differential structure of $\mathfrak{M}$.

If v and w are two vectors fields which do not commute, then the (1-2) becomes

$$R(v,w)z = -D_v D_w z + D_w D_v z + D_{[v,w]}z$$

II.2 - GEODESICS.

Given (parameterized) curve $\gamma: [a,b] \longrightarrow \mathfrak{M}$, its length is given by

$$\ell(\gamma) = \int_a^b \|\dot{\gamma}(t)\|\ dt \quad \text{where} \quad \|\dot{\gamma}(t)\| = \langle \dot{\gamma}(t), \dot{\gamma}(t) \rangle^{1/2}$$

The notion of length of a curve allows to induce a metric on $\mathfrak{M}$ as follows

$$d(x_0, x_1) = \inf \left\{ \ell(\gamma)\ |\ \gamma \in C^1([0,1], \mathfrak{M}),\ \gamma(0) = x_0;\ \gamma(1) = x_1 \right\}$$

Definition of the geodesics

A curve Γ is called geodesic if the distance of any two sufficiently close points on Γ is equal to the length of the portion of the curve between them; more precisely Γ is a geodesic if either

(i) $\ell(\gamma) = d(\gamma(a), \gamma(b))$ where γ: $[a,b] \longrightarrow \mathfrak{M}$ is a parameterization of Γ;

(ii) given any point $P_0 = \gamma(t_0)$ of Γ there is a neighborhood of t_0, $N = [a_0,b_0]$, such that $\gamma\big|_N$ satisfies (i).

A parameterized curve γ: $[a,b] \longrightarrow \mathfrak{M}$, is called (parameterized) geodesic if $\gamma([a,b])$ is a geodesic and it is parameterized with its arc-length.

A geodesic which satisfies (i) is called a minimal geodesic; since every geodesic is "locally minimal" by definition, its local properties can be studied restricting ourselves to minimal geodesics.

The energy functional

We recall the Holder inequality:

$$(2\text{-}1) \qquad \int_a^b \|\dot{\gamma}(t)\| \, dt \leq (b-a)^{1/2} \left[\int_a^b \|\dot{\gamma}(t)\|^2 \, dt \right]^{1/2}$$

Now, if γ is parameterized with its arc-length, $\|\dot{\gamma}(t)\|$ is constant and in (2-1) the equality holds; therefore, if γ is a minimal geodesic, it minimizes the integral

$$E(\gamma) = \frac{1}{2} \int_a^b \|\dot{\gamma}(t)\|^2 \, dt$$

among all the curves $\gamma \in C^1([0,1], \mathfrak{M})$ with fix extremes. Moreover, if a curve minimizes $E(\gamma)$ is a minimal geodesic provided that it is parameterized with its arc-length (and, as we will see, this fact always happens).

The functional $E(\gamma)$ is called "energy" functional. Since, it is easier to study than the functional $\ell(\gamma)$, we will use this functional to analyze the properties of geodesics.

The geodesics equation

Let $\gamma(t)$ be a minimal geodesic, $v(t)$ a vector valued function in E, $\sigma(t,s)$ a function of two variables defined in $[-1,1] \times [a,b]$ such that:

(2-2)
$$\begin{cases} \sigma(t,0) = \gamma(t) & \text{for every } t \in [a,b] \\[2mm] \partial_s \sigma(t,0) = v(t) & \text{with } v(t) \in T_{\gamma(t)}\mathfrak{M} \\[2mm] \sigma(0,s) = x_0 & \text{for every } s \in [-1,1] \\[2mm] \sigma(1,s) = x_1 & \text{for every } s \in [-1,1] \end{cases}$$

Since, the function $s \longmapsto E(\sigma(\cdot,s))$ has a minimum for $s = 0$,

$$\frac{d}{ds} E(\sigma(\cdot,s))\Big|_{s=0} = 0$$

Let us compute $\dfrac{d}{ds} E(\sigma(\cdot,s))$:

(2-3) $\dfrac{d}{ds} E(\sigma(\cdot,s))\Big|_{s=0} = \dfrac{1}{2} \dfrac{d}{ds} \displaystyle\int_a^b \|\dot{\sigma}(t,s)\|^2 \, dt\Big|_{s=0} =$

$$= \frac{1}{2} \int_a^b \partial_s \langle \dot{\sigma}(t,s),\dot{\sigma}(t,s) \rangle \Big|_{s=0} dt = \int_a^b \langle D_s \dot{\sigma}(t,0),\dot{\sigma}(t,0) \rangle \, dt =$$

$$= \int_a^b \langle D_s \partial_t \sigma(t,0),\dot{\gamma}(t) \rangle \, dt = \int_a^b \langle D_t \partial_s \sigma(t,0),\dot{\gamma}(t) \rangle \, dt =$$

$$= \int_a^b \langle D_t v(t),\dot{\gamma}(t) \rangle = \quad \text{(integrating by parts)}$$

$$= \int_a^b \langle v(t),D_t \dot{\gamma}(t) \rangle \, dt.$$

Since v has been chosen arbitrarily, we have that

(2-4) $D_t \dot{\gamma}(t) = 0$

Thus every minimal geodesic satisfy the equation (2-4); moreover since every geodesic is locally minimal, it follows that every geodesic satisfy (2-4). It is not difficult to see that also the converse is true, i.e. every curve which satisfies (2-4) is a geodesic.

The exponential map

Notice that the Cauchy problem

$$\begin{cases} D_t \dot{\gamma}(t) = 0 \\ \gamma(0) = x \\ \dot{\gamma}(0) = v \quad v \in T_{\gamma(t)}\mathfrak{M} \end{cases}$$

is always locally solvable. Its solution is denoted by

$$exp_x tv$$

and it is called exponential function. A Riemannian manifold is called geodesically complete if, for any given $(x,v) \in T\mathfrak{M}$, the map

$$t \longrightarrow exp_x tv$$

is defined for every $t \in \mathbb{R}$. Since a Riemannian manifold has a natural structure of metric space we can define also the usual notion of completeness (i.e. every Cauchy sequence is convergent). It is possible to prove that this two definitions of completeness are equivalent.

II.3 - THE MORSE THEORY FOR GEODESICS.

The space $\Omega^1(\mathfrak{M}; x_0, x_1)$.

The natural function space where to study

$$E(\gamma) = \int\limits_0^1 \|\dot{\gamma}(t)\|^2 \, dt$$

is

$$\Omega^1(\mathfrak{M}; x_0, x_1) =$$

$$= \left\{ \gamma \in W^1([0,1], \mathbb{R}^N) \mid \gamma(t) \in \mathfrak{M}; \ \gamma(0) = x_0; \ \gamma(1) = x_1 \right\}$$

where $W^1([0,1], \mathbb{R}^N)$ denotes the Sobolev space of vector valued functions having the first derivative square integrable.

Lemma *II.3.1. If* $\mathfrak{M}$ *is closed as submanifold of* $\mathbb{R}^N$, *then* $\Omega^1(\mathfrak{M}; x_0, x_1)$ *is a complete differentiable submanifold of* $W_0^1([0,1], \mathbb{R}^N)$ *and*

$$T_\gamma \Omega^1(\mathfrak{M}; x_0, x_1) =$$

$$\left\{ v \in W_0^1([0,1], \mathbb{R}^N) \mid \forall \ t \in [0,1], \ v(t) \in T_{\gamma(t)} \mathfrak{M} \right\}$$

Proof.See [Pa,63] or [Sc,69].

□

Lemma *II.3.2. The energy functional E is differentiable on* $\Omega^1(\mathfrak{M};x_0,x_1)$ *and we have*

$$dE(\gamma)[v] = \int_0^1 <\dot{\gamma}(t),D_t v(t)> \, dt \quad ; \quad v \in T_\gamma \Omega^1(\mathfrak{M};x_0,x_1)$$

Proof. It is a consequence of (2-3) and the properties of the Frechét derivative.

□

Lemma *II.3.3 If* $\mathfrak{M}$ *is complete, the energy functional E satisfies P.S. on* $\Omega^1(\mathfrak{M};x_0,x_1)$.

Proof. Let γ_n be a sequence of functions such that

(3-1) $$\int_0^1 <\dot{\gamma}_n(t),\dot{\gamma}_n(t)> \, dt \longrightarrow c$$

and

(3-2) $$\left| \int_0^1 <\dot{\gamma}_n(t),D_t v(t)> \, dt \right| \leq \varepsilon_n \|v\|$$

notice that $\|v\|$ denotes the norm of v in $W_0^1([0,1],\mathbb{R}^N)$; in fact we are using the metric of $W_0^1([0,1],\mathbb{R}^N)$.

By (3-1) we have that γ_n converges weakly in $W_0^1([0,1],\mathbb{R}^N)$ to some curve γ_0 and hence converges uniformly to γ_0. We have to prove that this convergence is strong in in $W_0^1([0,1],\mathbb{R}^N)$. To do this, set

$$v_n = P(\gamma_n)[\gamma_n-\gamma_0]$$

where $P(\gamma_n(t))$: $T_{\gamma_n(t)}\mathbb{R}^N \longrightarrow T_{\gamma_n(t)}\mathfrak{M}$ is the orthogonal projection. Clearly, $v_n(t) \in T_{\gamma_n(t)}\mathfrak{M}$. Moreover

$$\partial_t v_n(t) = \partial_t P(\gamma_n)[\dot{\gamma}_n - \dot{\gamma}_0] = P(\gamma_n)[\ddot{\gamma}_n - \ddot{\gamma}_0] + P'(\gamma_n)[\dot{\gamma}_n][\dot{\gamma}_n - \dot{\gamma}_0]$$

Now, since $\gamma_n \longrightarrow \gamma_0$ uniformly and $\dot{\gamma}_n$ is bounded in $L^2([0,1],\mathbb{R}^N)$, we have that $P'(\gamma_n)[\dot{\gamma}_n][\dot{\gamma}_n - \dot{\gamma}_0] \longrightarrow 0$ strongly in $L^2([0,1],\mathbb{R}^N)$. Thus

$$D_t v_n(t) = P(\gamma_n)\partial_t v_n(t) = P(\gamma_n)[\ddot{\gamma}_n - \ddot{\gamma}_0] + w_n$$

where $w_n \longrightarrow 0$ strongly in $L^2([0,1],\mathbb{R}^N)$. Thus, v_n is boumded in $W_0^1([0,1],\mathbb{R}^N)$ and by Lemma II.3.1, it beongs to $T_\gamma\Omega^1(\mathfrak{M};x_0,x_1)$. Using this information and the (3-2), replacing v with v_n, we get

$$\left| \int_0^1 \langle\dot{\gamma}_n, P(\gamma_n)[\ddot{\gamma}_n - \ddot{\gamma}_0] + w_n\rangle \, dt \right| \leq \varepsilon_n M_1$$

where M_1 is a suitable constant. Since w_n is infinitesimal and since $\dot{\gamma}_n(t) \in T_{\gamma_n(t)}\mathfrak{M}$,

(3-3)
$$\int_0^1 \langle\dot{\gamma}_n, \ddot{\gamma}_n - \ddot{\gamma}_0\rangle \, dt \longrightarrow 0$$

Since

$$\int_0^1 \langle\dot{\gamma}_0, \ddot{\gamma}_n - \ddot{\gamma}_0\rangle \, dt \longrightarrow 0$$

using (3-3), we have that

$$\int_0^1 \langle\ddot{\gamma}_n - \ddot{\gamma}_0, \ddot{\gamma}_n - \ddot{\gamma}_0\rangle \, dt \longrightarrow 0$$

and this concludes the proof.

$\square$

The Morse relations.

By virtue of the above lemmas, we can apply Morse theory to the functional E in order to get information on the geodesics on $\mathfrak{M}$.

Theorem *II.3.4 – Morse theory for geodesics. Set*

$$\mathfrak{G} = \left\{ \gamma \in \Omega^1(\mathfrak{M};x_0,x_1) \mid \gamma \ \text{is a geodesic} \right\}.$$

We have the following relations:

(i)
$$i_\lambda(\mathfrak{G}) = \mathcal{P}_\lambda(\Omega^1(\mathfrak{M};x_0,x_1)) + (1 + \lambda)\, Q_\lambda.$$

(ii)   *if a and b are not the energy of any geodesic in $\mathfrak{G}$, then*

$$i_\lambda(\mathfrak{G} \cap E_a^b) = \mathcal{P}_\lambda(E^b,E^a) + (1 + \lambda)\, Q_\lambda(a,b).$$

Moreover, if we assume the generic condition that all the geodesics in $\mathfrak{G}$ are non degenerate, (i) and (ii) read as follows

(iii) $$\sum_{\gamma \in \mathfrak{G}} \lambda^{m(\gamma)} = \mathcal{P}_\lambda(\Omega^1(\mathfrak{M};x_0,x_1)) + (1 + \lambda)\, Q_\lambda$$

(iv) $$\sum_{\gamma \in \mathfrak{G} \cap E_a^b} \lambda^{m(\gamma)} = \mathcal{P}_\lambda(E^b,E^a) + (1 + \lambda)\, Q_\lambda(a,b).$$

where $m(\gamma)$ is the Morse index of the functional E at γ.

Proof. Since the geodesics are the critical points of E and P.S. holds by Lemma II.3.3, (i) and (ii) are consequences of Theorem I.5.9, (iii) is a consequence of Theorem I.4.11 and (iv) is a consequence of Theorem I.2.3.

□

The first application of the above theorem concerns the existence and the multiplicity of geodesics. To do this is sufficient to know the homology (and hence the Poincaré polynomial) of $\Omega^1(\mathfrak{M};x_0,x_1)$. We have the following theorem

Theorem *II.3.5 – (i) if M is topologically trivial, then also $\Omega^1(\mathfrak{M};x_0,x_1)$ is topologically trivial; hence $P_\lambda(\Omega^1(\mathfrak{M};x_0,x_1)) = 1$.*

(ii) if M is not simply connected, then $H_0(\Omega^1(\mathfrak{M};x_0,x_1)) \neq 0$; hence $P_\lambda(\Omega^1(\mathfrak{M};x_0,x_1)) = 1 + at + Z(t)$ with $a > 0$.

(iii) if M is simply connected, then $H_q(\Omega^1(\mathfrak{M};x_0,x_1)) \neq 0$ for infinitely many q's; hence $P_\lambda(\Omega^1(\mathfrak{M};x_0,x_1))$ is a formal series.

Proof. (i) and (ii) are trivial. (iii) is a consequence of the Leray-Serre theorem [Se,51]; for the proof see also [Sc,69]

□

II.4 – THE INDEX THEOREM.

In this section, we will investigate the geometrical meaning of the Morse index $m(\gamma)$ of a geodesic γ and we will see that it is equal to a "geometrical" index $\mu(\gamma)$ which can be defined by mean of the Jacobi equation" which is the linearization of the geodesic equation.

The second variation of E.

Let γ be a geodesic joining x_0 and x_1. We know that the *Morse index* $m(\gamma)$ is the maximal dimension of a subspace of $T_\gamma \Omega^1$ on which the Hessian $E''(\gamma)$ is negative definite. Let us compute the Hessian of E.

Now, let $\xi(t)$ and $\xi'(t)$ be two tangent vector fields along a geodesic γ and let

$$\sigma: [0,1] \times \mathcal{U} \longrightarrow \mathfrak{M}$$

(where $\mathcal{U}$ is a neighborhood of O in $\mathbb{R}^2$) be a function of three variables t, s and r such that:

$$\sigma(t,o,o) = \gamma(t)$$
$$\partial_s \, \sigma(t,o,o) = \xi(t)$$
$$\partial_r \, \sigma(t,o,o) = \xi'(t)$$

The function

$$(s,r) \longmapsto \sigma(\,\cdot\,,s,r)$$

can be regarded as a function from $\mathcal{U}$ to Ω^1. Then the Hessian form of E at γ, is given by

$$H^E(\gamma)[\xi,\xi'] = \frac{\partial^2}{\partial s \partial r} \, E(\sigma(\,\cdot\,,s,r))\Big|_{s=0,r=0} =$$

$$= \frac{1}{2} \int_0^1 \frac{\partial^2}{\partial s \partial r} \, \langle \partial_t \sigma, \partial_t \sigma \rangle \, dt = \int_0^1 \frac{\partial}{\partial s} \, \langle D_r \partial_t \sigma, \partial_t \sigma \rangle \, dt =$$

$$= \int_0^1 \frac{\partial}{\partial s} \, \langle D_t \partial_r \sigma, \partial_t \sigma \rangle dt = \int_0^1 \langle D_t \partial_r \sigma, D_t \partial_s \sigma \rangle + \langle D_s D_t \partial_r \sigma, \partial_t \sigma \rangle \, dt =$$

$$= \int_0^1 \langle D_t \xi, D_t \xi' \rangle + \langle D_t D_s \partial_r \sigma, \partial_t \sigma \rangle - \langle R(\partial_r \sigma, \partial_t \sigma) \partial_r \sigma, \partial_t \sigma \rangle \, dt =$$

(where R is the Riemann curvature tensor)

$$= \int_0^1 <D_t\xi,D_t\xi'> - <R(\xi',\dot\gamma)\xi, \dot\gamma> \ dt + \int_0^1 <D_tD_s\xi, \dot\gamma> \ dt =$$

$$(since \ D_t \dot\gamma = 0)$$

$$= \int_0^1 <D_t\xi,D_t\xi'> - <R(\dot\gamma,\xi')\dot\gamma,\xi> \ dt + \int_0^1 \partial_t<D_s\xi, \dot\gamma> \ dt =$$

$$= \int_0^1 <D_t\xi,D_t\xi'> - <R(\dot\gamma,\xi')\dot\gamma,\xi> \ dt + \left[<D_s\xi, \dot\gamma>\right]_{t=0}^{t=1} =$$

$$= \int_0^1 <D_t\xi,D_t\xi'> - <R(\dot\gamma,\xi')\dot\gamma,\xi> \ dt.$$

So, we get the second variation formula

$$H^E(\gamma)[\xi,\xi'] = \int_0^1 <D_t\xi,D_t\xi'> - <R(\dot\gamma,\xi')\dot\gamma,\xi> \ dt.$$

Notice that $m(\gamma)$ is finite because $E''(\gamma)$ is a compact perturbation of the positive definite bilinear form

$$a(\gamma)[\xi,\xi'] = \int_0^1 <D_t\xi,D_t\xi'> \ dt$$

The Jacobi fields and the geometric index

For every geodesic γ, there is another index $\mu(\gamma)$, which we call "*geometric index*". We need some notations. For every $s \in \ ]0,1]$, we consider the functional

$$E_s(\gamma) = \frac{1}{2} \int_0^s <\dot\gamma,\dot\gamma> \ dt,$$

defined on the loop space $\Omega_s^1 = \Omega^1(\gamma_0,\gamma(s),M_0)$ of the curves joining x_0 and $\gamma(s)$ and parameterized in $[0,s]$. Moreover, let γ_s be the restriction of γ to $[0,s]$. Then γ_s is a critical point of E_s.

Definition II.4.1. *Let γ be a geodesic joining x_0 and x_1. A point $\gamma(s)$, $s \in (0,1]$, is said conjugate to x_0 along γ, if $E_s''(\gamma_s)$ is degenerate, i.e. there exists $\xi \in T_{\gamma_s}\Omega_s^1$, $\xi \neq 0$, such that*

$$E_s''(\gamma_s)[\xi,\xi'] = 0, \qquad \forall \xi' \in T_{\gamma_s}\Omega_s^1.$$

ξ is said Jacobi tangent field along γ_s.

The set of Jacobi tangent fields along γ_s is the kernel of $E_s''(\gamma_s)$ and is denoted by *Ker* $E_s''(\gamma_s)$.

The *multiplicity* of the conjugate point $\gamma(s)$ is the dimension of *Ker* $E_s''(\gamma_s)$. It is a finite number. Indeed, a vector field $\xi \in$ *Ker* $E_s''(\gamma_s)$ if and only if it solves the following system of linear ordinary differential equations:

$$\begin{cases} D_t^2\xi + R(\dot{\gamma},\xi)\dot{\gamma} = 0 \\ \xi(0) = \xi(s) = 0 \end{cases}$$

So, *Ker* $E_s''(\gamma_s)$ has dimension at most $n = \dim M_0$ (actually at most $n - 1$, see [Mi,63]). The n differential equations of the system of $D_t^2\xi + R(\dot{\gamma},\xi)\dot{\gamma} = 0$ are called "*Jacobi equations*".

On a geodesic, there is a finite number of conjugate points to its initial point. Indeed a point q is conjugate to a point p along a geodesic iff q is a critical value of the exponential map $\exp_p$.

Definition II.4.2. *The geometric index $\mu(\gamma)$ of a geodesic $\gamma:[0,1] \rightarrow M_0$ is the number of points $\gamma(s)$ conjugate to*

$\gamma(0)$, $s \in \,]0,1[$, *counted with their multiplicity i. e.*

$$\mu(\gamma) = \sum_{s \in]0,1[} \dim \, Ker \, E''_s(\gamma_s),$$

The index theorem

The Index theorem is one of the most important theorems in global Riemannian geometry.

Theorem II.4.3. *(Index theorem) For every geodesic γ, we have* m$(\gamma) = \mu(\gamma)$.

Proof. We refer to [Sc,69] or [Kl,82]). □

By the Index Theorem and Theorem II.3.4, we get

Theorem II.4.4. *If we assume the generic condition that all the geodesics in $\circledS$ are non degenerate, we have the following relations:*

(4-1) $$\sum_{\gamma \, \in \, \circledS} \lambda^{\mu(\gamma)} = \mathcal{P}_\lambda(\Omega^1(\mathfrak{M};x_0,x_1)) + (1 + \lambda) \, Q_\lambda$$

where Q_λ is a polynomial with positive integer coefficients and μ is the geometric index of γ.

C H A P T E R   III
AN APPLICATIONS TO SPACE-TIME GEOMETRY

One of the classical fields of application of the Morse theory, as we have seen in Chapter II, is the Riemannian geometry. However, in recent years, thanks to the development of the nonlinear analysis it has been possible to use variational methods also in Lorentzian geometry (i.e. the space-time geometry of the theory of General Relativity); we refer to [Be,Fo,ta] for a survey paper on this topic. In this chapter, we will present some of the main ideas which allow to apply the Morse theory to this field.

III.1 - INTRODUCTION

Lorentian geometry and General Relativity

Until the beginning of the 19th century the nature of the space as described by the Euclidean Geometry was considered an "absolute" truth. The discovery of non Euclidean geometries provide the first evidence that this point of view is too limitative. Later, at the beginning of the XXth century, the theory of General Relativity provided a "physical" evidence that the space, in the presence of matter, is not Euclidean. Moreover in General Relativity the space and the time are structurally joined: in this theory a gravitational field is described by assigning on a smooth 4-dimensional manifold $\mathfrak{M}$ a metric tensor (which we denote by $< , >_L$) having signature +++-. This means that, for all z in $\mathfrak{M}$, $< \, , \, >_L$ is a non degenerate symmetric bilinear form (on the tangent space $T_z\mathfrak{M}$ to $\mathfrak{M}$ at z) whose matrix representation has exactly one negative eigenvalue.

Such a metric $< \, , \, >_L$ is called *Lorentzian metric* and $\mathfrak{M}$ equipped with $<\cdot,\cdot>_L$ is called *Lorentzian manifold* or *space-time.* Any point in $\mathfrak{M}$ is called *event.* The geometry and the topology of the space-time $(\mathfrak{M}, < \, , \, >_L)$ are determined by the Einstein equations which connect the Ricci curvature and the metric tensor to the energy-matter distribution.

Geodesics in Lorentian manifolds

The aim of this chapter is to investigate on the possibility to construct a Morse theory for Lorentzian manifold following the lines of the "classical" Morse theory for Riemannian manifolds.

Before doing this, it is useful to recall some basic facts on Lorentzian geometry (see e.g.[Be,He,81], [Ha,El,73], [ON,83], [Pe,72] for extensive treatments).

The definitions of covariant derivative, curvature and geodesic can be transferred to Lorentzian metrics; of course $<\cdot,\cdot>$ needs to be replaced by $<\cdot,\cdot>_L$. The covariant derivative D is an operator on vector fields which satisfies (i), (ii), (iii) and (iv) of section II.1 where $<\cdot,\cdot>$ is replaced by $<\cdot,\cdot>_L$. The curvature tensor is defined in the same way. A geodesic on $\mathfrak{M}$ is a curve $\gamma = \gamma(s)$ solving the equation

$$D\dot{\gamma}(s) = 0 \qquad\qquad (1.1)$$

In terms of a local coordinates system with coordinates $z^1, \ldots, z^4$, a curve $\gamma = \gamma(s)$ determines four smooth real functions $z^1(s), \ldots, z^4(s)$. The equation (1.1) then takes the form

$$\ddot{z}^k + \sum_{i,j} \Gamma^k_{ij}(z^1, \ldots, z^4)\dot{z}^i\dot{z}^j = 0 \qquad (1.2)$$

where Γ^k_{ij} are smooth real functions on $\mathfrak{M}$ (Christoffel symbols), depending on the metric tensor. The Christoffel symbols replace the gravitational force of the Newtonian gravitational theory.

As in the Riemannian case, equation (1.1) is the Euler–Lagrange equation related to the "energy" functional

$$E(\gamma) = \int_0^1 \langle\dot{\gamma}(s), \dot{\gamma}(s)\rangle_L \, ds \qquad (1.3)$$

Notice that the name "energy" we give to (1.3) is not related to the physical meaning of this functional.

Causal type of the geodesics

If $\gamma = \gamma(s)$ is a geodesic, then the identity

$$\frac{d}{ds} \langle\dot{\gamma}(s), \dot{\gamma}(s)\rangle_L = 2 \langle D\dot{\gamma}(s), \dot{\gamma}(s)\rangle_L = 0$$

shows that $E(s) = \langle\dot{\gamma}(s), \dot{\gamma}(s)\rangle_L$ is constant along $\gamma(s)$. A geodesic $\gamma = \gamma(s)$ is called *space-like, light-like* or *time-like* if E is greater, equal or less than zero, respectively; this property is called the causal type of a geodesic.

In the theory of General Relativity, a timelike geodesic $\gamma = \gamma(s)$ is physically interpreted as the world line of a material particle under the action of a gravitational field. In this case the parameter s, (after an aprropriate rescaling such that $\langle \dot{\gamma}(s), \dot{\gamma}(s) \rangle_L = -1)$, is called *proper time* and it is interpreted as the time measured by a clock associated with the particle. A light-like geodesic is the world line of a light ray. The spacelike geodesics have a more subtle interpretation: for a suitable local observer, they represent "Riemannian" geodesics consisting of simultaneous events.

Before concluding this introduction we briefly indicate one of the main difficulties which occur in developing a Morse theory for geodesics in Lorentzian manifolds.

As already observed, the geodesics joining two given events z_0, z_1 in $\mathfrak{M}$ are the critical points $\gamma = \gamma(s)$ ($s \in [0,1]$) of the energy functional (1.3) on the infinite dimensional manifold $\Omega^1(\mathfrak{M}; z_0, z_1)$ consisting of the W^1-curves $\gamma(s)$, $s \in [0,1]$, on $\mathfrak{M}$ with $\gamma(0) = z_0$, $\gamma(1) = z_1$.

Unlike the situation for positive definite Riemannian metrics, the functional (1.3) is unbounded both from below and from above, since the Lorentzian metric $\langle \, , \, \rangle_L$ is indefinite. Moreover it is not difficult to see that a critical point γ of (1.3) is a saddle point with infinite Morse index (i.e. the Hessian form $E''(\gamma)$ of E at γ possesses infinitely many positive and infinitely many negative eigenvalues). This fact does not permit a direct definition of the Morse Index; moreover the "topology" of the sub-levels of E does not change when a critical value is "crossed".

III.2 - SOME EXAMPLES OF LORENTIAN MANIFOLDS.

We now give some examples of Lorentian manifolds.

The Minkowsky space time.

EXAMPLE III.2.1. The Minkowsky space time is the manifold $\mathbb{R}^4$ equipped with the flat Lorentzian metric:

$$(2\text{-}1) \qquad <\zeta,\zeta>_L = \zeta_1^2 + \zeta_2^2 + \zeta_3^2 - \zeta_4^2 \qquad \zeta = (\zeta_1,\zeta_2,\zeta_3,\zeta_4) \in \mathbb{R}^4$$

This is the space-time of Special Relativity which describes the situations in which the effects of the gravitational field are negligible.

The static space time.

EXAMPLE III.2.2. Let $(\mathfrak{M}_0,<\cdot,\cdot>)$ be a smooth, connected Riemannian manifold, dim $\mathfrak{M}_0 = n \geq 1$, $\mathfrak{M} = \mathfrak{M}_0 \times \mathbb{R}$. A *static metric* $<\cdot,\cdot>_L$ on $\mathfrak{M}$ is of the form

$$(2\text{-}2) \qquad\qquad <\zeta,\zeta>_L = <\xi,\xi> - \beta(x)\tau^2,$$

where $\gamma = (x,t) \in \mathfrak{M}$, $\zeta = (\xi,\tau) \in T_\gamma\mathfrak{M} \equiv T_x\mathfrak{M}_0 \times \mathbb{R}$, and $\beta : \mathfrak{M}_0 \longrightarrow \mathbb{R}$ is a smooth positive function on $\mathfrak{M}_0$. $(\mathfrak{M},<\cdot,\cdot>_L)$ is said *(standard) static space time.* From the physical point of view a static space-time describes a situation in which the matter is supposed to be motionless in some reference frame.

Space time of split-type

EXAMPLE III.2.3. Let $(\mathfrak{M}_0,<\cdot,\cdot>_R)$ be a smooth, connected Riemannian manifold. Consider the "time dependent" Lorentzian metric $<\cdot,\cdot>_L$ on $\mathfrak{M} = \mathfrak{M}_0 \times \mathbb{R}$ defined as follows:

$$(2\text{-}3) \qquad <\zeta,\zeta>_L = <A(x,t)\xi,\xi> - \beta(x,t)\tau^2,$$

where $\gamma = (x,t) \in \mathfrak{M}$, $\zeta = (\xi,\tau) \in T_\gamma\mathfrak{M} = T_x\mathfrak{M}_0 \times \mathbb{R}$, $A(x,t)$ is a positive linear operator defined on $T_x \mathfrak{M}_0$ and β is a smooth, positive scalar field on $\mathfrak{M}$. $\mathfrak{M}$, equipped with the Lorentzian metric (2.3), is called space-time of *split-type*.

Observe that, when both A and β do not depend on t, a space-time of split-type is static.

The class of the Lorentzian manifolds of split-type is quite large. In fact a celebrated result of Geroch [Ge,70]) shows that any *time-oriented* and *globally hyperbolic* Lorentzian manifold is of split-type. For the definition of time-orientability and global hyperbolicity we refer to [Be,He,81], [Ha,El,73], [Pe,72]. We recall that the global hyperbolicity is a notion introduced by Leray for the well posedness of the Cauchy problem.

III.3 - MORSE THEORY FOR LORENTIAN MANIFOLDS

As we have noticed in the introduction, the definitions of geodesics, conjugate points, Jacobi tangent fields, and geometrical index can be transferred to Lorentzian metrics. Thus it makes sense to ask ourselves if the analogous of theorem II.4.4 holds in some case. We restrict ourselves to Lorentzian manifolds of split-type. In this case the geodesics joining two points $z_0 = (x_0, t_0)$ and $z_1 = (x_1, t_1)$ are the critical points of the integral

(3.1) $E(\gamma) = E(x,t) =$

$$= \frac{1}{2} \int_0^1 \langle \dot{\gamma}, \dot{\gamma} \rangle_L \, ds = \frac{1}{2} \int_0^1 [\langle A(x,t)\dot{x}, \dot{x} \rangle + \beta(x,t)\dot{t}^2] \, ds,$$

defined on the loop space

$$\mathcal{Z} = \Omega^1(\mathfrak{M}, z_0, z_1) = \Omega^1(\mathfrak{M}_0, x_0, x_1) \times W^1(t_0, t_1),$$

where

$$W^1(t_0,t_1) = \left\{\, t \in W^1([0,1],\mathbb{R}) \,|\, t(0) = t_0,\ t(1) = t_1 \,\right\}.$$

$W^1(t_0,t_1)$ is a closed linear sub-manifold of $W^1([0,1],\mathbb{R})$ isomorphic to $W_0^1([0,1],\mathbb{R})$. For every $\gamma = (x,t) \in Z$, the tangent space is

$$T_\gamma Z = T_x \Omega^1 \times W_0^1([0,1],\mathbb{R}).$$

For every geodesic $\gamma = (x,t)$ and $\zeta = (\xi,\tau)$, $\zeta' = (\xi',\tau')$ $\in T_\gamma Z$, the hessian $E''(\gamma):T_\gamma Z \times T_\gamma Z \longrightarrow \mathbb{R}$ is given by

$$E''(\gamma)[\zeta,\zeta'] = \int_0^1 [<D_s\zeta,D_s\zeta'> - <R_L(\dot\gamma,\zeta)\dot\gamma,\zeta'>]ds,$$

where R_L is the Riemann curvature tensor with respect to the Lorentzian metric.

Morse theory for space-times of split-type

The Morse index $m(\gamma)$ of a geodesic is $+\infty$, because $E''(\gamma)$ is negative definite on the "time" directions $(0,\tau)$, $\tau \in W_0^1([0,1],\mathbb{R})$, thus the analogous of theorem II.3.4 does not make sense. However, the geometrical index $\mu(\gamma)$ is finite, as in the Riemannian case. Therefore, we we can try to use this index to construct a Morse theory and to see if the analogous of Theorem II.4.3 holds. It is quite surprising that this is the case for a large class of Lorentzian manifolds. In fact, we have the following theorem:

Theorem III.3.1. [Be,Fo,Ma,ta] – *Let $\mathfrak{M} = \mathfrak{M}_0 \times \mathbb{R}$ be a space-time of split-type. Assume that:*

(i) $(\mathfrak{M}_0, < , >_R)$ is a complete Riemannian manifold.

(ii) $v <\xi,\xi>_R \ \leq\ <A(x,t)\xi,\xi>_R \ \leq\ M <\xi,\xi>_R \ ;\ v \leq \beta(x,t) \leq M$

where ν and M are positive constants.

(iii) $\frac{\partial}{\partial t} A(x,t)$ and $\frac{\partial}{\partial t} \beta(x,t)$ are bounded.

(iv) $\displaystyle\limsup_{t \longrightarrow +\infty} <\frac{\partial}{\partial t} A(x,t)\xi,\xi>_R \leq 0$

   $\displaystyle\liminf_{t \longrightarrow +\infty} <\frac{\partial}{\partial t} A(x,t)\xi,\xi>_R \geq 0$

uniformly in $x \in \mathfrak{M}_0$ *and* $\xi \in T_x\mathfrak{M}_0$.

(v) $\frac{\partial^2}{\partial t^2} A(x,t), \frac{\partial^2}{\partial t^2} \beta(x,t)$ *are bounded.*

Under these assumptions, there is a Morse theory for the geodesics in in $\mathfrak{M}$. *More exactly, let* z_0, z_1 *be two points of* $\mathfrak{M}$ *and set*

$$\mathfrak{G} = \left\{ \gamma \in \Omega^1(\mathfrak{M};z_0,z_1) \mid \gamma \text{ is a geodesic} \right\}.$$

Then, if we assume the generic condition that all the geodesics in $\mathfrak{G}$ *are non degenerate, we have the following relations:*

(3-1) $\displaystyle\sum_{\gamma \in \mathfrak{G}} \lambda^{\mu(\gamma)} = \mathcal{P}_\lambda(\Omega^1(\mathfrak{M};z_0,z_1)) + (1 + \lambda) Q_\lambda$

where Q_λ is a polynomial with positive integer coefficients and μ is the geometric index of γ.

As in the Riemannian case, it is possible that some coefficients of the sum (3.1) are infinite cardinal numbers. In this case, the sum makes sense with the algebra of $\mathfrak{S}$.

Since the manifold

$$\left\{ t \in W^1([0,1],\mathbb{R}) \mid t(0) = t_0, \ t(1) = t_1 \right\}$$

is contractible to a point, $\Omega^1(\mathfrak{M};z_0,z_1)$ and $\Omega^1(\mathfrak{M}_0;x_0,x_1)$ are homotopically equivalent. Hence, $\mathcal{P}_\lambda(\Omega^1(\mathfrak{M};z_0,z_1)) =$

$\mathcal{P}_\lambda(\Omega^1(\mathfrak{M}_0;x_0,x_1))$. Then, if $\mathfrak{M}_0$ is simply connected and not topologically trivial, we deduce from *Th. II.3.5* that there are infinitely many geodesics joining any two non-conjugate points in $\mathfrak{M}$.

Morse theory for static space-times

The proof theorem III.3.1 is very technical and we refer to [Be,Fo,Ma,ta] for it. Here we consider the particular case of static space-times, i.e. we assume that $A(x,t)$ = *identity* and β does not depend on t. Thus the proof is much simpler and the main ideas will appear more clearly.

Theorem III.3.2. ([Be,Ma,92]) *Let* $(\mathfrak{M},<\cdot,\cdot>_L)$ *be a static Lorentzian manifold, such that*

   i) $(\mathfrak{M}_0,<\cdot,\cdot>)$ *is a complete riemannian manifold;*

  ii) $\exists M > 0$, *such that* $0 < \beta(x) \le M$.

    Then, the same result of theorem III.3.1 holds.

Remark III.3.3 Previous results about a Morse theory for geodesics in Lorentzian manifolds have been obtained in [Uh,75] (see also [Be,He,81] and its references). In [Uh,75] and in [Be,He,81], only causal geodesics (i.e. with $<\dot\gamma,\dot\gamma>_L < 0$) are considered. Therefore the Morse theory reflects only the topology of the "manifold" of causal curves. Moreover, in [Uh,75] also a Morse theory of light-like geodesics (i.e. with $<\dot\gamma,\dot\gamma>_L = 0$) joining a point with a given time-like curve is presented.

Remark III.3.4. If β is not bounded from above, theorem

III.3.2 does not hold. A counterexample is furnished by the Anti-De-Sitter space-time, see for instance [Pe,72].

III.4. PRELIMINARY LEMMAS

In this section we shall prove two lemmas which will be used in the proof of theorem III.3.2. We put ourselves in an abstract framework.

Let Λ be a Riemannian manifold modelled on a Hilbert space, let H be a closed affine space, isomorphic to a Hilbert space $\mathbb{H}$ (which can be identified with TH) and set $\mathcal{Z} = \Lambda \times H$.

In the applications, we will have

$$\mathcal{Z} = \Omega^1(\mathfrak{M};z_0,z_1),$$
$$\Lambda = \Omega^1(\mathfrak{M}_0;x_0,x_1),$$
$$H = W^1(t_0,t_1),$$
$$\mathbb{H} = W_0^1([0,1],\mathbb{R}).$$

Take a C^2-functional $f: \mathcal{Z} \longrightarrow \mathbb{R}$ and let $f'_t(x,t): \mathbb{H} \longrightarrow \mathbb{R}$ and $f'_x(x,t): T_x\Lambda \longrightarrow \mathbb{R}$ be the partial derivatives of f in a point $\gamma = (x,t)$ of $\mathcal{Z}$, which we can define in the following way:

$$f'_t(x,t)[\tau] = f'(x,t)[(0,\tau)], \text{ for every } \tau \in \mathbb{H},$$
$$f'_x(x,t)[\xi] = f'(x,t)[(\xi,0)], \text{ for every } \xi \in T_x\Lambda,$$

where $f'(x,t)$ denotes the Frechét differential of f in γ.

The "J" functional

Suppose now that the set $\{(x,t) \in \mathcal{Z} \mid f'_t(x,t) = 0\}$ is the graph of a smooth map $\phi: \Lambda \longrightarrow H$, so we have $f'_t(x,t) = 0$ iff $t = \phi(x)$. We consider now the restriction of f to the graph of ϕ, i.e. we consider the functional $J: \Lambda \longrightarrow \mathbb{R}$

defined by

$$J(x) = f(x, \phi(x)).$$

The first order variational principle

In order to look for the critical points of f, we have the following variational principle. It states that the critical points of f are on the graph of ϕ, and the research of the critical points of f is equivalent to the same problem for J.

Lemma III.4.1. *Let* $\gamma = (x,t) \in Z$, *then the following propositions are equivalent:*

a) *γ is a critical point of f;*

b) $\begin{cases} i) \ t = \phi(x); \\ \\ ii) \ x \text{ is a critical point of } J. \end{cases}$

The proof easily follows from the fact that for every $x \in \Lambda$ and for every $\xi \in T_x\Lambda$,

(4-1) $J'(x)[\xi] = f'_x(x, \phi(x))[\xi]$

The second order variational principle

Now we study the Hessian

$$f''(\gamma): T_\gamma Z \times T_\gamma Z \longrightarrow \mathbb{R} \quad \text{and} \quad J''(x): T_x\Lambda \times T_x\Lambda \longrightarrow \mathbb{R}.$$

We also consider the second partial derivatives

$$f''_{xx}(x,t): T_x\Lambda \times T_x\Lambda \longrightarrow \mathbb{R},$$
$$f''_{xt}(x,t) \equiv f''_{tx}(x,t) : T_x\Lambda \times \mathbb{H} \longrightarrow \mathbb{R}$$

and

$$f''_{tt}(x,t): \mathbb{H} \times \mathbb{H} \longrightarrow \mathbb{R},$$

defined by

$$f''_{xx}(x,t)[\xi,\xi'] = f''(x,t)[(\xi,0),(\xi',0)],$$
$$f''_{xt}(x,t)[\xi,\tau] = f''(x,t)[(\xi,0),(0,\tau)],$$
$$f''_{tt}(x,t)[\tau,\tau'] = f''(x,t)[(0,\tau),(0,\tau')].$$

We recall that f'', J'', f''_{xx} and f''_{tt} are symmetric bilinear forms.

We get the following "second order variational principle" about the kernels of $J''(x)$ and $f''(x,\phi(x))$.

Lemma III.4.2. *Let* $x_0 \in \Lambda$, $\gamma_0 = (x_0,\phi(x_0))$ *and* $\zeta = (\xi,\tau) \in T_{\gamma_0} Z$. *Suppose that* $f''_{tt}(\gamma_0)$ *is nondegenerate, i.e.*

$f''_{tt}(\gamma_0)[\tau,\tau'] = 0$ *for every* $\tau' \in \mathbb{H}$ *implies* $\tau = 0$.

Then the following propositions are equivalent:

a) $\zeta \in Ker\ f''(\gamma_0)$,

i.e. $f''(\gamma_0)[\zeta,\zeta'] = 0$, *for all* $\zeta' \in T_{\gamma_0} Z$;

b) $\begin{cases} i) \ \tau = \phi'(x_0)\xi; \\ ii) \ \xi \in Ker J''(x_0), \ i.e. J''(x_0)[\xi,\xi']=0, \ \text{for all } \xi \in T_{x_0}\Lambda, \end{cases}$

where $\phi'(x_0): T_{x_0}\Lambda \longrightarrow \mathbb{H}$ *is the differential of* ϕ *at* x_0.

Proof. a) $\Rightarrow$ b). Let $\zeta = (\xi,\tau) \in Ker\ f''(\gamma_0)$. First we prove i). From (4-1), for every $\xi' \in T_{x_0}\Lambda$ we get:

$$(4\text{-}2) \quad J''(x_0)[\xi,\xi'] = f''_{xx}(\gamma_0)[\xi,\xi'] + f''_{xt}(\gamma_0)[\xi,\phi'(x_0)\xi'].$$

Now, let $\tau' \in \mathbb{H}$, then we have:

$$(4\text{-}3) \qquad 0 = f''(\gamma_0)[(\xi,\tau),(0,\tau')] =$$
$$= f''_{xt}(\gamma_0)[\xi,\tau'] + f''_{tt}(\gamma_0)[\tau,\tau'].$$

Since for every $x \in \Lambda$, $f'_t(x,\phi(x))[\tau'] = 0$, differentiating with respect to x, we get:

(4-4) $f''_{xt}(\gamma_0)[\xi,\tau'] + f''_{tt}(\gamma_0)[\phi'(x_0)\xi,\tau'] = 0.$

From (4-3) and (4-4), we have

(4-5) $f''_{tt}(\gamma_0)[\tau,\tau'] = f''_{tt}(\gamma_0)[\phi'(x_0)\xi,\tau'].$

Since τ' is arbitrary and $f''_{tt}(\gamma_0)$ is nondegenerate, from (4-5) we deduce i). Now we prove ii). From (4-2) and part ii), for an arbitrary $\xi' \in T_{x_0} \Lambda$, we have

$$J''(x_0)[\xi,\xi'] = f''_{xx}(\gamma_0)[\xi,\xi'] + f''_{xt}(\gamma_0)[\xi,\phi'(x_0)\xi'] =$$
$$= f''_{xx}(\gamma_0)[\xi,\xi'] + f''_{xt}(\gamma_0)[\xi',\phi'(x_0)\xi] =$$
$$= f''(\gamma_0)[(\xi,\tau),(\xi',0)] = 0,$$

from which we deduce i).

b) $\Rightarrow$ a). Let $\xi \in Ker\ J''(x_0)$, and let $\tau = \phi'(x_0)\xi$. We prove that $\zeta = (\xi,\tau) \in Ker\ f''(\gamma_0)$. Indeed, for every $\zeta' = (\xi',\tau') \in T_{\gamma_0} \Lambda$, we have:

$$f''(\gamma_0)[\zeta,\zeta'] =$$
$$= f''_{xx}(\gamma_0)[\xi,\xi'] + f''_{xt}(\gamma_0)[\xi',\phi'(x_0)\xi] +$$
$$+ f''_{xt}(\gamma_0)[\xi,\tau'] + f''_{tt}(\gamma_0)[\phi'(x_0)\xi,\tau'] =$$
$$= J''(x_0)[\xi,\xi'] + f''_{xt}(\gamma_0)[\xi,\tau'] + f''_{tt}(\gamma_0)[\phi'(x_0)\xi,\tau'] =$$
$$= f''_{xt}(\gamma_0)[\xi,\tau'] + f''_{tt}(\gamma_0)[\phi'(x_0)\xi,\tau'].$$

From (4-4), we get

$$f''(\gamma_0)[\zeta,\zeta'] = 0,$$

so $\zeta \in Ker\ f''(\gamma_0)$.

$\square$

Remark III.4.3. The assumption on $f''_{tt}(\gamma)$ needs only to prove a) $\Rightarrow$ b).

Remark III.4.4. From Lemma III.4.2, we deduce that for every $\gamma = (x,\phi(x)) \in \Lambda$ such that $f''_{tt}(\gamma)$ is nondegenerate, the kernels of $f''(\gamma)$ and $J''(x)$ are isomorphic. The isomorphism is given by

$$(\xi,\tau) \longrightarrow \xi,$$

while the inverse isomorphism is given by

$$\xi \longrightarrow (\xi,\phi'(x)\xi).$$

$\square$

III.5. PROOF OF THE MORSE RELATIONS FOR STATIC SPACE-TIME

In this section we shall prove Theorem III.3.2. Fix a static Lorentzian manifold $(\mathfrak{M},<\cdot,\cdot>_L)$ and $z_0 = (x_0,t_0)$, $z_1 = (x_1,t_1)$ two points of $\mathfrak{M}$.

The "J" functional for static space-times

We can apply all the results of section 3 to the action integral. First of all, the set $\{(x,t)\,|\,E'_t(x,t) = 0\}$ is the graph of the smooth map $\phi\colon \Omega^1_\bullet \to W^1(t_0,t_1)$, defined by

$$\phi(x)(s) = t_0 + \frac{\Delta}{\displaystyle\int_0^1 \frac{1}{\beta(x(r))}dr} \cdot \int_0^s \frac{1}{\beta(x(r))}\,dr,$$

where $\Delta = t_1 - t_0$. The explicit form of ϕ follows from direct computations (see [Be,Fo,Gi,91]).

Moreover, for every $\gamma = (x,t) \in \mathcal{Z}$, $E''_{tt}(\gamma)$ is a nondegenerate bilinear form of $W^1_0([0,1],\mathbb{R})$. Indeed, for every $\tau \in W^1_0([0,1],\mathbb{R})$, we have

$$E''_{tt}(\gamma)[\tau,\tau] = - \int_0^1 \beta(x)\dot{\tau}^2 ds \leq - c_0 \|\tau\|^2,$$

where $c_0 = \underset{s \in [0,1]}{inf} \beta(x(s))$ and $\|\cdot\|$ is the usual norm of $W_0^1([0,1],\mathbb{R})$,

$$\|\tau\|^2 = \int_0^1 \dot{\tau}^2 ds.$$

Then, we deduce that $E''_{tt}(\gamma)$ is nondegenerate.

So, the functional J, as defined in the previous section makes sense. In this case it gets the following form:

$$(5\text{-}1) \quad J(x) = E(x,\phi(x)) = \frac{1}{2} \int_0^1 \langle \dot{x},\dot{x} \rangle ds - \frac{\Delta^2}{2} \frac{1}{\displaystyle\int_0^1 \frac{1}{\beta(x(s))} ds}.$$

Now, let x be a critical point of J and let $\gamma = (x,t)$, with $t = \phi(x)$. Then γ is a geodesic joining z_0 and z_1. Now, as in definition II.4.1, we consider the family of functionals

$$E_s(w) = \frac{1}{2} \int_0^s \langle \dot{w},\dot{w} \rangle dr,$$

defined on the path space

$$Z_s = \Omega^1(\mathfrak{M},z_0,\gamma(s)) = \Omega^1(\mathfrak{M}_0,x_0,x_1(s)) \times W^1(t_0,t_1(s)),$$

of the curves joining z_0 and $\gamma(s) \equiv (x_1(s),t_1(s))$, and parameterized in $[0,s]$, $s \in]0,1]$.

For every $s \in]0,1]$, E_s satisfies the assumptions of Theorems III.4.1 and III.4.2. In particular, the set $\{(y,t)|(E_s)'_u(u,t) = 0\}$ is the graph of the map

$$\phi_s: \Omega^1(\mathfrak{M}_0, x_0, x_1(s)) \longrightarrow W^1(t_0, t_1(s)),$$

given by

$$\phi_s(y)(r) = t_0 + \frac{\Delta_s}{s \displaystyle\int_0^1 \frac{1}{\beta(y(\rho))}d\rho} \cdot \int_0^r \frac{1}{\beta(y(\rho))}d\rho,$$

where $\Delta_s = t_1(s) - t_0$. So we set

$$J_s(y) = E_s(y, \phi_s(y)) = \frac{1}{2} \int_0^s \langle \dot{y}, \dot{y} \rangle dr - \frac{\Delta_s^2}{2} \frac{1}{s \displaystyle\int_0^1 \frac{1}{\beta(y(r))}dr}.$$

Now, recalling the definition II.4.1, we generalize the notion of conjugate point for a critical point of J.

Definition III.5.1 – *Let x be a critical point of J. A point $x(s)$, $s \in \,]0,1]$ is said conjugate to x_0 along x, if $J_s''(x_s)$ is degenerate, i.e. there exists $\xi \in T_{x_s}\Omega_s^1$, $\xi \neq 0$, such that*

(5-2) $\qquad J_s''(x_s)[\xi, \xi'] = 0, \qquad \forall \xi' \in T_{x_s}\Omega_s^1.$

The set of the vector fields which satisfies (5-2) is denoted with *Ker* $J''(x_s)$. The *multiplicity* of the conjugate point $x(s)$ is the dimension of *Ker* $J_s''(x_s)$. Finally the *geometric index* $\mu(x,J)$ is the number of conjugate points $x(s)$, $s \in \,]0,1[$, to x_0 along x, counted with their multiplicity.

The index theorem for static space-times

From Lemma III.4.2, we deduce the following

Theorem III.5.2. *Let* $\gamma = (x,t)$ *be a geodesic and* $s \in]0,1]$, *then we have:*

(i) $\zeta = (\xi,\tau) \in Ker\ E_s''(\gamma_s)$ *iff* $\tau = \phi_s'(x_s)\xi$ *and* $\xi \in Ker\ J_s''(x_s)$, *where* $\phi_s'(x_s)$ *is the differential of* ϕ_s *at* x_s;

(ii) $\gamma(s)$ *is conjugate to* $\gamma(0)$ *along* γ *iff* $x(s)$ *is conjugate to* $x(0)$ *along* x;

(iii) $\mu(\gamma,E) = \mu(x,J)$.

Proof. It follows from Lemma III.4.2 since, for every $s \in]0,1]$, $J_s(y) = E_s(y,\phi_s(y))$ and the assumptions of Lemma III.4.2 hold.

□

Remark III.5.3 – From Theorem III.5.2 we deduce that every conjugate point to $x(0)$ along x has finite multiplicity and x has finite geometrical index.

Now we state the analogous of the Index Theorem (cf. Th. II.4.3) for the functional J.

Theorem III.5.4. *(Index theorem for static space-times) Let* x *be a critical point of* J *and* $\gamma = (x,\phi(x))$. *We have:*

$$\mu(x,J) = m(x,J).$$

The proof is similar to that of the Index Theorem for the Riemannian action integral and it will be omitted.

From Theorems III.5.2 and III.5.4, we deduce the following

Corollary III.5.5 - *Let* $\gamma = (x,t)$ *be a critical point of* f, *then we have*

$$\mu(\gamma,E) = \mu(x,J).$$

Proof of the Morse relations for static space-times

Finally we shall prove Theorem III.3.2.

Proof of Th. III.3.2. Let $z_0 = (x_0,t_0)$ and $z_1 = (x_1,t_1)$ be two nonconjugate points of $\mathfrak{M}$. So the action integral E is a Morse function (i.e. has non degenerate critical points). From Theorem III.5.2, also x_0 and x_1 are nonconjugate for J. Hence, J is a Morse function.

Moreover, by assumption i) and ii), J is bounded from below and satisfies the Palais-Smale compactness condition; then by Theorem I.4.11, we get the Morse relations for the functional J:

$$\sum_{x \in Z_J} \lambda^{m(x,J)} = \mathcal{P}_\lambda(\Omega^1(\mathfrak{M}_0;x_0,x_1)) + (1 + \lambda)\, Q_\lambda,$$

where Z_J is the set of the critical points of J (it is at most a countable set, because J is bounded from below and is a Morse function). Then, by III.5.4, we get

$$(5.4) \qquad \sum_{x \in Z_J} \lambda^{\mu(x,J)} = \mathcal{P}_\lambda(\Omega^1(\mathfrak{M}_0;x_0,x_1)) + (1 + \lambda)\, Q_\lambda.$$

From the first variational principle (Lemma III.4.1), Theorem III.5.2.(iii) and the (5.4), we get

$$(5-5) \qquad \sum_{\gamma \in \mathfrak{G}} \lambda^{\mu(\gamma, E)} = \mathcal{P}_{\lambda}(\Omega^1(\mathfrak{M}_0; x_0, x_1)) + (1 + \lambda) Q_{\lambda}$$

where $\mathfrak{G}$ is the set of the geodesics joining z_0 and z_1. Now, since $W^1(t_0, t_1)$ is contractible to a point, $\Omega^1(\mathfrak{M}; z_0, z_1)$ and $\Omega^1(\mathfrak{M}_0; x_0, x_1)$ are homotopically equivalent. So, from (5-5), we deduce (3-1) and the proof of Theorem III.3.2 is complete.

□

C H A P T E R   IV
SOME APPLICATIONS TO A SEMILINEAR ELLIPTIC EQUATION

IV.1. INTRODUCTION

In this chapter we will be concerned with a semilinear elliptic equation. Even if this equation is simple, it presents a lot of very interesting phenomena which makes its study more important of the equation in itself. Moreover, since this equation is structurally simple, the technicalities are minimal.

The model problem.

The equation is the following one:

$$(1-1) \qquad \begin{cases} -\Delta u - \mu u + f(u) = 0 & in \ \Omega \\ u = 0 & in \ \partial\Omega \end{cases}$$

where μ is a real parameter, $\Omega \subset \mathbb{R}^N$, $N \geq 3$, is a smooth bounded domain and

$$f : \mathbb{R}^+ \longrightarrow \mathbb{R}$$

is a continuous function; if $f \in C^1$, the parameter μ allows to take $f'(0) = 0$ without any loss of generality.

The equation (1-1) is the Euler-Lagrange equation relative to the the "energy" functional

(1-2)
$$E(u) = \int_{\Omega} (\frac{1}{2} |\nabla u|^2 - \frac{\mu}{2} u^2 + F(u)) \, dx$$

where

$$F(s) = \int_0^s f(t) \, dt$$

Hence the solutions of (1-1) correspond to the critical points of (1-2) and in principle the Morse theory can be applied. In order to do this is necessary to define the appropriate functional setting.

Functional setting and differentiability.

We set

$$W_0^1(\Omega) = \left\{ u \in L^2(\Omega) \mid \nabla u \in L^2(\Omega) \right\}.$$

It is well known that $W_0^1(\Omega)$ is a Hilbert space if equipped with the scalar product

$$\langle u, v \rangle = \int_{\Omega} \nabla u \cdot \nabla v \, dx$$

and the relative norm

$$\|u\| = \sqrt{ \int_{\Omega} |\nabla u|^2 \, dx }$$

The space $W^{-1}(\Omega)$, which is the dual of $W^1_0(\Omega)$, can be identified with the completion of $L^2(\Omega)$ with respect to the norm

$$\|u\|_{W^{-1}(\Omega)} = max \left\{ \int_\Omega u\, \varphi\, dx \mid \varphi \in W^1_0(\Omega), \ \|\varphi\| = 1 \right\}$$

We now recall a well known result:

Theorem IV.1.1. *Suppose that*

(1-3) $|f(s)| \leq c(|s|^{p-1} + 1), \ c \in \mathbb{R}, \ p \in [1, \dfrac{2N}{N-2} \ ].$

Then E is a differentiable functional in $W^1_0(\Omega)$ *and*

$$dE : W^1_0(\Omega) \longrightarrow W^{-1}(\Omega)$$

has the form

$$dE(u) = -\Delta u - \mu u + f(u)$$

where $-\Delta$ *is the Laplace operator in the sense of distributions.*

Moreover, if f is of class C^1 *and*

(1-4) $|f'(s)| \leq c(|s|^{p-2} + 1), \ \ c \in \mathbb{R}, \ p \in [1, \dfrac{2N}{N-2} \ ],$

E is of class C^2, *and*

$$d^2E(u)[v,w] = \int_\Omega (\langle \nabla v, \nabla w \rangle - \mu\, vw + f'(u)uw) \, dx.$$

Proof. See e.g. [V,64].

□

Thus, the functional E is of class C^1 even if F is of class C^2 unless the (1-4) is satisfied. This assumption is too restrictive; in fact using the topological degree or the Ljiusternik-Schnirelmann theory, it is possible to get existence results without it. So the classical Morse theory is not adequate to study problem (1-1). However, using the results of section I.5, not only it is possible to use the Morse theory, but it provides better estimates on the number of solutions and sometimes on their qualitative properties.

Morse index for C^1-functionals.

If u is a critical point of E, by well known elliptic regularity results, $u \in L^\infty(\Omega)$, therefore, if f is of class C^1, the second variation of E

$$d^2 E(u)[v,w] = \int_\Omega (\langle \nabla v, \nabla w \rangle - \mu \, vw + f'(u)vw) \, dx.$$

is well defined on the critical points even if E is not twice Frecét differentiable. In fact

$$R(v) = E(u+v) - E(u) - dE(u)[v] - \frac{1}{2} d^2 E(u)[v,v]$$

is not $o(\|v^2\|)$ for $\|v\| \longrightarrow 0$. Nevertheless, the self adjoint realization in $L^2(\Omega)$ of the operator $-\Delta - \mu + f'(u_0)$ with domain $W_0^1(\Omega) \cap W^2(\Omega)$, is well defined.

Thus, it is possible to define $m(u_0)$ and $m^*(u_0)$ as the number of negative (non-positive) eigenvalues of L_{u_0}, the self adjoint realization of the operator $-\Delta - \mu + f'(u_0)$. The index $i_\lambda(u)$ is related to $m(u_0)$ and $m^*(u_0)$ in the obvious way as the following theorem shows.

Theorem IV.1.2 - *Let u_0 be a solution of (1-1). Suppose* that $f \in C^1$ and (1-3) is satisfied. Suppose that

(i) if L_{u_0}, is not singular, then $i_\lambda(u_0) = \lambda^{m(u_0)}$

(ii) if L_{u_0} is singular, but u_0 is isolated, then

$$i_\lambda(u_0,f) = \sum_{k=m(u_0)+1}^{k=m^*(u_0)-1} a_k \lambda^k,$$

(iii) if K is an isolated set of solutions of the (1-1) and $i_\lambda(K,f) = \sum_{k \geq 0} a_k \lambda^k$ with $a_k \neq 0$, then there exists $u \in K$ such that

$$m(u) \leq k \leq m^*(u).$$

Proof. It is an immediate consequence of Theorems I.5.11 and I.5.12.

□

Morse index and nodal regions.

We conclude this section stating a theorem which shows how the knowledge of the Morse index can be used e.g. to evaluate the number of the nodal regions of the solutions of (1.1).

Definition IV.1.3 - Let u be a continuous function defined in Ω. A nodal region for u is an open set $\omega \subseteq \Omega$ on which u does not change sign and such that $u\big|_{\partial\omega} \equiv 0$.

Theorem IV.1.4 - Let u be a solution of (1-1). Suppose that $f \in C^1$, $f(0) = 0$ and that it satisfies (1-3); moreover suppose that

(1-5) $f'(s)s^2 - f(s)s < 0$ for almost every

$s \in \mathbb{R}$.

Then

$$N(u) \leq m(u)$$

where $N(u)$ denotes the number of nodal regions of u.

Proof. Let $n = N(u)$ and let $\omega_1,\ldots,\omega_n$ be the nodal regions of u. Set

$$u_k = \begin{cases} u(x) & \text{if } x \in \omega_k \\ 0 & \text{if } x \notin \omega_k \end{cases}$$

and

$$V_n = Sp \left\{u_1,\ldots,u_n\right\}.$$

We claim that E is negative definite on V_n. Multiplying the equation (1-1) for u_k and integrating, we get

$$\int_\Omega (|\nabla u_k|^2 - \mu\, u_k^2 + f(u)u_k)\, dx = 0$$

and hence

$$\int_\Omega (|\nabla u_k|^2 - \mu\, u_k^2)\, dx = - \int_\Omega f(u)u_k^2\, dx.$$

Now take $v = \sum_{k=1}^{n} \lambda_k\, u_k \in V_n$; then, using the equation above,

$$d^2E(u)[v,v] = \sum_{k=1}^{n} \lambda_k^2 \int_\Omega (|\nabla u_k|^2 - \mu\, u_k^2 + f'(u)\, u_k)\, dx =$$

$$= \sum_{k=1}^{n} \lambda_k^2 \int_\Omega (f'(u)\, u_k^2 - f(u)u_k)\, dx < 0.$$

The inequality holds by the (1-5). Concluding we have

$$m(u) \geq dim \ V_n = N(u)$$

□

If f has a superlinear growth, the (1-5) is satisfied by functions of the type $f(u) = -|u|^{p-2}u$, $p > 2$, (cf. sections 4.3 and 4.4); if f has a sublinear growth, then (1-5) is satisfied by functions of the type $f(u) = -\ arctg\ u$. In the next sections we will see how to get existence results and estimates on the Morse index.

IV.2 - THE SUBLINEAR CASE.

In this section we will give existence results in the case in which the nonlinear term in (1-1) grows less than linearly; more precisely we assume that

(2-1) $|f(s)| \leq c(|s|^{p-1} + 1)$, $c \in \mathbb{R}$, $p \in [1, 2)$

The Box theorem for elliptic equations.

The first result which we have in this case is the following one:

Theorem IV.2.1. *Assume (2-1) and let* $\mu \in (\mu_k, \mu_{k+1})$, *where* μ_j *is the j-th eigenvalue of* $-\Delta$. *Then*
(a)   *the problem (1-1) has at least one solution* $\bar{u}$ *such that*

$$i_\lambda(\bar{u}) = \lambda^k + \mathcal{Z}(\lambda), \quad \mathcal{Z}(\lambda) \in \mathcal{S};$$

(b)   *if* f *is of class* C^1 *then*

$$m(\bar{u},E) \leq k \leq m^*(\bar{u},E);$$

The above theorem is an application of the Box theorem, I.6.8.; thus we have to verify its assumptions.

Lemma IV.2.2 - *Define*

$$\psi(u) = \int_{\Omega} F(u) \, dx$$

Then the I.(6-3) is satisfied.

Proof. Using the (2-1), the Holder and the Sobolev inequalities, for every $u,v \in W_0^1(\Omega)$, we have:

$$\psi'(u)[v] = \int_{\Omega} f(u)v \, dx \leq \int_{\Omega} c(|u|^{p-1} + 1)|v| \, dx \leq$$

$$\leq c \, \|u\|_{L^p}^{p-1} \cdot \|v\|_{L^p} + c \, \|v\|_{L^1} \leq k \, (\|u\|^{p-1} \cdot \|v\| + \|v\|).$$

Then, $\|\psi'(u)\| \leq k \, (\|u\|^{p-1} + 1)$

$\square$

Lemma IV.2.3 - *Under the assumptions of theorem IV.2.1, the functional E satisfies P.S. in bounded sets.*

Proof. Let u_n be a bounded PS-sequence i.e. a bounded sequence such $dE(u_n) \longrightarrow 0$ in $W^{-1}(\Omega)$. Then

(2-2) $\qquad -\Delta u_n - \mu u_n + f(u_n) = w_n$

where $w_n \longrightarrow 0$ in $W^{-1}(\Omega)$.

Since u_n is bounded in $W_0^1(\Omega)$, may be taking a subsequence, u_n is weakly convergent in $W_0^1(\Omega)$, and by Rellich theorem u_n is strongly convergent in $L^2(\Omega)$. Thus, also $- \mu u_n + f(u_n)$ is strongly convergent in $L^2(\Omega)$ and a fortiori in $W^{-1}(\Omega)$. Then, by the equation (2-2)

$$-\Delta u_n = \mu u_n - f(u_n) + w_n$$

with the right hand side strongly convergent in $W^{-1}(\Omega)$.
Therefore

$$u_n = (-\Delta)^{-1}(\mu u_n - f(u_n) + w_n)$$

is strongly convergent in $W_0^1(\Omega)$.

□

Proof of Theorem IV.2.1 - By Lemma *IV.2.2* and Lemma *IV.2.3*,
(a) is a consequence of Theorem I.6.8; (b) follows from Th.
IV.1.2.(iii).

□

The Three Solutions Theorem

The following theorem concerns the concept of
multiplicity introduced in section I.5.

Theorem IV.2.4. *Let f satisfy (2-1) and assume that*
$f(0) = 0$. *Then*

(a)   *if $f'(0)$ exists and $f'(0) \in (\mu_h, \mu_{h+1})$ with $h \neq k$, the*
         problem (2.1) has at least 3 solutions if counted
         with their multiplicity,

(b)   *moreover, if f is of class C^1, there are two*
         solutions u_1, u_2 of (2.1) (or one solution with
         multiplicity 2) such that

$$m(u_1, f) \leq k \leq m^*(u_1, f)$$

         and

$$m(u_2, f) - 1 \leq k \leq m^*(u_2, f) + 1.$$

Proof. (a) Since 0 is not degenerate, it is an isolated

critical point; then by theorem IV.1.2, $i_\lambda(0) = \lambda^h$. Then, denoting by K the set of the critical points of E,

$$i_\lambda(K) = \lambda^h + Z(\lambda), \quad (Z(\lambda) \in S),$$

We have to show that $Z(1) \geq 2$. By Theorem I.6.8, $i_\lambda(K) = \lambda^k + (1+\lambda)Q(\lambda)$; then

(2-4) $\qquad \lambda^h + Z(\lambda) = \lambda^k + (1+\lambda)Q(\lambda).$

Since $h \neq k$, $Q(\lambda) \neq 0$, then $Z(1) \geq 2$ and we get (a). (b) By the (2-4), we have that

$$Z(\lambda) = \lambda^k + \lambda^{k+1} + \text{other possible terms}$$

or

$$Z(\lambda) = \lambda^k + \lambda^{k-1} + \text{other possible terms}.$$

Then, (b) follows from Th. IV.1.2.(iii).

$\qquad\qquad\qquad\qquad\qquad\qquad\qquad\qquad\qquad\qquad\qquad$ □

Remark IV.2.5. Theorem IV.2.4 illustrates the concept of multiplicity and generalizes the Theorem of the 3 solutions of Amann (cf. [Am,82]).

If, in the proof of Theorem IV.2.1, we use the topological degree instead of the generalized Morse index, we need an assumptions assuring that the topological degree in 0 is different from the topological degree at infinity, i.e. $(-1)^h \neq (-1)^k$.

IV.3 - THE SUPERLINEAR CASE.

In this section we will give existence results in the case in which the nonlinear term in (1-1) grows more than linearly; more precisely we are concerned with the following problem:

$$(3\text{-}1) \quad \begin{cases} -\Delta u = f(u) & in \ \Omega, \\ \\ u = 0 & on \ \partial\Omega \end{cases}$$

where $\Omega \subset \mathbb{R}^N$, $N \geq 3$, is a smooth bounded domain and $f : \mathbb{R} \longrightarrow \mathbb{R}$ is a continuous function.

Moreover we will make the following assumptions:

(3-2) *there exists a > 0 such that, for every t > 0,*

$$|f(t)| \leq a + at^{p-1}$$

where a is a suitable positive constant and

$$p \in (1 \ , \ \frac{2N}{n-2})$$

(3-3)   *there exists* $\vartheta \in (0,1/2)$ *such that*

$$F(t) \leq \vartheta \ tf(t) \qquad\qquad t \geq 0$$

where, as usual

$$(3\text{-}4) \qquad F(t) = \int_0^t f(\tau)d\tau \qquad\qquad for \quad t \geq 0$$

$$(3\text{-}5) \qquad f(0) = 0 \ ; \qquad f'(0) = 0.$$

Moreover notice that (3-3) implies that

$$(3\text{-}6) \qquad f(t) \geq k \ t^{1/\vartheta - 1}, \qquad k > 0, t > 0$$

Lemma IV.3.1 – *If (3-2) and (3-3) hold, then* $E \in \mathcal{F}(W_0^1(\Omega))$.

Proof. It is well known that (3-2) and (3-3) imply that E satisfies P.S. (cf. e.g. [Am,Ra,73] or [Ra,86]). Then the conclusion follows from Theorem I.5.1.

□

Existence of infinitely many solutions.

By lemma IV.3.1, we can apply the Morse theory to the functional E and in particular the theorems of section I.6. We limit ourselves to apply Th. I.6.6 and get the following result.

Theorem IV.3.2 - *(i) if $f'(0)$ exists and $f'(0) \notin \sigma(-\Delta)$, i.e if it is not an eigenvalue of $-\Delta$, then the problem (3-1) has at least a solution different from 0.*
(ii) if f is odd f (3-1) has infinitely many solutions

Proof. Clearly E satisfies the assumptions of Theorem I.6.7. with

$$<Lx,x> = \int_{\Omega} |\nabla u|^{2} \, dx \quad \text{and} \quad \psi(x) = \int_{\Omega} F(u) \, dx.$$

Hence, E satisfies the assumptions of Th. I.6.6. Then, if $f'(0) \notin \sigma(-\Delta)$, (a) of Th. I.6.6 holds. If f is odd, E is even and (b) of Th. I.6.6 holds.

□

Remark IV.3.3 - The existence of infinitely many solutions of problem (3-1) has been proved by Ambrosetti and Rabinowitz with minimax methods [Am,Ra,73]. Our proof, based on Morse theory gives an extra information on the Morse index and hence, via Theorem IV.1.4, we get information on the nodal regions. In fact, combining Theorem IV.3.2.(ii) and Theorem IV.1.4, we get the existence of at least $h-k$ solutions with at most h nodal regions, where k is such that $f'(0) \in \,]\mu_{h},\mu_{h+1}[$ (μ_{h} is the h-th eigenvalue of $-\Delta$ with null boundary conditions).

IV.4. MORSE RELATIONS FOR POSITIVE SOLUTIONS

In this section we are concerned with a problem in which the topology plays a different role than in the previous situations. In fact, in the problem analyzed here, the number of solutions depend on the topology of Ω, the domain on which the equation is defined. This result is strictly related to the Morse theory. For this reason we shall treat it in details in the next sections of this chapter.

The model problem

The problem which we will study is the following:

$$(P_\varepsilon) \qquad \begin{cases} -\varepsilon\Delta u + u = f(u) & in\ \Omega, \\[2mm] u > 0 & in\ \Omega, \\[2mm] u = 0 & on\ \partial\Omega; \end{cases}$$

where $\varepsilon \in \mathbb{R}^+\backslash\{0\}$, $\Omega \subset \mathbb{R}^N$, $N \geq 3$, is a smooth bounded domain and

$$f : \mathbb{R}^+ \longrightarrow \mathbb{R}$$

is a $\mathcal{C}^{1,1}$-function.

We make the following assumptions on f:

(H_1) there exists $a > 0$ such that, for every $t > 0$,

$$|f(t)| \leq a + at^{p-1}$$
$$|f'(t)| \leq a + at^{p-2}$$

where a is a suitable positive constant and $p \in (2\ ,\ \dfrac{2N}{N-2}\ )$

(H_2) there exists $\vartheta \in (0,1/2)$ such that

$$F(t) \leq \vartheta\ tf(t) \qquad\qquad t \geq 0$$

where $F(t) = \int_0^t f(\tau)d\tau$ *for* $t \geq 0$

(H_3) *for every* $t > 0$, $\frac{d}{dt}\left(\frac{f(t)}{t}\right) > 0.$

(H_4) $f(0) = 0$; $f'(0) = 0.$

For what follows it is useful to extend f to $\mathbb{R}^-$ in the following way:

(H_5) $f(t) = 0$ *for* $t < 0.$

Moreover notice that (H_2) implies that

(H_2') $f(t) \geq k\, t^{1/\vartheta - 1}$, $k > 0, t > 0$

By theorem IV.3.2.(a), with $f(u)$ replaced by $f(u)-u$, (P_ε) has a positive solution. Actually, the existence of this solution could be proved also applying the Mountain Pass Theorem (Th. I.6.1). We want to prove that, if the topology of Ω is not trivial, and ε is sufficiently small, the problem (P_ε) has more than one solution.

In order to state the results, we consider the solutions of (P_ε) as the critical points of the energy functional

(4-1) $$E_\varepsilon(u) = \int_\Omega \left(\frac{1}{2}\,\varepsilon|\nabla u|^2 + \frac{1}{2}\,u^2 - F(u) \right)dx$$

where $F(t)$, for $t \geq 0$, is defined in (H_2) and, for $t \leq 0$, is 0. Notice that, by virtue of (H_1) and Th. IV.1.1, E_ε is a $\mathscr{C}^2$-functional on the Sobolev space $W_0^1(\Omega)$. If u is an isolated critical point of E_ε and $E_\varepsilon(u) = c$, its (polynomial) Morse index $i_\lambda(u)$ is well defined. We recall from section I.5 that the integer number $i_1(u)$ is called the multiplicity of u.

If u is a non-degenerate solution, its multiplicity is 1. If the multiplicity of u is n, a generic $\mathscr{C}^2$-small perturbation splits u into at least n non-degenerate solutions.

The main result

Now, we can state a theorem which gives a particular kind of Morse relations for the positive solutions of our problem.

IV.4.1 - Theorem. [Be,Ce,ta] *Suppose that f satisfies the assumptions (H_1), ...,(H_4). Moreover suppose that*

(i) $\varepsilon \in (o,\varepsilon^*]$ *where ε^* is a suitable positive constant,*

(ii) *the set $\mathcal{K}$ of nontrivial solutions of problem (P_ε) is discrete;*

then

$$(4\text{-}2) \quad \sum_{u \in \mathcal{K}} i_\lambda(u) = \lambda \, \mathcal{P}_\lambda(\Omega) + \lambda^2 \left[\mathcal{P}_\lambda(\Omega) - 1\right] + \lambda(1+\lambda) \, Q(\lambda)$$

where $Q(\lambda)$ is a polynomial with non-negative integer coefficients.

Let us remark that Theorem 4.1 implies that the problem (P_ε) has at least $2 \, \mathcal{P}_1(\Omega) - 1$ solutions if they are counted with their multiplicity. Of course, if Ω is topologically trivial, then $\mathcal{P}_1(\Omega) = 1$, and the above theorem does not give any extra information. When Ω is topologically rich, we obtain good information on the solutions of (P_ε).

The proof of Theorem IV.4.1 is quite involved and we shall dedicate to it the next sections, but first we will see an application.

A particular case

An example of this is given by the following corollary.

*IV.4.2 - **Corollary**. Let A and C_i, (i = 1,..,k) be contractible open non-empty sets in $\mathbb{R}^N$, smooth and bounded; suppose that*

$$\overline{C}_i \cap \overline{C}_j = \emptyset, \ (i,j = 1,..,k, \ i \neq j)$$

$$\overline{C}_i \subset A, \ (i = 1,..,k)$$

and set

$$\Omega = A \setminus \bigcup_i \overline{C}_i.$$

Then, there exists $\varepsilon^* > 0$ such that, for any $\varepsilon \in (0,\varepsilon^*]$, the problem (P_ε) has at least 2k + 1 solutions, if they are counted with their multiplicity.

Moreover, if the solutions are non-degenerate, k of them have Morse index N, k of them have index N + 1, and one (the Mountain Pass Solution) has index 1.

Proof. The assertion is a consequence of the theorem IV.4.1 and the fact that

(4-3)
$$\mathcal{P}_\lambda(\Omega) = 1 + k\lambda^{N-1}$$

The computation of $\mathcal{P}_\lambda(\Omega)$ is an easy application of algebraic topology techniques which we will show for completeness. Using the excision property and the fact that C_i's are ENR, we have:

$$H_q(A,\Omega) \cong H_q(\bigcup_i \overline{C}_i, \bigcup_i \partial C_i) \cong \bigoplus_{i=1}^{k} H_q(\overline{C}_i, \partial C_i) \cong \bigoplus_{i=1}^{k} H_q(B_N, \partial B_N).$$

Hence $\mathcal{P}_\lambda(A,\Omega) = k\lambda^N$.

From the exactness of the following sequence,

$$\longrightarrow H_q(A) \xrightarrow{j_q} H_q(A,\Omega) \xrightarrow{\partial_q} H_{q-1}(\Omega) \xrightarrow{i_{q-1}} H_{q-1}(A) \longrightarrow$$

it follows that, for q = N,

$$\dim\left[H_{N-1}(\Omega)\right] = \dim\left[H_N(A,\Omega)\right] = k;$$

for q ≥ 2 and q ≠ N

$$\dim\left[H_{q-1}(\Omega)\right] = \dim\left[H_q(A,\Omega)\right] = 0;$$

moreover, since Ω is connected,

$$\dim\left[H_0(\Omega)\right] = 1.$$

Concluding, we get (4-3).

□

IV.5 - THE FUNCTIONAL SETTING

In this section we will state the functional setting in which to study the problem (P_ε).

The manifold M_ε

We set

$$(5-1) \quad J_\varepsilon(u) = E_\varepsilon'(u)[u] = \int_\Omega \left(\varepsilon|\nabla u|^2 + u^2 \right) dx - \int_\Omega f(u)u \, dx$$

IV.5.1 - **Lemma.** *Suppose that $J_\varepsilon(u) = 0$ and that $u \neq 0$. Then*

$$J_\varepsilon'(u)[u] < 0$$

Proof. First notice that $J_\varepsilon(u) = 0$ and $u \neq 0$ imply

$$meas\left\{x \in \Omega \mid u(x) > 0\right\} > 0,$$

in fact, for every $u \leq 0$, by (H_5) and $J_\varepsilon(u) = 0$, we have that

$$\int_\Omega \left(\varepsilon|\nabla u|^2 + u^2 \right) dx = 0$$

and hence $u = 0$ a.e.

Now,

$$J_\varepsilon'(u)[u] = 2 \int_\Omega \left(\varepsilon|\nabla u|^2 + u^2 \right) dx - \int_\Omega \left(f'(u)u^2 + f(u)u \right) dx =$$

$$= \int_\Omega \left(f(u)u - f'(u)u^2 \right) dx;$$

therefore the claim follows by (H_3).

□

We now set

$$(5\text{-}2) \quad \mathscr{S} = \left\{ v \in W_0^1(\Omega) \mid \; \|v\| = 1 \right\} \setminus \left\{ v \in W_0^1(\Omega) \mid v \leq 0 \text{ a.e.} \right\}$$

(where $\|\cdot\|$ denotes the norm of $W_0^1(\Omega)$). Clearly $\mathscr{S}$ is a smooth manifold of codimension 1. Moreover define

$$(5\text{-}3) \qquad\qquad M_\varepsilon = \left\{ u \in W_0^1(\Omega) \mid u \neq 0 \text{ and } J_\varepsilon(u) = 0 \right\}$$

*IV.5.2 – **Lemma.** M_ε is a manifold diffeomorphic to $\mathscr{S}$ and the diffeomorphism $\psi_\varepsilon : \mathscr{S} \longrightarrow M_\varepsilon$ is of class $\mathscr{C}^{1,1}$. Moreover there exist $h_\varepsilon > 0$ and $k_\varepsilon > 0$ such that for every $u \in M_\varepsilon$ the following relations hold true*

$(5\text{-}4)$ $\qquad\qquad\qquad\qquad \|u\| \geq h_\varepsilon$

$(5\text{-}5)$ $\qquad\qquad\qquad\qquad E_\varepsilon(u) \geq k_\varepsilon$

Proof. For every $\bar{v} \in \mathscr{S}$, let $\xi_\varepsilon(\bar{v})$ be the positive number which realizes the maximum of the function $\lambda \longrightarrow E_\varepsilon(\lambda \bar{v})$ defined on $\mathbb{R}^+$. $\xi_\varepsilon(\bar{v})$ is well defined. In fact, $\max\limits_{\lambda \in \mathbb{R}^+} E_\varepsilon(\lambda \bar{v})$ is achieved since, by (H_1) 0 is a local strict minimum of $E_\varepsilon(u)$ and by (H_2'),

$$\lim_{\lambda \longrightarrow \infty} E_\varepsilon(\lambda \bar{v}) = -\infty.$$

Moreover the uniqueness of $\xi_\varepsilon(\bar{v})$ follows observing that, by

(H_3), $\frac{1}{\lambda} f(\lambda \bar{v})\bar{v}$ is a strictly increasing function of λ and

$$0 = \frac{\partial}{\partial \lambda} E_\varepsilon(\lambda \bar{v}) = \lambda \int_\Omega \left(\varepsilon |\nabla \bar{v}|^2 + \bar{v}^2 \right) dx - \int_\Omega f(\lambda \bar{v})\bar{v} \, dx$$

implies

$$\int_\Omega \left(\varepsilon |\nabla \bar{v}|^2 + \bar{v}^2 \right) dx = \frac{1}{\lambda} \int_\Omega f(\lambda \bar{v})\bar{v} \, dx \ .$$

Clearly $\xi_\varepsilon(\bar{v})\bar{v} \in M_\varepsilon$. Thus M_ε is the graph of the function $\psi_\varepsilon : \mathscr{S} \longrightarrow M_\varepsilon$ defined by

$$\psi_\varepsilon(v) = \xi_\varepsilon(v)v$$

By the implicit function theorem and lemma IV.5.1, ψ_ε and ξ_ε are functions of class $\mathscr{C}^{1,1}$; (5-4) is a consequence of the definition of the function ξ_ε and of the behavior of E_ε. Lastly let us consider $u \in M_\varepsilon$, then using (H_2) and (5-4) we obtain

$$E_\varepsilon(u) = \int_\Omega \left(\frac{1}{2} \varepsilon |\nabla u|^2 + \frac{1}{2} u^2 - F(u) \right) dx \ \geq$$

$$\geq \frac{1}{2} \int_\Omega \left(\varepsilon |\nabla u|^2 + u^2 \right) dx - \vartheta \int_\Omega f(u)u \, dx =$$

$$= (1/2 - \vartheta) \int_\Omega \left(\varepsilon |\nabla u|^2 + u^2 \right) dx \geq (1/2 - \vartheta) \, \varepsilon \, \|u\|^2 \geq k_\varepsilon.$$

□

The constrained critical points.

Even if it is not necessary for the proof of Th. IV.4.1, in this subsection we state a variational principle which is the analogous of the first order variational principle presented in section III.4.

IV.5.3 - **Lemma.** *The following statements are equivalent:*

(i) *u is a critical point of* E_ε;

(ii) *u is a critical point of* E_ε *constrained on* M_ε;

Proof. (i) $\Rightarrow$ (ii) *is immediate since* $E_\varepsilon'(u) = 0 \Rightarrow u \in M_\varepsilon$.

(ii) $\Rightarrow$ (i); *let* u_0 *be a critical point of* E_ε *constrained on* M_ε. *Then there exists* $\lambda \in \mathbb{R}$ *such that*

$$E_\varepsilon'(u_0) - \lambda J_\varepsilon'(u_0) = 0.$$

Thus,

$$J_\varepsilon(u_0) = E_\varepsilon'(u_0)[u_0] = \lambda J_\varepsilon'(u_0)[u_0].$$

This equality and lemma IV.5.1 imply that $\lambda = 0$.

$\square$

By standard arguments, we have that the Palais - Smale condition holds for both the free functional E_ε and the functional E_ε constrained on M_ε, i.e. we have the following Lemma.

IV.5.4 - **Lemma.** (i) *if* $u_n \in W_0^1(\Omega)$ *is a sequence such that*

(5-6) $\|\nabla E_\varepsilon(u_n)\| \longrightarrow 0$ *and* $E_\varepsilon(u_n)$ *is bounded*

then u_n *has a converging subsequence.*

(ii) *if* $u_n \in M_\varepsilon$ *is a sequence such that*

$$(5\text{-}7) \qquad \left\| \nabla E_\varepsilon(u_n) - \frac{(\nabla E_\varepsilon(u_n), \nabla J_\varepsilon(u_n))}{\|\nabla J_\varepsilon(u_n)\|^2} \nabla J_\varepsilon(u_n) \right\| \longrightarrow 0$$

and $E_\varepsilon(u_n)$ *is bounded then* u_n *has a converging subsequence in* M_ε.

IV.6 – SOME HARD ANALYSIS.

Since we want to prove Th. IV.4.1 in full detail, some "hard analysis" is now necessary.

From now on , for any function $u \in W^1_0(\mathcal{D})$ $(\mathcal{D} \subseteq \mathbb{R}^N)$, we denote with the same symbol its extension to $\mathbb{R}^N$ obtained setting $u = 0$ outside of $\mathcal{D}$.

Moreover, for any $u \in W^1_0(\mathcal{D})$ $(\mathcal{D} \subseteq \mathbb{R}^N)$, we denote with the symbols E_ε, J_ε, the objects corresponding to the ones we have defined by (5-3) and (5-1) for $u \in W^1_0(\Omega)$. Also, for any $\mathcal{D} \subseteq \mathbb{R}^N$, $M_\varepsilon(\mathcal{D})$ is the submanifold of $W^1_0(\mathcal{D})$ defined as in (5-3).

The number m(ε,ρ)

We set

(6-1)
$$m(\varepsilon,\mathcal{D}) = \inf \left\{ E_\varepsilon(u) \, , \, u \in M_\varepsilon(\mathcal{D}) \right\}$$

Notice that $m(\varepsilon,\mathcal{D})$ is well defined by (5-5) of Lemma IV.5.2. Moreover, whenever $\mathcal{D} \subset \mathbb{R}^N$ is bounded, the infimum is achieved since E_ε satisfies (PS) on $M_\varepsilon(\mathcal{D})$ (Lemma IV.5.4). If

$$\mathcal{D} = B_\rho(y) = \left\{ x \in \mathbb{R}^N : |x\text{-}y| < \rho \right\}$$

the number $m(\varepsilon,B_\rho(y))$ does not depend on y; thus, for any y $\in \mathbb{R}^N$, we set

(6-2) $m(\varepsilon,\rho) = m(\varepsilon,B_\rho(y))$.

Moreover, it is a trivial fact that

(6-3) $\rho_1 < \rho_2 \;\Rightarrow\; m(\varepsilon,\rho_1) > m(\varepsilon,\rho_2)$

If $\mathcal{D} = \mathbb{R}^N$, then (PS) for E_ε fails, however the following result holds:

*IV.6.1 - **Lemma.*** $m(\varepsilon,\mathbb{R}^N)$ *is achieved by a positive function radially symmetric* $u(r)$ *where* r *is the radial coordinate.* $u(r)$ *is decreasing in* r *and has the asymptotic behavior*

$$\lim_{r \to \infty} r^{(N-1)/2} e^r u(r) = \eta_1 > 0$$

(6-4)

$$\lim_{r \to \infty} r^{(N-1)/2} e^r u'(r) = \eta_2 > 0$$

Proof. If we restrict our consideration to the subspace $W_r^1(\mathbb{R}^N)$ of $W^1(\mathbb{R}^N)$ consisting of functions radially symmetric about the origin, the embedding

$$j : W_r^1(\mathbb{R}^N) \longrightarrow L^p(\mathbb{R}^N), \; p \in (2, \frac{2N}{N-2}),$$

is compact [St,77]; so the functional E_ε satisfies the (P.S.) condition on $M_\varepsilon(\mathbb{R}^N) \cap W_r^1(\mathbb{R}^N)$ and

$$\inf\left\{ E_\varepsilon(u) \mid u \in M_\varepsilon(\mathbb{R}^N) \cap W_r^1(\mathbb{R}^N)\right\}$$

is achieved. Then in order to prove that $m(\varepsilon,\mathbb{R}^N)$ is achieved, it suffices to show that, for any $u \in M_\varepsilon(\mathbb{R}^N)$, there exists $w \in M_\varepsilon(\mathbb{R}^N) \cap W_r^1(\mathbb{R}^N)$ such that $E_\varepsilon(w) \le E_\varepsilon(u)$. Let us denote by u^* the Schwartz symmetrized function about the origin of u and set $w = t^* u^*$

where $t^* = \xi_\varepsilon \left(\dfrac{u^*}{\|u^*\|} \right) \|u^*\|^{-1}$. Then, using the Riesz

inequality and the properties of the spherical rearrangements, we obtain

$$E_\varepsilon(w) = (1/2) \ (t^*)^2 \cdot \int_{\mathbb{R}^N} \left(\varepsilon |\nabla u^*|^2 + |u^*|^2 \right) dx - \int_{\mathbb{R}^N} F(t^* u^*) \ dx \leq$$

$$\leq (1/2) \ (t^*)^2 \cdot \int_{\mathbb{R}^N} \left(\varepsilon |\nabla u|^2 + |u|^2 \right) dx - \int_{\mathbb{R}^N} F(t^* u) \ dx =$$

$$= E_\varepsilon(t^* u) \leq \max_{t \in \mathbb{R}^+} E_\varepsilon(tu) = E_\varepsilon(u)$$

Then, the estimates (6-4) follow from a well known theorem of Gidas, Ni, Nirenberg [Gi,Ni,Ni,80].

□

The barycentre

In the following, $W^1_{comp}(\mathbb{R}^N)$ will denote the subspace of $W^1(\mathbb{R}^N)$ of functions whose support is compact.

For any $u \in W^1_{comp}(\mathbb{R}^N)$ we shall consider

$$\beta(u) = \frac{\displaystyle\int_{\mathbb{R}^N} x \cdot |\nabla u(x)|^2 \ dx}{\displaystyle\int_{\mathbb{R}^N} |\nabla u(x)|^2 \ dx}.$$

The functional β: $W^1_{comp}(\mathbb{R}^N) \ ///\!\!\to \ \mathbb{R}^N$. is called the barycentre of u.

The number $m^*(\varepsilon, \gamma)$

We define, for every $\rho > 0$ and every $\gamma > 1$,

(6-5) $m^*(\varepsilon, \rho, \gamma) \equiv$

$$\equiv \inf \left\{ E_\varepsilon(u) \mid u \in M_\varepsilon(B_{\gamma\rho}(0) \backslash B_\rho(0)), \ \beta(u) = 0 \right\}.$$

Let us point out that the number $m^*(\varepsilon,\rho,\gamma)$ does not change if we move the center of the balls and $\beta(u)$ to any other point $x \in \mathbb{R}^N$.

It is clear that

$$m^*(\varepsilon,\rho,\gamma) > m(\varepsilon,\mathbb{R}^N)$$

Now we set

(6-6) $$m^*(\varepsilon,\gamma) \equiv \inf_{\rho>0} m^*(\varepsilon,\rho,\gamma)$$

Estimates on $m(\varepsilon,\rho)$ and $m^*(\varepsilon,\gamma)$

IV.6.2 - **Lemma.** *The relation*

(6-7) $$\lim_{\rho \to +\infty} m(1,\rho) = m(1,\mathbb{R}^N)$$

holds.

Proof. Let us denote by $\Psi \in W_r^1(\mathbb{R}^N)$ a positive function spherically symmetric about the origin, such that

$$E_1(\Psi) = m(1,\mathbb{R}^N) \ , \ \Psi \in M_1(\mathbb{R}^N)$$

and consider the function $w_\rho \in W^1(B_\rho(0))$ defined by

$$w_\rho(x) = \zeta_\rho(x)\psi(x)$$

where $\zeta_\rho(x): \mathbb{R}^N \longrightarrow [0,1]$ is a C^∞-function defined by

$$\zeta_\rho(x) = \tilde{\xi}\left(\frac{|x|}{\rho}\right)$$

$\tilde{\xi} : \mathbb{R}^+ \cup \{0\} \longrightarrow [0,1]$ being a decreasing C^∞-function such that

$$\tilde{\xi}(t) = \begin{cases} 1 & t \leq \dfrac{1}{2} \\ 0 & t \geq 1 \end{cases}.$$

Put

$$u_\rho(x) = t_\rho \, w_\rho(x) \quad \text{where } t_\rho = \xi_1 \left(\frac{w_\rho}{\|w_\rho\|} \right) \|w_\rho\|^{-1}.$$

Then $u_\rho \in M_1(B_\rho(0))$ and $E_1(u_\rho) \geq m(1,\rho) > m(1,\mathbb{R}^N)$. Hence to prove the assertion it is sufficient to show that

(6-8)

$$\text{(a)} \quad \int_{\mathbb{R}^N \backslash B_{\rho/2}(0)} f(\Psi)\Psi \, dx = o(1/\rho)$$

$$\text{(b)} \quad \int_{\mathbb{R}^N \backslash B_{\rho/2}(0)} F(\Psi) \, dx = o(1/\rho)$$

(6-9)

$$\int_{\mathbb{R}^N} \left(|\nabla(\Psi - w_\rho)|^2 + |\Psi - w_\rho|^2 \right) dx = o(1/\rho).$$

The limit (6-8)(a) follows from the fact that, if ρ is large enough, (6-4) and (H_1) imply

$$0 \leq \int_{\mathbb{R}^N \backslash B_{\rho/2}(0)} f(\Psi)\Psi \, dx \leq$$

$$\leq \int_{\mathbb{R}^N \backslash B_{\rho/2}(0)} (a\Psi + a\Psi^{p+1}) \, dx \leq$$

$$\leq k_1 \int_{\mathbb{R}^N \backslash B_{\rho/2}(0)} \left(\frac{1}{e^{|x|} |x|^{(N-1)/2}} \right) dx +$$

$$+ k_2 \int_{\mathbb{R}^N \backslash B_{\rho/2}(0)} \left(\frac{1}{e^{|x|} |x|^{(N-1)/2}} \right)^{p+1} dx =$$

$$= o\left(\frac{1}{\rho}\right) \quad \text{as } \rho \longrightarrow +\infty.$$

Analogously we prove that

$$0 \le \int_{\mathbb{R}^N \setminus B_{\rho/2}(0)} F(\Psi) \, dx \le$$

$$\le \int_{\mathbb{R}^N \setminus B_{\rho/2}(0)} (b\Psi + b\Psi^{p+1}) \, dx = o(1/\rho)$$

Also, using (6-4), if ρ is large enough, we obtain

$$\int_{\mathbb{R}^N} \left(|\nabla(\Psi - w_\rho)|^2 + |\Psi - w_\rho|^2 \right) dx \le$$

$$\le k_3 \int_{\mathbb{R}^N \setminus B_{\rho/2}(0)} |\nabla\Psi|^2 dx + k_3 \int_{\mathbb{R}^N \setminus B_{\rho/2}(0)} |\Psi|^2 dx \le$$

$$\le k_4 \int_{\mathbb{R}^N \setminus B_{\rho/2}(0)} \left(\frac{1}{e^{|x|} |x|^{(N-1)/2}} \right)^2 dx = o\left(\frac{1}{\rho}\right).$$

$\square$

IV.6.3 - **Lemma.** *The inequality*

(6-10) $$m^*(1,\gamma) > m(1,\mathbb{R}^N)$$

holds for any fixed $\gamma > 1$.

Proof. It is obvious that $m^*(1,\gamma) \geq m(1,\mathbb{R}^N)$. To prove the strict inequality, we argue by contradiction and we suppose that the equality holds. In this case there exists a sequence ρ_n such that

(6-11) $m^*(1,\rho_n,\gamma) \longrightarrow m(1,\mathbb{R}^N)$ for $n \longrightarrow + \infty$

We can exclude at once that $\{\rho_n\}$ is bounded. In fact if it were bounded by L, we should have

$$m^*(1,\rho_n,\gamma) \geq m(1,\gamma L) > m(1,\mathbb{R}^N).$$

So we can assume that (6-11) holds with $\rho_n \uparrow +\infty$. Then there exists a sequence of functions $\{u_n\}$ such that

$$u_n \in W^1_0(B_{\gamma\rho_n}(0)\backslash B_{\rho_n}(0)), \; u_n \neq 0, \; \beta(u_n) = 0 \;,$$

$$\frac{1}{2}\int_{B_{\gamma\rho_n}(0)\backslash B_{\rho_n}(0)}\left(|\nabla u_n|^2 + u_n^2\right)dx +$$

$$- \int_{B_{\gamma\rho_n}(0)\backslash B_{\rho_n}(0)} F(u_n)dx \longrightarrow m(1,\mathbb{R}^N).$$

$$\int_{B_{\gamma\rho_n}(0)\backslash B_{\rho_n}(0)}\left(|\nabla u_n|^2 + u_n^2\right)dx = \int_{B_{\gamma\rho_n}(0)\backslash B_{\rho_n}(0)} f(u_n)u_n \, dx$$

On the other hand it is known that any minimizing sequence in $W^1_0(\mathbb{R}^N\backslash B_{\rho_1}(0))$ has the form

(6-12) $w_n(x) + \Psi(x - y_n)$

where $\{w_n(x)\} \subset W^1(\mathbb{R}^N)$ is a sequence going strongly to 0 in $W^1(\mathbb{R}^N)$, $\{y_n\} \subset \mathbb{R}^N$ is such that $|y_n| \longrightarrow +\infty$ and $\Psi \in W^1(\mathbb{R}^N)$ is a positive function, spherically symmetric about the origin, such that $E_1(\Psi) = m(1,\mathbb{R}^N)$, $\Psi \in M_1(\mathbb{R}^N)$. Thus, in particular, u_n has the form (6-12).

Since any regular solution φ of $-\Delta u + u = f(u)$ in $\mathbb{R}^N$

satisfies the following Pohozaev type inequality

$$\int_{\mathbb{R}^N} |\nabla\varphi|^2 dx = \frac{2N}{N-2} \int_{\mathbb{R}^N} \left[-\frac{\varphi^2}{2} + F(\varphi) \right] dx$$

we have

$$\int_{\mathbb{R}^N} |\nabla\Psi|^2 \, dx = N \, m(1,\mathbb{R}^N) = b > 0.$$

Since $\|w_n\|_{W^1(\mathbb{R}^N)} \longrightarrow 0$, it follows that

$$\int_{B_{\rho_n/4}(y_n)} |\nabla(w_n(x) + \Psi(x-y_n))|^2 dx \xrightarrow[n\longrightarrow+\infty]{} b$$

Set $C_n = B_{\gamma\rho_n}(0) \backslash B_{\rho_n}(0)$; clearly

$$\int_{C_n} |\nabla u_n|^2 dx =$$

$$= \int_{C_n \cap [B_{\rho_n/4}(y_n)]} |\nabla u_n|^2 dx + \int_{C_n \backslash [B_{\rho_n/4}(y_n)]} |\nabla u_n|^2 dx$$

and

$$(6\text{-}13) \qquad 0 \le \int_{C_n \backslash [B_{\rho_n/4}(y_n)]} |\nabla u_n|^2 dx \le$$

$$\le \int_{\mathbb{R}^N \backslash [B_{\rho_n/4}(y_n)]} |\nabla(w_n(x)+\Psi(x-y_n))|^2 dx \xrightarrow[n\longrightarrow+\infty]{} 0$$

$$(6\text{-}14) \qquad \int_{C_n \cap [B_{\rho_n/4}(y_n)]} |\nabla u_n|^2 dx =$$

$$= \int_{C_n \cap [B_{\rho_n/4}(y_n)]} |\nabla(w_n(x) + \Psi(x-y_n))|^2 dx \xrightarrow[n\longrightarrow+\infty]{} b;$$

thus $C_n \cap [B_{\rho_n/4}(y_n)] \neq \emptyset$ and $|y_n| > (3/4)\, \rho_n$ for n big enough.

Now

$$\frac{\displaystyle\int_{C_n \cap [B_{\rho_n/4}(y_n)]} x\,|\nabla u_n|^2\,dx}{\displaystyle\int_{C_n \cap [B_{\rho_n/4}(y_n)]} |\nabla u_n|^2\,dx} =$$

$$= y_n + \frac{\displaystyle\int_{C_n \cap [B_{\rho_n/4}(y_n)]} (x - y_n)\,|\nabla u_n|^2\,dx}{\displaystyle\int_{C_n \cap [B_{\rho_n/4}(y_n)]} |\nabla u_n|^2\,dx}$$

so

$$\left| \frac{\displaystyle\int_{C_n \cap [B_{\rho_n/4}(y_n)]} x\,|\nabla u_n|^2\,dx}{\displaystyle\int_{C_n \cap [B_{\rho_n/4}(y_n)]} |\nabla u_n|^2\,dx} \right| \geq \rho_n/2.$$

On the other hand

$$\left| \frac{\displaystyle\int_{C_n \setminus [B_{\rho_n/4}(y_n)]} x\,|\nabla u_n|^2\,dx}{\displaystyle\int_{C_n \setminus [B_{\rho_n/4}(y_n)]} |\nabla u_n|^2\,dx} \right| < \gamma \rho_n$$

Since $\beta(u_n) = 0$, we have

$$\int_{C_n \cap [B_{\rho_n/4}(y_n)]} x|\nabla u_n|^2 dx + \int_{C_n \setminus [B_{\rho_n/4}(y_n)]} x|\nabla u_n|^2 dx = 0,$$

then

$$(\rho_n/2) \cdot \left| \int_{C_n \cap [B_{\rho_n/4}(y_n)]} |\nabla u_n|^2 dx \right| \le$$

$$\le \left| \int_{C_n \cap [B_{\rho_n/4}(y_n)]} x|\nabla u_n|^2 dx \right| =$$

$$= \left| \int_{C_n \setminus [B_{\rho_n/4}(y_n)]} x|\nabla u_n|^2 dx \right| <$$

$$< \gamma\rho_n \left| \int_{C_n \setminus [B_{\rho_n/4}(y_n)]} |\nabla u_n|^2 dx \right|$$

and using (6-13) and (6-14), we obtain:

$$(\rho_n/2) \cdot (b + o(1)) \le \gamma\rho_n \cdot o(1)$$

and this inequality gives

$$b/(2\gamma) < o(1)$$

which is a contradiction.

□

IV.6.4 – Lemma. *For any $\gamma > 1$, there exists $\overline{R} \equiv \overline{R}(\gamma) > 0$ such that*

(6-15) $$m(1,R) < m^*(1,R,\gamma)$$

for every $R \ge \overline{R}(\gamma)$.

Proof. From lemmas IV.6.2 and IV.6.3, we deduce that there exists $\bar{R} > 0$ such that for every $R \geq \bar{R}$

$$m(1,R) < m^*(1,\gamma).$$

Then, (6-15) holds since, by definition,

$$m^*(1,R,\gamma) \geq m^*(1,\gamma)$$

for every R.

$\square$

IV.6.5 – **Corollary.** *For every $\rho > 0$ and $\gamma > 1$, there exists $\bar{\varepsilon} \equiv \bar{\varepsilon}(\gamma,\rho)$, such that the relation*

(6-16) $m(\varepsilon,\rho) < m^*(\varepsilon,\rho,\gamma)$

holds for every $\varepsilon \in (0,\bar{\varepsilon}]$.

Proof. We assert that $\bar{\varepsilon} = (\rho/\bar{R})^2$ where $\bar{R}$ is the number found in lemma IV.6.4. Indeed it is possible to define a one to one map

$$T \; : \; W_0^1(B_\rho(0)) \longrightarrow W_0^1(B_{\rho/\sqrt{\varepsilon}}(0))$$

by $T(u) = u_\varepsilon(x) \equiv u(\sqrt{\varepsilon}x)$ for every $u \in W_0^1(B_\rho(0))$. Then a simple calculation shows that

$$\int_{B_{\rho/\sqrt{\varepsilon}}(0)} \left(|\nabla u_\varepsilon|^2 + u_\varepsilon^2\right) dx - \int_{B_{\rho/\sqrt{\varepsilon}}(0)} f(u_\varepsilon)u_\varepsilon \, dx =$$

$$= \varepsilon^{-N/2} \left[\int_{B_\rho(0)} \left(\varepsilon|\nabla u|^2 + u^2\right) dx - \int_{B_\rho(0)} f(u)u \, dx \right];$$

and

$$(1/2)\int_{B_{\rho/\sqrt{\varepsilon}}(0)} \left(|\nabla u_\varepsilon|^2 + u_\varepsilon^2\right) dx - \int_{B_{\rho/\sqrt{\varepsilon}}(0)} F(u_\varepsilon) \, dx =$$

$$= \varepsilon^{-N/2}\left[(1/2)\int_{B_\rho(0)} \left(\varepsilon|\nabla u|^2 + u^2\right) dx - \int_{B_\rho(0)} F(u) \, dx \right].$$

So

$$m(\varepsilon,\rho) \; = \; \varepsilon^{N/2} \, m(1,\rho/\sqrt{\varepsilon}) \; .$$

Analogously it is easy to verify that

$$m^*(\varepsilon,\rho,\gamma) \; = \; \varepsilon^{N/2} \, m^*(1,\rho/\sqrt{\varepsilon},\gamma) \; ;$$
$$m(\varepsilon,\mathbb{R}^N) \; = \; \varepsilon^{N/2} \, m(1,\mathbb{R}^N) \; .$$

Hence if $\rho/\sqrt{\varepsilon} \geq \overline{R}$, that is if $\varepsilon \leq (\rho/\overline{R})^2$, the relation (6-16) follows from (6-15).

$\square$

IV.7. THE PHOTOGRAPHY METHOD

The photography method consists in finding a pairs of maps φ_1 and φ_2 as in theorem I.2.2.(vii) with $(X,A) = \Omega$ and where (Y,B) is a suitable subset of our function space; in our case we have $(Y,B) = (M_\varepsilon{}^c,\varnothing)$ where $M_\varepsilon{}^c$ is given by

$$M_\varepsilon{}^c = \left\{ u \in M_\varepsilon \;\middle|\; E_\varepsilon(u) \leq c \right\}.$$

Thus, the theorem I.2.2.(vii), allows to estimate the topology of $M_\varepsilon{}^c$ by the topology of Ω (cf. Th. IV.7.3). The name of this method comes from the fact that $\varphi_1(\Omega)$ is seen as a photography of Ω in $M_\varepsilon{}^c$.

Estimates on the range of β

Now let us develop the technical part. In what follows, without any loss of generality, we shall assume $0 \in \Omega$. Moreover, we denote by $r > 0$, a number such that the sets

$$\Omega^+ = \left\{ x \in \mathbb{R}^N \;\middle|\; d(x,\Omega) \leq r \right\}$$

$$\text{and } \Omega^- = \left\{ x \in \Omega \;\middle|\; d(x,\partial\Omega) \geq 2r \right\}$$

are homotopically equivalent to Ω and $B_r(0) \subset \Omega$. Finally we fix

$$\gamma = \frac{diam\ \Omega}{r}.$$

*IV.7.1 - **Lemma**. There exists ε^* such that*

(7-1) $u \in M_\varepsilon,$ $E_\varepsilon(u) \leq m(\varepsilon,r)$ $\Rightarrow$ $\beta(u) \in \Omega^+$

for every $\varepsilon \in (0,\varepsilon^*]$.

*Proof.*First of all we observe that by the choice of r,

$$m(\varepsilon,\Omega) < m(\varepsilon,r)$$

for any $\varepsilon > 0$.

Thus the set of the functions verifying the condition on the left side of (7-1) is non-empty. Also let us notice that our choice of r and γ implies that

$$\gamma r = diam\ \Omega.$$

Now let ε^* be the number, depending of course on γ and r, which satisfies the assertion of Corollary IV.6.5.

In order to prove (7-1), let us suppose $\varepsilon \in (0,\varepsilon^*]$ and let $u^* \in M_\varepsilon$ be a function such that $E_\varepsilon(u^*) \leq m(\varepsilon,r)$.

We argue by contradiction and we assume that $x^* = \beta(u^*) \notin \Omega^+$. Then $\Omega \subset B_{diam\Omega}(x^*) \backslash B_r(x^*) = B_{\gamma r}(x^*) \backslash B_r(x^*)$.

Therefore

$$m^*(\varepsilon,r,\gamma) = \inf\Bigg\{\int_{B_{\gamma r}(x^*)\backslash B_r(x^*)} \left[\frac{1}{2}\,\varepsilon|\nabla u|^2 + \frac{1}{2}\,u^2 - F(u)\right] dx :$$

$$\int_{B_{\gamma r}(x^*)\backslash B_r(x^*)} \left[\varepsilon|\nabla u|^2 + u^2 - f(u)u\right] dx = 0;$$

$$u \in W_0^1(B_{\gamma r}(x^*)\backslash B_r(x^*));\quad \beta(u) = x^*\Bigg\} \leq E_\varepsilon(u^*) \leq m(\varepsilon,r).$$

that contradicts, by our choice of ε, (6-16).

□

The map ϕ_ε

For any $\varepsilon > 0$, let us define the operator

$$\phi_\varepsilon \; : \; \bar{\Omega} \; \longrightarrow \; W^1_0(\Omega)$$

by

$$\left[\phi_\varepsilon(y)\right](x) \; = \; \begin{cases} u_\varepsilon(|x-y|) & \forall \, x \in B_{2r}(y) \\ \\ 0 & \forall \, x \in \Omega \setminus B_{2r}(y) \end{cases}$$

where $u_\varepsilon(|x|)$ is a positive function, radially symmetric about the origin, such that $u_\varepsilon \in M_\varepsilon(B_{2r}(0))$, $E_\varepsilon(u_\varepsilon) = m(\varepsilon, 2r) < m(\varepsilon, r)$.

Note that ϕ_ε is continuous and that

$$\phi_\varepsilon(y) \in M_\varepsilon \quad , \quad \beta\left[\phi_\varepsilon(y)\right] = y.$$

IV.7.2 – **Lemma.** *Let $\varepsilon^* > 0$ be as in lemma 4.1, then for every $\varepsilon \in (0, \varepsilon^*]$*

$$\beta\left(M_\varepsilon^{m(\varepsilon, r)}\right) \subseteq \Omega^+ \quad , \quad \phi_\varepsilon(\bar{\Omega}^-) \subset M_\varepsilon^{m(\varepsilon, r)} \; ;$$

and

$$\beta \circ \phi_\varepsilon = j$$

where $j : \bar{\Omega}^- \longrightarrow \Omega^+$ denotes the embedding map, i.e.,

$$j(x) = x, \quad \forall \, x \in \bar{\Omega}^-$$

Proof. The proof is an immediate consequence of the relation (7-1) and the definition of ϕ_ε.

$\square$

IV.7.3 - **Theorem.** *Let* ε^* *be as in lemma 4.1 ; then for any* $\varepsilon \in (0,\varepsilon^*]$

$$(7\text{-}2) \qquad\qquad \mathcal{P}_t(M_\varepsilon{}^{m(\varepsilon,r)}) = \mathcal{P}_t(\Omega) + \mathcal{Z}(t),$$

where $\mathcal{Z}(t)$ *is a polynomial with non-negative coefficients.*

Proof. Since Ω, Ω^- and Ω^+ are homotopically equivalent, the conclusion follows from Lemma IV.7.2 and Th.I.2.2.(vii).

□

IV.8. THE TOPOLOGY OF THE STRIP.

By Theorem IV.7.3, we know that $M_\varepsilon{}^{m(\varepsilon,r)}$ has a topology at least as rich as the topology of Ω. Now using this estimate and Th.I.2.2.(vii), we can compute the Poincaré polynomial of suitable strips of $W_0^1(\Omega)$ and to prove Th. IV.4.1.

Computation of $\mathcal{P}_\lambda(E_\varepsilon{}^{m(\varepsilon,r)}, E_\varepsilon{}^\delta)$

First, let us compute the homology of the strip $E_\varepsilon{}^c \backslash E_\varepsilon{}^\delta$.

VI.8.1 - **Lemma.** *Let* ε^* *and* ε *be as in Th. IV.8.1,* $\delta \in (o, k_\varepsilon/2)$ *(k_ε is defined in lemma IV.5.2) and let $c \in (\delta, +\infty]$ be a noncritical level of E_ε; then*

$$(8\text{-}1) \qquad\qquad \mathcal{P}_\lambda(E_\varepsilon{}^c, E_\varepsilon{}^\delta) = \lambda \, \mathcal{P}_\lambda(M_\varepsilon{}^c).$$

Proof. Take two open neighborhoods U and V of M_ε, $\bar{U} \subset V$, and let χ be a C^1-function which is 1 in U and 0 out of V. Then define

$$F(u) = \nabla E_\varepsilon(u) - \chi(u) \frac{(\nabla E_\varepsilon(u), \nabla J_\varepsilon(u))}{\|\nabla J_\varepsilon(u)\|^2} \nabla J_\varepsilon(u)$$

and let be $\eta(t,u)$ be flow relative to the Cauchy problem

$$\begin{cases} \dfrac{d}{dt} \eta(t,u) = - \dfrac{F(u)}{1 + \|F(u)\|} \\[2mm] \eta(0,u) = u \end{cases}$$

It is easy to check that the Cauchy problem is well posed and $\eta(t,u)$ is defined for every $t \in \mathbb{R}$ and every $u \in W_0^1(\Omega)$. Moreover, if V is sufficiently small and $\nabla E_\varepsilon(u) \neq 0$, then

$$\frac{d}{dt} E_\varepsilon(\eta(t,u)) < 0.$$

Now set

$$W = \left\{ u \in (E_\varepsilon)_a^b \mid \forall\, t \geq 0,\ \eta(t,u) \in (E_\varepsilon)_a^b \right\}.$$

Notice that, by the construction of our flow, we have that

(8-2) $$W = (E_\varepsilon)_a^b \cap M_\varepsilon = M_\varepsilon^b$$

Hence M_ε^b is a manifold of codimension 1, positively invariant in $(E_\varepsilon)_a^b$. So, by the theorem I.3.8.(x), with $\mathcal{M} = \mathrm{int}\,\overline{(E_\varepsilon)_a^b}$ and $\bar{N} = M_\varepsilon^b$,

$$\mathcal{P}_\lambda(E_\varepsilon^c, E_\varepsilon^\delta) = I_\lambda((E_\varepsilon)_a^b) = \lambda\, I_\lambda(M_\varepsilon^b)$$

□

VI.8.2 – Corollary. *Let ε^*, ε and δ be as in lemma IV.5.2; then*

(8-3) $$\mathcal{P}_\lambda(E_\varepsilon^{m(\varepsilon,r)}, E_\varepsilon^\delta) = \lambda \mathcal{P}_\lambda(\Omega) + \lambda Z(\lambda)$$

(8-4) $$\mathcal{P}_\lambda(W_0^1(\Omega), E_\varepsilon^\delta) = \lambda \mathcal{P}_\lambda(M_\varepsilon) = \lambda$$

where $\mathcal{Z}(\lambda)$ is a polynomial with non–negative integer coefficients

Proof. As we have already observed in the proof of Th. IV.5.1, we can assume without loss of generality that $m(\varepsilon,r)$, is not a critical level of E_ε. Then (8-3) follows from (7-2) and (8-1). (8-4) follows from (8-2) and the fact that M_ε is contractible, so *dim* $H_k(M_\varepsilon) = 1$ if $k = 0$ and *dim* $H_k(M_\varepsilon) = 0$ if $k \neq 0$.

$\square$

Computation of $\mathcal{P}_\lambda(W_0^1(\Omega), E_\varepsilon^{m(\varepsilon,r)})$

Next we will compute the homology of the strip $W_0^1(\Omega) \backslash E_\varepsilon^{m(\varepsilon,r)}$

VI.8.3 – **Lemma.** *Let ε^* and ε be as in lemma IV.5.2; then*

(8-5) $$\mathcal{P}_\lambda(W_0^1(\Omega), E_\varepsilon^{m(\varepsilon,r)}) = \lambda^2 \left[\mathcal{P}_\lambda(\Omega) + \mathcal{Z}(\lambda) - 1 \right]$$

where $\mathcal{Z}(\lambda)$ is a polynomial with non–negative integer coefficients

Proof. Let δ be as in Corollary *VI.8.2* and consider the exact sequence,

$$\rightarrow H_k(W_0^1(\Omega), E_\varepsilon^\delta) \xrightarrow{\ j_k\ } H_k(W_0^1(\Omega), E_\varepsilon^{m(\varepsilon,r)}) \xrightarrow{\ \partial_k\ }$$

$$\xrightarrow{\ \partial_k\ } H_{k-1}(E_\varepsilon^{m(\varepsilon,r)}, E_\varepsilon^\delta) \xrightarrow{\ i_{k-1}\ } H_{k-1}(W_0^1(\Omega), E_\varepsilon^\delta) \longrightarrow$$

Thus, for $k > 2$,

$$dim \left[H_k(W_0^1(\Omega), E_\varepsilon^{m(\varepsilon,r)}) \right] = dim \left[H_{k-1}(E_\varepsilon^{m(\varepsilon,r)}, E_\varepsilon^\delta) \right]$$

For $k = 2$, we have

$$\rightarrow H_2(W_0^1(\Omega),E_\varepsilon^\delta) \xrightarrow{\ j_2\ } H_2(W_0^1(\Omega),E_\varepsilon^{m(\varepsilon,r)}) \xrightarrow{\ \partial_2\ }$$

$$\xrightarrow{\ \partial_2\ } H_1(E_\varepsilon^{m(\varepsilon,r)},E_\varepsilon^\delta) \xrightarrow{\ i_1\ } H_1(W_0^1(\Omega),E_\varepsilon^\delta) \longrightarrow$$

since, i_1 is an isomorphism,

$$H_2(W_0^1(\Omega),E_\varepsilon^{m(\varepsilon,r)}) = j_2\left[H_2(W_0^1(\Omega),E_\varepsilon^\delta)\right] = 0.$$

For $k = 1$, we have

$$\xrightarrow{\ \partial_2\ } H_1(E_\varepsilon^{m(\varepsilon,r)},E_\varepsilon^\delta) \xrightarrow{\ i_1\ } H_1(W_0^1(\Omega),E_\varepsilon^\delta) \xrightarrow{\ j_1\ }$$

$$H_1(W_0^1(\Omega),E_\varepsilon^{m(\varepsilon,r)}) \xrightarrow{\ \partial_1\ }$$

$$H_1(W_0^1(\Omega),E_\varepsilon^{m(\varepsilon,r)}) = 0.$$

Moreover

$$H_0(W_0^1(\Omega),E_\varepsilon^{m(\varepsilon,r)}) = 0.$$

By the above formulas and $(8\text{-}1)$, we conclude that

$$P_\lambda(W_0^1(\Omega),E_\varepsilon^{m(\varepsilon,r)}) = \lambda\left[P_\lambda(E_\varepsilon^{m(\varepsilon,r)},E_\varepsilon^\delta) - \lambda\right] =$$

$$= \lambda^2\left[P_\lambda(\Omega) + Z(\lambda) - 1\right].$$

$\square$

Proof of the main results

*VI.8.4 – Lemma. Let ε^* be as in lemma IV.5.1 ; suppose
that*

(i) $\varepsilon \in (o,\varepsilon^*]$

*(ii) the set K of nontrivial solutions of problem (P_ε)
is discrete;*

then

$(5\text{-}6)$
$$\sum_{u \in \mathfrak{C}_1} i_\lambda(u) = \lambda\, P_\lambda(\Omega) + \lambda\, Z(\lambda) + (1+\lambda)\, Q_1(\lambda)$$

$(5\text{-}7)$
$$\sum_{u \in \mathfrak{C}_2} i_\lambda(u) = \lambda^2\left[P_\lambda(\Omega) + Z(\lambda) - 1\right] + (1+\lambda)\, Q_2(\lambda)$$

where

$$\mathfrak{C}_1 = \left\{ u \in \mathcal{K} \mid \delta < E_\varepsilon(u) \leq m(\varepsilon,r) \right\}$$

and

$$\mathfrak{C}_2 = \left\{ u \in \mathcal{K} \mid E_\varepsilon(u) > m(\varepsilon,r) \right\}$$

Proof. It follows from theorem I.5.9

□

Now we are ready to prove theorem IV.4.1

Proof of theorem IV.4.1 - Choose ε^* and ε as in lemma VI.8.3. Since E_ε does not have any non-zero solution below the level δ, $\mathcal{K} = \mathfrak{C}_1 \cup \mathfrak{C}_2$, then

$$\sum_{u \in \mathcal{K}} i_\lambda(u) = \sum_{u \in \mathfrak{C}_1} i_\lambda(u) + \sum_{u \in \mathfrak{C}_2} i_\lambda(u)$$

The conclusion follows by lemma IV.5.7.

□

REFERENCES

[Am,82] Amann H., *A note on degree theory for gradient mappings,* Proc. Amer. Math. Soc. <u>85,</u> (1982), 591-597.

[Am,Ra,73] Ambrosetti A., Rabinowitz P. H., *Dual variational methods in critical point theory and applications,* J. Funct. Anal., **44**, (1973), 349-381.

[Av,63] Avez A., *Essais de geometrie Riemannienne hyperbolique globale. Application a la Relativite' Generale,* Ann. Inst. Fourier.132,(1963), 105-190.

[Ba,Be,Fo,83] Bartolo P., Benci V., Fortunato D., *Abstract critical point theorems and applications to some nonlinear problems with "strong resonance" at infinity'*, Journal of Nonlinear Anal. T.M.A., **7**, (1983), 981-1012.

[Be,He,81] Beem J.K., Herlich E., *Global Lorentzian Geometry*. New York-Basel, Marcel Dekker Inc. 1981.

[Be,82] Benci V., *On critical point theory for indefinite functionals in the presence of symmetries,* Trans. Amer. Math. Soc. 274, (1982), 533-572.

[Be,87] Benci V., *Some applications of the generalized Morse-Conley theory,* Conferenze del Seminario di Matematica dell' Università di Bari, 218, Laterza, (1987).

[Be,91] Benci V., *A new approach to the Morse-Conley theory and some applications,* Ann. Mat. Pura Appl. (iv), <u>158</u>, (1991), 231-305.

[Be,Ce,91] Benci V., Cerami G., The effect of the domain topology on the number of positive solutions of nonlinear elliptic problems, Arch.Rational Mech.Anal. 114 (1991), 79-93

[Be,Ce,ta] Benci V., Cerami G., *Multiple positive solutions of some elliptic problems via the Morse theory and the domain topology,* to appear.

[Be,Fo,86] V.Benci, D.Fortunato, *Subharmonic solutions of prescribed period for non-autunomous differential equations,* in "Recent advances in Hamiltonian Systems", Dell'Antonio and D'onofrio ed., World Scientific Publishing, Singapore (1987), 1-52.

[Be,Fo,89] V.Benci-D.Fortunato, *A remark on the nodal regions of the solutions of some superlinear elliptic equations,* Proc. Roy. Soc. Edimb. <u>111A,</u> (1989), 123-128.

[Be,Fo,90] Benci V., Fortunato D., *Existence of geodesics for the Lorentz metric of a stationary gravitational field.* Ann. Inst. H. Poincare'. Analyse nonlineaire.7,(1990),27-35.

[Bè,Fo,ta] Benci V., Fortunato D., *Geodesics on space-time manifolds: a variational approach,* to appear.

[Be,Fo,Gi,91] Benci V., Fortunato D., Giannoni F. *On the*

existence of multiple geodesics in static space-time. Ann. Inst. H.Poincare'. Analyse nonlineaire. **8** ,(1991),79-102.

[Be,Fo,Gi,92] Benci V., Fortunato D., Giannoni F. *On the existence of geodesics in static Lorentz manifolds with singular boundary.* Ann. Scuola Norm. Sup. Pisa. **19** , (1992).

[Be,Ma,92] Benci V., Masiello A., *A Morse Index for geodesics in static Lorentz manifolds,* Math. Annalen.,293, (1992),433-442.

[Be,Fo,Ma,ta] Benci V., Fortunato D., Masiello A., to appear

[Be,Gi,92] V.Benci, F.Giannoni, *On the existence of closed geodesics on noncompact Riemannian manifolds,* Duke Mathematical Journal, (1992).

[Be,Gi,92] V.Benci, F.Giannoni, *Morse theory for functionals of class* C^1, C. R. Acad. Sci. Paris, 315, Serie I, (1992), 883-888.

[Be,Ma,92] Benci V., Masiello A., *A Morse index for geodesics in static Lorentz manifolds,* Math. Annalen, **293**, (1992), 433-442.

[Be,Ra,79] V. Benci, P. H. Rabinowitz - *Critical point theorems for indefinite functionals* , Inv. Math., **52**, (1979), 336-352.

[Bo,82] Bott R., *Lectures on Morse theory old and new.* Bull. Am. Math. Soc.,6,(1982),331-358.

[Ch,86] Chang K.C - Infinite dimensional Morse theory and its applications, Seminaire de Mathematiques Supérieures, 97, Presses de L'Université de Montréal, (1986)

[Co,78] C.Conley, *Isolated invariant sets and the Morse index,* C.B.M.S., Reg. Conf. Ser. in Math., 38, Amer. Math. Soc., Providence, RI, (1978).

[Co,Ze,84] C.Conley, E. Zahnder, *Morse type index theory for flows and periodic solutions For Hamiltonian equations,* Comm. Pure Appl. Math, 27, (1984), 211-253.

[De,ta] G. F. Dell' Antonio, *Variational calculus and stability of periodic solutions of a class of Hamiltonian systems,* t.a.

[De,DO,Ek,ta] G. F. Dell' Antonio, B:M: D'Onofrio, I. Ekeland, *Les systémes hamiltoniens convexes et pairs ne son pas ergodiques en gèneéral, CRAS,* to appear.

[Do,72] A.Dold, *Lectures on Algebraic Toplogy,* Springer-Verlag, Berlin-Heidelberg-New-York, (1972).

[Ek,84] I. Ekeland, *Une théorie de Morse pour les systèmes hamiltoniens convexes,* Analyse non linéaire, 1, (1984), 19-78.

[Ek,90] I. Ekeland, *Convexity Methods in Hamiltonian Mechanics,* Springer-Verlag, Berlin-Heidelberg-New-York, (1980).

[Ge,70] Geroch R.,*Domains of dependence.* J.Math.Phys.11,(970),437-449
[28] Hawking S. W., Ellis G. F.R., *The large scale structure of space-time.* Cambridge University Press (1973)

[Gi,Ni,Ni,80] Gidas B - Ni W. M.- Nirenberg L. - *Symmetry of positive solutions of nonlinear elliptic equations in* $\mathbb{R}^N$. Mathematical Analysis and Applications - Part A - Advances in Mathematics Supplementary studies. Vol. 7A, 369-402.

[Gr,Ha,81] M.J.Greenberg-J.R.Harper, *Algebraic Topology - a first course,* Addison-Wesley P.C., Redwood City, CA, (1981).

[Ho,85] H.Hofer, *A geometric description of the neighbourhood of a critical point given by the Mountain Pass Theorem,* J.London Math. Soc. (2), 31, 566-570, (1985).

[Kl,82] Klingenberg W.,*Riemannian Geometry.* Berlin-New York Walter de Gruiter (1982).

[Ma,Pr,75] A.Marino-G.Prodi, *Metodi perturbativi nella teoria di Morse,* Boll. U.M.I. (4) 11, suppl. fasc. 3, (1975), 1-32.

[Ma,90] Masiello A., *Metodi variazionali in geometria*

lorenziana. Tesi Dott. Ricerca. Univ. Pisa (1990).

[Mi,63] Milnor J.,*Morse Theory.* Ann. Math. Studies **51**,(1963),Princeton University Press.

[ON,83] O' Neill, *Semiriemannian Geometry with application to Relativity,* Academic Press Inc., New York-London, 1983.

[Pa,63] R.S.Palais, *Morse theory on Hilbert manifolds,* Topology, <u>2</u>, (1963), 299-340.

[Pa,66,a] R.S.Palais, *Lusternik-Schnirelmann theory on Banach manifolds,* Topology, <u>5</u>, (1966), 115-132.

[Pa,66,b] R.S.Palais, *Homotopy theory of infinite demensional manifolds,* Topology, <u>5</u>, (1966), 1-16.

[Pa,Sm,64] R.S.Palais, S, Smale, *A Generaliztion of Morse theory,* Bull. A.M.S., <u>70</u>, (1964), 165-172.

[Pe,72] Penrose R., *Techniques of differential topology in Relativity,* Conf. Board Math. Sci. **7**,(1972),S.I.A.M.

[Ra,78,a] P.H.Rabinowitz, *Some minimax theorems and applications to nonlinear partial differential equations,* Nonlinear Analysis, a collections of papers in honor of Erich Rothe, Academic Press, New York, (1978), 161-177.

[Ra,78,b] P.H.Rabinowitz, *Some critical point theorems and applications to semilinear elliptic partial differential equations,* Ann. Sc. Norm. Sup. Pisa, (4), 5, (1978), 215-223.

[Ra,86] P.H.Rabinowitz, *Minimax methods in critical point theory with applications to differential equations* C.B.M.S., Reg. Conf. series in Math, **65,** Amer. Math . Soc., Providence, RI, 1986.

[Sc,69] Schwartz J.,*Nonlinear functional analysis,* Gordon and Breach, New York, 1969.

[Se,51] Serre, *Homologie Singulière des Espaces Fibrès,* Ann. Math. <u>54</u>, (1951).

[Se,70] Seifert H.J., *Global connectivity by time-like geodesics,* Z. Naturefor. 22a, (1970), 1356.

[St,77] Strauss W., *Existence of solitary waves in higher dimensions*, Comm. Math. Phys. 55, (1977), 149-162.

[St,90] Struwe M., *Variational Methods: applications to nonlinear partial differential equations and Hamiltonian systems,* Springer-Verlag, Berlin, Heidelberg, New-York, 1990.

[Uh,75] Uhlenbeck K., *A Morse theory for geodesics on a Lorentz manifold*, Topology **14**, 69-90, 1975

[V,64] Vainberg M. M., *Variational methods for the study of nonlinear operators* , Holden Day, (1964).

Applications of Singularity Theory to the Solutions of Nonlinear Equations

James Damon

Department of Mathematics
University of North Carolina
Chapel Hill, NC 27514

The past twenty-five years has seen major strides in our understanding of the solutions to nonlinear ordinary and partial differential equations. This progress has built on earlier foundational work which was often concerned with, for example, establishing the existence of at least one solution for a nonlinear equation or determining the parameter values at which a solution bifurcates. These questions have been refined to ones involving more precise information about the solutions. This includes the precise properties of the set of solutions such as: the exact description of the bifurcation of equilibrium points and limit cycles for ordinary differential equations, the multiplicity of solutions to nonlinear elliptic equations, the number of periodic solutions appearing from perturbations of Hamiltonian systems, and the nature of the branching of such solutions. At the same time, the methods lead to increased knowledge about the nature of the solutions to nonlinear equations, including symmetry properties of solutions to nonlinear PDE's as well as properties of steady state solutions to evolution equations such as reaction–diffusion equations.

These developments arise from recent refinements in the topological analysis of nonlinear operators, which have largely involved variational methods based on variants of Morse theory, various notions of degree theory, and bifurcation theory, including equivariant versions of these theories. In each case, these methods can be thought of as either: extentions to the infinite dimensional case of methods used for mappings between finite dimensional spaces, or as methods for actually reducing questions about nonlinear operators to ones about associated mappings between finite dimensional spaces.

Many of these problems such as solutions for nonlinear Dirichlet problems or periodic solutions for various nonlinear oscillations can be represented as the solutions of an equation involving a nonlinear Fredholm operator between infinite dimensional function spaces. Then the classical Lyapunov–Schmidt procedure allows us to determine the local

---

[1] Partially supported by a grant from the National Science Foundation

solution set as that of an associated nonlinear mapping between finite dimensional spaces. If we use the "Full Lyapunov-Schmidt Reduction", we can furthermore represent the nonlinear operator locally as an -infinite dimensional unfolding-, i.e. an infinite dimensional parametrized family of finite dimensional mappings. Often the original operator contains special information relevant to the problem; examples are: distinguished roles played by certain parameters, symmetries, constraints, or special properties of the solutions such as being sections of bundles with special properties. This information is often retained in the Lyapunov-Schmidt reduction.

An important question for nonlinear analysis is how readily we can determine the properties of the associated finite dimensional mappings, preserving the special conditions when present, to yield conclusions for the original nonlinear operators. The study of the properties of smooth mappings between finite dimensional manifolds is exactly the domain of singularity theory. Inspired by the earlier work of Morse and Whitney, Thom envisioned a general theory for smooth mappings between manifolds analogous to that provided by Morse theory for real valued functions on manifolds. Thom's ideas for such a theory were explained in the "Thom-Levine Notes" [ThL] as well as in several fundamental papers on the (topological) classification of smooth mappings [Th2], [Th3], [Th4]. In addition, Thom has often been described as a catalyst in the fundamental work of Malgrange and Whitney on the Malgrange preparation theorem and the Whitney stratification theorem for complex analytic varieties.

However, it was Mather who, in his series of fundamental papers, brought Thom's ideas to full fruition going well beyond what Thom envisioned by completely developing the theory of the structural stability of C^∞-mappings.

Independently, V.I. Arnol'd, with his deep interest in differential equations and more generally applied mathematics, developed a school in Moscow devoted to the application of ideas from singularity theory to various questions from applied mathematics such as the behavior of caustics of wave fronts, properties of oscillatory integrals, bifurcation in differential equations, obstacle problems, etc. Because we shall concentrate on the analysis of nonlinear operators, we shall placing greater emphasis on the applications of the results of Thom and Mather rather than those of Arnol'd and his coworkers.

During the past twenty five years these fundamental ideas and results of Thom and Mather in singularity theory have sparked an explosion in our understanding of smooth nonlinear mappings in a variety of contexts. Workers have used Mather's fundamental results as a paradigm for developing modified theories applicable in varied settings. We shall describe these recent dramatic advances, explaining how these results translate into significant results for the local behavior of nonlinear operators and pointing to further potential applications for nonlinear problems.

Our overall aim here is to survey the results of singularities during this period which are applicable to the solutions of nonlinear equations. Hopefully our explanations will be of value to workers in areas other than singularities; so that someone in an area such as nonlinear analysis will see the applicability of the fundamental ideas of singularity theory for nonlinear problems. Besides recounting some of the developments as they historically occurred, providing a platform for later developments, we have also gone into considerable depth in several examples (especially §4) to illustrate how exactly the theorems and ideas are applied.

Specifically, in §2 and 3 we shall highlight the key results of Mather's theory of $C\infty$-stability of mappings, including: stability, finite determinacy, and the universal unfolding theorem as presented by Martinet. We will indicate in §4 how the local results extend to variants of the theory, which introduce an equivalence using groups of diffeomorphisms preserving special conditions such as those already described. The essence of these results is their reduction to infinitesimal conditions which are algebraic in nature. We indicate general principles for the applicability of such methods. The applications of the theory will be illustrated for two main examples: imperfect bifurcation theory, including symmetries, and periodic solutions of Hamiltonian systems.

The discussion in § 2, 3 and 4 would make it appear that in singularity theory workers have a compulsion to write down algebraic expressions for tangent spaces of virtually anything. In fact, there is probably some truth to this. Unable to entirely resist this urge, this author hopes that an interested reader will nonetheless find some justification for this, especially in the examples of §4. The reader is encouraged to freely skip over such details, but hopefully still follow the general ideas being described.

Much more concrete questions are considered in § 5 and 6, even though underlying them are the algebraic results from the earlier sections. In §5, we shall describe how certain nonlinear operators can be represented as -infinite dimensional versions- of finite dimensional stable mappings, and how the global finite dimensional theory of Mather leads quite generally to such an explicit local description. The finite dimensional representations allow us to draw conclusions about the multiplicities of solutions for nonlinear operators.

In §6, we contrast the known results for computing multiplicities of solutions in the complex case versus the more difficult real case, which is the important case for concrete applications. There are two approaches for the real case: computing the local degree using a theorem of Eisenbud–Levine , and in certain cases explicitly computing the number of solutions for stable germs as the algebraic dimension of its local algebra via a result of Galligo and the author. We describe how these results naturally lead to further generalizations for various branching phenomena in bifurcation theory including equivariant bifurcation, and equivariant degrees. Here the local results overlap with the global results obtained by workers in nonlinear analysis.

We shall conclude in §7 with a description of how the preceding results have been further extended to allow topological equivalence of smooth mappings. The presence of moduli, which parametrize continuous change under differentiable equivalence, forces us to consider such weaker topological equivalences. We indicate how these problems have been lurking in the background in even the simplest examples of the earlier sections. The topological analogues of Mather's results which have been developed are applied to these cases, actually simplifying the applicability for certain nonlinear problems.

The author would like to express his gratitude to Professors Alfonso Vignoli and Michele Matzeu for their very generous hospitality and infectious enthusiasm which made the conference at Frascati as thoroughly enjoyable as it was mathematically stimulating (and for their patience in awaiting the completion of this paper).

Contents

§1   **The Full Lyapunov–Schmidt Reduction**

Many problems can be reduced to the solution of an equation involving a nonlinear operator between spaces of functions which incorporate boundary conditions and differentiability conditions in the definition of the spaces. Two standard examples are :

(1.1)   the nonlinear Dirichlet problem on a domain $\Omega \subset \mathbb{R}n$ with smooth boundary.

$$\Delta u + h(u,x) \qquad = \ g(x) \qquad \text{on } \Omega$$
$$u| \, \partial\Omega \qquad\qquad = \ 0 \qquad \text{on } \partial\Omega.$$

For this problem we seek solutions to the equation $F(u) = g$, for the operator

$$F(u) \ = \ \Delta u + h(u,x) \ : \ C_0^{2+\alpha}(\bar\Omega) \ \longrightarrow \ C^{\alpha}(\bar\Omega)$$

(with the usual notation for Holder spaces of functions with "0" denoting those functions satisfying the boundary condition).

(**1.2**) <u>periodic solutions for forced nonlinear oscillations</u>

$$x" \ + \ \lambda x + P(x,x',t) \ = \ q(t)$$

where both $P(x,x',t)$ and $q(t)$ are $2\pi/\lambda$ periodic in t. This includes Van der Pol, Lienard, and Duffing equations among others. For simplicity in what follows we let $\lambda = 1$. Then, the solutions are those for the equation $F(x) = q$, where

$$F(x) \ = \ x" + x + P(x,x',t) \ : \ C_{2\pi}^{2}(\mathbb{R}) \ \longrightarrow \ C_{2\pi}^{0}(\mathbb{R})$$

where $C_{2\pi}^{k}(\mathbb{R})$ denotes the C^k 2π–periodic functrions on $\mathbb{R}$.

Now, we consider such nonlinear operators $f : E \longrightarrow F$, with E and F denoting appropriate Banach spaces of functions. Ideally we would like to understand the structure of such nonlinear operators by "visualizing" exactly how the operator f, viewed now as a nonlinear mapping, would map the space E to the space F. This "infinite dimensional vision" would allow us to see: how parts of F were covered multiple times by the image of E; which regions of F were entirely missed by the image of E ; and how these regions were divided by separating (nonlinear) subspaces. To see what we have in mind, we consider a finite dimensional example

(1.3) $\qquad$ $f(x, y) = (x^2 - y^2 + \varepsilon \cdot x, 2xy - \varepsilon \cdot y)$ $\qquad$ for small ε

This is the map $z \mapsto z^2 + \varepsilon \cdot \bar{z}$ (considered by Arnol'd) expressed in real coordinates. The critical set, where the derivative df is not invertible is given by $x^2 + y^2 = \varepsilon^2/4$. Its image is the hypocycloid shown in fig. 1.4 (a). We see that circles of radius less than $\varepsilon/2$ map homeomorphically to a closed curve inside the hypocycloid while those of radius greater than $\varepsilon/2$ map to increasingly larger pretzel shaped curves as shown in fig. 1.4 (b). With a bit of effort we can furthermore see that inside the hypocycloid there are four preimages of each point, outside there are two preimages, and on the nonsingular points of the curve there are three (the exact justification for this, without explicitly solving the equations, follows from the local classification for stable singularities (§3) plus global information given by the degree of the mapping (§6)).

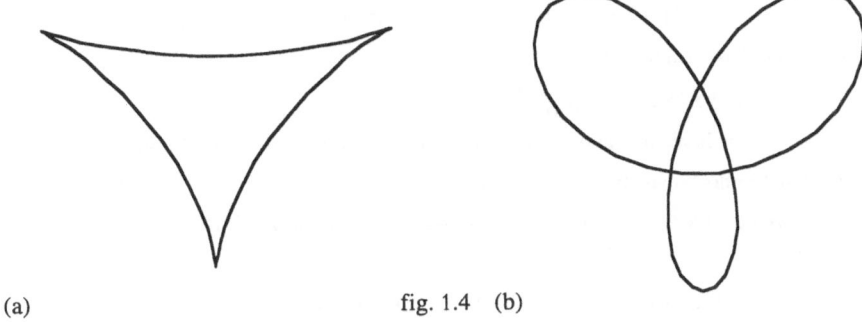

(a) $\qquad\qquad\qquad\qquad\qquad\qquad\qquad\qquad$ fig. 1.4 $\quad$ (b)

$\qquad$ It would be extremely valuable to have similar "vision" for nonlinear mappings between infinite dimensional spaces. A first step is to determine the local behavior of the mapping f in a neighborhood of a point of $x_0 \in E$.

$\qquad$ Suppose $f(x_0) = y_0$. If we are interested in the solutions of the equation $f(x) = y_0$ for x near x_0, then we are led to the classical Lyapunov–Schmidt reduction to a corresponding equation involving a finite dimensional mapping (see e.g. [N]). For this reduction we consider an f which is C^r ($1 \leq r \leq \infty$) in a neighborhood of x_0 (in the Frechet differentiable sense), and which is Fredholm so that $df(x_0) : E \longrightarrow F$ is a Fredholm operator. Thus, $F_1 = \text{image}(df(x_0))$ is a closed subspace of finite codimension with finite dimensional complement F_0, and $\ker(df(x_0)) = E_0$ is finite dimensional with closed complement E_1. By the open mapping theorem $df(x_0)| E_1$ maps E_1 invertibly to F_1.

$\qquad$ Classically the Lyapunov–Schmidt reduction is applied by first decomposing

$$y_0 = P(y_0) + (I-P)(y_0) = y_1 + y_2,$$

where $P: F \to F_1$ denotes projection along F_0, and similarly $x_0 = x_1 + x_2$. Then,

$$f(x) = y_0 \qquad \text{iff} \qquad P \circ f(x) = y_1 \quad \text{and} \quad (I-P) \circ f(x) = y_2.$$

We may then apply the implicit function theorem to $\tilde{f} = P \circ f$ so that in a neighborhood of x_0, $\tilde{f}^{-1}(y_1) = \text{graph}(g)$ for $g: U \to E_1$ with U a neighborhood of x_1 in E_0. Then, if $f_0(x') = (I-P) \circ f(x', g(x'))$, $f_0: U \to F_0$; and via g, $f^{-1}(y_0)$ is identified with $f_0^{-1}(y_2)$. Hence, the determination of the solution set for $f(x) = y_0$ has been locally reduced to a problem for the finite dimensional mapping f_0.

This provides a glimpse of the local structure of f from $f^{-1}(y_0)$. However, if we back up a bit we can do considerably better and obtain the

"*Full Lyapunov–Schmidt Reduction*".

Consider $\Phi: E_0 \oplus E_1 \to E_0 \oplus F_1$ defined by

$$\Phi(x', x'') = (x', \tilde{f}(x',x'')) \qquad \text{for} \quad x = (x', x'') \in E_0 \oplus E_1.$$

A simple calculation shows that

$$d\Phi(x_0) = \text{id}_{E_0} \oplus df(x_0)$$

so that $d\Phi(x_0)$ is invertible and so the inverse function theorem applies to Φ; hence, Φ is a C^r-local diffeomorphism and Φ^{-1} is defined and C^r in a neighborhood U' of $\Phi(x_0)$.

If we consider $f_1 = f \circ \Phi^{-1}: U' \to F$, we see that it has the form

(1.5) $$f_1(x',u) = (\bar{f}_1(x', u), u) \qquad \text{for} \quad (x', u) \in E_0 \oplus F_1.$$

(this u is quite distinct from the function u in (1.1)).

In fact, from the definition of Φ, we can compute the F_1-component of $f \circ \Phi^{-1}$ by

$$P \circ (f \circ \Phi^{-1}) = (P \circ f) \circ \Phi^{-1} = (P \circ \Phi) \circ \Phi^{-1} = P.$$

Now we have considerably more information about f. First, by the definition of f_1, Φ locally maps $f^{-1}(y_0)$ to $f_1^{-1}(y_0)$. Hence, we still can recover the local structure of $f^{-1}(y_0)$ from f_1. In fact, since $y_0 = y_1 + y_2$, if we let $f_0(x') = \bar{f}_1(x', y_2)$ and $U'' = U' \cap E_0$ then $f_0: U'' \to F_0$ and Φ locally maps $f^{-1}(y_0)$ to $f_0^{-1}(y_1)$ (and this f_0 is the same as above). However, we still have the complete local structure of f since f_1 only differs from it by a local diffeomorphism. Moreover, we see that this form allows us to begin visualizing the infinite dimensional mapping, at least locally. We begin with the value $u = y_2$ and obtain the mapping f_0 ; then as we vary u (which we can think of as a parameter varying in the parameter space F_1) we obtain the perturbations $\bar{f}_1(x', u)$ of f'_0. The mapping f_1 simultaneously keeps track of the pertubations as well as the parameter values u. In the finite dimensional case a mapping given in the form (1.5) is called an *unfolding* of the initial mapping f_0.

Usually we will apply translations sending 0 to x_0 and y_0 to 0 so we may always assume that $f(0) = 0$ and view f as being defined locally near 0. Of course, after translation, a simple operator such as

$$f(x(t)) = x'' + x + a_1(t) \cdot x^2 : C_{2\pi}^2(\mathbb{R}) \longrightarrow C_{2\pi}^0(\mathbb{R}) \quad \text{with } a_1(t) \in C_{2\pi}^0(\mathbb{R})$$

would become

$$f_1(x(t)) = x'' + a_2(t) \cdot x + a_1(t) \cdot x^2 + \quad \text{(where } a_2(t) = 1 + 2a_1(t) \cdot x_0(t)\text{)}.$$

This may considerably complicate the local analysis; for example, even if a_1 is constant, the first derivative of the operator f_1 may be a time dependent operator, yielding a Mathieu-type differential equation. This is the intrinsic nature of such nonlinear operators.

Moreover, for local considerations, if we have two distinct mappings which agree in a neighborhood of 0 then we do not wish to distinguish between them. Local mappings under this equivalence of agreeing on some neighborhood of the point 0 in question are referred to as germs at the point. We denote the *germ of* f *at* 0 by f: E, 0 $\rightarrow$ F, 0. When f is an unfolding we represent it in the form $f(x, u) = (\bar{f}(x, u), u)$ with $\bar{f}(x, 0) = f_0(x)$ and say it is an unfolding of f_0.

This special form considerably simplifies deducing properties of the mapping; as an example, an immediate corollary is the infinite dimensional analog of the constant rank theorem for finite dimensional mappings.

Corollary 1.6 (Constant Corank Theorem): *Suppose that* f: E, 0 $\rightarrow$ F, 0 *is a* C^r *Fredholm mapping which has constant corank* = k *in a neighborhood of* 0. *Then there exists a neighborhood* U' *of* 0 *such that* f(U') *is a* C^r *submanifold of* F *of co-dimension* k.

Proof: After the Lyapunov–Schmidt reduction, we may assume that f has the form of an unfolding $f(x, u) = (\bar{f}(x, u), u)$ where for $(x, u) \in E_0 \oplus F_1$, with $\dim_{\mathbb{R}} E_0 = k$ and $\bar{df}(0) \equiv$

0. As df has constant corank = k, $d_x\bar{f}(x,u) \equiv 0$ for all (x, u) in a neighborhood U of 0. However, by standard calculus, this implies that $\bar{f}$ is independent of x, and so is a function of u on this neighborhood. Thus, the image of f on a neighborhood $U' = U_1 \times U_2 \subset U$ is $\{(\bar{f}(u), u) : u \in U_2\}$ which is the graph of the C^r function $\bar{f}$.   □

For the general purpose of understanding the structure of f, we can always replace f by f_1 so we may as well assume that f itself has the form of an unfolding. There are several natural questions about the unfolding f which lead to the subject of singularity theory proper. These include the following.

i)   Versality : What can we say about the types of perturbations which are possible starting from a given f_0?

ii)   Stability : In particular, there is an infinite dimensional space for the parameters u. Are there circumstances when we can effectively say that certain of these parameters are redundant (in some sense)?

iii)   Classification : When can we deduce the specific form which f must take?

iv)   Determinacy : When can we add higher order terms to f beginning with some order s and know that we have not changed the local structure of f near 0?

v)   Computability : How can we use derivative computations involving f to obtain information about the preceding questions, i.e. can we deduce answers to i) – iv) from finite Taylor expansion information about f_0 and f ?

Remark: It is in the final question v) that the subtlety of the Lyapunov–Schmidt reduction comes into play. The unfolding structure only becomes apparent when we have obtained $f_1 = f_0 \circ \Phi^{-1}$. To perform derviative computations on f_1 we must use the chain rule for $d^j(f_0 \circ \Phi^{-1})$ in terms of the derivatives $d^k(f)$ and $d^\ell(\Phi^{-1})$ with k and $\ell \leq j$. In turn, by the chain rule applied to $d^\ell(\Phi \circ \Phi^{-1})$, the derivatives $d^\ell(\Phi^{-1})$ can be expressed in terms of $d^m(\Phi)$ with $m \leq \ell$. Then, we can compute

$$(1.7) \qquad d^m(\Phi)(0) = \begin{cases} \mathrm{id}_{E_0} \oplus P \circ df(0) & m = 1 \\ \\ P \circ d^m f(0) & m > 1. \end{cases}$$

However, the formulae for the compositions can become quite complicated (see e.g. [Fr] and [Rg]). We shall see that singularity theory does allow us to answer the above questions for a particular problem using finite derivative information; however, our effectiveness in using it may be limited by our computational ability.

§2   Mather's Theory of C^∞ - Stability of Mappings - Global Theory

We have hinted in the preceding discussion that to "visualize", at least locally, nonlinear operators viewed as infinite dimensional mappings we would have to understand the perturbations of mappings between finite dimensional spaces. We describe Mather's theory of C^∞-stability of mappings, indicating how it exactly allows us to answer the questions raised earlier in applying the Lyapunov–Schmidt procedure. It as well provides the general scheme for the many extensions of this theory which we will consider later.

To understand why the theory has taken its particular form, we begin by recalling several key elements of Morse theory, concentrating on the properties of Morse functions. Although the basic facts about Morse theory are well-known, we wish to stress its role as a precursor for the general theory of stable mappings.

The truly remarkable fact about Morse theory is that so much information can be gained about a manifold M from information provided by an "appropriately chosen" smooth function f: M → ℝ. Contrary to our expectations, "appropriately chosen" does not mean that f has to be chosen from among a very select set of functions; but rather that "almost any function" will have the desired properties. It is this surprising fact that almost any function will have the especially desirable properties that is so remarkable; so we proceed to explain exactly what we mean.

First, we mention that even some smooth functions can have truly bizarre and undesirable properties. A theorem of Whitney [Wh1] asserts that given any closed subset $A \subset \mathbb{R}^n$, there is a smooth function f: $\mathbb{R}^n \to \mathbb{R}$ with $f^{-1}(0) = A$. Hence, from the point of view of the Lyapunov–Schmidt procedure we could conceivably have arbitrarily bad solution sets for nonlinear operators. Yet, in general we don't; so how do we explain this?

To do so, we begin by recalling the key ideas from Morse theory.

Definition : A smooth function f: $M \to \mathbb{R}$ is a *Morse function* if

i)   every critical point of f is *nondegenerate*; by this we mean that for any choice of local coordinates $(x_1, \dots, x_n)$ about a critical point x_0, the Hessian matrix

$$H_f = \left(\frac{\partial^2 f}{\partial x_i \partial x_j} (x_0) \right) \text{ is nonsingular}$$

ii)   the critical values of distinct critical points are distinct.

Remark : If M is not compact then f is usually required to be proper (i.e. f^{-1}(compact) is compact) and we will suppose so. Also, condition ii) is not necessary for many applications so a the function also satisfying ii) is sometimes called an *excellent Morse function*.

         Several properties of Morse functions underly their usefulness.

Local Normal Form : Every critical point of a Morse function is referred to as a Morse singularity. Any such can be put into a standard form as a result of the Morse Lemma.

Morse Lemma: *If x_0 is a nondegenerate critical point of f: $M \to \mathbb{R}$ then there is a local coordinate chart (φ, U) about x_0 such that in these local coordinates f has the form*

$$f \circ \varphi^{-1}(x_1, \dots, x_n) = -\sum_{j=1}^{r} x_j^2 + \sum_{j=r+1}^{n} x_j^2 + c$$

(*where* $c = f(x_0)$). *Then, r is called the index of the critical point.*

         We recall that the essential point of the Morse lemma is that the specific nature of the local normal form for a Morse singularity allows us to determine the change in topology of the level sets of f as we pass a critical point. We know that for noncritical values $a < b$ we can obtain $M^b = \{x \in M: f(x) \leq b\}$ from M^a by adjoining cells whose dimensions are the indices of the critical points with values between a and b. This leads to estimates for the Betti numbers of the relative homology of (M^b, M^a) and in the case of compact M for the Betti numbers of M itself in terms of the indices of the critical points of a Morse function. Moreover, by the work of Witten (and implicit in earlier work of Smale), from the gradient flow of a Morse function (for "most" metrics) it is possible to construct a cell complex whose homology is exactly that of M.

With such rich information contained in Morse functions, it is, in a way, somewhat surprising that they are so plentiful.

<u>Density</u> : The set of Morse functions are dense in $C^\infty(M, \mathbb{R})$ (or in the case that M is not compact, in the space of proper functions $C^\infty_{pr}(M, \mathbb{R})$ with the Whitney C^∞-topology rather than the regular C^∞-topology).

In fact, for a properly embedded submanifold $M \subset \mathbb{R}^m$, there are two standard ways to obtain Morse functions.

1) Given $x' \in \mathbb{R}^m$, we define the distance–squared function $f_{x'}(x) = \| x-x' \|^2$. Then, for "almost all" $x' \in \mathbb{R}^m$, the restiction of $f_{x'}$ to M is a Morse function. In fact, as explained in Milnor [Mi], the bad x' are the focal points of M and the set of such points can be identified as the critical set of a smooth function and hence, by Sard's theorem , forms a set of measure zero.

2) Second, for a line $L \subset \mathbb{R}^m$ we let f_L denote orthogonal projection of $\mathbb{R}^m$ onto L. Then, for "almost all" lines L, the restriction of f_L to M is a Morse function. This function is a "height function on M"; hence, all submanifolds have Morse functions analogous to the height function on the standardly embedded torus.

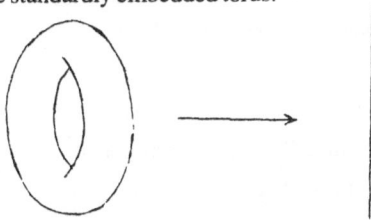

Thus, density ensures that even badly behaved functions such as a the Whitney–type functions mentioned earlier can be arbitrarily closely approximated by Morse functions so the bad properties will disappear.

Finally, despite the key importance of the Morse lemma and the density for Morse functions we state, in what almost seems an afterthought, one final property of Morse functions.

<u>Stability</u> : Any Morse function f has a neighborhood U in $C^\infty(M, \mathbb{R})$ (or $C^\infty_{pr}(M, \mathbb{R})$) such that if $g \in U$ then there exist diffeomorphisms φ of M and ψ of $\mathbb{R}$ so that $g =$

$\psi \circ f \circ \varphi^{-1}$. Furthermore, the φ and ψ may be chosen to depend continuously on g.

Hence, the structure of f remains unchanged under sufficiently small perturbations. This is what is meant by the "structural stability" of Morse functions.

Thom's Program

Ideally one would like to have an analogous theory for the structure of generic, i.e. most, smooth mappings between any two smooth manifolds. We might expect that the principal ideas present in Morse theory might apply more generally. This would include: i) characterization of generic mappings by local and global properties; ii) local normal forms for the local singularities; iii) finite determination of the local singularities by finite parts of their Taylor expansions, and iv) density and stability of the generic mappings. Evidence that such a general result might be true was provided by Whitney, who established just such a theory for mappings from the plane to the plane [Wh2] as well as for mappings from $\mathbb{R}^n$ to $\mathbb{R}^m$ with $m \geq 2n-1$ [Wh3].

Thom proposed that a version of the theory should be valid for smooth mappings between arbitrary manifolds. Such a theory should be built around a general notion of stability of mappings. Quite generally we let $C^\infty_{pr}(N, P)$ denote the space of proper smooth mappings f: N $\rightarrow$ P provided with the Whitney C^∞-topology (if N is compact this is just $C^\infty(N, P)$ with the usual C^∞-topology).

Definition 2.1 : A smooth mapping f: N $\rightarrow$ P is said to be C^∞-*stable* if there exists a neighborhood $\mathcal{U}$ of f in $C^\infty_{pr}(N, P)$ such that if $g \in \mathcal{U}$ then there exist diffeomorphisms φ of N and ψ of P such that

$$g = \psi \circ f \circ \varphi^{-1}$$

It is not required that (φ, ψ) depend continuously on g. If we only require that φ and ψ be homeomorphisms, then f is said to be *topologically stable*.

While this seems like a very natural notion extending the corresponding property for Morse functions and those mappings of Whitney, there remains the question of how useful a notion it is. For example, how can we verify the stability of a particular f, and how can we characterize stability via local and global properties as is done in Morse theory?

To start we can rephrase the condition for stability in terms of the orbit of a group action. Following Mather we let $\mathcal{A}(N, P) = \text{Diff}(N) \times \text{Diff}(P)$ be the group of pairs of diffeomorphisms. It acts on $C^\infty{}_{pr}(N, P)$ by

$$(\varphi, \psi) \cdot f \;=\; \psi \circ f \circ \varphi^{-1}$$

Two mappings in the same orbit are said to be ($\mathcal{A}$)-equivalent (or alternately left–right equivalent). For a fixed f we have the corresponding orbit map

(2.2)
$$\alpha_f : \mathcal{A}(N, P) \longrightarrow C^\infty{}_{pr}(N, P)$$
$$(\varphi, \psi) \longmapsto \psi \circ f \circ \varphi^{-1}$$

The C^∞-stability of f is equivalent to the $\mathcal{A}(N, P)$-orbit of f, which is the image of α_f, containing a neighborhood of f. If we could apply the implicit function theorem, then a sufficient condition for this would be surjectivity of the derivative $d\alpha_f$ at the identity element. Unfortunately, both of the spaces $\mathcal{A}(N, P)$ and $C^\infty{}_{pr}(N, P)$ are Frechet spaces so the usual implicit function theorem fails. There is the stronger Nash–Moser implicit function theorem; however, efforts to apply to this situation even more refined versions of the Nash–Moser theorem, such as that due to Hamilton [Ha], have been unsuccessful. For this reason it is all the more remarkable that Mather was able to prove what is effectively an implicit function theorem for this case; namely, $d\alpha_f$ (id) being surjective, i.e. "infinitesimal stability", implies that Image(α_f) contains a neighborhood of f, i.e. stability.

The fundamental discovery by Mather, that not only made this theorem possible but also allowed him to go on and answer the other questions regarding stable mappings, is the crucial role played by algebra, versus the emphasis in functional analysis on the topologies of the function spaces. To see how this appears, we recall how to compute the tangent spaces.

Computation of the Tangent spaces for Spaces of Mappings

Let $f_t : N \to P$ denote a smooth family of mappings such that $f_0 = f$. Then, we compute the tangent vector to this curve of mappings at f to be $\dfrac{\partial f_t}{\partial t}\Big|_{t = 0}$. For fixed x, this is a tangent vector at t = 0 to the curve $\gamma(t) = f_t(x)$ in P. This defines a mapping $\zeta : N \to TP$ with the property that $\pi \circ \zeta = f$, where π denotes projection $\pi : TP \to P$. Such ζ is a "vector field along f"; it indicates the direction of infinitesimal change at f(x) to obtain the deformation f_t. The space of such vector fields is denoted by $\theta(f)$. In the case $f = id_M$, we obtain the vector fields on M, and use the notation $\theta(M)$. As Diff(M) is open in $C^\infty_{pr}(M, M)$, this is the tangent space to Diff(M) at id_M. Then, to compute $d\alpha_f$, we let φ_t and ψ_t be families of diffeomorphisms with $\varphi_0 = id_N$ and $\psi_0 = id_P$ so that

$$\frac{\partial \varphi_t}{\partial t}\Big|_{t=0} = \xi \in \theta(N) \quad \text{and} \quad \frac{\partial \psi_t}{\partial t}\Big|_{t=0} = \eta \in \theta(P)$$

then, by a simple calculation using the chain rule (see [Ma1–II]) we obtain

(2.3) $$d\alpha_f(\xi, \eta) = \frac{\partial}{\partial t}\left(\psi_t \circ f \circ \varphi_t^{-1} \right)\Big|_{t=0}$$

$$= \frac{\partial}{\partial t}(\psi_t)\Big|_{t=0} \circ f \circ \varphi_0^{-1} + d\psi_0 \circ df\left(\frac{\partial}{\partial t}(\varphi_t^{-1})\Big|_{t=0}\right)$$

$$= \eta \circ f - df(\xi)$$

Thus,

$$d\alpha_f : \theta(N) \oplus \theta(P) \longrightarrow \theta(f)$$

(2.4) $$(\xi, \eta) \longmapsto \eta \circ f - \xi(f)$$

where $\xi(f)$ ($= df(\xi)$) denotes the directional derivative with respect to the vector field ξ. We will refer to this as the *infinitesimal orbit map* and we say that f is *infinitesimally stable* if $d\alpha_f$ is surjective. We can describe $d\alpha_f$ via composition with df and f in the following diagram.

$$
\begin{array}{ccc}
 & df & \\
TN & \longrightarrow & TP \\
\xi \uparrow \pi & & \pi \uparrow \eta \\
\downarrow & f & \downarrow \\
N & \longrightarrow & P
\end{array}
$$

We observe that both $\theta(N)$ and $\theta(f)$ are modules over the ring $C^\infty(N)$ and similarly for $\theta(P)$ over $C^\infty(P)$; $d\alpha_f \mid \theta(N)$ is then a $C^\infty(N)$-module homomorphism, while $d\alpha_f \mid \theta(P)$ is a module homomorphism over the ring homomorphism $f^* : C^\infty(P) \longrightarrow C^\infty(N)$ defined by pull–back $f^*(h) = h \circ f$.

Remark : Not all actions by groups of diffeomorphisms have a nice algebraic structure. For example, Diff(M) acting on $C^\infty(M, M)$ by conjugation $\varphi \cdot f = \varphi \circ f \circ \varphi^{-1}$ has for its infinitesimal orbit map $\xi \mapsto -\xi(f) + \xi \circ f$, which does not preserve the module structure. In part, this is the reason that conjugation does not yield to such a complete analysis as does $\mathcal{A}$-equivalence.

Then, Mather's fundamental theorem "infinitesimal stability implies stability" becomes

Theorem (Mather [Ma1–II]) : *If* $f \in C^\infty{}_{pr}(N, P)$ *is infinitesimally stable then there exists a neighborhood* $\mathcal{U}$ *of* f *and a continuous mapping*

$$\mathcal{U} \longrightarrow \mathcal{A}(N, P)$$
$$g \longmapsto (\varphi_g, \psi_g)$$

sending $f \mapsto (id_N , id_P)$ *such that*

$$g = \psi_g \circ f \circ \varphi_g{}^{-1} \qquad \textit{for all } g \in \mathcal{U}.$$

Infinitesimal Methods

Because Mather's proof does not rely upon an implicit function theorem and because the ideas in the proof play an equally crucial role in answering the questions for the local theory, we indicate an outline of how the theorem is proven.

The Key Ideas are:

1) proving triviality of families of mappings;

2) reducing triviality to an infinitesimal equation for triviality;

3) using algebraic methods to solve the infinitesimal equation; and

4) in the case of stability, constructing families to sufficiently near mappings in a continuous manner and proving that the infinitesimal equations can be solved for all sufficiently near mappings.

It is almost impossible to directly prove that two mappings f_0 and f_1 are $(\mathcal{A})$-equivalent. However, suppose that these mappings can be joined by a homotopy, i.e. by a family of mappings f_t. Then,we can try to prove instead that the family f_t is trivial, i.e. there are families of diffeomorphisms φ_t and ψ_t such that $\varphi_t \circ f_t \circ \psi_t \equiv f_0$ so that f_t is equivalent to the constant family $f_0 \times \text{id}$. Then we conclude that f_1 is equivalent to f_0.

On first glance, this looks harder than just directly proving that f_1 is equivalent to f_0; however,as Thom observed [ThL], it has the considerable advantage that it can be proved by infinitesimal methods and is the underlying idea in his notion of "homotopic stability".

To describe the infinitesimal reduction, we observe that giving a family f_t is equivalent to giving an unfolding $f: N \times I \longrightarrow P \times I$ of the form $f(x, t) = (\bar{f}(x, t), t)$ where $f_t(x) = \bar{f}(x, t)$ and $I = [0, 1]$. Likewise families of diffeomorphisms φ_t and ψ_t correspond to unfoldings $\varphi(x, t) = (\bar{\varphi}(x, t), t)$ and $\psi(y, t) = (\bar{\psi}(y, t), t)$ of diffeomorphisms. These unfoldings differ from those obtained via Lyapunov–Schmidt reduction because they are defined globally on the interval rather than for just small values of the parameter t. Then, f is said to be an $(\mathcal{A})$-trivial unfolding if there are unfoldings of diffeomorphisms φ and ψ so that $\psi \circ f \circ \varphi^{-1} = f_0 \times \text{id}$; this is just a restatement that f_t is a an $(\mathcal{A}-)$ trivial family. Then by differentiating the first factor of this equation with respect to t, we are repeating the calculation of (2.3) except that we do not evaluate at $t = 0$ and now f is also time – dependent. We obtain

$$(2.5) \qquad \frac{\partial}{\partial t}(\bar{\psi} \circ \bar{f} \circ \bar{\varphi}^{-1}) \; = \; d\psi \, (\, \eta(f) + \frac{\partial \bar{f}}{\partial t} - \xi(f) \,) \circ \varphi^{-1}$$

where

$$(2.6) \qquad \frac{\partial \bar{\varphi}}{\partial t} \; = \; \xi \circ \varphi \; , \quad \frac{\partial \bar{\psi}}{\partial t} \; = \; \eta \circ \psi, \quad \text{and} \quad \xi \in \theta(\pi_N) \text{ and } \eta \in \theta(\pi_P)$$

are "time dependent" vector fields lying over the projections $\pi_N: N \times I \to N$ and $\pi_P: P \times I \to P$. The equations in (2.6) state exactly that the families of diffeomorphisms φ and ψ are the flows generated by the time dependent vector fields ξ and η.

If we extend the action of $\mathcal{A}$ to unfoldings f, then (2.4) again gives the correct formula for the infintesimal orbit map except now it is applied to "time dependent" vector fields. By reversing the argument used to deduce (2.5), Thom–Levine [ThL] give the infinitesimal criteria for triviality ("homotopic triviality" in their terminology).

Theorem 2.7 : *Given an unfolding* f: N × I ⟶ P × I. *If there exist time dependent vector fields* η *and* ξ *such that*

(2.8)
$$-\frac{\partial \bar{f}}{\partial t} = \eta(f) - \xi(\bar{f})$$

then f *is a trivial unfolding of* f_0.

The line of reasoning outlined above then proceeds in the case of stability by first establishing 4). This follows from two general results about spaces of mappings and the integrability of vector fields established by Mather.

Local Pathwise Contractibility (see [Ma1–II§7]) :

Given $f_0 \in C^\infty{}_{pr}(N, P)$ there exists a neighborhood $\mathcal{U}$ of f_0 and a continuous map $\gamma^* : \mathcal{U} \longrightarrow C^\infty{}_{pr}(N \times I, P)$ such that for $f_1 \in \mathcal{U}$, $\gamma^*(f_1)(x, 0) = f_0(x)$ and $\gamma^*(f_1)(x, 1) = f_1(x)$.

Local integrability (see [Ma1–II§7]) :

Given a smooth manifold M (possibly with boundary and corners), there exists a neighborhood O of 0 in $\theta(\pi_M)$, where again $\pi_M: M \times I \to M$ denotes projection, and a continuous map $\theta: O \longrightarrow C^\infty{}_{pr}(M \times I, M)$ such that if $\theta(\zeta) = \varphi(x, t)$ then $\varphi(x, t) = (\bar{\varphi}(x, t), t)$ is the flow generated by ζ, i.e.

$$\frac{\partial \bar{\varphi}}{\partial t}(x, t) = \zeta \circ \varphi(x, t)$$

Then, the local pathwise contractibility provides a continuous construction of homotopies from the fixed f_0 to near f_1. Also, the local integrability provides for the continuous dependence of the flows on the time dependent vector fields. The remaining *step 3) is the crucial one* requiring algebraic results to solve the infinitesimal equations. It is here that for f(x, t) = (f̄(x, t), t) the unfolding with $\bar{f} = \gamma^*(f_1)$ we must solve equation (2.8) with the time dependent vector fields ξ and η depending continuously on f_1. If we let

$\Delta(f_1) = -\dfrac{\partial \bar{f}}{\partial t}$, then this requires the solution of the lifting problem in fig. 2.9.

$$\theta(\pi_N) \oplus \theta(\pi_P)$$

$$\downarrow d\alpha_f$$

fig. 2.9 Δ

$$\mathcal{U} \xrightarrow{\quad\quad} C^\infty_{pr}(N \times I, TP)$$

Mather solves this problem for sufficiently small $\mathcal{U}$ using his Global Division Theorem. It is the global analogue of the local Malgrange division theorem, from which follows the Malgrange preparation theorem, the basis for all of the local results to be discussed next. However, we will not discuss the lifting problem for 2.9 here (see [Ma1-II,§6])

§3   Mather's Local Theory as Paradigm

At first glance, Mather's elegant infinitesimal characterization of stability seems to be very noncomputible. Part of Mather's remarkable insight into these questions was his realization that infinitesimal stability could be understood in terms of : a) local algebraic conditions on the mapping at its singular points together with b) global transversal behavior of the image of the singular set. These seemingly disparate notions are brought together through Mather's geometric characterization of stability in terms of multi-transversality (in the jet bundle).

A good analogy is with the development of topology. Initially point set topology dealt with basic topological properties such as connectedness, compactness, separation properties, metrization, dimension, etc. While these properties are fundamental in the study of topology, they clearly missed many of the basic features of spaces. A major breakthrough occurred when topology was algebraicized. It was the ability to compute the algebraic invariants of spaces, especially manifolds, that has lead to the enormous developments and applications of topology to most areas of mathematics. In a similar vein, Mather's work algebraicized local singularity theory. Unlike the earlier work of Morse and Whitney which determined the local behavior for stable mappings without providing a systematic approach. Mather did just that by providing algebraic methods [Ma1,III–VI] for answering questions regarding:

1) local stability, algebraically characterizing the germs of stable mappings;

2) finite   determinacy, determining for a given germ, a level ℓ, such that the germ is equivalent to the germ defined by its ℓ-th order Taylor expansion;

3) classification, providing algebraic criteria for deciding when two germs are equivalent;

4) normal forms, beginning from algebraic data which classifies stable germs and giving an algorithm for constructing standard models (up to equivalence);

5) universal unfoldings, finding finite parameter families of germs which describe (up to equivalence) all possible small perturbations of a given germ.

Remark : We should note that although Mather first worked out a theory of universal unfoldings for right equivalence, this work remained unpublished [Ma2]. A definitive treatment of right equivalence was given by Wasserman [Ws1], and more importantly Martinet in [Mar] gave the unfolding theory for both $\mathcal{A}$ and $\mathcal{K}$-equivalence in a form which has served as the prototype for subsequent extensions of the theory. Also, earlier Tougeron [Tg] had independently obtained results for finite determinacy for $\mathcal{K}$ and $\mathcal{R}$-equivalence.

If we look back to the questions posed at the end of §1 when we considered the Lyapunov–Schmidt reduction, we see that the questions which Mather answers for finite dimensional map germs are essentially the same local ones we were asking for the infinite dimensional nonlinear operators. It should come as no surprise that the answers which we describe in this section, together with the extensions in the next two sections allowing various special conditions will allow us to draw ample conclusions about the local structure of nonlinear operators.

We now describe the algebraic techniques and results which answer the above questions. First we deduce local consequences of the infinitesimal stability condition. Let $f: N \longrightarrow P$ be infinitesimally stable. Let $\Sigma(f) = \{x \in N: rk(df(x)) < dimP\}$ denote the critical set of f. A consequence of infinitesimal stability and the properness of f is that if $y \in P$ then $f^{-1}(y) \cap \Sigma(f) = \{x_1, \dots, x_\ell\}$ is finite. Then, infinitesimal stability implies that given germs of vector fields $\zeta_i : N, x_i \longrightarrow T_yP$ along f, there exist germs of vector fields $\xi_i : N, x_i \longrightarrow T_{x_i}N$ and $\eta : P, y \longrightarrow T_yP$ such that

$$(3.1) \qquad\qquad \zeta_i \; = \; -\xi_i(f) + \eta \circ f \qquad \text{for } 1 \le i \le \ell$$

(basically the local ζ_i can be extended to a global ζ and the ξ_i and η are obtained by restricting the global solutions of $\zeta = d\alpha_f(\xi, \eta)$)).

Then (3.1) can be rephrased in terms of the local action of the group $\mathcal{A}$ on germs f: $N, x_i \longrightarrow P, y$. We let $\theta(N)_{x_i}$ denote the germs of vector fields at x_i and similarly for $\theta(P)_y$. Then (3.1) implies:

i) the infintesimal orbit map for germs

$$(3.2) \qquad\qquad d\alpha_f : \theta(N)_{x_i} \oplus \theta(P)_y \longrightarrow \theta(f)_{x_i}$$

is surjective for each i, and (can be shown to imply)

ii) the subspaces $\{d_{x_i}f(T_{x_i}N)\}$ intersect transversally in T_yP (i.e. any $d_{x_i}f(T_{x_i}N)$ together

To transform i) into a useful local algebraic criterion, we concentrate on a single germ f_0 and introduce local coordinates so that we may assume $f_0 : \mathbb{R}^n, 0 \longrightarrow \mathbb{R}^p, 0$. We use local coordinates $\mathbf{x} = (x_1, \dots, x_n)$ for $\mathbb{R}^n$ and $\mathbf{y} = (y_1, \dots, y_p)$ for $\mathbb{R}^p$. When we consider an unfolding $f : \mathbb{R}^{n+q}, 0 \longrightarrow \mathbb{R}^{p+q}, 0$ of f_0, we use local coordinates $\mathbf{u} = (u_1, \dots, u_q)$ for $\mathbb{R}^q$ and write $f(\mathbf{x}, \mathbf{u}) = (\bar{f}(\mathbf{x}, \mathbf{u}), \mathbf{u})$.

A Few Algebraic Preliminaries

The importance of algebra for infinitesimal methods is that it provides a systematic method to work with "higher order terms". Recall that if $h: U \longrightarrow \mathbb{R}$ and $0 \in U$, then h is of order $\geq k$ near 0, usually denoted $| h(x) | = O(\| x \|^k)$, if there exists a constant $C > 0$ such that $| h(x) | \leq C \| x \|^k$ in a neighborhood of 0. We can easily algebraically describe such terms of order $\geq k$.

Using the local coordinates just described above, we let C_x denote the ring of germs of smooth functions $h : \mathbb{R}^n, 0 \longrightarrow \mathbb{R}$, which has a maximal ideal m_x consisting of germs vanishing at 0, and similarly for C_y (and m_y) on $\mathbb{R}^p$. We denote the corresponding ring on $\mathbb{R}^{n+q}$ by $C_{x,u}$ (with $m_{x,u}$), etc. If m_x^k denotes the k-th power of m_x, then by what is often called Hadamard's lemma, m_x^k is generated as an ideal by monomials in $x_1, \dots, x_n$ of degree k; then m_x^k consists of the "higher order terms" of degree $\geq k$.

Although working with infinitely flat smooth germs may seem to make algebraic computations impossible, there are several nice features about C_x which overcome this. By Hadamard's lemma $C_x / m_x^k \simeq \mathbb{R}[x_1, \dots, x_n] / m_n^k$ (where $\mathbb{R}[x_1, \dots, x_n]$ is the usual polynomial ring and m_n is the ideal generated by $x_1, \dots, x_n$). Thus, as a vector space C_x / m_x^k has a finite basis consisting of the monomials of degree $< k$.

We can also express the modules of vector fields as modules over these rings. We represent the germs of vector fields along the germ f_0 by $\theta(f_0)$ which is the finitely generated C_x-module $C_x \{ \frac{\partial}{\partial y_1}, \dots, \frac{\partial}{\partial y_p} \}$ generated by the constant vector fields $\{ \frac{\partial}{\partial y_1}, \dots, \frac{\partial}{\partial y_p} \}$. We also write $\theta(\mathrm{id}_{\mathbb{R}^n}) = \theta_n = C_x \{ \frac{\partial}{\partial x_1}, \dots, \frac{\partial}{\partial x_n} \}$ and similarly for θ_p. When the number of generators is understood we may abbreviate $\theta_n = C_x \{ \frac{\partial}{\partial x_i} \}$. We can likewise form e.g. $m_x^k \cdot \theta(f_0)$ which consists of finite sums of terms which are products of elements of $\theta(f_0)$ by elements of m_x^k; these are again terms of order $\geq k$ among the vector fields along f_0.

Let $J^\ell(n, p)$ denote the jet space consisting of the ℓ-jets of germs $g : \mathbb{R}^n, 0 \longrightarrow \mathbb{R}^p, 0$, which are essentially just the ℓ-th order Taylor expansions, and write $j^\ell(g)$ for the ℓ-jet of g. There is the natural identification (by representing ℓ-jets by ℓ-th degree polynomials)

$$(3.3) \qquad J^\ell(n, p) \; \simeq \; m_x \cdot \theta(f_0)/m_x^{\ell+1} \cdot \theta(f_0) \; \simeq \; m_x\{\frac{\partial}{\partial y_i}\}/ m_x^{\ell+1}\{\frac{\partial}{\partial y_i}\}$$

$$(g_1, ..., g_p) \; \longleftrightarrow \; \sum g_i \frac{\partial}{\partial y_i}$$

We wish to be able to reduce the infinite complexity associated with smooth germs to considerations involving lower order terms. The first reason that we can sometimes do this is Nakayama's Lemma, which we state in a form suggesting further extensions. We let R denote a local ring with maximal ideal m over a field k, e. g. R is a ring such as C_x over the field $\mathbf{k} = \mathbb{R}$ with a unique maximal ideal m_x. Suppose that $\alpha: N \longrightarrow M$ is an R-module homorphism between finitely generated R-modules, with $M_0 \subset M$ a finitely generated R-submodule.

Nakayama's Lemma :
1)   if $M_0 \subseteq \alpha(N) + mM_0$ then $M_0 \subseteq \alpha(N)$
2)   if $\dim_k(M/\alpha(N)) = r < \infty$ then there exists an ℓ such that $m^\ell.M \subseteq \alpha(N)$ (in fact, $\ell = r$).

Remark (What Nakayama's lemma really says): for example, in 1) suppose $\alpha : N \to M_0$ then if we can express the generators of M_0 as images of elements of N *modulo higher order terms of* M_0 then all elements of M_0 are images of elements of N; also, 2) ensures that if $\alpha(N)$ has finite codimension in M then $\alpha(N)$ contains all "higher order terms" of M (of order $\geq \ell$).

For example, ideals in the local ring C_x are C_x-modules. From the germ $f_0 = (f_{01}$, ... , $f_{0p})$ we can associate the ideal $I(f_0)$ generated by the germs f_{01}, ... , f_{0p}, and the algebra $Q(f_0) = C_x/I(f_0)$, which is called the *local algebra* of f_0. We make several observations. First, although the definition of $I(f_0)$ appears to depend on the local coordinates chosen for $\mathbb{R}^p$, it is intrinsic as it equals the ideal $f_0^* m_y \cdot C_x$ where $f_0^*: C_y \to C_x$ is the ring homomorphism $f_0^*(h) = h \circ f_0$. Second, if $f(x, u) = (\bar{f}(x, u), u)$ is an unfolding of f_0, then since $\bar{f}(x, 0) = f_0(x)$, $I(f)$ is easily seen to equal the ideal in $C_{x,u}$ generated by $I(f_0)$ and m_u so that $Q(f) \simeq Q(f_0)$. Hence, we do not change the local algebra when we replace a germ by an unfolding of it. Third, by Nakayama's lemma, if $\dim_{\mathbb{R}}(Q(f_0)) = \ell < \infty$ then $m_x \ell \subseteq I(f_0)$, i.e. $I(f_0)$ contains all higher order terms of degree $\geq \ell$. Hence, $Q(f_0)$ can be represented as a quotient of a polynomial ring. We shall see that $Q(f_0)$ plays an important role in the classification of stable germs, in the construction of normal forms, and even in the determination of the number of local solutions to the equation $f_0(x) = y$ (see §6).

Returning to the local question of infinitesimal stability, we want to rewrite (3.2) in a simpler form. In doing this we want to take advantage of being able to represent f as an unfolding of a germ f_0. Thus, we begin with $\mathbf{x}'$ and $\mathbf{y}'$ denoting local coordinates for N and P, and with $n' = \dim N$ and $p' = \dim P$. Then, (3.2) becomes

(3.4) $d\alpha_f : \theta_{n'} \oplus \theta_{p'} \longrightarrow \theta(f)$

As we remarked in the global case, $\theta_{n'}$ and $\theta(f)$ are both $C_{x'}$-modules while θ_p is a $C_{y'}$-module and the restriction $d\alpha_f \mid \theta_{p'}$ is a module homomorphism over f^*. Hence, although we would like to establish the surjectivity of $d\alpha_f$, we can not use any of the results involving Nakayama's lemma which only concerns modules over a single ring. We could view the $C_{x'}$-modules as $C_{y'}$-modules via f^*; however, we would no longer know that they are finitely generated so Nakayama's lemma still could not be used. This type of problem in related situations was recognized as being equivalent to the smooth version of the classical Weierstrass preparation theorem (which superficially looks quite different). The smooth version is considerably more difficult and follows from the smooth local division theorem of Malgrange [Mal].

Malgrange Preparation Theorem :   If M is a finitely generated $C_{x'}$-module then it is a finitely generated $C_{y'}$-module iff $\dim_{\mathbb{R}}(M/m_{y'} \cdot M) < \infty$ (and $\{\varphi_1, \dots, \varphi_s\}$ generate M as a $C_{y'}$-module iff they span $M/m_{y'} \cdot M$ as a vector space).

Note :   Here and in what follows we often suppress the ring homomorphism f^* to simplify notation, so we write e.g. $m_{y'} \cdot M$ instead of $f^* m_{y'} \cdot M$.

A consequence is that the results stated in Nakayama's lemma also hold for homomorphisms such as $d\alpha_f$. To state the conclusion, we now apply the Lyapunov-Schmidt reduction and assume that f is represented as an unfolding of the germ f_0 so $f(x, u) = (\bar{f}(x, u), u)$ (with $n' = n + q$ and $p' = p + q$). We let $x' = (x, u)$ and $y' = (y, u)$; then by the preparation theorem (applied to $\theta(f)/d\alpha_f(\theta_{n'}))$, (3.4) is surjective iff

(3.5) $d\alpha_f(\theta_{n'} \oplus \theta_{p'}) + m_{y'} \cdot \theta(f) = \theta(f)$

We proceed to simplify the condition (3.5). First, observe that $d\alpha_f(m_y \cdot \theta_{p'}) \subset m_{y'} \cdot \theta(f)$. On dividing both sides by $m_{y'} \cdot \theta(f)$, using the special form for f, and recalling that dividing by $m_u \cdot \theta(f)$ will remove the unfolding variables u, we obtain after some algebraic simplification

$$(3.6) \qquad C_x\{\frac{\partial f_0}{\partial x_1}, ..., \frac{\partial f_0}{\partial x_n}\} + f_0^* \, m_y \cdot C_x\{\frac{\partial}{\partial y_1}, ..., \frac{\partial}{\partial y_p}\} + \langle \partial_1 f, ..., \partial_q f, \frac{\partial}{\partial y_1}, ..., \frac{\partial}{\partial y_p}\rangle$$

$$= C_x\{\frac{\partial}{\partial y_1}, ..., \frac{\partial}{\partial y_p}\}$$

where $\partial_j f = \frac{\overline{\partial f}}{\partial u_j}\big|_{u=0}$ and $\langle \varphi_1, ..., \varphi_s\rangle$ denotes the vector space spanned by $\varphi_1, ..., \varphi_s$.

We see that the first two summands are just $d\alpha_f(\theta_{n'}) + m_{y'} \cdot \theta(f)$ after we divide by $m_u \cdot \theta(f)$. Although this still looks quite complicated we note that now all modules are C_x-modules so that Nakayama's lemma can be applied. Also, by the Lyapunov–Schmidt reduction, f has no linear

terms so neither does f_0; hence, $\frac{\partial f_0}{\partial x_i}$ and $\partial_j f \in m_x \cdot \theta(f_0)$. Thus, (3.6) may be equivalently written in the abbreviated form

$$(3.7) \qquad C_x\{\frac{\partial f_0}{\partial x_i}\} + f_0^* \, m_y \cdot \theta(f_0) + \langle \partial_1 f, ..., \partial_q f\rangle = m_x \cdot \theta(f_0)$$

In fact, by 2) of Nakayama's lemma, (3.7) implies

$$m_x^{q+1} \cdot \theta(f_0) \subseteq C_x\{\frac{\partial f_0}{\partial x_i}\} + f_0^* \, m_y \cdot \theta(f_0)$$

and conversely a more subtle use of Nakayama's lemma implies that if (3.7) is only true modulo $m_x^{q+2} \cdot \theta(f_0)$ then it is true. Hence, observing that $f_0^* \, m_y \cdot \theta(f_0)$ is the direct sum $I(f_0)^{(p)}$ of p

copies of $f_0^* \, m_y = I(f_0)$ and letting $L = C_x\{\frac{\partial f_0}{\partial x_i}\}$, we conclude

Theorem (Verification Criterion [Ma1-IV]) : *Let* $f : \mathbb{R}^{n+q}, 0 \longrightarrow \mathbb{R}^{p+q}, 0$ *be an unfolding of* f_0 *which has rank 0, i.e.* $df_0(0) = 0$. *Then,* f *is infinitesimally stable iff*

$\{\partial_1 f, \dots, \partial_q f\}$ *spans the (finite dimensional) quotient space*

(3.8) $N(f_0) = m_x^{(p)}/(L + I(f_0)^{(p)} + (m_x^{q+2})^{(p)})$.

Remark: It follows that if f is infinitesimally stable then so is $f \times id_{\mathbb{R}^k}$.

Note : It is convenient to shorten the expression infinitesimally stable germ to just stable germ. There is a local notion for stability for germs which also makes this literally correct.

Normal Forms

This provides an algorithm for constructing normal forms for stable germs. Given f_0 : $\mathbb{R}^n,0 \longrightarrow \mathbb{R}^p,0$ with $df_0(0) = 0$, we let $\{\varphi_1, \dots, \varphi_q\}$ be a set of germs whose images span $N(f_0)$. Then

(3.9) $f(x, u) = (f_0(x) + \sum_{i=1}^{q} u_i \cdot \varphi_i, u_1, \dots, u_q)$

is a stable germ and moreover, $Q(f) \simeq Q(f_0)$. Such an f is called a *stable unfolding* of f_0. Given a stable germ f, which is represented as an unfolding of f_0, the verification criterion allows us to construct a simpler unfolding f_1 between the same spaces as f and with the same local algebra $Q(f) \simeq Q(f_0) \simeq Q(f_1)$. What is the relation between the stable germs f and f_1? Mather's classification theorem provides the answer.

Classification Theorem [Ma1-IV] : *Let* f , g: $\mathbb{R}^n,0 \longrightarrow \mathbb{R}^p,0$ *be stable germs such that* $Q(f) \simeq Q(g)$ *then f and g are* $(\mathcal{A})$-*equivalent.*

Hence, all stable germs can be represented (up to equivalence) by normal forms.

Examples (3.10):

1) Morin singularities: Beginning with $f_0(x) = x^{n+1}$, we see that L is the ideal generated by $(n+1)x^n$ and so $N(f_0) = m_x/m_x^n$ is spanned by $\{x, x^2, \dots, x^{n-1}\}$ so that

$$f(x, u) = (x^{n+1} + \sum_{i=1}^{n-1} u_i \cdot x^{n-i}, u_1, \dots, u_{n-1})$$

is infinitesimally stable as is $f \times id_{\mathbb{R}^k}$. It follows by the classification theorem that any stable germ g: $\mathbb{R}^m,0 \longrightarrow \mathbb{R}^m,0$ with $dim_{\mathbb{R}}(ker(dg(0)) = 1$ has local algebra $Q(g) \simeq \mathbb{R}[x]/(x^s)$ for some s, and hence, is equivalent to $f \times id_{\mathbb{R}^k}$ for some Morin singularity f.

In the case m = 2, this gives the two stable singularities obtained by Whitney [Wh], the standard "cusp singularity" $f(x, u) = (x^3 + ux, u)$ and the "fold" $f(x, u) = (x^2, u)$.

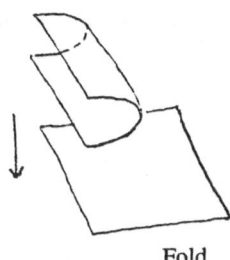
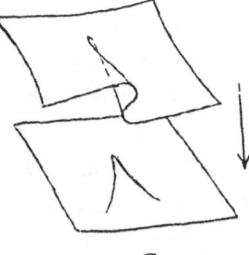

Fold   Cusp

2) "Whitney Umbrella" : From $f_0 : \mathbb{R},0 \longrightarrow \mathbb{R}^2,0$ defined by $f_0(x) = (x^2, 0)$, we obtain

$$\frac{\partial f_0}{\partial x} =$$

$(2x, 0)$ so $N(f_0) = (0) \oplus m_x/m_x^2$ is spanned by $(0, x)$ and $f(x, u) = (x^2, ux, u)$ is the stable unfolding $f : \mathbb{R}^2,0 \longrightarrow \mathbb{R}^3,0$ with image shown below. Note $Q(f) \simeq Q(f_0) \simeq \mathbb{R}[x]/(x^2)$, which is also the local algebra for the fold singularity. Hence, very distinct stable mappings between different dimensional spaces can have the same local algebra.

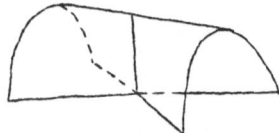

WhitneyUmbrella

Remark : From the point of view of nonlinear operators, f is an overdetermined system of equations; however, we see that it still has very stable well–determined singular behavior.

3) Consider $f_0 : \mathbb{R}^2,0 \longrightarrow \mathbb{R}^2,0$ defined by $f_0(x, y) = (x^2 + \varepsilon y^2, xy)$ with $\varepsilon = \pm 1$. Then,

$$\frac{\partial f_0}{\partial x} = (2x, y) \text{ and } \frac{\partial f_0}{\partial y} = (2\varepsilon y, x) \text{ and } N(f_0) = (m_{x,y}/m_{x,y}^2)^{(2)}/\langle (2x, y), (2\varepsilon y, x) \rangle \text{ and is}$$

spanned by
e.g. $(x, 0)$ and $(0, y)$. Thus,

$$f(x, y, u, v) = (x^2 + \varepsilon y^2 + ux, xy + vy, u, v)$$

is the stable unfolding. In the case $\varepsilon = -1$, f_0 is equivalent to the complex map $z \mapsto z^2$ and the perturbation $z^2 + t \cdot \bar{z}$ considered in §1 corresponds to $(u, v) = (t/2, -t/2)$.

Suppose $f_0 : \mathbb{R}^n,0 \longrightarrow \mathbb{R}^p,0$ is a germ with $n \leq p$. In this case, it can be shown that

$N(f_0)$ is finite dimensional iff $Q(f_0)$ is; then f_0 is said to be a *finite map germ*. Using the verfication criterion, we can explicitly answer the determinacy question for this case, . We first make a definition which formalizes the idea that adding higher order terms does not change the germ up to equivalence.

Definition 3.11: A germ $f : \mathbb{R}^n, 0 \longrightarrow \mathbb{R}^p, 0$ is ℓ-*determined* if given $g : \mathbb{R}^n, 0 \longrightarrow \mathbb{R}^p, 0$ with $j^\ell(f) = j^\ell(g)$ then f and g are ($\mathcal{A}$)-equivalent. A germ is *finitely-determined* if it is ℓ-determined for some ℓ.

Mather has given a general result for the finite determinacy of stable germs [Ma1 – IV].

Theorem 3.12: *Suppose* $f : \mathbb{R}^n, 0 \longrightarrow \mathbb{R}^p, 0$ *is a stable germ, then it is* $p+1$-*determined.*

In the case $n \leq p$, this can be somewhat strengthened by a direct application of the verification criterion and the classification theorem.

Corollary 3.13: *Suppose* $f_0 : \mathbb{R}^n, 0 \longrightarrow \mathbb{R}^p, 0$ *is a finite map germ of rank 0 (i.e.* $df_0(0) = 0$) *with stable unfolding* $f : \mathbb{R}^{n+q}, 0 \longrightarrow \mathbb{R}^{p+q}, 0$ *such that* $m_x{}^\ell \cdot Q(f_0) = 0$, *then,* f *is* ℓ-*determined.*

Outline of Proof (of corollary): By the classification theorem, it is enough to prove that g is also infinitesimally stable and $Q(g) \simeq Q(f)$. Let $h = f-g \in m_{x,u}{}^{\ell+1} \cdot \theta(f)$. To apply the Lyapunov–Schmidt reduction to $g = f-h$, we only have to change the coordinate functions of f by terms in $m_{x,u}{}^{\ell+1}$; hence, g is $\mathcal{A}$-equivalent to

(3.14) $\qquad g_1(x, u) = (\bar{f}(x, u) + h_1(x, u), u) \qquad$ with $h_1 \in m_{x,u}{}^{\ell+1} \cdot \theta(\bar{f})$.

It is sufficient to prove the result for g_1, which is an unfolding of $g_0(x) = f_0(x) + h_1(x, 0)$. Then, $m_x{}^\ell \cdot Q(f_0) = 0$ implies

(3.15) $\qquad\qquad m_x{}^\ell \subseteq I(f_0), \qquad$ and hence $\qquad m_x{}^\ell \cdot \theta(f_0) \subseteq I(f_0)^{(p)}$.

Nakayama's lemma and the form of g_0, together with (3.14) and (3.15), imply both $I(g_0) = I(f_0)$ and $N(g_0) = N(f_0)$. Thus, $Q(g_1) \simeq Q(g_0) \simeq Q(f_0) \simeq Q(f)$. Also, (3.14) implies

$$\partial_i g_1 \equiv \partial_i f \mod m_x{}^\ell \cdot \theta(f_0)$$

Thus, as $N(g_0) = N(f_0)$, by the verification criterion, g_1 is infinitesimally stable. $\square$

For example 3.10, Morin singularities defined from x^{n+1} are $n+1$ determined by theorem 3.12 or corollary 3.13. For 2) and 3) we use the stronger corollary 3.13; for 2)

$m_x^2 \cdot Q(f_0) = 0$ so the Whitney umbrella map f is 2-determined, and for 3) $m_{x,y}^3 \cdot Q(f_0) = 0$ hence the stable unfolding f is 3-determined (in fact, a more refined analysis shows it is 2-determined).

If at this point we look back on the list of results we wanted to consider we see that for stable germs we have reasonably complete answers, except that for given n and p we would still have to work out which algebras would give rise to stable germs between these dimensions. Also, we haven't really addressed how the germs fit together to give global mappings. We shall only briefly comment on this in §7, since it isn't relevant to the local questions we are considering.

We have yet to discuss how what we have learned so far from the investigation of stable germs and mappings applies to nonstable germs. For example, if a nonlinear mapping has a parameter, as we vary the parameter we will ocassionally pass unstable behavior; however, surprisingly this process of encountering unstable behavior can itself be made stable! We have seen something of this in the construction of the stable unfolding; however, it seemed very formal and algebraic without other motivation. We now clear up these remaining points and show how they point to answers for the remaining questions.

Groups of Equivalences Acting on Spaces of Germs

We have seen that we can consider a local version of $\mathcal{A}$-equivalence for germs. Let $C(n, p)$ denote the space of smooth germs $f_0 : \mathbb{R}^n, 0 \longrightarrow \mathbb{R}^p, 0$, and $\mathcal{D}_n$ denote the group of germs of smooth diffeomorphisms $\varphi : \mathbb{R}^n, 0 \longrightarrow \mathbb{R}^n, 0$. Then, $\mathcal{A} = \mathcal{D}_n \times \mathcal{D}_p$ acts on $C(n, p)$ as in the global case. Given $f_0 \in C(n, p)$, we can compute the infinitesimal orbit map $d\alpha_{f_0}$. However, it is slightly different from (3.4). This is because $\mathcal{A}$ now fixes 0 so that the vector fields in $T\mathcal{A}$ vanish at 0; likewise the germs in $C(n, p)$ always send $0 \mapsto 0$ so the vector fields along f_0 that we obtain from "curves" in $C(n, p)$ must vanish at 0. However, because the local action is still the same, the infintesimal orbit space map will just be the restriction of that in (3.4). The tangent spaces and the infintesimal orbit space map for the local action are given by

$$T\mathcal{A} = m_x \cdot \theta_n \oplus m_y \cdot \theta_p \qquad \text{and} \qquad TC(n, p) = m_x \cdot \theta(f_0)$$

(3.16)
$$d\alpha_{f_0} : m_x \cdot \theta_n \oplus m_y \cdot \theta_p \longrightarrow m_x \cdot \theta(f_0)$$
$$(\xi, \eta) \longmapsto \eta \circ f - \xi(f)$$

Perturbations via Unfoldings

There is a way to understand (3.4) from a local viewpoint. We note that the preceding computation of the tangent space is not well suited for considering perturbations since the distinguished role of 0 is preserved. Instead let $f(x, u) = (\bar{f}(x, u), u)$ denote an unfolding of f_0. Then, when $u_0 \neq 0$, $\bar{f}(., u_0)$ need not send 0 to 0 and so the special role of 0 is lost. For example, for $\bar{f}(x, u) = x^3 - ux$, when $u > 0$ the singular point at 0 splits into two critical points at $x = \pm\sqrt{(u/3)}$; and the distinguished role of 0 disappears. Hence, instead of considering families in $\mathcal{A}$ acting on $C(n, p)$, we consider the action of unfoldings. We let $\mathcal{A}_{un}(q)$ denote the group of unfoldings on q-parameters $u = (u_1, ... , u_q)$. This group acts on unfoldings on q-parameters u, $f(x, u) = (\bar{f}(x, u), u)$, of germs $f_0 \in C(n, p)$. For a 1-parameter unfolding f, since $\bar{f}(x, u)$ is not

required to send 0 to 0 when $u \neq 0$. Thus, $\frac{\partial \bar{f}}{\partial u}\big|_{u=0}$ will be a vector field along f_0, but it

need not vanish at 0. This leads to "extended tangent spaces" for both $\mathcal{A}$ and $C(n, p)$, obtained by differentiating 1-parameter unfoldings and evaluating at $u = 0$. Furthermore, given $f_0 \in C(n, p)$, $\mathcal{A}_{un}(1)$ will act on the trivial unfolding $f_0 \times id_{\mathbb{R}}$. It gives rise to an orbit map α_{f_0}. We now obtain

$$T\mathcal{A}_e = \theta_n \oplus \theta_p \qquad \text{and} \qquad TC(n, p)_e = \theta(f_0)$$

and $d\alpha_{f_0}$ is none other than that given by (3.4); it exactly captures the perturbations allowing the origin to move. The regular tangent spaces $T\mathcal{A}$ and $TC(n, p)$ differ from these by just the constant vector fields. We define the (extended) tangent spaces to the orbit of f by

$$T\mathcal{A} \cdot f = d\alpha_{f_0}(T\mathcal{A}) \qquad \text{and} \qquad T\mathcal{A}_e \cdot f = d\alpha_{f_0}(T\mathcal{A}_e).$$

The finite determinacy and unfolding theorems are stated in terms of these tangent spaces.

Before doing so we recall one more fundamental observation made by Mather [Ma1–III–

IV]; namely, in (3.7) the module $C_x\{\frac{\partial f_0}{\partial x_i}\} + f_0^* m_y \cdot \theta(f_0)$ which appears on the left hand side

is itself the extended tangent space for another very basic group action. The group $\mathcal{K}$, called *contact equivalence*, consists of $H \in \mathcal{D}_{n+p}$ such that there is an $h \in \mathcal{D}_n$ so that $H \circ i = i \circ h$ and $\pi \circ H = h \circ \pi$ where $i(x) = (x, 0)$ is the inclusion $i : \mathbb{R}^n \hookrightarrow \mathbb{R}^{n+p}$ and $\pi(x, y) = x$ is the projection $\pi : \mathbb{R}^{n+p} \to \mathbb{R}^n$. Then, $\mathcal{K}$ acts on $C(n, p)$ by

$$(h(x), H \cdot f(x)) = H(x, f(x)) \qquad \text{i.e.} \qquad graph(H \cdot f) = H(graph(f)).$$

Germs are $\mathcal{K}$-*equivalent* if they lie in a common orbit of the $\mathcal{K}$-action. This group also extends to a group of unfoldings $\mathcal{K}_{un}(q)$ acting on q-parameter unfoldings. Here the group consists of unfoldings of H and h, with H preserving $\mathbb{R}^n \times \{0\} \times \mathbb{R}^q$, acting in the same way on unfoldings f.

This equivalence, despite its apparently strange definition captures the ambient equivalence of $f_0^{-1}(0)$ for $f_0 : \mathbb{R}^n, 0 \longrightarrow \mathbb{R}^p, 0$ (it sends graph(f_0) to graph(H·f_0) and preserves $\mathbb{R}^n \times \{0\}$, but the intersection graph(f_0) $\cap$ ($\mathbb{R}^n \times \{0\}$) is exactly $f_0^{-1}(0)$). Furthermore, Mather shows that f and g : $\mathbb{R}^n, 0 \longrightarrow \mathbb{R}^p, 0$ are contact equivalent iff there is a $\varphi \in \mathcal{D}_n$ so that $\varphi^*(I(f)) = I(g)$. This implies Q(f) $\simeq$ Q(g); and if both algebras are finite dimensional, the implication reverses.

Quite remarkably, we are coming full circle from our beginning with the Lyapunov–Schmidt reduction. There we explicitly emphasized the full reduction to study the entire mapping and not just the zero-set. The classification theorem implies that for stable germs that the zero-set (with some associated algebraic data added on) determines the germ up to $\mathcal{A}$-equivalence!

A computation for $\mathcal{K}$ [Ma1–III] shows that for α_{f_0} denoting the orbit map for $\mathcal{K}$-equivalence

$$d\alpha_{f_0} : T\mathcal{K}_e \left(= C_x \left\{\frac{\partial}{\partial x_j}\right\} + m_y \cdot C_{x,y} \left\{\frac{\partial}{\partial y_i}\right\}\right) \longrightarrow T\mathcal{C}(n, p)_e = \theta(f_0)$$

$$(\xi, \eta) \longmapsto -\xi(f_0) + \eta \circ \tilde{f}_0$$

where $\tilde{f}_0 : \mathbb{R}^n, 0 \longrightarrow \mathbb{R}^{n+p}, 0$ is defined by $\tilde{f}_0(x) = (x, f_0(x))$. Then, again letting the (extended) tangent space be defined by $T\mathcal{K} \cdot f_0 = d\alpha_{f_0}(T\mathcal{K})$ (or $T\mathcal{K}_e \cdot f_0 = d\alpha_{f_0}(T\mathcal{K}_e)$) we obtain

$$T\mathcal{K} \cdot f_0 = m_x \left\{\frac{\partial f_0}{\partial x_j}\right\} + f_0^* m_y \cdot C_x \left\{\frac{\partial}{\partial y_i}\right\}$$

$$T\mathcal{K}_e \cdot f_0 = C_x \left\{\frac{\partial f_0}{\partial x_j}\right\} + f_0^* m_y \cdot C_x \left\{\frac{\partial}{\partial y_i}\right\}.$$

In addition to $\mathcal{A}$ and $\mathcal{K}$, there are several other basic groups which Mather considered; namely, $\mathcal{R} = \mathcal{D}_n$ and $\mathcal{L} = \mathcal{D}_p$ acting on $C(n, p)$, and C which is the subgroup of $H \in \mathcal{K}$ for which $h = \mathrm{id}_{\mathbb{R}^n}$. The action of $\mathcal{R}$ is referred to as "right equivalence" and that of $\mathcal{L}$, "left equivalence". These additional groups have tangent spaces $T\mathcal{G}f_0$ (or $T\mathcal{G}_e \cdot f_0$) which are the appropriate summands of the tangent spaces for $\mathcal{A}$ and $\mathcal{K}$ e.g. $T\mathcal{R}_e \cdot f_0 = C_x$ $\left\{ \frac{\partial f_0}{\partial x_i} \right\}$. Likewise, we can introduce the *normal spaces*

$$N\mathcal{G}f_0 = \theta(f_0)/T\mathcal{G}f_0 \qquad \text{and} \qquad N\mathcal{G}_e \cdot f_0 = \theta(f_0)/T\mathcal{G}_e \cdot f_0$$

For any of these groups $\mathcal{G}$ we have the straight-forward modification of (3.11) to the notion of finite $\mathcal{G}$-determinacy. A germ f_0 is said to have *finite $\mathcal{G}$-codimension* if $\dim_{\mathbb{R}}(N\mathcal{G}f_0) < \infty$. These two notions are related. First, if f is ℓ-determined for $\mathcal{G}$-equivalence and $h \in m_x^{\ell+1} \cdot \theta(f_0)$, then $f_t = f_0 + t \cdot h$ lies in the orbit $\mathcal{G}f_0$. Thus, $h = \dfrac{\partial f_t}{\partial t}\Big|_{t=0}$

$\in T\mathcal{G}f_0$ and so $m_x^{\ell+1} \cdot \theta(f_0) \subset T\mathcal{G}f_0$, or f_0 has finite $\mathcal{G}$-codimension. Then, for $\mathcal{G} = \mathcal{A}$, $\mathcal{K}$, $\mathcal{R}$, $\mathcal{L}$, or C, Mather proves the much more difficult converse [Ma1–III].

Finite Determinacy Theorem [Ma1–III]: *A germ $f_0 : \mathbb{R}^n, 0 \longrightarrow \mathbb{R}^p, 0$ is finitely $\mathcal{G}$-determined iff it has finite $\mathcal{G}$-codimension*

In the case of $\mathcal{G} = \mathcal{K}$, $\mathcal{R}$, or C, $T\mathcal{G}f_0$ is a C_x-module, and using Nakayama's lemma an explicit order of determinacy can be given based on the smallest integer r such that $m_x^r \cdot \theta(f_0) \subseteq T\mathcal{G}f_0$; in each of these cases f_0 will be r+1-determined for $\mathcal{G}$-equivalence. In the case of $\mathcal{A}$-equivalence, the results are considerably harder and require subtle use of the preparation theorem. For example, (see also [GaDu])

Theorem [Gaf1] : *Suppose for $f_0 : \mathbb{R}^n, 0 \longrightarrow \mathbb{R}^p, 0$ that $m_x^r \cdot \theta(f_0) \subseteq T\mathcal{A}_e \cdot f_0$ and that $m_x^s \cdot \theta(f_0) \subseteq T\mathcal{K}_e \cdot f_0$ then f_0 is r+s-determined for $\mathcal{A}$-equivalence.*

In general, such estimates are not the best possible; this can be partially dealt with using a result of Bruce-DuPlessis-Wall [BWD], which we describe further below. Another difficulty in the case of nonlinear operators is the precise computation of sufficiently many derivatives to apply criteria for finite determinacy for $\mathcal{A}$-equivalence. In the case of stability, we will apply the preceding results in §5.

Consequences for Classification of Germs

First, we explain what we mean by a classification. We can not possibly classify all germs even up to some equivalence relation. Instead we try to classify all germs up to a given codimension m. This means that we can find a list of germs such that every germ in the list has $\mathcal{G}$-codimension (or possibly $\mathcal{G}_e$-codimension) $\leq$ m and every germ of codimension $\leq$ m is $\mathcal{G}$-equivalent to one of the germs on the list. The m that is used varies from case to case and depends on the number of parameters in a problem as well as the patience of the person carrying out the computations.

Then, finite determinacy also plays a crucial role in the classification of germs beyond that of specifying the order of the Taylor expansion determining the germ,. We explain this further. Any of the preceding groups $\mathcal{G}$ have an induced action of $\mathcal{G}^{(\ell)}$ = {the ℓ-jets of elements of $\mathcal{G}$} on finite jets $J^\ell(n, p)$ (e.g. for $\mathcal{A}$, $(j^\ell(\varphi), j^\ell(\psi)) \cdot j^\ell(g) = j^\ell(\psi \circ g \circ \varphi^{-1})$). This group $\mathcal{G}^{(\ell)}$ is a finite dimensional Lie group. If f_0 is ℓ-determined for $\mathcal{G}$-equivalence, then so is any germ g with ℓ-jet $j^\ell(g)$ $\mathcal{G}^{(\ell)}$-equivalent to $j^\ell(f_0)$; for if $j^\ell(\sigma \cdot g)$ = $j^\ell(f_0)$ for $\sigma \in \mathcal{G}$, then the ℓ-determinacy of f_0 implies $\sigma \cdot g$, and hence g itself, is $\mathcal{G}$-equivalent to f_0. In particular it is true for $j^\ell(g)$ viewed as a polynomial; so g and $j^\ell(g)$ are both $\mathcal{G}$-equivalent to f_0, and hence to each other. Thus, the $\mathcal{G}$-orbit $\mathcal{G} \cdot f_0$ is determined by the $\mathcal{G}^{(\ell)}$-orbit of $j^\ell(f_0)$. However, determining the orbit of a Lie group is analogous to the triviality results already considered. Like so much else it follows by a lemma of Mather [Ma1–IV§3]

Mather's Geometric Lemma : *Let* G *be a finite dimensional Lie group acting smoothly on a manifold* M *(without boundary). A sufficient condition that a curve* γ : [0, 1] $\longrightarrow$ M *belong to a single G-orbit is that for all* t $\in$ [0, 1]:
i) $\gamma'(t) \in T_{\gamma(t)}M$,   and   ii) $\dim_{\mathbb{R}} T_{\gamma(t)}M$ *is independent of* t.

By (3.3) $T\mathcal{G}^{(\ell)} \cdot f_0$ can be identified with $T\mathcal{G} \cdot f_0 / m_x^\ell \cdot \theta(f_0)$ so that results analogous to those already considered suffice to classify germs up to $\mathcal{G}$-equivalence. For example, let f_t = f +th where $j^2(h) = 0$ and

$$f_0(x, y) \; = \; (x^2 + \varepsilon y^2, xy).$$

Then, f_0 is 3-$\mathcal{K}$-determined by an earlier remark. Also, for each t

$$T\mathcal{K}^{(3)} \cdot f_t = (m_{x,y}{}^2)^{(2)} + \langle (2x,y), (2\varepsilon y, x) \rangle \quad \mathrm{mod} \ (m_{x,y}{}^3)^{(2)}$$

so by Mather's geometric lemma $j^3(f_t)$ belongs to a single $\mathcal{K}^{(3)}$-orbit, and so all f_t are $\mathcal{K}$-equivalent. Now we can return to (3 of example 3.10. If we replace f by $F_t = f + tg$ where $j^2(g) = 0$, then after applying Lyapunov–Schmidt reduction to F_t, we obtain unfoldings of germs of the form f_t. Thus, by the preceding argument $Q(F_1) \simeq Q(f)$. Also, both f and F_1 are stable so they are equivalent and f is 2-$\mathcal{A}$-determined.

There is a significant improvement in Mather's Geometric Lemma which allows Bruce–DuPlessis–Wall to provide a very effective method for establishing the order of determinacy. The idea is that if $j^\ell(g) = j^\ell(f)$ then to produce an element $\sigma \in G$ which makes g G-equivalent to f it may be possible to use a special σ, e.g. $j^r(\sigma) = $ id for some r. Hence, it may be enough to replace G by a subgroup $\mathcal{U}$ which has special properties.

We describe their result for the important special cases where $G = \mathcal{A}, \mathcal{K}, \mathcal{R}, \mathcal{L}$, or C. An important class of subgroups are the unipotent subgroups $\mathcal{U}$ which are also "strongly closed subgroups", (this latter condition means that $\mathcal{U}$ contains some $G_s \overset{\mathrm{def}}{=} \{\sigma \in G : j^s(\sigma) = \mathrm{id}\}$. Essentially, for the group $\mathcal{U}$ to be unipotent it is sufficient that the group $\mathcal{U}^{(1)}$ of 1–jets in $\mathcal{U}$ should be unipotent [BDW§1], which means that there is a choice of vector space basis for $J^1(n, p) \simeq \mathrm{Hom}(\mathbb{R}^n, \mathbb{R}^p)$ such that the matrix representation for the action of each $j^1(\sigma)$ is upper triangular. For example, for $G = \mathcal{R}$ it is sufficient to find local coordinates so that $d\psi(0)$ is upper triangular for each $\psi \in \mathcal{U}$.

Then, an optimal estimate for determinacy in terms of these unipotent subgroups is given by the following.

Theorem [BDW] : *Suppose that G has a strongly closed subgroup $\mathcal{U}$ with $\mathcal{U}^{(1)}$ unipotent. If $f_0 : \mathbb{R}^n,0 \longrightarrow \mathbb{R}^p,0$ has $m_x{}^{r+1} \cdot \theta(f_0) \subseteq T\mathcal{U} \cdot f_0$ then f_0 is r–determined for G–equivalence*

For the earlier example
$$f(x, y, u, v) = (x^2 + \varepsilon y^2 + ux, xy + vy, u, v)$$

we can use for $\mathcal{U}$ the subgroup of $\mathcal{A}$ consisting of (φ, ψ) such that $(j^1(\varphi)(0), j^1(\psi)(0)) =$ (id, id), i.e. $\mathcal{A}_1$. Then, a calculation using the stability of f shows

$$m_{x,y,u,v}{}^3 \cdot \theta(f) \subseteq T\mathcal{U} \cdot f = m_{x,y,u,v}{}^2 \left\{ \frac{\partial \bar{f}}{\partial x}, \; \ldots, \frac{\partial \bar{f}}{\partial v} \right\} + m_{z,u,v}{}^2 C_x \left\{ \frac{\partial}{\partial z_i}, \frac{\partial}{\partial u}, \frac{\partial}{\partial v} \right\}$$

Thus, by the theorem, f is 2–determined for $\mathcal{A}$-equivalence. We also saw this earlier using a more involved argument involving stability.

Versal Unfoldings

Having seen how we can classify germs for a number of different equivalences, we now answer the remaining question of how we can determine all possible small perturbations of a germ f_0. In this we follow Martinet [Mar2] and concentrate on the cases $G = \mathcal{A}$ and $\mathcal{K}$(or Wasserman for $\mathcal{R}$). The basic idea is that possible small perturbations of f_0 are described by the possible unfoldings of f_0. To describe all possible small perturbations via a finite parameter unfolding, we must define a relation among unfoldings so that we can decide when one contains all of the perturbations described by the other. Let $f : \mathbb{R}^{s+q}, 0 \to \mathbb{R}^{t+q}, 0$ and $g : \mathbb{R}^{s+r}, 0 \to \mathbb{R}^{t+r}, 0$ be unfoldings of f_0, with $f(x, u) = (\bar{f}(x, u), u)$ and $g(x, v) = (\bar{g}(x, v), v)$. For unfoldings on the same parameters, we have described for $\mathcal{A}$ and $\mathcal{K}$-equivalence the corresponding action on unfoldings. We extend this to different parameter spaces via the notion of a *mapping of unfoldings*. A mapping from g to f consists of a germ $\lambda : \mathbb{R}^r, 0 \to \mathbb{R}^q, 0$ such that $\lambda^* f(x, v) = (\bar{f}(x, \lambda(v)), v)$ is G-equivalent as an unfolding to g. We denote it $\lambda : (g, r) \to (f, q)$. Then, the unfolding f of f_0 is said to be G-versal if for any other unfolding $g : \mathbb{R}^{s+r}, 0 \to \mathbb{R}^{t+r}, 0$ of f_0, there is a mapping of unfoldings $\lambda : (g, r) \to (f, q)$. Note this captures exactly the idea that every perturbation $\bar{g}(x, v)$ of f_0 can be realized from f (up to G-equivalence). Then, Martinet (and Wasserman for $\mathcal{R}$) gives the following simple infinitesimal characterization of versality.

Unfolding Theorem [Mar2] [Ws1]: *Let* $f(x, u) = (\bar{f}(x, u), u)$ *be an unfolding of* f_0, *where* $f : \mathbb{R}^{s+q}, 0 \to \mathbb{R}^{t+q}, 0$. *For* $G = \mathcal{A}$, $\mathcal{K}$, *or* $\mathcal{R}$, *the following are equivalent*

i)   *f is* G-versal
ii)   *f is infinitesimally* G-versal i.e.

$$T\mathcal{G}_e \cdot f_0 + \langle \partial_1 f, \dots, \partial_q f \rangle \;=\; \theta(f_0)$$

(*where as earlier in* (3.6) $\partial_j f = \dfrac{\overline{\partial f}}{\partial u_j}\big|_{u=0}$).

There are several immediate corollaries which Martinet deduces regarding uniqueness of versal unfoldings. Infinitesimal versality requires that the images of $\{\partial_1 f, \dots, \partial_q f\}$ span $N\mathcal{G}_e \cdot f_0$; then a $\mathcal{G}$–versal unfolding (f, q) is said to be *miniversal* if q = $\dim_{\mathbb{R}} N\mathcal{G}_e \cdot f_0$. Next, we say two unfoldings of f_0, (f, q) and (g, q) are $\mathcal{G}$–*equivalent* if there exist mappings of unfoldings $\lambda : (g, q) \to (f, q)$ and $\sigma : (f, q) \to (g, q)$ such that the composition in either order is the identity (i.e. λ is a germ of a diffeomorphism).

Corollary 3.17 : For $\mathcal{G} = \mathcal{A}, \mathcal{K}$, or $\mathcal{R}$,

i)   *A germ f_0 has a $\mathcal{G}$-versal unfolding iff* $\dim_{\mathbb{R}} N\mathcal{G}_e \cdot f_0 = q < \infty$

ii)   *Any two $\mathcal{G}$-miniversal unfoldings of f_0 are $\mathcal{G}$-equivalent*

iii)   *Any $\mathcal{G}$-versal unfolding (g, r) of f_0 is $\mathcal{G}$-equivalent to* (f × $\mathrm{id}_{\mathbb{R}^{r-q}}$, r) *for a miniversal unfolding* (f, q).

As consequences of this theorem, we can repeat the construction of (3.9) and let $\{\varphi_1, \dots, \varphi_q\}$ be a set of germs whose images form a basis for $N\mathcal{G}_e \cdot f_0$. Then, a $\mathcal{G}$-*miniversal unfolding* of f_0 is given by

(3.18)
$$f(x, u) \;=\; \Big(f_0(x) + \sum_{i=1}^{q} u_i \cdot \varphi_i, \, u_1, \dots, u_q\Big)$$

The corollary implies that any $\mathcal{G}$-miniversal unfolding is $\mathcal{G}$-equivalent to (3.18), and any $\mathcal{G}$-versal unfolding is $\mathcal{G}$-equivalent to a product of (3.18) with an identity map.

Furthermore, this has as a consequence a relation between stability and $\mathcal{K}$-versality. We directly see that (3.7) characterizing the stability of the germ f(x, u) = ($\overline{f}$(x, u), u), viewed as an unfolding of f_0, is equivalent to the $\mathcal{K}$-versality of the unfolding F with $\overline{F}$(x, u, v) = $\overline{f}$(x, u)–v (here v ∈ $\mathbb{R}^p$). Moreover, if we let $\mathcal{V} = \overline{F}^{-1}(0)$ then $\mathcal{V}$ is a smooth submanifold and the projection (x, u, v) $\mapsto$ (u, v) : $\mathcal{V} \longrightarrow \mathbb{R}^{p+q}$ is none other than f. This is the underlying reason for the classification theorem for stable germs and allowed Martinet to give a direct proof using only the unfolding theorem. This projection map $\mathcal{V} \longrightarrow \mathbb{R}^{p+q}$ is the also the versal deformation of V = $f_0^{-1}(0)$ in the sense of algebraic geometry. Also, it follows that those germs f_0 which have stable unfoldings are exactly the finitely $\mathcal{K}$-determined map germs. These are also referred to as germs of *finite singularity type*; if n ≤

p these are exactly the finite map germs.

Example 3.19 (splitting lemma):

As an illustration of the usefulness of the versality theorem, we reconsider a germ f $: \mathbb{R}^m, 0 \to \mathbb{R}, 0$ with a degenerate critical point at 0 so the Morse lemma doesn't apply. What can still be said about the form we can obtain for f after a change of coordinates in $\mathbb{R}^m$? Suppose that $d^2f(0)$ has rank n. We may assume, for example, that $d^2f(0)| \mathbb{R}^n$ is nondegenerate. If we let $f_0(x_1, \ldots, x_n) = f(x_1, \ldots, x_n, 0, \ldots, 0)$ then f_0 has a nondegenerate critical point as a germ $f_0 : \mathbb{R}^n, 0 \to \mathbb{R}, 0$. Thus by the Morse lemma, there is a local change of coordinates φ on $\mathbb{R}^n, 0$ so that

$$(3.20) \qquad f_0 \circ \varphi^{-1}(x_1, \ldots, x_n) = -\sum_{j=1}^{r} x_j^2 + \sum_{j=r+1}^{n} x_j^2$$

Hence, composing with $\varphi \times id_{\mathbb{R}^{m-n}}$ we might as well assume f has this form. We let x' denote $(x_1, \ldots, x_n)$, x'' denote $(x_{n+1}, \ldots, x_m)$, and $Q(x')$ denote the right hand side of (3.20). By a straightforward calculation using the versality theorem, we see that the $\mathcal{R}$-versal unfolding of $f_0(x') = Q(x')$ is simply $F(x', u) = (f_0(x') + u, u)$. However, f defines an unfolding of f_0 on the parameters x''. Thus, the versality theorem implies there is a smooth germ $\lambda : \mathbb{R}^{m-n}, 0 \to \mathbb{R}, 0$ and a germ of a diffeomorphism $\psi : \mathbb{R}^m, 0 \to \mathbb{R}^m, 0$ of the form $\psi(x', x'') = (\bar{\psi}(x', x''), x'')$ such that

$$f \circ \psi^{-1}(x', x'') = f_0(x') + \lambda(x'').$$

By the properties of f, it follows that λ has a critical point at 0 and $d^2\lambda(0) \equiv 0$. Thus, we obtain an extension of the Morse lemma which is usually called the splitting lemma.

Splitting Lemma : *Suppose a germ* f $: \mathbb{R}^m, 0 \to \mathbb{R}, 0$ *has a critical point at 0 with* rk $d^2f(0) = n$, *then f is right –equivalent to a germ of the form*

$$-\sum_{j=1}^{r} x_j^2 + \sum_{j=r+1}^{n} x_j^2 + h(x_{n+1}, \ldots, x_m)$$

where $dh(0) = d^2h(0) = 0$ *(furthermore, h can be shown to be unique up to right equivalence).*

Remark : There is a slight variant of $\mathcal{R}$-equivalence which allows the addition of a constant to the function; this is sometimes called $\mathcal{R}^+$-equivalence (basically it is the relative values of critical values that are of interest, not their absolute values). The analysis of the $\mathcal{R}^+$-versal unfoldings of germs $f_0 : \mathbb{R}^n, 0 \longrightarrow \mathbb{R}, 0$ of $\mathcal{R}^+_e$-codimension ≤ 4 constitutes Thom's "elementary catastrophe theory" [Th1]. This includes the descriptions of the generic transitions which occur in a 1-parameter family joining distinct Morse functions, including normal forms for the "birth" and "death" of critical points (usually called A_2-singularities). The quadratic part of the function in the splitting lemma has no effect in catastrophe theory; so the splitting lemma is the reason that only unfoldings of certain standard functions in one and two variables are needed for elementary catastrophe theory.

Versality and Transversality

We conclude this section by briefly explaining how we can understand geometrically the versality theorem. For open sets $U \subseteq \mathbb{R}^n$ and $V \subseteq \mathbb{R}^p$, we can consider the jet space $J^\ell(U, V)$ consisting of ℓ-jets of germs g: $U, x \longrightarrow Y, y$. It is a trivial fiber bundle over $U \times V$ with fiber $J^\ell(n, p)$. Hence, by (3.3), for $f_0 : \mathbb{R}^n, 0 \longrightarrow \mathbb{R}^p, 0$ we can express the tangent space algebraically

(3.21)
$$T_{j^\ell(f_0)} J^\ell(U, V) \simeq \theta(f_0)/m_x^{\ell+1} \cdot \theta(f_0) \oplus \mathbb{R}^n.$$

Because each of the groups $\mathcal{G}$ we have been considering have induced actions as Lie groups $\mathcal{G}^{(\ell)}$ on $J^\ell(n, p)$, we can form "bundles of orbits" in $J^\ell(U, V)$ of the form: $\tilde{\mathcal{A}}^{(\ell)} \cdot g = U \times V \times \mathcal{A}^{(\ell)} \cdot g$ for $\mathcal{A}$, $\tilde{\mathcal{K}}^{(\ell)} \cdot g = U \times \{y\} \times \mathcal{K}^{(\ell)} \cdot g$ for $\mathcal{K}$, etc. so that in (3.21)

(3.22)
$$T_{j^\ell(f_0)} \tilde{\mathcal{G}}^{(\ell)} \cdot f_0 + \langle \frac{\partial f_0}{\partial x_1}, ..., \frac{\partial f_0}{\partial x_n} \rangle \simeq T\mathcal{G}_e \cdot f_0$$

For f_0: $U \longrightarrow V$ we can define the jet map $j^\ell(f_0) : U \longrightarrow J^\ell(U, V)$ sending $x \mapsto j^\ell(f_0)(x)$, the ℓ-jet of f_0 at x; and for an unfolding f: $U \times W \longrightarrow V \times W$ with $W \subseteq \mathbb{R}^q$ we have $j_x^\ell(f) : U \times W \longrightarrow J^\ell(U, V)$ sending $(x, u) \mapsto j^\ell(f(.,u))(x)$. This latter map has derivative

(3.23) $dj^\ell(f)(0)(\frac{\partial}{\partial x_i}) = j^\ell(\frac{\partial f_0}{\partial x_i})$ and $dj^\ell(f)(0)(\frac{\partial}{\partial u_j}) = j^\ell(\frac{\partial f}{\partial u_j}|_{u=0})$

Finally, we see that $j_x^\ell(f)$ is transverse to $\tilde{\mathcal{G}}^{(\ell)} \cdot f_0$ at 0 iff

$$\mathrm{dj}_x{}^{\ell}(f)(T_{(0,0)}U \times W) + T_j{}^{\ell}(f_0)(0) \; \tilde{\mathcal{G}}^{(\ell)} \cdot f_0 \;=\; T_j{}^{\ell}(f_0)(0) \; J^{\ell}(U, V)$$

From the identification (3.3) together with (3.22) and (3.23), this becomes

$$(3.24) \qquad\qquad T\mathcal{G}_e \cdot f_0 + <\partial_1 f, \dots, \partial_q f> \;\equiv\; \theta(f_0) \quad \mathrm{mod} \quad m_x{}^{\ell+1} \cdot \theta(f_0)$$

Thus, versality implies, via (3.24), transversality of the jet map; and conversely using the preparation theorem, it can be shown that transversality to the bundle of $\mathcal{G}^{(\ell)}$-orbits for sufficiently large ℓ implies infinitesimal versality and hence versality. However, the Thom Transversality theorem (see §7 or also e.g. [GG,chap. 2]) implies that transversality to such bundles of orbits is "generic behavior"; thus, we should normally expect to see such versal behavior in families. In particular, if we are looking at k-parameter families then generically we should only expect to encounter versal families of codimension k germs, i.e. as asserted earlier, passing through nonstable germs in families occurs in a stable fashion. In fact, this is correct as stated in "low codimensions" by the Thom transversality theorem; however, eventually we encounter moduli which parametrize continuous change in equivalence classes. Then, we are forced to weaken the equivalence to a topological version to still recover a correct version of this claim as described in §7.

In the next section, we consider how these results serve as models for numerous variations of the theory when special conditions are added. Again we would like to have a local theory which takes into account the special conditions; and yet establishes the basic theorems of singularity theory, namely, the unfolding and determinacy theorems, stability, and a method for classifying germs.

§4   Singularity Theory with Special Conditions

It was recognized quite early on (at the beginning of the 70≈s) by people who wanted to apply singularity theory to physical problems that often the problems did not fit into the theories which had been developed up to that point by Thom and Mather [Th1],[Ma1]. This includes the structural stability already discussed as well as elementary catastrophe theory which, as we mentioned, concerns the study of real–valued functions and their deformations under right equivalence. For example, Thom in his book -Structural Stability and Morphogenesis- proposed applications of elementary catastrophe theory to explain a variety of phenomena in physics and biology. Zeeman further explored applications such as to the equations for the heartbeat [Ze] (see also [PS]). Arnold also used the classification in a very unexpected way to classify the generic structure of caustics appearing in wave–front evolution via projections of Lagrangian submanifolds [A2]. The work of Arnold on caustics and Zeeman on the heartbeat both showed that applications may require additional information to be incorporated into standard singularity theory. This leads to the next question.

QUESTION: Can versions of Mather's results be developed which take into account special features of a problem such as: symmetries in physical problems, special roles played by certain variables or parameters, constraint conditions, etc?

One important example of such additional information is the special role that a variable such as time may play. Wasserman [Ws2] obtained a modified version of catastrophe theory in which there is a distinguished unfolding parameter, usually denoting time. Then, with Mather's local theory and Martinet's unfolding theorem for $\mathcal{A}$ and $\mathcal{K}$ as paradigms, workers began producing a flood of applications. Golubitsky and Schaeffer [GS1] recognized that for bifurcation problems the distinguished parameter could be incorporated as part of the data for the problem rather than treating it as a special unfolding parameter. For a compact symmetry group G, Gerald Schwarz [Sz] proved the analogue of the Hilbert theorem for the ring of smooth G-invariant functions, i.e. there are a finite number of G-invariant polynomials p_i so that any G-invariant smooth function is a smooth function of the p_i. This provides the basis for applications involving symmetries (see also Bierstone [Bi]). The use of symmetries for bifurcation questions was proposed by Sattinger [Sa2] and considerably developed through the work of Golubitsky-Schaeffer, see e.g. [GS3] [GSS].

Just how generally can the question be answered? We list some of the special conditions which singularity theory can incorporate, allowing it to be remarkably versatile for analyzing models for nonlinear problems.

(4.1)   Special   Conditions

1)   Distinguished Parameters :   Problems may have a set of parameters distinguished in various ways. For example, bifurcation problems may have a distinguished parameter [GS1] [GS2], a subspace of parameters [La],[Iz1] [Pr] [LL], or sequential parameters for a system modeled by various inputs, outputs and controls [DS]; these are examples of a general scheme for allowing various subsets of parameters given by a "ladder of mappings " [D14-I,§8].

For example, for $f_0(x, \lambda)$: $\mathbb{R}^{n+1},0 \longrightarrow \mathbb{R}^n,0$, imperfect bifurcation equivalence as introduced by Golubitsky–Schaeffer is defined via the action of a subgroup $\mathcal{K}^*$ of $\mathcal{K}$ consisting of those (H, h) for which H acts linearly on y for each fixed (x, λ) and h(x, λ) = (h$_1$(x, λ), h$_2$(λ)) so that h preserves the projection onto the λ–axis and hence the special role played by λ.

2)   Symmetries :   Mappings and equivalences can preserve symmetries [P], [Be], [Rg],[Iz2], [Rb], [D1]. The importance of symmetries for bifurcation problems was stressed in [Sa2] and developed in [Sa1], and via singularity theory in [GS2] and [GSS]. This includes: symmetry–breaking, as e.g. [GSt] or see [GSS] for numerous applications; symmetry–breaking in equations [GS4], [D1], or more generally any example in 1) with a symmetry group added [D3] or [D1].

    If G is a compact Lie group which has representations on $\mathbb{R}^n$ and $\mathbb{R}^p$, then we consider G–equivariant mappings f$_0$: $\mathbb{R}^n$,0 $\longrightarrow$ $\mathbb{R}^p$,0, so f$_0$($\sigma \cdot$x) = $\sigma \cdot$f$_0$(x) for $\sigma \in$ G, and G–equivariant equivalences. By the theorem of Schwarz [Sz], the ring $C_x{}^G$ of G–invariant germs on $\mathbb{R}^n$ (i.e. h($\sigma \cdot$x) = h(x) for all $\sigma \in$ G) consists of the smooth functions in the polynomial invariant functions; and C(n, p)^G the space of G–equivariant mappings is a finitely generated $C_x{}^G$–module by an argument in Poenaru [P].

3)   Special Bifurcations :   Bifurcation may be considered not just for zero–sets but for matrices [A3], sets of group orbits [Ge1,2], or constraint sets [D2]. Here the equivalences of germs f$_0$: $\mathbb{R}^n$,0 $\longrightarrow$ $\mathbb{R}^p$,0 form a subgroup of $\mathcal{K}$ which acts via a special equivalence on $\mathbb{R}^p$.

4)   Compositions :   For mappings given as compositions, the equivalences may preserve the composition structure [Ba], [Du1], [Du2], [FM]. For example, for f$_0$: $\mathbb{R}^n$,0 $\longrightarrow$ $\mathbb{R}^p$,0 given as a composition f$_0$ = h$_0$◦g$_0$ with g$_0$: $\mathbb{R}^n$,0 $\longrightarrow$ $\mathbb{R}^m$,0 and h$_0$: $\mathbb{R}^m$,0 $\longrightarrow$ $\mathbb{R}^p$,0, the equivalence is given by germs of diffeomorphisms (φ, χ, ψ) acting via

$$(g_0, h_0) \mapsto (\chi \circ g_0 \circ \varphi^{-1}, \psi \circ h_0 \circ \chi^{-1}).$$

5)   Boundaries and Singular Subspaces :   For points on the boundary of a region, the equivalences can be required to preserve the region (and its boundary), even allowing singular boundaries [Si2], [A1], [BR], [BG2,3], [G], [Ly], [Ta].

6)   Constraint Spaces : If the mappings are restricted to singular constraint spaces [Gal1] [Lê] [D14-I,§8], the equivalences can be restricted to these singular spaces, even allowing the constraint spaces to deform [D4].

7)   Periodic solutions of Hamiltonian Systems : The behavior of perturbations from equilibrium can be analyzed via associated equivariant mappings [Dm] [vM] [MRS1, 2].

For a survey of these and still other examples see e.g. [A4], [Wa2], [D1a], and [D14-I,§8].

On the other hand, singularity theory, in the form we are describing here, does not apply directly to a number of situations including: conjugation of mappings, symplectic equivalence for Hamiltonian systems, nonstable Lagrangian projections [Za], "divergent diagrams" in the sense of Dufour [Du3] [Du4], nor to solutions to partial differential equations [DD], [Iz3], [D5]. For these questions, either nonstandard methods must be employed or the methods must be applied to obtain partial answers about special features for problems.

What are simple guidelines which can be used for deciding when singularity theory in this standard form can be applied? We shall explain the criteria which suffices. In such circumstances, we will then be able to answer a number of questions which were answered in the previous section for the classical notions of equivalence, including:

1) stability : giving criteria for ensuring a germ is stable (under deformations);

2) versality : giving infinitesimal criteria for the existence of a versal unfolding;

3) determinacy: characterizing finite determinacy in terms of the finite codimension of the germ;

4) classification: determining when two germs are equivalent.

APPLYING SINGULARITY THEORY TO PROBLEMS :

To model nonlinear problems via singularity theory, we must specify three ingredients : the models, an equivalence relation on the models which captures the desired behavior, and a collection of allowable perturbations.

Models : They are given by a space of germs $\mathcal{F}$ which is either a linear or affine subspace of the space of germs $C(n, p) = \{ f : \mathbf{k}^n, 0 \longrightarrow \mathbf{k}^p, 0 \}$, which now may denote the space of smooth or real analytic germs if $\mathbf{k} = \mathbb{R}$ or holomorphic germs if $\mathbf{k} = \mathbb{C}$. We do this to emphasize that unlike many results for mappings, the results we describe are equally valid for the real or complex cases.

Equivalence Relation : On this space $\mathcal{F}$ there is an equivalence relation induced by the action of a group of germs of diffeomorphisms $\mathcal{G}$. The space of germs and the group are chosen to model the problem under consideration, taking into account special conditions such as those already described.

Perturbations : The allowable perturbations of a germ $f_0 \in \mathcal{F}$ are given by a space of unfoldings of f_0, i.e. germs $f(x,u) = (\bar{f}(x,u),u)$ with $f(x,0) = f_0$ and f belonging to an (affine) space of allowable unfoldings $\mathcal{F}_{un}(q)$ (here u denotes local coordinates for $\mathbf{k}^q$). For $\mathcal{G}$, we must extend the equivalence to unfoldings $\mathcal{G}_{un}(q)$ acting on $\mathcal{F}_{un}(q)$. This is not intrinsically given; however, there it usually an obvious extension to unfoldings. When we refer to $\mathcal{G}$, it will be understood that the $\mathcal{G}_{un}$ are included.

An example is the action of $\mathcal{A}$ on $C(n, p)$ extending to an action on unfoldings f by $(\varphi', \psi') \cdot f = \psi' \circ f \circ \varphi'^{-1}$ for φ' and ψ' unfoldings of germs of diffeomorphisms. We are principally interested in the case when $\mathcal{G}$ is a subgroup of $\mathcal{A}$ or $\mathcal{K}$. In fact, all of the equivalences described in 1) – 7) are given by groups which can be realized as subgroups of $\mathcal{A}$ or $\mathcal{K}$. Hence, we can denote elements as elements of either $\mathcal{A}$ or $\mathcal{K}$. We will usually use $\mathcal{A}$ for illustration.

Once certain basic results are established for these two groups, these results will often be inherited by naturality for a large collection of subgroups. Part of our goal is to describe these subgroups. On the other hand we might also ask why intrinsically should A and K be so important? This we can also answer. The largest group of diffeomorphisms of kn+p,0 which carries graphs of germs to graphs of germs is the -gauge group- K which consists of $H \in Dn+p$, with an associated $h \in Dn$ as for the contact group, except it is not required to preserve the subspace kn × {0}. Any interesting group which maps germs to germs will be a subgroup of K . Then, we have the inclusions

$$\mathcal{K}(n, p) \subset \tilde{\mathcal{K}}(n, p) \subset \mathcal{A}(n, n+p) \subset \tilde{\mathcal{K}}(n, n+p)$$

where the middle inclusion is given by $H \mapsto (h, H)$ and arises by identifying a germ $f \in C(n, p)$ with a germ of a section of the trivial vector bundle $\mathbf{k}^{n+p}$ over $\mathbf{k}^n$ with fiber $\mathbf{k}^p$, namely, $\tilde{f}(x) = (x, f(x))$. These inclusions naturally extend to the corresponding groups of unfoldings (unlike the inclusion $\mathcal{A}(n, p) \subset \mathcal{K}(n, p)$ which does not, as unfoldings in $\mathcal{A}_{un}$ do not preserve 0 when $u \neq 0$).

Thus, we can always suppose we are considering subgroups of $\mathcal{A}$ or $\mathcal{K}$.

Tangent Spaces and the Infinitesimal Orbit Map :

First, for $\mathcal{G}$ or $\mathcal{F}$ we can compute the tangent spaces and extended tangent spaces exactly as in §3, by differentiating families or unfoldings. For $\mathcal{G}$ we compute the tangent space at id while for $\mathcal{F}$ it doesn't matter as $\mathcal{F}$ is linear or affine. Likewise we can compute tangent spaces for $\mathcal{G}_{un}$ or $\mathcal{F}_{un}$ which will be vector fields depending on the parameters u.

For example, for the group $\mathcal{K}^*$ given in 1) of (4.1) defining imperfect bifurcation equivalence, we observe that for an equivalence of one parameter unfoldings, we only allow change of coordinates for λ to depend on (λ, u). Hence, when we compute tangent spaces we obtain

$$(4.2) \qquad T\mathcal{K}^*_e \;=\; C_{x,\lambda}\left\{\tfrac{\partial}{\partial x_j}\right\} + C_{x,\lambda}\left\{y_j \cdot \tfrac{\partial}{\partial y_i}\right\} + C_\lambda\left\{\tfrac{\partial}{\partial \lambda}\right\}$$

Again, the action of $\mathcal{G}$ on f_0 induces an orbit map

$$\alpha_{f_0} : \mathcal{G} \longrightarrow \mathcal{F}$$
$$(\varphi, \psi) \mapsto (\varphi, \psi) \cdot f_0$$

The derivative of this map induces a map.

$$d\alpha_{f_0} : T\mathcal{G}_e \longrightarrow T\mathcal{F}_e$$

There is an analogous map for unfoldings f, which is exactly the u–parametrized version of that for f_0.

$$d\alpha_f : T\mathcal{G}_{un,e} \longrightarrow T\mathcal{F}_{un,e}$$

The restrictions to $T\mathcal{G}$ or $T\mathcal{G}_{un}$ give the maps on tangent spaces of the groups (here the number of unfolding parameters q is to be understood). Again the images represent the tangent spaces to the orbit, $T\mathcal{G}_e \cdot f_0$ (or $T\mathcal{G} f_0$), $T\mathcal{G}_{un,e} \cdot f$, etc and we can define as in §3 the corresponding normal spaces $N\mathcal{G}_e \cdot f_0 = T\mathcal{F}_e / T\mathcal{G}_e \cdot f_0$, etc. For example, for an unfolding $f(x, u) = (\bar{f}(x, u), u)$,

$$T\mathcal{K}^*_{un,e}\cdot f \;=\; C_{x,\lambda,u}\!\left\{\frac{\overline{\partial f}}{\partial x_i}\right\} + \overline{f}^* \, m_y \, C_{x,\lambda,u}\!\left\{\frac{\partial}{\partial y_j}\right\} + C_{\lambda,u}\!\left\{\frac{\overline{\partial f}}{\partial \lambda}\right\} \;\subseteq\; C_{x,\lambda,u}\!\left\{\frac{\partial}{\partial y_j}\right\}$$

All of the groups we have considered (even those for which standard singularity theory will not be applicable) are defined by geometric conditions. Consequently, there are two general properties which all of them share.

(4.3) 1) <u>Naturality under Pull-backs</u>: Change of parameters still takes an allowable unfolding into another one. Specifically, if $f \in \mathcal{F}_{un}(q)$ and $\lambda: : k^r,0 \longrightarrow k^q,0$, then the pull-back unfolding (as defined in §3) $\lambda^* f \in \mathcal{F}_{un}(r)$; and similarly for $\mathcal{G}_{un}$.

(4.3) 2) <u>Exponential Map</u>: Integrating a (parametrized) vector field representing an element of the tangent space of $\mathcal{G}$ or $\mathcal{G}_{un}(q)$ gives a germ of a one-parameter family in the group. More generally, let $\mathcal{G}_{eq}(q)$ denote equivalences of unfoldings obtained by composing elements of $\mathcal{G}_{un}(q)$ with changes of coordinates in k^q. Then, the corresponding exponential map (for $\mathcal{A}$ or $\mathcal{K}$) restricts to

$$\exp: T\mathcal{G}_{un,e}(q) \oplus \theta_q \longrightarrow \mathcal{G}_{eq}(q+1).$$

(we end up in $\mathcal{G}_{eq}(q+1)$ because the time parameter becomes an unfolding parameter)

It is not really surprising that general properties such as those above are satisfied by many subgroups including those such as conjugation, symplectic equivalence, etc., for which standard singularity theory doesn't apply. What is also reasonably clear is that no general results about the action of such a group are possible without first obtaining precise information about the properties of the tangent spaces and the infinitesimal orbit map between them. From the viewpoint of analysis, this information would take the form of the topology on the spaces (Banach, Frechet, etc.) and the behavior of the infinitesimal orbit map; however, as we have alluded to earlier, the key for singularity theory is the algebraic structure on the tangent spaces and the algebraic properties of the infinitesimal orbit map.

Algebraic structure of the Tangent Spaces and the Infinitesimal Orbit Map

We have seen in various examples that the tangent spaces are sums of modules possibly over several different rings. These rings are connected by various homomorphisms induced by germs. For $\mathcal{A}$ it is $f_0^*: C_y \rightarrow C_x$; for $\mathcal{K}^*$ it is the inclusion

π^*: $C_\lambda \rightarrow C_{x,\lambda}$ where $\pi(x, \lambda) = \lambda$ is projection; etc. Furthermore, the infinitesimal orbit map "behaves well" ,i.e. on each summand of the tangent spaces it is a module homomorphism over a ring homomorphism. It is this property which we wish to capture.

Formally we consider a finite collection of such rings $\{R_\alpha\}_{\alpha \in \mathcal{D}}$ for ($\mathcal{D}, \leq$) a partially ordered index set. We have homomorphisms $\varphi_{\alpha\beta}$: $R_\alpha \rightarrow R_\beta$ exactly when $\alpha \leq \beta$, satisfying the compatibility condition $\varphi_{\beta\gamma} \circ \varphi_{\alpha\beta} = \varphi_{\alpha\gamma}$ ($\varphi_{\alpha\alpha}$ = id). We refer to such a collection $\{R_\alpha\}$ as a system of rings and consider $\{R_\alpha\}$-modules M to be a direct sum of modules $\oplus M_\alpha$ with each M_α an R_α-module and M is finitely generated iff each M_α is. A homomorphism of such modules ψ: M $\rightarrow$ N will be one whose restriction to summands $\psi_{\alpha\beta}$: $M_\alpha \rightarrow N_\beta$ is a module homomorphism over $\varphi_{\alpha\beta}$. An important but unpleasant fact of life is that such homomorphisms completely mix up the different module structures so the image has no simple structure.

The rings that are allowed are those such as C_x, $C_{x,\lambda}$, or their equivariant versions $C_{x,\lambda}{}^G$. We also allow their u-parametrized analogues such as $C_{x,\lambda,u}$, denoted $R_{\alpha,u}$. Such rings are contained in a general class of "differentiable or analytic algebras" introduced by Malgrange [Mg]. We do not give the precise definition here; but we will refer to them simply as DA algebras.

Lastly, the partial ordering on $\mathcal{D}$ is important; it is said to be "adequately ordered" if any $\alpha \in \mathcal{D}$ has at most one immediate successor in the ordering. This means that many rings in the collection may map to a given ring R_α; however, it maps to at most one other ring immediately following it in the ordering. Dufour [Du2] shows that if this condition is violated then the standard application of singularity theory fails. In all 7 of the classes of examples the rings satisfy this condition.

Then, the allowable algebraic structure on the tangent spaces takes the following form.

(4.3) 3) <u>Tangent Space Condition</u> : The tangent spaces $T\mathcal{G}_e$ and $T\mathcal{F}_e$ (and $T\mathcal{G}$ and $T\mathcal{F}$) are finitely generated modules over a system of DA-algebras $\{R_\alpha\}$ (as are $T\mathcal{G}_{un,e}$ and $T\mathcal{F}_{un,e}$ over the system of u-parametrized rings $\{R_{\alpha,u}\}$). For germs f_0 and unfoldings f, $d\alpha_{f_0}$ and $d\alpha_f$ are module homomorphisms over these systems of rings.

The final condition is that the algebraic structure can be used to represent the jet space structure. If m_α denotes the maximal ideal of R_α , we let $\{m_\alpha{}^\ell\}\cdot$M denote

$\oplus m_\alpha^\ell \cdot M_\alpha$ (i.e. the "ℓ-th order terms" in each summand). Then, via (3.3), $T\mathcal{F}/\{m_\alpha^{\ell+1}\}\cdot T\mathcal{F}$ can be identified with the image of $\mathcal{F}$ in the ℓ-jet space; this holds as well for $\mathcal{F}_{un}$.

(4.3) 4) <u>Filtration Condition</u> : $\mathcal{G}_{un}$ preserves the filtration $\{m_\alpha^\ell\}\cdot T\mathcal{F}_{un}$ and induces an action on the quotient $T\mathcal{F}_{un}/\{m_\alpha^\ell\}\cdot T\mathcal{F}_{un}$.

This condition reflects the fact that the group action, even allowing unfoldings, acts at the level of jets, which is essentially a consequence of the chain rule for ℓ-jets. Hence, of these four conditions it is mainly 3) which separates the allowable groups from those which aren't.

Definition : A subgroup $\mathcal{G}$ of $\mathcal{A}$ or $\mathcal{K}$ which satisfies (4.3) 1) – 4) will be said to be a *geometric subgroup of $\mathcal{A}$ or $\mathcal{K}$.*

It is for such groups (provided the algebras are adequately ordered) that the basic theorems of singularity theory are valid and for which the above questions can be answered for $\mathcal{G}$ and $\mathcal{F}$ (see [D1 a or b]).

In applying singularity theory, we either must be able to model the problem using such a geometric subgroup or alternatively concentrate on a part of the problem for which such a group can be used. This is exactly the philosophy that has proven so successful; however, as in Arnol'd's work on caustics, it may require very clever introduction of singularity theory. Alternately, one must find an inventive way to overcome problems with the failure of traditional methods as is done in Dufour's work on divergent diagrams .

The preceding examples provide many illustrations of the applicability. We indicate here two such. Before doing so we should at least state the form which the general theorems take. They are based on a local analogue of the homotopic triviality that we described in §3. This fundamental idea is that the criterion for $\mathcal{G}$-triviality of an unfolding f, which is essentially Martinet's reduction criterion [Mar2], is induced by the naturality and exponential properties to apply to any subgroup satisfying 1) and 2).

Proposition 4.4 (Infinitesimal Reduction Criterion) : *Suppose that $\mathcal{G}$ acting on $\mathcal{F}$ is a geometric subgroup of $\mathcal{A}$ or $\mathcal{K}$. Let $f \in \mathcal{F}_{un}(q)$ be an unfolding of f_0, and let f_1 be* *an unfolding on q–1 parameters defined by*

$$\bar{f}_1(x, u_1, \ldots, u_{q-1}) = \bar{f}(x, u_1, \ldots, u_{q-1}, 0).$$

If

$$\frac{\partial \bar{f}}{\partial u_q} \in C_u\{\frac{\partial \bar{f}}{\partial u_1}, \ldots, \frac{\partial \bar{f}}{\partial u_{q-1}}\} + T\mathcal{G}_{un,e} \cdot f$$

then f is a trivial extension of f_1, *in that as an unfolding it is* $\mathcal{G}$*-equivalent to* $f_1 \times id_k$.

Once this criterion is established, the algebra available from the tangent space condition, together with an extension of the Malgrange preparation to homomorphisms over systems of rings yield *Mather's Algebraic Lemma* (see e.g. [D1a,§3]) which provides the analogues of Nakayama's lemma for the infinitesimal orbit maps. This allows us to establish the reduction criterion for certain types of unfoldings (generally following the line of reasoning of Mather and Martinet), and obtain the two main theorems: the finite determinacy theorem and the universal unfolding theorem.

For example, to prove that f_0 is ℓ-determined for $\mathcal{G}$-equivalence we show that if $j^\ell h(0) = 0$ then $f_0 + h$ is $\mathcal{G}$-equivalent to f_0 by considering the unfolding $f(x, t) = (f_0 + th, t)$. It is enough by the compactness of $I = [0, 1]$ to prove that the germ of f at $(0, t_0)$ is locally trivial at each point $t_0 \in I$. This is proven using the reduction lemma (showing in addition that the vector field vanishes along $\{0\} \times I$).

For either theorem we let $\mathcal{G}$ *be a geometric subgroup of* $\mathcal{A}$ *or* $\mathcal{K}$ *acting on* $\mathcal{F}$ *with an adequately ordered system of DA–algebras.*

Theorem (determinacy theorem) : *The germ* $f_0 \in \mathcal{F}$ *is finitely* $\mathcal{G}$*-determined iff it has finite* $\mathcal{G}$*-codimension i.e.* $\dim_k(N\mathcal{G}f_0) < \infty$.

As in §3 we say an unfolding (f, q) of f_0 is $\mathcal{G}$-versal if for any other unfolding (g, r), there is a mapping of unfoldings λ: (g, r) $\longrightarrow$ (f, q), i.e. there is a germ $\lambda : k^r, 0 \rightarrow k^q, 0$ such that $\lambda^* f(x,v) = (\bar{f}(x,\lambda(v)), v)$ is $\mathcal{G}$-equivalent to g.

Theorem (versality theorem) : *For an unfolding* $f \in \mathcal{F}_{un}$ *of* f_0 *the following are equivalent:*

i)   f *is* $\mathcal{G}$-*versal*

ii) $T\mathcal{G}_e \cdot f_0 + \langle \partial_1 f, \ldots, \partial_q f \rangle = T\mathcal{F}_e$ *(infinitesimal versality)*

iii)   *any unfolding* f_1 *of* f_0 *which extends* f *is a* $\mathcal{G}$-*trivial extension (i.e.* f_1 *is* $\mathcal{G}$-*equivalent as an unfolding to* f × id *).*

Note: If f is $\mathcal{G}$-versal then q is an upper bound for $\mathcal{G}_e$-codim(f_0).

Just as in §3, there is uniqueness of versal unfoldings, etc. We also deduce a simple characterization of stability under deformations of a germ f_0, i.e. any unfolding f of f_0 is a $\mathcal{G}$-trivial unfolding.

Corollary 4.5 (stability under deformations) : *Let* $\mathcal{G}$ *be a geometric subgroup of* $\mathcal{A}$ *or* $\mathcal{K}$ *with an adequately ordered system of DA-algebras. A germ* f_0 *is stable under deformations iff* $T\mathcal{G}_e \cdot f_0 = T\mathcal{F}_e$.

In fact, by the versality theorem $T\mathcal{G}_e \cdot f_0 = T\mathcal{F}_e$ implies that f_0 is its own versal unfolding. Hence, any unfolding f of f_0 is $\mathcal{G}$-equivalent to f_0 × id.   □

The only remaining question regards the classification problem, which involves determining specific orders of determinacy, and applying Mather's geometric lemma. We illustrate this with two detailed examples, which illustrate the preceding theorems.

Example (4.6)   Hopf Bifurcation with Symmetry

We consider the bifurcation from equilibrium for the equation

(4.7) $\dot{x} = F(x, \lambda)$

where $F : \mathbb{R}^n \times \mathbb{R} \longrightarrow \mathbb{R}^n$ is equivariant with respect to a nontrivial action of O(2) on $\mathbb{R}^n$ and F(0, 0) = 0. If dF(0, 0) has a pair of purely imaginary eigenvalues then generically one has classical Hopf-bifurcation. In certain cases there can arise degenerate situations in which there is a pair of purely imaginary double eigenvalues. This can lead to the bifurcation of the equilibrium point into tori. Certain special cases of the generic and codimension one behavior has been considered by various workers (see [GR] for a discussion of these results). Golubitsky and Roberts [GR] apply an equivariant version of imperfect bifurcation equivalence which takes into account the O(2)–action to systematically account for the earlier results as well as extending them to allow further degeneracies.

We consider periodic solutions of (4.7). By change of time coordinate we may assume the double eigenvalues are $\pm i$. By a reduction using the center manifold theorem, it is enough to consider small periodic solutions of

(**4.8**) $\dot{x} = f(x, \lambda)$

where now $f : \mathbb{R}^4 \times \mathbb{R}, 0 \longrightarrow \mathbb{R}^4, 0$ is equivariant with respect to a action of O(2) on $\mathbb{R}^4 \simeq \mathbb{C}^2$ by $e^{i\theta} \cdot (z_1, z_2) = (e^{i\theta} \cdot z_1, e^{-i\theta} \cdot z_2)$ for $e^{i\theta} \in SO(2)$ together with $-I$ acting by interchanging the z_i. First, the linear part of (4.8) is invariant under an additional S^1–symmetry of phase shift φ acting on $\mathbb{C}^2$ by multiplication by $e^{i\varphi}$. By applying Birkhoff normal form, f can be made to commute with this extra S^1–action up to arbitrarily high degree (and in this situation, for the purposes of studying small amplitude periodic solutions, it can be replaced by f which exactly commutes, see [GR,§2]).

This situation is further simplified using a reduction of Swift [Sw]:

1)   such $O(2) \times S^1$ =equivariant functions are of the form

(**4.9**) $f(z_1, z_2, \lambda) = g_1 \cdot (z_1, z_2) + g_2 \cdot \delta \cdot (z_1, -z_2)$

where the g_i are smooth $\mathbb{C}$–valued functions of λ, $N = |z_1|^2 + |z_2|^2$, and $\Delta = \delta^2$ where $\delta = |z_1|^2 - |z_2|^2$;

2)   if $z_1 = xe^{i\psi_1}$ and $z_2 = ye^{i\psi_2}$ then with f of the form (4.9), equation (4.8) decouples into separate equations for the amplitudes (x, y) and the phases (ψ_1, ψ_2); and

3)   the equations for the amplitudes take the form

(**4.10**) $\dot{x} + (p + r\delta)x = 0$

                                                        $\dot{y} + (p - r\delta)y = 0$

Where p and r are smooth functions of λ, $N = x^2 + y^2$, and $\Delta = \delta^2$ where $\delta = x^2 - y^2$. These equations are now invariant under the standard action of the dihedral group D_4 on $\mathbb{R}^2$ defined by $\sigma(x,y) = (x,-y)$ and $\tau(x,y) = (y,x)$ for generators σ and τ (extended to $\mathbb{R}^3$ with coordinates (x,y,λ) by letting D_4 act trivially on the third factor).

Thus, the original problem of bifurcation of the equilibrium point has been reduced to the bifurcation of the equilibrium point of the system (4.10). Following Golubitsky–

Roberts [GR], this can be analyzed using a special case of equivariant bifurcation equivalence $\mathcal{K}^{G*}$, where the imperfect bifurcation equivalence in (4.1) preserves a group action of G. We consider the set $\mathcal{F}$ of smooth equivariant germs $f_0: \mathbb{R}^3, 0 \longrightarrow \mathbb{R}^2, 0$. The ring of D_4-invariant germs on $\mathbb{R}^3$, denoted by $C_{x,y,\lambda}{}^{D_4}$, is generated by λ, $N = x^2 + y^2$, and $\Delta = \delta^2$ where $\delta = x^2 - y^2$. If we let

$$\zeta_1(x,y,\lambda) = (x,y) \qquad \text{and} \qquad \zeta_2(x,y,\lambda) = (\delta x, -\delta y)$$

then by [GR] any D_4-equivariant germ $f_0 \in \mathcal{F}$ may be uniquely written

(4.11) $$f_0 = p(N,\Delta,\lambda) \cdot \zeta_1 + r(N,\Delta,\lambda) \cdot \zeta_2.$$

which is abbreviated $f_0 = [p, r]$.

Question: Which forms (4.11) do we expect to see stably or in families with few parameters ?

The answer requires: the classificaton of low codimension germs under D_4-equivariant bifurcation equivalence; determination of normal forms for these germs; and construction of versal unfoldings. At this point, we are going to explain how such calculations sre carried out. *Readers who do not want to "get their hands dirty" with these calculations are encouraged to proceed directly to the results summarized in* (4.17) *for the first few low codimension germs.*

The action of $\mathcal{K}^{D_4*}$ on $\mathcal{F}$ consists of pairs (S,h) with S: $\mathbb{R}^3, 0 \longrightarrow GL(\mathbb{R},2)$, which is D_4-equivariant for the action of D_4 on $GL(\mathbb{R},2)$ by conjugation, and h: $\mathbb{R}^3, 0 \longrightarrow \mathbb{R}^3, 0$ a germ of a diffeomorphism of the form $h(x,y,\lambda) = (h_1(x,y,\lambda), h_2(\lambda))$ (if h is just a germ of a diffeomorphism we would have instead equivariant contact equivalence $\mathcal{K}^G$). Then $\mathcal{K}^{G*}$ acts on $\mathcal{F}$ by

$$(S,h) \cdot f_0(x,y,\lambda) = S(x,y,\lambda) \cdot f_0 \circ h \qquad \text{(here } \cdot \text{ denotes matrix multiplication).}$$

Then, the tangent space to $\mathcal{K}^{D_4*}$ is computed in [GR] to be

(4.12) $$T\mathcal{K}^{D_4*}{}_e = C_{x,y,\lambda}{}^{D_4}\{\zeta_1, \zeta_2, \eta_1, \dots, \eta_4\} \oplus C_\lambda\{\tfrac{\partial}{\partial\lambda}\}$$

In this case the system of rings is $p^*: C_\lambda \longrightarrow C_{x,y,\lambda}{}^{D_4}$. Since both ζ_i vanish at 0,

$T\mathcal{K}^{D_4*}$ only differs by the absence of the single term $\frac{\partial}{\partial\lambda}$.

The ζ_i are as just defined above and the vector fields η_i are defined by $\eta_i(x, y) =$

$S_i \cdot (x,y)^T$ where · denotes matrix multiplication, $(\ )^T$ denotes transpose, and the S_i denote the matrices

(4.13) $\quad \begin{bmatrix} 1 & 0 \\ 0 & 1 \end{bmatrix} \quad \begin{bmatrix} x^2 & xy \\ xy & y^2 \end{bmatrix} \quad \begin{bmatrix} -x^2 & xy \\ xy & -y^2 \end{bmatrix} \quad \begin{bmatrix} 0 & x^3y \\ xy^3 & 0 \end{bmatrix}$

Then $S_i \cdot f_0$ ($= d\alpha_{f_0}(\eta_i)$) denotes the vector field along f_0. These vector fields become, respectively:

(4.14) $\quad [\,p, r\,]$, $[\,Np - r\Delta, 0\,]$, $[\,0, p - Nr\,]$, and $\quad [(N^2 - \Delta) \cdot p, -(N^2 - \Delta) \cdot r\,]$.
Also,

$$d\alpha_{f_0}(\zeta_1) = \zeta_1(f_0) = [\,p + 2Np_N + 4\Delta p_\Delta, \; 3r + 2Nr_N + 4\Delta r_\Delta]$$

(4.15)

$$d\alpha_{f_0}(\zeta_2) = \zeta_2(f_0) = [-2\Delta p_N - 4N\Delta p_\Delta + \Delta r, \; p - 2Nr - 2\Delta r_N - 4N\Delta r_\Delta\,].$$

(here e.g. p_N denotes the partial derivative of p with respect to N).

The simplest example beyond the case $p(0) \neq 0$ (which is without bifurcation) is the case $p(0) = 0$; but generically $\varepsilon = r(0) \neq 0$ and $p = \alpha N + \beta \Delta + \gamma \lambda + \text{h.o.t.}$ Then, by (4.14) and (4.15), we obtain modulo higher order terms in each coordinate

$S_1 \cdot f_0 \equiv [\,\alpha N + \beta \Delta + \gamma \lambda, \varepsilon\,]$, $\quad S_2 \cdot f_0 \equiv [-\varepsilon \Delta, 0\,]$, $\quad \zeta_1(f_0) \equiv [\,3\alpha N + 4\beta \Delta, 3\varepsilon\,]$
Provided $\gamma \neq 0$, we obtain $[\Delta, 0]$, $[\lambda, 0]$, and $[\,\alpha N, \varepsilon\,]$ (modulo h.o.t.)
Also,

$$rS_3 \cdot f_0 - (p-Nr)S_1 \cdot f_0 \equiv [\,(\alpha N + \beta \Delta + \gamma \lambda)((\alpha - \varepsilon)N + \beta \Delta + \gamma \lambda), 0].$$

If $\alpha \neq \varepsilon, 0$, we obtain $[\,N^2, 0]$ modulo ($m_{N,\Delta,\lambda} \cdot \{\Delta, 0], [\lambda, 0]\}$). Then, Nakayama's Lemma implies (since $C_{x,y,\lambda}{}^D 4 = C_{N,\Delta,\lambda}$)

$$[\,m_{N,\Delta,\lambda}{}^2, m_{N,\Delta,\lambda}\,] \subseteq C_{x,y,\lambda}{}^D 4\{\zeta_1, \eta_1, \eta_2, \eta_3\}.$$

Also, the remaining terms $S_4 \cdot f_0$ and $\zeta_2(f_0)$ belong to the left hand side while

$$\frac{\partial f_0}{\partial \lambda} \equiv [\,\gamma, \varphi] \quad (\text{modulo h.o.t.}).$$

Hence, if $\gamma \neq 0$, $\alpha \neq \varepsilon, 0$, we obtain

(4.16) $T\mathcal{K}^{D_4*}{}_e \cdot f_0 \; = \; < [\Delta, 0], [\lambda, 0], [\alpha N, \varepsilon], [\gamma, 0] > \; \oplus \; [m_{N, \Delta, \lambda}{}^2 , m_{N, \Delta, \lambda}]$

while for $T\mathcal{K}^{D_4*} \cdot f_0$, we remove $[\gamma, 0]$.

From (4.16) f_0 has $\mathcal{K}^{D_4*}$-codimension = 2 with complement to $T\mathcal{K}^{D_4*} \cdot f_0$ spanned by $\{ [N, 0], [1, 0] \}$. Thus, it is finitely $\mathcal{K}^{D_4*}$-determined; however, we do not yet have an order of determinacy. However, by Mather's geometric lemma , $j^\ell f_0$ is $\mathcal{K}^{D_4*(\ell)}$-equivalent to $[\alpha N + \varepsilon'\lambda, \varepsilon]$ for $\ell \geq 2$. Here we may arrange ε, $\varepsilon' = \pm 1$, with $\alpha \neq 0$, ε; however, we can not reduce α to a fixed value by a change of coordinates. In fact, the versal unfolding is given by $[(\alpha + u)N + \varepsilon'\lambda, \varepsilon]$ (because $[1, 0]$ is in the extended tangent space). Since $[N, 0]$ does not belong to the orbit tangent space for any α, it follows that α is a "modulus" parametrizing continuous change in equivalence class. We shall see in §7 how to collapse this to a finite classification by using topological $\mathcal{K}^{D_4*}$-equivalence.

This is just the beginning of the classification. It continues as follows:

(4.17)

Condition	Normal Form $(\varepsilon, \varepsilon', \varepsilon'' = \pm 1)$		Unfolding Terms
$p(0) = 0$	$[\alpha N + \varepsilon'\lambda, \varepsilon]$	$\alpha \neq 0, \varepsilon$	$[N, 0]$,
$p(0) = 0, p_N(0) = 0$	$[\varepsilon''N^2 + \varepsilon'\lambda, \varepsilon]$		$[N, 0]$
$p(0) = 0, p_N(0) = r(0)$	$[\varepsilon''\Delta + \varepsilon N + \varepsilon'\lambda, \varepsilon]$		$[N, 0]$
$p(0) = 0, r(0) = 0$	$[m\Delta + \varepsilon N + \varepsilon'\lambda, \varepsilon N]$	$m \neq 0$	$[0, 1], [\Delta, 0]$

etc.

We see consequences of the classification for the bifurcation of the equilibrium as pointed out by Golubitsky-Roberts. For (4.10) $f_0 = [p, r]$. If $(x_0, y_0) \in f_0{}^{-1}(0)$ then the set of points in $\mathbb{R}^4$ with amplitudes = (x_0, y_0) will form a torus if neither = 0, a circle if exactly one of x_0 or y_0 = 0, or a point if both = 0. By examining the normal forms, the form of the bifurcation (and even the stability in the dynamical systems sense) can be determined. For example, from (4.10) the equations become

$$(p + r\delta)x \; = \; 0 \qquad\qquad (p - r\delta)y \; = \; 0.$$

For $[\alpha N + \varepsilon'\lambda, \varepsilon]$ with say $x \neq 0$, $y = 0$, then $p + r\delta = 0$ yields the branch of periodic solutions $(\alpha + \varepsilon)x^2 + \varepsilon'\lambda = 0$. If $x \neq 0$, $y \neq 0$, then $p = 0$ and $r\delta = 0$, giving a branch of 2-tori with $x = \pm y$ and $2\alpha x^2 + \varepsilon'\lambda = 0$. Golubitsky-Roberts go on to determine the bifurcation behavior including transitions between the different types of behavior for these and higher codimension behavior.

The complete classification up to codimension 2 is quite involved and is considerably simplified using arguments which apply the Bruce-DuPlessis-Wall unipotency results (see [Gaf2]).

Example (4.18):   Periodic Solutions of Hamiltonian Systems

Let $H : \mathbb{R}^{2n}, 0 \longrightarrow \mathbb{R}, 0$ be a smooth Hamiltonian function with $(p, q) = (p_1, \dots p_n, q_1, \dots q_n)$ denoting coordinates for $\mathbb{R}^{2n}$. Suppose $dH(0) = 0$ and for simplicity that the quadratic part of H has the form

$$(4.19) \qquad\qquad\qquad H_2 = \sum_{i=1}^{n} \lambda_i(q_i^2 + p_i^2).$$

This is the "semisimple" form. In general the quadratic part can be more complicated having "nilpotent" terms; we refer to [VM] for a discussion of the work of Cushman, Sanders, Van der Meer, etc. on this more general case.

One problem is to understand the small oscillations which are solutions to the Hamiltonian system

$$(4.20) \qquad \frac{dH}{dq} = \frac{\partial H}{\partial p} \qquad\qquad \frac{dH}{dp} = -\frac{\partial H}{\partial q}$$

If in (4.20) we had H_2 in place of H then there are independent oscillations in each (q_i, p_i) plane of period $2\pi/\lambda_i$. Lyapunov proved that a similar result continues to hold for H provided a nonresonance condition holds, namely, no two λ_i are rational multiples. Under this assumption, the Lyapunov center theorem asserts there are n families of periodic solutions (parametrized by the energy level H near 0) whose periods are close to $2\pi/\lambda_i$. These are the nonlinear normal modes. This need not hold in the presence of resonance; and considerable work by Weinstein [We], Moser [Mo], and Rabinowitz [Ra] [FR] has been devoted to determining conditions which ensure that part of this result remains true.

Their work centers on the signature of H_2 restricted to resonance subspaces V_α, which are the coordinate subspaces of those q_i and p_i for which the λ_i are rational multiples of a number α (where, say, $\alpha > 0$ denotes the maximal number such that all λ_i associated to V_α are integer multiples of α). We summarize the key results by the following fundamental theorem.

Theorem 4.21:

1) (Weinstein [We]) *If* H_2 *is positive definite then the conclusions of Lyapunov's theorem still hold, i.e. there are n families of periodic solutions paranetrized by the energy level* $H = \varepsilon$.

2) (Moser [Mo]) *Suppose instead that* H_2 *restricted to a resonance subspace* V_α *(of dim = 2r, say) is positive definite. Then, there are r families of periodic solutions corresponding to the* λ_i *associated to this subspace.*

3) (Fadell and Rabinowitz [FR]) *Thirdly, suppose instead that* H_2 *restricted to a resonance subspace* V_α *has signature 2m. Then, there are* |m| *families of periodic solutions having periods close to* $T = 2\pi/\alpha$.

A basic question is to determine how the higher order terms of H influence the periodic behavior of solutions. We indicate how Duistermaat, Van der Meer, and Montaldi-Roberts-Stewart have applied singularity theory to begin obtaining answers. First, because the key role is played by the behavior of H on resonance subspaces, we assume that there is only one such subspace, i.e. all λ_i are integer multiples of a maximal α > 0. By a change of time coordinate we may assume that the λ_i are half-integers; and as a reminder, we denote them by $k_i/2$. To study the effects of higher order terms of H, we use Birkhoff normal form. To the quadratic part H_2 is associated the Hamiltonian vector field X_{H_2} which induces an S^1 -action on $\mathbb{R}^{2n}$, with $e^{i\theta}$ acting by multiplication by $e^{ik_j\theta}$ on z_j = $q_j + i \cdot p_j$. We may then try to modify H by symplectic change of coordinates so that it commutes with this S^1 -action. This may be carried out to arbitrarily high order k, which is referred to as Birkhoff normal form to this order. However, this cannot in general be applied to all of H (each time it is applied to bring the next higher order terms into normal form the neighborhood on which it is valid may shrink and approach 0 as $k \to \infty$; this is the small divisor problem). This is exactly an example where singularity theory cannot be applied in its standard form. However, after k steps we may assume H has the form

(4.22) $$H = H_2 + H_k + \tilde{H}$$

where H_k denotes terms of order > 2 and $\leq$ k, with $H_2 + H_k$ in Birkhoff normal form.

Then, we ask whether the form of H_k allows us to deduce additional information about the periodic solutions of (4.20). The work on this has taken two directions and in each case the information is supplied by singularity information about an auxillary map.

Energy–Momentum Map (Duistermaat [Dm] and Van der Meer [VM]) :

For this method there are really two steps: one is to replace H by an $\hat{H}$ which is invariant under the S^1 -action such that there is a diffeomorphism taking the periodic orbits of the system for H to those for $\hat{H}$. To see that finding such an $\hat{H}$ is possible we refer to [Dm,§2]; it is usually called the Moser–Weinstein Reduction. Secondly, given an S^1 - invariant H, we would like to deduce information from only part of it. This second step uses singularity theory. We consider the S^1 -invariant map $f = (H_2 , H) : \mathbb{R}^{2n} ,0 \longrightarrow \mathbb{R}^2,0$. It is still not possible to put f into acceptable normal form using symplectic equivalence on $\mathbb{R}^{2n}$. However, the reduction also shows that the periodic solutions are given by critical S^1 -orbits for f, i.e. where the map f has rank < 2. This critical set is preserved under arbitrary S^1-equivariant diffeomorphisms of $\mathbb{R}^{2n},0$. Thus, it is sufficient to investigate the classification of such germs under S^1 -equivariant $\mathcal{A}$-equivalence (with trivial S^1 -action on $\mathbb{R}^2$).

Duistermaat analyzes the case of n = 2 and $H_2 = k(q_1^2 + p_1^2) + \ell(q_2^2 + p_2^2)$ (with k and ℓ relatively prime). He classifies the lowest codimension germs and determines their versal unfoldings to describe the detuning from resonance. He only excludes the 1:−1 resonance case (where nilpotent terms in H_2 can occur) treated by Van der Meer [VM]. For $0 < \ell < |k|$, the algebra of S^1-invariant functions is generated by

$$\rho_1 = |z_1|^2 = q_1^2 + p_1^2 \qquad \rho_2 = |z_2|^2 = q_2^2 + p_2^2$$

and $\quad \psi = Re(z_1^{|k|} \bar{z}_1^{\ell}) \qquad\qquad \chi = Im(z_1^{|k|} \bar{z}_1^{\ell}).$

Then, if e.g. $\ell + |k| > 4$ then if the Birkhoff normal form is

$$H = k\rho_1 + \ell\rho_2 + (a/2)\rho_1^2 + b\rho_1\rho_2 + (c/2)\rho_2^2 + d\psi + e\chi + h.o.t.$$

where d or e $\neq 0$ and ρ_1 and ρ_2 are "assigned" degree 1 and ψ and χ degree 2. Then, provided the coefficients a, b and c satisfy certain conditions, f is S^1 -equivariant $\mathcal{A}$- equivalent to the $\hat{f}$ obtained from

(4.23) $\hat{H} = k\rho_1 + \ell\rho_2 + (a/2)\rho_1^2 + b\rho_1\rho_2 + (c/2)\rho_2^2 + \psi.$

Morover, the versal unfolding of f is given by

(4.24) $$\hat{H} + u_1\rho_2 + u_2\rho_1\rho_2.$$

The parameter u_2 is again a modulus indicating that this parameter can be removed for most values by using topological equivalence. Using these normal forms, Duistermaat is able to explicitly determine the periodic solutions obtained as well the behavior arising from detuning by the unfolding parameters (see appendix of [Dm]). An analogous result is obtained for $\ell = 1$ and $k = \pm 2$. However, interestingly, for $\ell = 1$ and $k = \pm 3$, the germ can be given an analogous normal form but it has codimension 3 and exhibits special behavior.

<u>Symmetric Hamiltonian Systems and A Nondegeneracy Criterion</u> (Montaldi–Roberts–Stewart [MRS1,2])

Suppose now the Hamiltonian H is invariant under the action of a symmetry group Γ, which is a compact Lie group acting symplectically $\mathbb{R}^{2n},0$. An example is the planar spring pendulum (see fig 4.25) in which the potential energy is invariant under reflection about the vertical axis. Then, H can still be placed in Birkhoff normal form in such a way

(fig 4.25)

that it is still Γ–invariant. However, the symmetry will often force H_2 to have repeated eigenvalues so there is resonance. In addition, it may force the absence of certain lower order terms in H. For example in (4.22) H_k may consist of terms of degree k.

Montaldi–Roberts–Stewart [MRS1] show that an analogue of the Moser–Weinstein results continue to hold in the presence of a symmetry group (in fact, they also allow time reversal; however, for simplicity we ignore that in the statement of the theorems). On a resonance subspace V_α there is an induced action of $\Gamma \times S^1$ (since Γ acts symplectically, it commutes with the S^1-action). Let $\Sigma \subseteq \Gamma \times S^1$ denote an isotropy subgroup, i.e. for some $x \in V_\alpha$, $\Sigma = \{\sigma \in \Gamma \times S^1: \sigma \cdot x = x\}$. Let Fix$(\Sigma, V_\alpha)$ denote the subspace of points fixed by Σ. The Equivariant version of Moser–Weinstein becomes

Theorem 4.26 [MRS1] : *Suppose that H_2 is nondegenerate and restricted to V_α is positive definite, then for each isotropy subgroup Σ there exist m_α families of periodic solutions parametrized by the energy level $H = \varepsilon$ with periods near $2\pi/\alpha$ and symmetry group containing Σ, where $m_\alpha = (1/2)\dim \mathrm{Fix}(\Sigma, V_\alpha)$.*

In addition, they are also able to obtain a determinacy theorem which implies that the information about periodic orbits for the Hamiltonian system for h in terms of the system for $\hat{H} = H_2 + H_k$. where $\hat{H}$ is in Birkhoff normal form to degree k but H_k only consists of terms of degree k. For simplicity we suppose H is fully resonant ($V_\alpha = \mathbb{R}^{2n}$), and let J denote the standard symplectic involution on $\mathbb{R}^{2n}$.

Definition : H is said to be *nondegenerate* if the associated $\Gamma \times S^1$ -equivariant map
(4.27) $$\Psi : \mathbb{R}^{2n} \times \mathbb{R} \times \mathbb{R}, 0 \longrightarrow \mathbb{R}^{2n}, 0$$

$$((p, q), \lambda, \tau) \mapsto J \cdot d_x(-\tau H_2 + H_k) + \lambda d_x H_2$$

is finitely determined for $\mathcal{K}^{\Gamma \times S^1}$ -equivalence ($d_x(\cdot)$ denotes the derivative with respect to (p, q)).

A geometric interpretation of this condition is given [MRS2,§ 4] in terms of any S^1 -orbit of $\Psi^{-1}(0)$ being a nondegenerate critical orbit for $\hat{H} \mid H_2 = c$, and the nonsingularity of the associated infinitesimal Floquet operator. Then, the finite determinacy theorem takes the form

Theorem 4.28 [MRS2] : *Suppose that H as above is nondegenerate, Then there exists a $\Gamma \times S^1$ -equivariant homeomorphism germ $\mathbb{R}^{2n} \times \mathbb{R}, 0 \longrightarrow \mathbb{R}^{2n} \times \mathbb{R}, 0$ which maps the periodic orbits of H to those of $\hat{H}$, preserving the symmetry groups.*

Remark : An interesting point about the proof is that unlike the Lyapunov– Schmidt type reduction used in the Moser-Weinstein theorem, this result uses an infinite-dimensional version of the splitting principle (§3) due to Magnus (see [MRS2, §2]).

In the case of the spring pendulum which does not exactly obey Hooke's law, the potential energy is a nonquadratic function $\sigma(\ell)$ of the length ℓ. With certain normalizations (and invariance under the vertical reflectional symmetry $q_2 \mapsto -q_2$) the Hamiltonian H may be written in the form

$$H(p, q) = (1/2)(p_1{}^2 + p_2{}^2) + (\omega^2 q_1{}^2 + q_2{}^2 + a_1 q_1{}^3 + b_1 q_1{}^4 + b_2 q_1{}^2 q_2{}^2) + \dots$$

where the terms in q denote the potential energy. The coefficients ω^2, a_1, and b_i can be written in terms of the Taylor coefficients of σ: $\omega^2 = \sigma''(0)$, $d = \sigma'''(0)$, and $f = \sigma^{iv}(0)$. If $\omega = 1$ it is in 1 : 1 resonance; terms up to 4-th order can be written in the $\mathbb{Z}_2 \times O(2)$ – invariant form (which includes the reflectional symmetry , S^1-action, and time reversal)

(4.29) $$H = N + (\alpha_1 N^2 + \alpha_2 P + \beta_1 NQ + \beta_2 Q^2)$$

with $N = |z_1|^2 + |z_2|^2$, $P = |z_1|^2 \cdot |z_2|^2$, and $Q = \mathrm{Re}(z_1 \bar{z}_2)$ (here the z_i are complex coordinates representing the (p, q)).

The nondegeneracy is expressed in terms of linear relation among the α_i and β_i, which can, in turn, be expressed in terms of d and f. Hence, theorem 4.28 allows us to deduce the possible periodic behaviors by reducing consideration to the Hamiltonian (4.29). The nondegeneracy conditions decompose the (d, f) – plane into regions exhibiting various periodic behavior. Furthermore, symmetry properties are applied [MRS2, §5] to determine for each region the symmetry of the periodic orbits under the reflectional and time reversal symmetries and show that depending on the coefficients there may exist 2, 3, or 4 different symmetry types for the periodic solutions giving rigorous justification for earlier results of Kummer (see [MRS§5]).

The Moral from the Examples :

These detailed examples illustrate two key principles.

First, singularity theory is not a sledgehammer which one uses to crush a problem, but rather a finely honed tool which is used to extract useful and important information about a problem. This requires that the user first lay the proper ground-work, so it can be applied. This involves reducing the original problem to one amenable to singularity theory, as in the symmetric Hamiltonian–Hopf case reducing to D_4-equivariant bifurcation for the amplitude equations, or in the Hamiltonian case to the energy momentum map or the associated map (4.27).

Secondly, although the results obtained may only concern one aspect of the problem, these results are exact and are not based on formal arguments which may lack rigor and at times imply misleading results. The basic theorems are applied to reduce to computible normal forms such as (4.17), (4.23), (4.24), and (4.29). From these normal forms we can both analyze and understand the behavior occurring for a problem including the possible perturbations of such behavior.

We continue in this direction in the next sections when we deduce information regarding the exact local structure of nonlinear operators and the number of solutions to nonlinear problems.

§5   The Structure of Nonlinear Fredholm Operators

We have now indicated how the results on stability of mappings allow us to describe the local structure of mappings between finite dimensional manifolds. We have seen in §4 how this refinements of the theory allow us to deduce important consequences for nonlinear problems. We go further and ask whether the finite dimensional structure theorems can help in determining exactly the local structure for nonlinear Fredholm operators viewed as mappings? Lets first consider two results which suggest that such a general result might be lurking about.

Example 5.1   The forced undamped Duffing equation :

We seek 2π-periodic solutions of the equation

$$(5.1) \qquad \frac{d^2x}{dt^2} + x + \lambda_1 x + x^3 \ = \ \lambda_2 \cos(t)$$

In analogy with the Van der Pol equation, we might expect that this equation also has isolated periodic solutions. To determine the periodic solutions Hale and Rodriguez [HR] consider the nonlinear operator

$$(5.2) \qquad F(x) \ = \ \frac{d^2x}{dt^2} + x + \lambda_1 x + x^3 \quad : C^2_{2\pi}(\mathbb{R})^{\#} \longrightarrow C^0_{2\pi}(\mathbb{R})^{\#}$$

where $C^k_{2\pi}(\mathbb{R})^{\#}$ denotes the C^k 2π-periodic even functrions on $\mathbb{R}$.

Regarding the nature of the solutions, they prove (see [HR] or [CH,chap. 8])

Theorem 5.3 : *There is a neighborhood $\mathcal{U}$ of 0 in $C^2_{2\pi}(\mathbb{R})^\#$ and a neighborhood U*

of (0, 0) in $\mathbb{R}^2$ such that for $(\lambda_1, \lambda_2) \in U$, all 2π-periodic solutions to (5.1) sufficiently

close to 0 are even functions. Moreover, there is a curve γ in U, approximately given by

the cusp $\lambda_1^3 + c\lambda_2^2 = 0$ (for appropriate $c \neq 0$), which divides the complement $U\backslash\gamma = U_1$

$\cup\ U_2$ such that: for $(\lambda_1, \lambda_2) \in U_1$ there are three 2π-periodic solutions to (5.1) in $\mathcal{U}$

while if $(\lambda_1, \lambda_2) \in U_2$ there is exactly one 2π-periodic solutions to (5.1) in $\mathcal{U}$ (on

$\gamma\backslash\{0\}$ there are two solutions).

A related result was proven by Ambrosetti and Prodi [AP] (and their forthcoming Camb. Univ. Press monograph) concerning a nonlinear Dirichlet problem on a domain Ω with smooth boundary.

(5.4)
$$\Delta u + h(u) = g(x) \quad \text{on } \Omega$$
$$u| \partial\Omega = 0 \quad \text{on } \partial\Omega$$

where $h(u)$ is function (at least C^3) which satisfies $h(0) = 0$, $h''(t) > 0$ all t, and asymptotically satisfies $0 < \lim_{t \to -\infty} f'(t) < \lambda_1 < \lim_{t \to \infty} f'(t) < \lambda_2$, where $0 < \lambda_1 < \lambda_2$ $< ...$ represent the eigenvalues for Δ on Ω. This problem can now be described in terms of the operator

$$F(u) = \Delta u + h(u) : C_0^{2+\alpha}(\bar{\Omega}) \longrightarrow C^\alpha(\bar{\Omega})$$

in the usual notation for Holder spaces of functions with "0" denoting the space of functions satisfying the boundary condition. Then, they show

Theorem 5.5 : *There is a codimension 1 submanifold $\Sigma \subset C^\alpha(\bar{\Omega})$ which divides the*

complement into two components $U_1 \cup U_2$ such that: for $g(x) \in U_1$ there are exactly

two solutions to (5.2), if $g(x) \in U_2$ there are no solutions, while if $g(x) \in \Sigma$ there is

exactly one solution.

This is similar to the local result of Hale-Rodriguez in that it decomposes a space into regions where the number of solutions are specified. However, we note that the result for the Duffing equation is only local and, in fact, only gives the result for a special class of functions, namely, $\lambda_2 \cos(t)$. The Ambrosetti–Prodi result does describe the global decomposition of the space of functions; however, the decomposition has a simpler form than that for the Duffing equation.

Berger and Church [BC] recognized that the mapping in the Ambrosetti–Prodi result as an "infinite-dimensional fold", i.e. the operator F is differentiably equivalent as a mapping to $s(x) \times id_E$ where $s(x) = x^2$, E is a Banach space and there are global diffeomorphisms φ and ψ such that the diagram commutes

$$C_0^{2+\alpha}(\bar\Omega) \xrightarrow{\;F\;} C^\alpha(\bar\Omega)$$

(5.6) $\varphi \downarrow$ $\qquad$ $\psi \downarrow$

$$\mathbb{R} \oplus E \xrightarrow{\;s \times id_E\;} \mathbb{R} \oplus E$$

Berger, Church, and Timourian went on to show that there is a local analog of the Ambrosetti–Prodi result for the nonlinear Dirichlet problem with cubic nonlinearity

(5.7) $\qquad\qquad\qquad\qquad \Delta u + \lambda u - u^3 = g(x) \qquad$ on Ω

$\qquad\qquad\qquad\qquad\qquad\quad u | \partial\Omega = 0 \qquad\quad$ on $\partial\Omega$

For the mapping

(5.8) $\qquad\quad F(u, \lambda) = (\Delta u + \lambda u - u^3, \lambda) : C_0^{2+\alpha}(\bar\Omega) \times \mathbb{R} \longrightarrow C^\alpha(\bar\Omega) \times \mathbb{R}$

they proved [BCT1][BCT2]

Theorem 5.9 : *The nonlinear mapping (5.8) is locally smoothly equivalent to the infinite-dimensional cusp map $g \times id_E$ where $g(x, y) = (x^3 + yx, y)$ is the standard cusp map.*

In fact, if one looks back at the Hale and Rodriguez result for the Duffing equation, one sees in the parameter space the telltale cusp which strongly suggests that there is a cusp mapping present as well. Given that these two model mappings are both stable mappings, we ask the obvious question whether there are general criteria for deciding whether a nonlinear operator can be represented locally or globally as a stable mapping?

In certain special cases it has proven to be the case that operators are globally infinite dimensional fold or cusp maps; such examples have occurred for certain Ricatti equations considered by McKean–Scovel [MSc]. See Cafagna [Ca] and especially Church–Timourian [CT] for the known examples of global representability of nonlinear operators. However, in general the models for these stable singularities are local and they will not globally describe a nonlinear operator (see e.g. [Mc]).

It is still quite reasonable to expect that nonlinear operators can be described locally by infinite-dimensional versions of stable mappings. We will see that this is not only the case, but indeed it will follow from a slightly strengthened version of Mather's global theorem "infinitesimal stability implies stability".

We return to the case of a smooth mapping $f: N \rightarrow P$ between finite dimensional manifolds. We recall from Mather's theorem that the mapping $\mathcal{U} \longrightarrow \mathcal{A}(N, P)$ sending $g \longmapsto (\varphi_g, \psi_g)$ was continuous. We would like to know more information about its smoothness properties. This can be provided by an addendum to Mather's theorem. For it, we let $M \subset N$ be a codimension 0 compact submanifold of N (we allow M to possibly have boundaries and corners). Also, we let K and L be codimension 0 compact submanifolds of N and P with $M \subset \text{int}(K)$ and $f(M) \subset \text{int}(L)$.

Theorem 5.10 (Addendum to Mather's Theorem, see [D6]): *If the mapping* $f|M : M \rightarrow P$ *is infinitesimally stable, then there exists a neighborhood* $\mathcal{U}$ *of* $f \mid M$ *in* $C^\infty(M, P)$ *and a continuous mapping*

$$H = (H_1, H_2): \mathcal{U} \longrightarrow \mathcal{A}(N, P) = \text{Diff}(N) \times \text{Diff}(P)$$

which satisfies the conclusions of Mather's theorem (*with* $\varphi_g = H_1(g)$ *and* $\psi_g = H_2(g)$), *namely,* $H(f|M) = (\text{id}_N, \text{id}_P)$ *and*

$$g = (\psi_g \circ f \circ \varphi_g^{-1})|M \quad \text{for all } g \in \mathcal{U}.$$

Moreover, the composition of each H_i *with the restriction maps yields smooth mappings of Frechet spaces*

$$\mathcal{U} \xrightarrow{H_1} \text{Diff}(N) \xrightarrow{r_1} C^\infty(M, N) \qquad \mathcal{U} \xrightarrow{H_2} \text{Diff}(P) \xrightarrow{r_2} C^\infty(L, P)$$

(*here the smoothness is in the Gateaux differentiable sense, see e.g.* [Ha]).

Now given a germ of a nonlinear mapping f: E,0 $\longrightarrow$ F,0 in the form of an unfolding f(x, u) = ($\bar{f}$(x, u), u) where E = $E_1 \oplus G$, F = $F_1 \oplus G$, with E_1 and F_1 finite dimensional, x $\in E_1$, u $\in G$, and $\bar{f}$(x, u) $\in F_1$.

Definition : f: E,0 $\longrightarrow$ F,0 is *infinitesimally stable* if there exists a finite dimensional subspace $G_1 \subset G$ (with closed complement G_2) such that f | : $E_1 \oplus G_1,0 \longrightarrow F_1 \oplus G_1,0$ is infinitesimally stable.

By a result known as the "openness of versality " (see e.g. [Te] or [Ma V]), it follows that there is an open neighborhood U of 0 in $E_1 \oplus G_1$ such that $f_1 = f | : U \longrightarrow F_1 \oplus G_1$ is globally infinitesimally stable (the closure $\bar{U}$ can be chosen to be a compact manifold with boundary). Then, $\bar{f}$ defines a smooth mapping $U_1 \to C^\infty(\bar{U} , F_1 \oplus G_1)$. Thus, applying the addendum to Mather's theorem allows us to prove

Theorem 5.11 [D6]: *If the germ of the mapping f: E,0 $\longrightarrow$ F,0 is infinitesimally stable, then f is C^∞-equivalent to $f_1 \times id_G$ (i.e. there are germs of diffeomorphisms* φ: E,0 $\to$ *E,0 and* ψ: *F,0 $\to$ F,0 such that f = $\psi \circ (f_1 \times id_G) \circ \varphi^{-1}$).*

Remark : We can define Q(f) = $Q(f_1)$, if e.g. f is infinitesimally stable then it can be shown that Q(f) is well-defined, independent of f_1.

This result shows that all of the local results already described are consequences of this general result. We derive several immediate corollaries for nonlinear operators, which are the infinite dimensional analogues of the local results in §3.

Corollary 5.12 (local stability): *Suppose that* f: E,0 $\longrightarrow$ F,0 *is an infinitesimally stable germ with E, F, and H Banach spaces. If g : E$\oplus$H, 0 $\longrightarrow$ F$\oplus$H, 0 is an unfolding of f, then g is equivalent to f $\times$ id$_H$.*

Corollary 5.13 (finite determinacy): *Suppose that*   f: E,0 $\longrightarrow$ F,0 *is an infinitesimally stable germ with, associated finite dimensional stable germ f_1. If f_1 is ℓ-determined for $\mathcal{A}$-equivalence, e.g. if $m_x{}^\ell \cdot Q(f_1) = 0$, then f is ℓ-determined in the sense that if h: E,0 $\longrightarrow$ F,0 with $\| h \| = O(\| x \|^{\ell+1})$ for x $\in$ E, then f+ h is equivalent to f.*

Corollary 5.14 (stable unfolding): *Suppose that* f: E,0 ⟶ F,0 *is a germ with associated finite dimensional germ* f_0 *which is finitely $\mathcal{K}$-determined. Then, there is a finite dimensional unfolding F of f such that F is equivalent to* $f_1 \times id_H$ *where* f_1 *is a stable unfolding of* f_0.

To apply these results beginning with a nonlinear Fredholm operator f: E ⟶ F, we first must apply the full Lyapunov–Schmidt reduction to obtain the operator $f \circ \Phi^{-1}$ which has the form of an unfolding. As mentioned in §1, to determine the infinitesimal stability of $f \circ \Phi^{-1}$ we must compute the derivatives $d^j(f \circ \Phi^{-1})$.

Example 5.15 : Morin Singularities

Suppose f: E,0 ⟶ F,0 has Fredholm index 0 and K = ker(df(0)) with $\dim_\mathbb{R}(K) = 1$, and K spanned by h. We let P denote projection onto Im(df(0)) and $f_2 = P \circ f$. For f to be smoothly equivalent to an infinite dimensional version of a Morin singularity (one of (3.10) , we must have:

(5.16) i) $d^j(f \circ \Phi^{-1})(h, h, ..., h) \in Im(df(0))$ for $1 \le j \le n$;

ii) $d^{n+1}(f \circ \Phi^{-1})(h, h, ..., h) \notin Im(df(0))$; and

iii) there are $g_k \in Im(df(0))$ for $1 \le k \le n-1$ so that
$d^j(f \circ \Phi^{-1})(g_k, h, ..., h) \in Im(df(0))$ for $k \le j$ but $\notin Im(df(0))$ for $j = k+1$.

In the case that df(0) is self adjoint, h also spans an L^2-complement to Im(df(0)) so the conditions that g is or is not in Im(df(0)) is given by an integral condition $\int_\Omega gh = 0$ or $\ne 0$.

Nonlinear Dirichlet Problem Reconsidered:

Consider the nonlinear Dirichlet problem on a (connected) bounded domain Ω with smooth boundary.

(5. 17) $\Delta u - \lambda_1 u + a_1(x)u^2 + b(x, u) = g(x)$ on Ω

$u | \partial\Omega = 0$ on $\partial\Omega$,

where λ_1 is the first eigenvalue of Δ with eigenfunction h and $b(x, u) = b'(x, u)u^3$ for smooth a_1, b, and b'. Then, in order for

(5. 18) $f(u) = \Delta u - \lambda_1 u + a_1(x)u^2 + b(x, u) : C_0^{2+\alpha}(\bar{\Omega}) \longrightarrow C^\alpha(\bar{\Omega})$

To locally define a fold map near 0, it is sufficient that $d^2(f \circ \Phi^{-1})(h, h) \notin \text{Im}(df(0))$ (here and below all derivatives are taken at 0). Applying the formulas for higher derivatives of compositions

$$d^2(f \circ \Phi^{-1})(d\Phi, d\Phi) \; = \; d^2f(\,,\,) \; - \; (df \circ (d\Phi)^{-1} \, d^2\Phi(\,,\,))$$

and $d^j\Phi = d^jf$ for $j > 1$. Since $d\Phi(h) = h$,

$$d^2(f \circ \Phi^{-1})(h, h) \; \equiv \; d^2f(h, h) \; = \; a_1(x) \cdot h^2 \mod \text{Im}(df(0))$$

Thus, $d^2(f \circ \Phi^{-1})(h, h) \notin \text{Im}(df(0)) = \text{Im}(\Delta - \lambda_1) = L^2$-complement to $\langle h \rangle$ iff

(5. 19) $$\int_\Omega a_1(x) \cdot h^3 \neq 0.$$

Thus, (5.19) gives a sufficient condition for a local fold. For example, since it is well-known that the first eigenfunction h may be chosen to be positive, this recovers the local version of Ambrosetti and Prodi [AP] and Berger–Church [BC] that

$$f(u) = \Delta u - \lambda_1 u + u^2 \; : \; C_0^{2+\alpha}(\bar{\Omega}) \longrightarrow C^\alpha(\bar{\Omega})$$

defines a local fold near $u = 0$.

If (5.19) fails, then we can continue using (5.16) to see if higher order Morin Singularities are present; this is carried out for cusps in [BCT] [BCT2] and for higher order singularities in [D6]. We can see using the criteria in (5.16) that , for example, the forced Duffing equation in (5.1) defines a cusp singularity

(5.20) $$F : C_{2\pi}^2(\mathbb{R})^\# \times \mathbb{R}, \, (0, 0) \longrightarrow C_{2\pi}^0(\mathbb{R})^\# \times \mathbb{R}, \, (0, 0)$$

$$(x(t), \lambda) \longmapsto (\frac{d^2 x}{dt^2} + x + \lambda x + x^3 \, , \, \lambda)$$

Remark : The conditions in (5.16) for n = 1 or 2 give the fold and cusp; the conditions are equivalent to those of Berger–Church–Timourian for those two local normal forms (although their result also applies when there is only finite differentiability of the operator).

Nonlinear Oscillations :

Consider the nonlinear equation

(5.21) $$\frac{d^2 x}{dt^2} + x + a_1(t) \cdot x^2 \; = \; b(t)$$

where both $a_1(t)$ and $b(t)$ are continuous functions periodic of period 2π. Then, the operator corresponding to the equation is

$$\textbf{(5.22)} \qquad f(x(t)) \;=\; \frac{d^2 x}{dt^2} + x + a_1(t){\cdot}x^2 \;:\; C^2_{2\pi}(\mathbb{R}),\, 0 \;\longrightarrow\; C^0_{2\pi}(\mathbb{R}),\, 0$$

This time we do not obtain a Morin singularity since $\dim_{\mathbb{R}}\ker(df(0)) = 2$. However, a calulation shows that the classification and infinitesimal stability of f is determined by conditions on an *open dense* set of values for the first four Fourier coefficients of $a_1(t)$, $\{a_0,\ \dots,\ a_4,\ b_1,\ \dots,\ b_4\}$ for $\sum a_n{\cdot}\cos(nt) + b_n{\cdot}\sin(nt)$. Several special cases are:

i)   if $(a_1, b_1) \neq (0, 0)$, $(a_3, b_3) = (0, 0)$, and $a_2 \neq b_2$ or

ii)   if $(a_1, b_1) = (0, 0)$, $(a_3, b_3) \neq (0, 0)$, and $a_2, b_2 \neq 0$ then

f in (5.22) is infinitesimally stable with local algebra isomorphic to $\mathbb{R}[x, y]/(x^2 + \varepsilon y^2, xy)$ where $\varepsilon = 1$ in case i) and -1 in case ii). Thus, by (5.11) and (3.10), f is smoothly equivalent to

$$(x^2 + \varepsilon y^2 + ux,\, xy + vy,\, u,\, v) \times \text{id}$$

Remark : Not only does the the normal form for (5.22) mark a departure from those obtained earlier; but it is the simplest example of the stable germs needed to describe infinitesimally stable operators (5.23). A version of this example is also contained in the forthcoming Ambrosetti–Prodi monograph.

$$\textbf{(5.23)} \qquad f(x(t)) \;=\; \frac{d^2 x}{dt^2} + x + P(x, x',t) \;:\; C^2_{2\pi}(\mathbb{R}),\, 0 \;\longrightarrow\; C^0_{2\pi}(\mathbb{R}),\, 0$$

In general, we need *all possible stable germs* $f : \mathbb{R}^n,\, 0 \longrightarrow \mathbb{R}^n,\, 0$ (any n) with $\dim\mathrm{Ker}(df(0)) = 2$ (see [D7]). We shall see consequences of this for determining the number of periodic solutions for nonlinear oscillations in §6.

Final Remarks : Unlike the previous section, the results described here give exactly the local structure of nonlinear operators. By constrast, the results of §4 include extra structure pertaining to the operator, but they only apply to the restriction to finite dimensional subspaces. In fact, there is a version of "infinitesimal stability implies stability" for global subgroups of $\mathcal{A}$. These subgroups satisfy global analogues of the conditions for geometric subgroups in §4 [MD,chap 4]. These conditions include "natural conditions" such as the local pathwise contractibility and local integrability described in §2, together with conditions on the tangent spaces, which are global analogues of the tangent space condition from §4. This then allows the application of such global results to the local structure of nonlinear operators. Hence, the properties for operators with extra structure do hold locally for the operator without restricting to finite dimensional subspaces.

§6   Multiplicities of Solutions to Nonlinear Equations

Suppose nonlinearities are introduced in an linear equation whose Fredholm index is zero. Then what commonly occurs is that the positive dimensional space of solutions to the equation is replaced by a discrete set of solutions. This is evident from the examples of nonlinear Dirichlet problems and nonlinear oscillations already mentioned. We shall describe several results from singularity theory which help determine the number of such solutions.

Of course, the Lyapunov–Schmidt procedure reduces the infinite dimensional problem to a finite dimensional one; unfortunately that still leaves us a long way from determining even the number of solutions. In fact, counting the number of solutions to an equation $f(x) = y$ for a smooth mapping $f: \mathbb{R}^n \to \mathbb{R}^n$ is notoriously hard. Consequently, we shall only be concerned with the number of solutions for a local equation. Two such questions are: 1) for a germ $f_0: \mathbb{R}^n, 0 \to \mathbb{R}^n, 0$ determine the number of solutions to $f_0(x) = y$ for x and y sufficiently close to 0; and 2) given a perturbation f_t of f_0, determine the number of solutions to $f_t(x) = 0$ with x near 0 and t sufficiently small.

We begin with holomorphic mappings $f: \mathbb{C}^n \to \mathbb{C}^n$ where some striking results serve as our starting point. Classically, there is Bezout's theorem which is a natural generalization of the fundamental theorem of algebra to a statement about the number of solutions for n polynomial equations in n variables.

Theorem (Bezout's Theorem see e.g.[GH,chap. 6]): *Suppose the map* $f: \mathbb{C}^n \to \mathbb{C}^n$ *is defined by n homogeneous polynomials* $f_i(z_1, ..., z_n)$ *of positive degrees* d_i, $i = 1, ..., n$. *If f has 0 as an isolated zero, i.e.* $f^{-1}(0) = 0$, *then the number of solutions to the equation* $f(z) = w$, *counting multiplicities, is* $d = d_1 d_2 \cdots d_n$. *Furthermore, for almost all w (i.e. for w not belonging to the discriminant of f, which is a closed algebraic subset of* $\mathbb{C}^n$*) there are exactly d distinct solutions.*

In a sense, Bezout's theorem is really a local result. The homogeneity of the coordinate functions of f can be restated in modern terminology by saying that f is equivariant with respect to $\mathbb{C}^*$-actions on each $\mathbb{C}^n$. The multiplicative group $\mathbb{C}^* = \mathbb{C}\backslash\{0\}$ acts on the source $\mathbb{C}^n$ by $t\cdot(z_1, ...,z_n) = (t\cdot z_1, ...,t\cdot z_n)$, and on the target $\mathbb{C}^n$ by $t\cdot(w_1, ...,w_n) = (t^{d_1}\cdot w_1, ...,t^{d_n}\cdot w_n)$. Then, the equivariance of f just states that f commutes with the group actions, $f(t\cdot z) = t\cdot f(z)$. The key geometric consequence of this action is that the properties of f are constant along $\mathbb{C}^*$-orbits, and the orbits approach 0 as $t \to 0$. Hence, the number of solutions to $f(z) = w$ is the same as for $f(z) = t\cdot w$. Thus, the local behavior of f determines it globally. Thus, the statement is really about the germ $f : \mathbb{C}^n, 0 \longrightarrow \mathbb{C}^n, 0$.

Remark 6.1 : This notion of homogeneity naturally extends to the case where a more general action on the source space $t\cdot(z_1, ...,z_n) = (t^{a_1}\cdot z_1, ...,t^{a_n}\cdot z_n)$, with all $a_i > 0$. Then, an equivariant f is said to be *weighted homogeneous*; this corresponds to each f_i being a sum of monomials

$$z^\alpha = z_1^{\alpha_1} z_2^{\alpha_2}...z_n^{\alpha_n}, \quad \text{with} \quad \text{wt}(z^\alpha) \stackrel{\text{def}}{=} \sum a_i\cdot\alpha_i = d_i$$

Surprisingly, normal forms which occur in low codimension for a variety of problems are weighted homogeneous far more often than we have any right to expect; in addition, their properties are far more amenable to examination.

The second observation about Bezout's theorem is that the number d is really the complex dimension of the algebra $Q(f) = \mathcal{O}_n/(f_1, ...,f_n) \simeq \mathbb{C}[z_1, ..., z_n]/(f_1, ..., f_n)$, where $\mathcal{O}_n$ denotes the ring of holomorphic germs on $\mathbb{C}^n$ at 0. In this form, the result need no longer be restricted to homogeneous f, but was proven by Palomodov [Pa] to be valid for any holomorphic germ with an isolated zero.

Theorem 6.2 (Palomodov [Pa]): *Suppose the holomorphic map germ f:* $\mathbb{C}^n, 0 \longrightarrow \mathbb{C}^n, 0$ *has 0 as an isolated zero, i.e.* $f^{-1}(0) = 0$. *Define*

$$\delta(f) \stackrel{\text{def}}{=} \dim_{\mathbb{C}}\mathcal{O}_n/(f_1, ...,f_n) .$$

Then, there is a representative of f, still denoted by f: $U \longrightarrow V$ *such that for* $w \in V$, *the number of solutions* $z \in U$ *to the equation* $f(z) = w$, *counting multiplicities, is* $\delta(f)$. *Furthermore, for almost all* $w \in V$ *there are exactly* $\delta(f)$ *distinct solutions.*

As a first step in determining the multiplicities of solutions to nonlinear equations in the real case, we might hope to extend this local result. There are two problems with this: first the number of solutions to a real equation f(x) = y is not the same for almost all y; and second, simple examples show that the number of real solutions is often much smaller than the number for the complex case. These problems have been long recognized and one method to avoid them is to count the solutions x with signs ±1 depending on whether df(x) preserves or reverses orientation. This yields the degree of a mapping. In its global form it has the a number of important properties, one of the most important being its homotopy invariance (see e.g. [N]). It is this property which makes it a principal tool for proving the existence of solutions to nonlinear equations. Typically one constructs a homotopy to a solvable linear equation (with a unique solution) and prove via a priori estimates that the degree is invariant under the homotopy so the degree for the nonlinear equation is nonzero.

To conclude additional information about the numbers of solutions we need either: more refined methods for computing the degree or an alternate method for computing the exact number of solutions. With Palamodov's theorem as a starting point, we describe how to partially answer these questions in the local case. We consider the case of a finite map germ f: $\mathbb{R}^n,0 \longrightarrow \mathbb{R}^n,0$. Recall from §3 that this means $\dim_{\mathbb{R}} Q(f) < \infty$, and has as a consequence that f has representative, still denoted f: $U \longrightarrow V$ which has an isolated 0 at 0.

Definition 6.3 : To define the *local degree* of f: $\mathbb{R}^n,0 \longrightarrow \mathbb{R}^n,0$ we choose a representative f: $U \longrightarrow V$ which has an isolated 0 at 0. Then, there is a possibly smaller neighborhood V' of 0, such that for a regular value $y \in V'$, the number

$$\sum \text{sgn}(x') \qquad \text{summed over } x' \in f^{-1}(y) \cap U$$

is independent of y and is called the local degree of f, deg(f). Here, as usual, sgn(x') = ±1 depending on whether df(x') preserves or reverses orientation.

We also define for any finite map germ f: $\mathbb{R}^n,0 \longrightarrow \mathbb{R}^p,0$ with n ≤ p the local multiplicity [DGal].

Definition 6.4 : For the map germ f: $\mathbb{R}^n,0 \longrightarrow \mathbb{R}^p,0$ we choose a representative f: $U \longrightarrow$ V which again has an isolated 0 at 0. Then, we define the *local (real) multiplicity* of f, denoted m(f), to be the maximum k such that given any smaller neighborhood V' of 0, there is a $y \in V'$ so that

$$k = \text{card}(f^{-1}(y) \cap U) \qquad \text{(card(S) denotes the cardinality of the set S).}$$

Remark : If n < p then, we are in the "overdetermined case" and for general y there are no solutions; nonetheless, we shall see that for stable germs the maximal number of solutions can still be determined.

For holomorphic germs viewed as smooth mappings with $\mathbb{C}^n \simeq \mathbb{R}^{2n}$ (and p = n) these two notions give the same result. For the real case they differ considerably, with deg(f) taking into account the cancellation of pairs of points with opposite orientations and retaining a local form of the homotopy invariance property, while m(f) acknowledges the varying number of solutions possible and concerns itself with the maximum such.

Example 6.5 Morin Singularities :

We saw in (3.10) that

$$f(x_1, ...,x_n) = (x_1^{n+1} + x_2 x_1^{n-1} + ... + x_n x_1 , x_2, ..., x_n)$$

defines a stable germ f: $\mathbb{R}^n,0 \to \mathbb{R}^n,0$. Then, calculations can be made "by hand" in this case to see: deg(f) = 1 if n is even and 0 if n is odd, while m(f) = n + 1. We note that in only half of these cases could the degree be used to detect solutions for y near 0, and it effectively says nothing about the number of solutions. Nevertheless, both of these invariants can be extremely useful and we will describe how we can systematically make these calculations.

The basic problem is to understand what happens as we go from the complex to the real case. If f: $\mathbb{R}^n,0 \to \mathbb{R}^n,0$ is polynomial or real analytic, then we can form its complexification $f_{\mathbb{C}}: \mathbb{C}^n,0 \to \mathbb{C}^n,0$. Then,

$$\delta(f) = \dim_{\mathbb{R}} Q(f) = \dim_{\mathbb{C}} Q(f_{\mathbb{C}}) = \delta(f_{\mathbb{C}}).$$

Since m(f) $\leq$ m($f_{\mathbb{C}}$) = $\delta(f_{\mathbb{C}})$ (by Palomodov's theorem), we conclude m(f) $\leq$ δ(f). In fact, this is true quite generally for any smooth finite map germ [GG,chap. 7].

Proposition 6.6 : *If* f: $\mathbb{R}^n,0 \to \mathbb{R}^p,0$ *is a smooth finite map germ, then* m(f) $\leq$ δ(f).

Since deg(f) is generally smaller than m(f) $\leq$ δ(f), we ask how much smaller. The answer to this was discovered by Eisenbud-Levine [EL] (a version was also obtained by Himshashvili [Hi]). They show that deg(f) only depends on Q(f). Moreover, Eisenbud-Levine give a simple formula for it using Grothendieck's local duality [GH] [Hr]. Since

$\dim_{\mathbb{R}} Q(f) < \infty$, we may choose polynomial generators for I(f). So we may as well assume f itself is polynomial, and for algebraic purposes we identify f with its complexification. We also let Jac(f) denote the Jacobian determinant of f.

Theorem (Grothendieck local duality): *Suppose the holomorphic map germ* f: $\mathbb{C}^n$,0 $\longrightarrow$ $\mathbb{C}^n$,0 *has 0 as an isolated zero, i.e.* $f^{-1}(0) = 0$. *Then, there is a linear functional* φ: Q(f) $\longrightarrow$ $\mathbb{C}$ *with* $\varphi(Jac(f)) \neq 0$ *such that for any such* φ, *multiplication in* Q(f) *induces a nonsingular pairing*

$$Q(f) \times Q(f) \longrightarrow \mathbb{C}$$
$$(g, h) \mapsto \varphi(g.h).$$

In the real case, $Q(f_{\mathbb{C}})$ is the complexication of Q(f) and Jac(f) is real so we may choose φ: Q(f) $\longrightarrow$ $\mathbb{R}$ with $\varphi(Jac(f)) > 0$. Thus, the pairing on the real algebra Q(f) is a nondegenerate symmetric bilinear form. Over the reals we have an additional invariant for such bilinear forms, namely the signature; this is the difference of the dimensions of the maximal subspaces on which the pairing is positive definite (respectively negative definite). We denote this signature by sig(Q(f)). Then, Eisenbud–Levine give the following formula for deg(f).

Theorem (Eisenbud–Levine [EL]): *For the smooth finite map germ* f: $\mathbb{R}^n$,0 $\longrightarrow$ $\mathbb{R}^n$,0, *let* φ: Q(f) $\longrightarrow$ $\mathbb{R}$ *be a linear functional with* $\varphi(Jac(f)) > 0$. *Then,*
$$deg(f) = sig(Q(f)).$$

We next see quite generally how we can determine m(f) provided the algebra Q(f) is not too complicated. Recall from §3 that stable germs classified by their local algebras. Not surprisingly, these algebras play a key role in determining m(f).

Theorem 6.7 ([DGal]): *For infinitesimally stable finite map germs* f: $\mathbb{R}^n$,0 $\longrightarrow$ $\mathbb{R}^p$,0

i)   m(f) *only depends on* Q(f) *in the strong sense that if* g: $\mathbb{R}^m$,0 $\longrightarrow$ $\mathbb{R}^q$,0 *is another stable finite map germ between different spaces but with* Q(f) $\simeq$ Q(g), *then* m(f) = m(g).

ii)   *if* $\dim_{\mathbb{R}}(df(0)) \leq 2$ *(i.e.* Q(f) $\simeq$ $\mathbb{R}[x, y]/I$) *or* Q(f) *belongs to the class of discrete algebra types (see §7) then*

$$m(f) = \delta(f).$$

Remark 6.8 : Mather's classification theorem for stable germs ensures i) of the theorem in the case that f and g are between the same spaces. However, it is nontrivial when f and g are between different dimensional spaces and the proof uses the unfolding representation of stable germs from their local algebras. For example, the Whitney umbrella g: $\mathbb{R}^2, 0 \longrightarrow \mathbb{R}^3, 0$ defined by $g(x_1, x_2) = (x_1^2, x_2 x_1, x_2)$ is stable with local algebra $Q(g) \simeq \mathbb{R}[x]/(x^2)$ which is the same as for the fold $f(x_1, x_2) = (x_1^2, x_2)$. Thus, $m(g) = m(f) = 2$. Although generically there are no solutions to the equation $g(x) = y$, there are y for which the maximum number occurs and it is exactly the same as the maximum number for the fold f. Hence, overdetermined equations defined by stable germs can have for certain values the same maximum number of solutions as for equations of Fredholm index = 0 with the same local algebra!

Example 6.9 : The examples in (6.5) can be computed using these results, but instead let's return to the examples

(6.10) $\qquad f(x, y, u, v) = (x^2 + \varepsilon y^2 + ux, xy + vy, u, v)$ $\qquad$ with $\varepsilon = \pm 1$.

They define stable mappings $f : \mathbb{R}^4, 0 \longrightarrow \mathbb{R}^4, 0$, with $Q(f) \simeq \mathbb{R}[x, y]/(x^2 + \varepsilon y^2, xy)$. A vector space basis for Q(f) is given by $\{1, x, y, x^2\}$ with $Jac(f) = 2(x^2 - \varepsilon y^2) \equiv 4x^2$ in Q(f). Thus, we may choose $\varphi : Q(f) \longrightarrow \mathbb{R}$, with $\varphi = 0$ on 1, x, and y, while $\varphi(x^2) = 1$. Then, 1 and x^2 are "dually paired" under multiplication and contribute nothing to the signature. Since $y^2 \equiv -\varepsilon x^2$, the Eisenbud–Levine theorem implies deg(f) = the signature on the subspace $\langle x, y \rangle = 2$ if $\varepsilon = -1$ and $= 0$ if $\varepsilon = 1$. In either case, they are stable germs with dimKer(df(0)) = 2; hence $m(f) = \delta(f) = 4$.

(6.11) $\qquad$ More generally, consider the stable unfolding f of $f_0(x, y) = (x^a + \varepsilon y^b, xy)$, with $\varepsilon = \pm 1$. These germs include the previous family and provide a doubly infinite family in Mather's classification [Ma1-VI].

$\qquad$ Quite generally for any unfolding f of f_0, $\deg(f) = \deg(f_0) = sig(Q(f_0))$. Because f_0 is weighted homogeneous, the calculations considerably simplify. Under multiplication, weights add; also, $Jac(f_0)$ is weighted homogeneous and has the highest weight of a nonzero term in $Q(f_0)$. It follows that only the "middle weight part" of $Q(f_0)$, i.e. terms of

$(1/2)$weight$(Jac(f_0))$ contribute to the signature (this simple observation is even more useful for bifurcation problems [D9] and to follow).

A basis for $Q(f_0)$ is given by $\{1, x, ..., x^a, y, ..., y^{b-1}\}$. If wt$(x) = b$ and wt$(y) = a$ then wt$(Jac(f_0)) = ab$. Thus, by the Eisenbud–Levine theorem, if both a and b are odd, there are no middle weight terms so deg$(f) = $ sig$Q(f_0) = 0$; if exactly one of a or b is even, deg(f) $= \pm 1$; and if both a and b are even, a calculation of the pairing on $\langle x^{(a/2)}, y^{(b/2)} \rangle$ shows deg$(f) = 1-\varepsilon$. Again, the information obtained is quite limited and is essentially independent of how large a and b are; but it applies even when f is not a stable unfolding. By contrast, theorem 6.7 implies when f is stable that that $m(f) = \delta(f) = a+b$, giving precisely the maximum number of roots.

Remark : Using these calculations we can return to example (1.3)

$$f(x, y) = (x^2 - y^2 + \varepsilon \cdot x, 2xy - \varepsilon \cdot y) \qquad \text{for small } \varepsilon$$

which is the map $z \mapsto z^2 + \varepsilon \cdot \bar{z}$. By example (6.9) we know the maximal number of solutions is 4. Also, near the cusps of the hypocycloid, f is equivalent to a Whitney cusp so the local degree $= 1$; and for the inside points there are locally 3 solutions, while outside there are 1. However, the local degree of f (when $\varepsilon = 0$) is 2 and the global degree can be shown not change for (small) ε. Thus, there must be a fourth solution for inside points and a second for outside points. This justifies the claims for that example.

Consequences for Periodic Solutions of Nonlinear Oscillations:
(6.12) We saw these germs in (6.10) occurring in §5 in the local description of nonlinear oscillations

$$\frac{d^2x}{dt^2} + x + a_1(t) \cdot x^2 = b(t).$$

Hence, we can conclude from the calculation of the local multiplicity that when the generic conditions on the Fourier coefficients of a(t) ensuring stability are satisfied then there are forcing terms b(t) arbitrarily close to 0 such that this equation has 4 periodic solutions near 0. Furthermore, this is the maximum number that can occur; and by the 2–determinacy of the germs in (6.10), this number is does not change even if we add terms $h(x, t) \cdot x^3$ to the equation (see [D6, §3]).

This number 4 is somewhat unexpected given the local results of Ambrosetti–Prodi from §5. However, much more is true. Suppose we consider instead the equation

$$(6.13) \qquad \frac{d^2 x}{dt^2} + x + P(x, x', t) = b(t)$$

where now P is a polynomial of degree 2n in x and x'. A number of results for various P imply that we can expect about n periodic solutions for appropriate forcing functions b(t). Already the preceeding shows this to be much less than really possible. In fact, in [D7] it is shown that there are P of degree 2n (only involving linear terms in x') such that the associated operator in (5.23) is infinitesimally stable with local algebra $\simeq \mathbb{R}[x, y]/(x^{n+1}, y^{n+1})$. Hence, by theorem 6.7 there are small forcing terms b(t) for which (6.13) will have $(n+1)^2$ periodic solutions.

Thus, given N there are nonlinear P of degree N and forcing terms b(t) such that the number of periodic solutions to (6.13) is of order N^2.
(Calculations suggest there may be exactly N^2).

Further Considerations regarding the Local Multiplicity:

Unlike the Eisenbud–Levine result, theorem 6.7 still leaves much unanswered; however, we cannot expect ii) of the theorem to hold in full generality. A result of Iarrobino [Ia1] implies, via the arguments in [DGal], that under very general conditions on (n, p), there are stable germs in the complex case f: $\mathbb{C}^n, 0 \to \mathbb{C}^p, 0$ with n < p for which m(f) < δ(f). Since for a real stable polynomial germ f, its complexification $f_\mathbb{C}$ satisfies m(f) $\leq$ m($f_\mathbb{C}$), this also holds for real stable germs. Hence, we can not hope to completely extend 6.7.

The result of Iarrobino actually concerns properties of the Hilbert Scheme for "thick points" in algebraic geometry yet these results, appropriately applied, have consequences for this question about stable germs (see also [DGal2]). Recently, Behnke–Mond–Pellikan have shown that this failure also occurs for germs f: $\mathbb{C}^n, 0 \to \mathbb{C}^{n+1}, 0$. However, as pointed out in [DGal2], theorem 6.7 equally well applies to complex stable germs. In light of the theorem being equally valid in the real and complex case, we may reasonably expect the following to be true.

Conjecture : Suppose a stable finite map germ f: $\mathbb{R}^n,0 \rightarrow \mathbb{R}^p,0$ has a complexification $f_{\mathbb{C}}$ then

$$m(f) = m(f_{\mathbb{C}}).$$

Unfortunately this is not even known to be generally true for n = p and $\dim\mathrm{Ker}(df(0)) \geq 3$.

Because these questions are still very much open we indicate how such results are proven , hopefully inspiring a reader to carry these investigations further. They are based on the use of "standard bases" for ideals. These were introduced by Hironaka as a method to analyze the remainders when polynomials are divided by generators of an ideal (see e.g. [Gal2], [BGal], and [Brc] or [Ia2]). It turns out that these same ideas were introduced by Grobner. The algorithm for computing them introduced by Buchberger [Bu] has made them an important tool for many problems in computer algebra, and are now standardly referred to as "Grobner bases".

Once part i) of the theorem is established, to show that $m(f) = \delta(f)$ for a stable germ f: $\mathbb{R}^n,0 \rightarrow \mathbb{R}^p,0$ it is sufficient to establish it for any other stable germ g with $Q(g) \simeq Q(f)$. Arguments in [DGal] show that, in light of proposition 6.6, it is sufficient to find a deformation of a set of generators $\{g_1, \dots , g_\ell\}$ for I(f), say $\{G_{1t}, \dots , G_{\ell t}\}$, so that for small t $\neq 0$ there are at least $\delta(g)$ solutions $G_{1t} = \dots = G_{\ell t} = 0$ in a small neighborhood of 0.

We illustrate the construction of such a deformation for $(x^2 + \varepsilon y^2 , xy)$, $\varepsilon = \pm 1$. The (reverse lexicographical) ordering of the monomials of $\mathbb{R}[x, y]$ is given by $x^{a_1}y^{b_1} < x^{a_2}y^{b_2}$ if $b_1 < b_2$ or $b_1 = b_2$ and $a_1 < a_2$. To each element h of an ideal I, we assign a lattice point $(a, b) \in \mathbb{N}_+^2$ corresponding to the monomial $x^a y^b$ of lowest order appearing in h. The set of such lattice points form a region closed under addition by $\mathbb{N}_+^2$. Such a region is referred to as the stairs of the ideal. For the example in question we obtain fig. 6. 14.   The

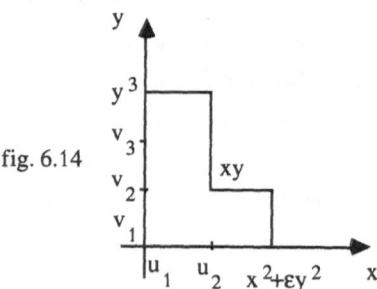

fig. 6.14

standard (or Grobner) basis for the ideal is a set of generators of I corresponding to the vertices of the stairs (in fact, a basis can be constructed for other orderings on the monomials). In this case, they are $(x^2 + \varepsilon y^2, xy, y^3)$. We construct a deformation of these generators, consider

(6.15)

$$G_{1u,v}(x, y) = (x-u_1)(x-u_2) + \varepsilon(y-v_1)(y-v_2)$$
$$G_{2u,v}(x, y) = (x-u_1)(y-v_3)$$
$$G_{3u,v}(x, y) = (y-v_1)(y-v_2)(y-v_3)$$

Then we count solutions to $G_{u,v}(x, y) (= (G_1, G_2, G_3)) = 0$:

for $y = v_3$ sufficiently small, there are two solutions $x = u_1', u_2'$;

for $x = u_1$, there are two more solutions $y = v_1, v_2$.

Thus, for f in (6.10), $m(f) = 4$ as asserted. However, the specific form of the stairs and the generators for the standard basis has a significant influence on whether this argument can be successfully applied (see [BGal] and [DGal]). For example, from the proof of theorem 6.7 it follows that if f is a stable germ and I can generated by monomials then $m(f) = \delta(f)$ (see also [Ia2]). As a consequence we obtain a rough lower bound for $m(f)$. If f: $\mathbb{R}^n,0 \rightarrow \mathbb{R}^p,0$ is a polynomial representative for a stable germ, we let $\tilde{I}(f)$ denote the ideal generated by the monomials in the equations for f and let $\tilde{Q}(f) = \mathbb{R}[x_1, ..., x_n]/\tilde{I}(f)$.

Corollary 6.16 : *For infinitesimally stable finite map germs* f: $\mathbb{R}^n,0 \rightarrow \mathbb{R}^p,0$,

$$m(f) \geq \tilde{\delta}(f) \stackrel{def}{=} \dim_{\mathbb{R}}\tilde{Q}(f).$$

Remark : In the opposite direction, Khovanski [Kh] has discovered in the real case more precise upper bounds (better than $\delta(f)$) for the number of solutions of a system of equations in terms of the number of monomials appearing in them, without regard to questions of stability.

Bifurcation, Equivariant Bifurcation, and Equivariant Degrees

Suppose we begin with a finite map germ f_0: $\mathbb{R}^n,0 \rightarrow \mathbb{R}^n,0$ and allow it to bifurcate via a perturbation $F(x, \lambda)$: $\mathbb{R}^{n+1},0 \rightarrow \mathbb{R}^n,0$. The singularity theory for analyzing this situation has already been discussed in §4. While questions of stability, determinacy, unfoldings and classification were answered, no methods were provided to derive geometrical information about F. Basically, we can find the normal form, and then by ad hoc methods determine geometric information from the normal form. In light of the preceding discussion, it would be especially desirable to have analogous algebraic results for bifurcation problems.

As λ is varied, the points of $F^{-1}(0)$ will lie on branches of curves. Actually what we think of topologically as a branch of a curve is really only one half of a branch in the algebraic sense. To avoid confusion about which we refer to, we use "half-branch" to denote the topological sense. There are several main questions we naturally ask regarding these half-branches: 1) how many half-branches are there; 2) how many of the half-branches correspond to the cases of $\lambda > 0$ and < 0; and 3) how many half-branches satisfy $\det(d_xF) > 0$ and < 0 ? To see the value of the third question observe that if $F = \operatorname{grad}_x(H)$ and $H(x, \lambda)$ is a Morsification of a function $h(x) = H(x, 0)$, then at (x, λ), $\det(d_xF) = (-1)^{\operatorname{ind}(H)}$ with $\operatorname{ind}(H)$ the Morse index of $H(., \lambda)$ at the critical point x. Quite generally we can ask for the index of d_xF, which is the number of negative eigenvalues, i.e. $\operatorname{ind}(H)$ in the gradient case.

If moreover, F is G-equivariant for a compact Lie group G acting linearly on $\mathbb{R}^n$, then $F^{-1}(0)$ will be a union of G-orbits. If G is finite then we still have (half-)branches of a curve with an action of G on them. If not, then in general $F^{-1}(0)$ will have higher dimensional components. In either case, the additional information of the G-action on $F^{-1}(0)$ provides that much more knowledge about F and its properties, provided this can be computed.

First consider the case without a group action. There are two possible ways to obtain Levine-Eisenbud type results. One is to reduce a calculation to the local degree of an associated map germ. Then, the Levine-Eisenbud result gives a signature formula involving the local algebra of the associated germ. The second possibility is to directly prove that an invariant can be computed using a signature formula for some local algebra.

The first method was used by Aoki-Fukuda-Nishimura [AFN1] and [AFN2] who obtained signature formulas answering questions 1) and 2). Let $\operatorname{Jac}(F)$ denote $\det(d_xF)$. Define map germs F_1, $F_2 : \mathbb{R}^{n+1}, 0 \longrightarrow \mathbb{R}^{n+1}, 0$ by $F_1 = (F, \lambda \cdot \operatorname{Jac}(F))$ and $F_2 = (F, \operatorname{Jac}(F))$. By our assumption on F, they are finite map germs. Also, each F_i has local algebra $Q(F_i) = C_{x,\lambda}/I(F_i)$.

Theorem 6.17 ([AFN2]): *Consider* $F(x, \lambda) : \mathbb{R}^{n+1}, 0 \longrightarrow \mathbb{R}^n, 0$, *with* $f_0(x) = F(x, 0)$ *a finite map germ and* $\text{Jac}(F) = \det(d_x F) \neq 0$ *on* $F^{-1}(0) \backslash \{0\}$. *Then,*

i)   *the number of half-branches of* $F^{-1}(0)$ $=$ $2 \cdot \text{sig} Q(F_1)$

ii)   *(number of half-branches with* $\lambda > 0$) - (*number of half-branches with* $\lambda < 0$)

$$= 2 \cdot \text{sig} Q(F_2)$$

This result provides exactly the information asked for in two of the questions and the degree of $(F(x, \lambda), \lambda) : \mathbb{R}^{n+1}, 0 \longrightarrow \mathbb{R}^{n+1}, 0$ provides the third. We shall further see that these results can be extended to take into account a group of symmetries, and to benefit from special properties of F such as weighted homogeneity.

Equivariant Bifurcation and Equivariant Degrees

We are interested in two related questions: first, for a G-equivariant germ $F : \mathbb{R}^n, 0 \longrightarrow \mathbb{R}^n, 0$, with $F^{-1}(0) = \{0\}$ can we define a G-equivariant degree; and second, what invariants can we define for G-equivariant germs $F : \mathbb{R}^{n+1}, 0 \longrightarrow \mathbb{R}^n, 0$, which provide information about the branching of F ?

A number of workers have considered ways to define an equivariant degree for F for various groups G. Just as the local degree for F can be defined as an element of $\pi_{n-1}(S^{n-1})$, so too a global G-degree for an equivariant map $F : \Omega \, (\subset \mathbb{R}^n) \longrightarrow \mathbb{R}^p$ has been defined by Ize-Massabo-Vignoli [IMV] as an element in G-equivariant homotopy groups of spheres. This is shown [IMV2] to yield in the special case of S^1-equivariance the Fuller index and the S^1-degree for gradient maps of Dancer. Because this will be covered in the report of Ize, we concentrate on results from singularity theory and their relation with equivariantbifurcation.

In particular, for the remainder of this section we shall concentrate on the case of finite groups G, which will have very little overlap with the above results.

With the action of the group G there are two related additional pieces of information: the action of G on a branch and the isotropy subgroups G_x of points x on branches, i.e. $G_x = \{ \sigma \in G : \sigma \cdot x = x \}$. For bifurcation, the action of G simplifies the determination of branching. We would hope to obtain G-equivariant analogues of the local degree and the results in theorem 6.17.

The simplest example is the equivariant branching lemma of Vanderbauwhede and Cicogna (see [GSS, chap.13]) which provides sufficient conditions for the existence of branching in terms of the behavior of f I Fix(H) where H is a maximal isotropy subgroup of G and Fix(H) is the fixed point space of H. A conjecture was proposed by Golubitsky that for absolutely irreducible representations of G on $\mathbb{R}^n$ with trivial extension to the last factor of $\mathbb{R}^{n+1}$, the branches had maximal isotropy subgroups. Counterexamples were found by Chossat and Lauterbach (see [GSS, chap.13]) and even for finite groups by Field and Richardson [FRc1]. Thus, there is a need to introduce refined methods to understand submaximal equivariant branching.

There is a related question considered by Field and Richardson [FRc2, I II] concerning how far along in the Taylor expansion of G-equivariant germs we must go to ensure that generically germs exhibiting bifurcation do so stably so that normal forms may be determined and analyzed. Specifically, let d denote the smallest integer so that there is an equivariant homogeneous polynomial of degree d which is not of the form h·id for a G-invariant function h. Consider equivariant germs with d jets of the form

$$(6.18) \qquad j^d(F) \ (= \ F_0) \ = \ R + Q + \lambda \cdot id \qquad \text{with } R = h \cdot id \ \text{ and } \ Q \text{ not of this form.}$$

When can we specify a dense open subset of such F which are stable under small perturbation (but remaining of this form) ? Here stability either refers to stability of the topological type of the set $F^{-1}(0)$ with G-action (weak stability) or with additional information of the directions of branching and the index of $d_x F$ (stability). The latter condition is important for F thought of as a family of vector fields, with its index of hyperbolicity at its zeros.

When an open dense set can be found, Field and Richardson use the very unfortunate terminology (from our viewpoint) that the problem is *(weakly) d-determined*, which conflicts with standard terminology of singularity theory.

Their answer is expressed in terms of the phase vector field

$$P_Q(u) \ \overset{def}{=} \ Q(u) - \langle Q(u), u \rangle \cdot u \qquad \text{for} \ \ u \in S^{n-1}.$$

Let $\alpha(Q,u) = \langle Q(u), u \rangle$.

Theorem 6.19 [FRc2, I] :   *Suppose that all zeros of the phase vector field* P_Q *are simple (resp. hyperbolic). Then*

i) $F^{-1}(0)$ *and* $F_0^{-1}(0)$ *are topologically G-equivariant equivalent (and there is a*

G-equivariant bijective correspondence with the zeros of P_Q *)*;

ii)   *if* $\alpha(Q,z) \neq 0$ *for all z which are zeros of* P_Q *then*

    a)   F *is weakly stable (resp. stable);*

    b)   *the topological equivalence preserves* $\mathrm{sign}(\lambda)$ *(resp.* $\mathrm{sign}(\lambda)$ *and* $\mathrm{index}(d_x f)$*)*

Often the degree d is sufficiently small and the truncated form sufficiently simple that they are able to apply various results from group theory to deduce information about branching. Also, the form that Field-Richardson obtain bears a close relation with the weighted homogeneous germs that we considered earlier. We describe this relation and state a result which extends the Eisenbud-Levine and Aoki-Fukuda-Nishimura theorems to give specific information regarding the representation of G on the set of half-branches. Montaldi-Van Straten have shown that most formulas involving a signature, such as those in Theorem 6.19, could be obtained from a general formula obtained by them for counting the number of branches of a real curve singularity [MvS]. It was shown in [D8] how to extend this result to the G-equivariant case for finite groups. Using this we describe how to obtain the G-representations on the half-branches for bifurcation germs.

Given $f = (f_1, \ldots, f_p) : \mathbb{R}^n, 0 \to \mathbb{R}^p, 0$, we say it is *semi-weighted homogeneous* if we can assign weights $\mathrm{wt}(x_i) = a_i > 0$ so that with $f_{0i} = \mathrm{in}(f_i)$ denoting the nonzero terms of f_i of lowest weight $(= d_i)$, $f_0 = (f_{01}, \ldots, f_{0p})$ defines a germ of finite $\mathcal{K}$-codimension. Alternately, we can define semi-weighted homogeneous germs for equivariant bifurcation equivalence, if instead we require that f_0 has finite codimension for this equivalence.

(6.20)   For example, the germ in (6.18) with $R = 0$, $F_0 = Q + \lambda \cdot \mathrm{id}$ is weighted homogeneous with $\mathrm{wt}(x_i) = 1$ and $\mathrm{wt}(\lambda) = d-1$ and $\mathrm{in}(F) = F_0$. If F_0 has finite codimension for one of the equivalences, then F is semi-weighted homogeneous for the equivalence. If F_0 has finite $\mathcal{K}$-codimension, then by the change of coordinates $\lambda \mapsto \lambda + R$, $F + R$ is $\mathcal{K}$-equivalent to a germ with d-jet $= F_0$. Thus, for describing the branches without regard to $\mathrm{sign}(\lambda)$ we again can reduce to a semi-weighted homogeneous germ with initial part $= F_0$.

We suppose $F(x, \lambda) : \mathbb{R}^{n+1}, 0 \to \mathbb{R}^n, 0$ is a G-equivariant semi-weighted homogeneous germ with initial part F_0, such that $F_0(x, 0) = f_0(x) : \mathbb{R}^n, 0 \to \mathbb{R}^n, 0$ is a finite map germ. We consider three algebras : the *local algebra* $Q(f_0)$; the *bifurcation algebra* defined above $\mathcal{B}(F_0) = C_{x,\lambda}/(F_1, ..., F_n, \det(d_x F))$, with $F_1, ..., F_n$ denoting the coordinate functions of F; and the *Jacobian algebra* $\mathcal{J}(F) = C_{x,\lambda}/J(F)$, where $J(F)$ denotes the ideal generated by $F_1, ..., F_n$, together with the $n \times n$ minors of the Jacobian matrix $d_{(x,\lambda)}F$.

We relate the induced representation of G on each of these algebras with the permutation representations of G on various collections of half-branches. Let B_h denote the set of half branches of $F^{-1}(0)$. Suppose $B' \subset B_h$ denotes a subset on which G acts. We can associate the permutation representation on the vector space V' with basis $\{e_b : b \in B'\}$ with action $\sigma \cdot e_b = e_{\sigma \cdot b}$. It is well-known that the complex character $\chi_{\mathbb{C}}$ of this representation uniquely determines it. Thus, it is sufficient to determine $\chi_{\mathbb{C}}$. Unfortunately, it is not these complex characters which can be explicitly computed, but rather the (characteristic 2) modular characters. These are $\mathbb{C}$-valued characters associated to characteristic 2 representations (see e.g. [D8,§4] for a brief summary of their properties); but they can be most simply described as the usual complex character restricted to $G°$, the set of odd order elements of G. We denote the character by $\chi = \chi_{\mathbb{C}}| G°$. For B_h itself we use V_{hp} and χ_{hp}.

Second, to incorporate other information about branches, suppose that by some method we have attached signs $\varepsilon(b)$ to the half-branches $b \in B_h$ in such a way that $\varepsilon(b) = \varepsilon(\sigma \cdot b)$ for all $b \in B_h$ and $\sigma \in G$. If $\lambda \neq 0$ on any branch, then examples are $\varepsilon(b) = \text{sgn}(\lambda)$ or $\text{sgn}(\det(d_x F))$. Then, we may define virtual representations on the half branches: we decompose $B_h = B_{he}^+ \cup B_{he}^-$, corresponding to the sign of ε. Associated to $B_h^\pm$ are permutation representations $V_{he}^\pm$ with modular characters $\chi_{he}^\pm$ and the virtual representation $V_{he} = V_{he}^+ - V_{he}^-$ and *virtual modular character* $\chi_{he} = \chi_{he}^+ - \chi_{he}^-$. For $\varepsilon(b) = \text{sign}(\lambda)$ we use V_{hb} and χ_{hb}, and for $\text{sgn}(\det(d_x F))$, V_{hd} and χ_{hd}.

While the half-branch characters are more natural to use for the actual investigation of singularities, the characters for representations on branches are algebraically more intrinsic and provide cleaner formulas. In stating the results we use these latter modular characters χ_ε which are related to the half branch characters by $\chi_{he} = 2\chi_\varepsilon$ (basically because each branch has two half branches, although this would not be correct for the usual complex characters).

The formulas for computing the modular characters will be based on the G-signature of the multiplication pairings on the algebras (see e.g. prop. 1.1 of [D8]). This strengthens the form of the usual signature when there is a symmetric bilinear form φ on a vector space V which is G-invariant, i.e. $\varphi(g \cdot v, g \cdot w) = \varphi(v, w)$. Then, there are maximal dimensional G-invariant subspaces $V_\pm$ on which φ is positive (resp. negative) definite; and any other choices will be isomorphic as G-representations. Thus, if $\chi_\pm$ denote their modular characters, then the G-signature $\mathrm{sig}_G(\varphi) = \chi_+ - \chi_-$ (or we use $\mathrm{sig}_G(V)$ if φ is fixed).

The three algebras $\mathcal{J}(F_0)$, $\mathcal{B}(F_0)$, and $Q(f_0)$ all have multiplication pairings to the nonzero top weight, so as earlier, we can restrict the pairing to the middle weight parts. Let

$$s = \sum_{i=1}^{n} d_i - \sum_{j=1}^{n} a_j .$$

Then, the middle weights for the algebras are given by Table 6.20.

Table 6.20

ALGEBRA		MIDDLE WEIGHT
Jacobian algebra	$\mathcal{J}(F_0)$	$s_p = s - \mathrm{wt}(\lambda)$
Bifurcation algebra	$\mathcal{B}(F_0)$	$s_b = s - (1/2)\mathrm{wt}(\lambda)$
Local algebra	$Q(f_0)$	$s_d = (1/2)s$

To denote the middle weight parts of these algebras we use the notation that the weight m part of the algebra A will be denoted by A_m. If m is not an integer, then $A_m = (0)$. Then, we can compute the modular characters as follows.

Theorem 6.21: *Suppose that* $F : \mathbb{R}^{n+1},0 \longrightarrow \mathbb{R}^n,0$ *is G-equivariant and semi-weighted homogeneous with initial part* F_0. *Then,*

1) $$\chi_p = 1 + \mathrm{sig}_G(\mathcal{J}(F_0)_{s_p})$$

2) $$\chi_d = \mathrm{sig}_G(Q(f_0)_{s_d})$$

If moreover F is semi-weighted homogeneous for bifurcation equivalence then

3) $$\chi_b = \mathrm{sig}_G(\mathcal{B}(F_0)_{s_b})$$

Remarks : These formulas also allow us to obtain information without computing a signature! If $\tilde{\chi}(G)$ denotes the ring of virtual modular characters, then $\tilde{\chi}(G) / 2\tilde{\chi}(G) \simeq (\mathbb{Z}/2\mathbb{Z})^r$, where r = the number of odd order conjugacy classes of G. Since $\mathrm{sig}_G(\varphi) \equiv \chi_+ + \chi_-$ mod $2\tilde{\chi}(G)$, We obtain a $(\mathbb{Z}/2\mathbb{Z})^r$-invariant just by analyzing the action of G on the middle weight part of the local algebras. It can be shown that

$$\mathcal{A}(F_0)_{s_p} \simeq Q(F_0)_{s_p} \qquad \text{and} \qquad \mathcal{B}(F_0)_{s_b} \simeq Q(F_0)_{s_b}$$

as G-representations $(Q(F_0) = C_{x,\lambda}/(F_{01}, ..., F_{0n})$ is not finite dimensional but any weighted part is). Thus, representation information on the half branches, including $\mathrm{sign}(\lambda)$ or $\mathrm{sgn}(\det(d_x F))$ is determined just from the representations on $Q(F_0)_m$ and $Q(f_0)_m$ for appropriate m. As discussed above, the form determined by Field-Richardson is often semi-weighed homogeneous and allows one to read off the appropriate weights. Then, for example, the calculations on $Q(F_0)_m$ actually provide a method [D9,§4] for detecting submaximal branching for finite groups.

These results are related with the G-equivariant degree referred to earlier. $\mathrm{sig}_G(Q(f_0))$ computes the G-degree as the character for the permutation representations of G_y on $f_0^{-1}(y)$ [D8,§2]. In fact, Gusein-Zade [GZ] has shown that to improve this formula to obtain a G-degree as a regular complex character, one has to determine the representation on imaginary points. Thus, the modular character seems to give the optimum information in the real case without involving complex solutions. It raises a basic question.

Question : How can one define a G-degree as a character of a representation for the case of compact Lie groups G; and how does this relate to the definition of the degree as an element of the stable homotopy groups of spheres by Ize-Massabo-Vignoli ?

§7   The Theory for Topological Equivalence

Given how successfully we have answered the questions about the local properties of nonlinear mappings for various notions of smooth equivalence, we might find it rather strange that the question of topological equivalence of mappings is even brought up. We have been able to work with topological invariants such as the local degree and local multiplicity, as well as various invariants for bifurcation theory, using invariants defined from smooth equivalence.

Despite this, since the beginning of our "excursion", there has been lurking just beyond our view, basic questions which can only be addressed via topological equivalence. These questions all stem from a fundamental problem in any application of singularity theory, namely, the presence of moduli which parametrize continuous change in the G-equivalence classes in $\mathcal{F}$. Often the G-orbits in $\mathcal{F}$ locally form a foliation of a subspace of $\mathcal{F}$. By taking a section to these orbits we obtain a modulus describing a continuous change in the G-equivalence class.

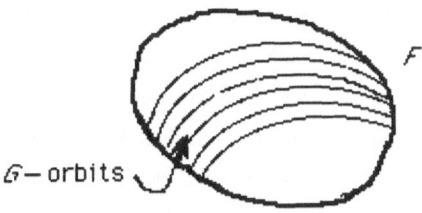

$G-$orbits F

We have alluded to the appearance of moduli in several earlier examples. To see the necessity of allowing topological equivalence to handle moduli, consider a standard simple example

(7.1) $$f_t(x, y) \ = \ xy(y-x)(y-tx)$$

For fixed t it defines a smooth germ $f_t : \mathbb{R}^2, 0 \rightarrow \mathbb{R}, 0$ and $f_t^{-1}(0)$ is a union of four lines. If f_t and $f_{t'}$ were right equivalent, say $f_{t'} = f_t \circ \varphi$, then both φ and its derivative $d\varphi(0)$ would send the four lines to the other four lines. However, associated to four lines through the origin in $\mathbb{R}^2$ is a number called the cross ratio (a line through the origin in $\mathbb{R}^2$ can be identified with a point in a projective line, and the cross ratio is a classical invariant of projective geometry associated to four such points). Since a linear transformation such as $d\varphi(0)$ must preserve the cross ratio, and the cross ratio varies in a continuous fashion with t, we conclude that for most t and t', the germs f_t and $f_{t'}$ are <u>not</u> right equivalent.

Thus, we see geometrically that t is a modulus, parametrizing continuous change in the right equivalence class. However, it is clear that there is a homeomorphism φ of $\mathbb{R}^2$ sending four distinct lines through 0 to another four distinct lines (and it can be proven that there is such a φ so that $f_{t'} = f_t \circ \varphi$). Thus, to reduce the uncountable collection of classes given by $\{ f_t : t \in \mathbb{R} \}$ to a finite list we must change to the topological version of right equivalence.

This same phenomena occurs repeatedly: in symmetric Hamiltonian–Hopf bifurcation, we saw the appearance of a modulus α in the simplest example in (4.17) (this modulus is detected algebraically rather than geometrically); there is a modulus in the normal forms obtained for the energy momentum map (4.23); Theorem 4.28 asserts a homeomorphism (rather than diffeomorphism) between the set of periodic orbits and a model set (also due to any moduli present); Theorem 6.21 expressing the branching phenomena for bifurcation for semi–weighted homogeneous germs in terms of their initial parts depends on topological equivalence to overcome moduli; and finally the stability result of Field–Richardson , theorem 6.19, likewise in given terms of the topological classification of the branching behavior to avoid moduli problems.

Even for the original question of stability of mappings, the problem with moduli has been skirted over to allow an unbroken development of the smooth theory. When we outlined the key results of Morse theory we emphasized the density of Morse functions allowing any function to be arbitrarily closely approximated by a Morse function. We never returned to the question of density in any further discussion. There is a very good reason why– it does not hold in general for stable mappings. There are certain dimensions (n, p) such that the set of stable mappings will not be dense among smooth mappings f: N $\longrightarrow$ P with dim N = n and dim P = p.

For example, this has relevance for determining the possible local structure of nonlinear operators f : E,0 $\longrightarrow$ F,0 of Fredholm index 0. Already for nonlinear osciliations (5.23) with P cubic of the form $P(x, x', t) = b_1(t) \cdot x^3 + b_2(t) \cdot x^2 \cdot x'$, moduli occur and topological stability is used (see [D6,§8]). Also, when dimKer(df(0)) = 3, generically there are finite dimensional models which are topologically stable but not stable.

Thom was aware of this failure of density and recognized it was due to the presence of moduli [ThL,chapIII]. Again, the work of Thom [Th4], along with major refinements by Mather [Ma3], [Ma4], proved that density held in all dimensions when stability was replaced by topological stability. Moduli also prevents map germs from being finitely determined "in general" by results of Mather [Ma5]. Again an approach to overcome this problem using topological determinacy was outlined by Thom [Th5]. A proof was first given by

Varchenko [Va] using different methods, and later by DuPlessis [DP] using the stratification theory of Thom–Mather.

How do we deal with questions raised by the intrusion of moduli into the problems we have discussed? How must these be taken into account in trying to determine the local behavior of nonlinear operators?

Basic questions for Topological Equivalence ;

1)   In trying to provide models for nonlinear phenomena, the "stable models" may really only be topologically stable, so how do we determine and construct such local models?

2)   Often information in a problem is of a topological nature so it is not really necessary to obtain smooth equivalence between germs to deduce that they carry the same information. How can one determine when two germs are topologically equivalent. for example, if a germ is finitely determined at some unspecified order k, can we find a lower well–determined order at which the germ is topologically determined?

3)   For problems involving a finite number of parameters, can we determine which parameters in versal unfoldings are topologically redundant?

4)   The preceding refer to "topological equivalence" without reference to which equivalence. We have seen in §4 there are numerous different equivalence relations defined by groups of germs of diffeomorphisms G. For each there is a corresponding notion of topological G-equivalence. We would like answers to the preceding which applies to each such group.

5)   We have been using the notion of "topological equivalence" in a rather cavalier manner because it is the most natural notion in the absence of differentiability. In fact, the types of topological equivalence we use will have differentiability properties. Our spaces will be decompose into finite unions of manifolds (called strata) and the homeomorphisms will be differentiable on each stratum. The question is : how few strata can we use to keep as much differentiability as possible?

Stability and Transversality

To understand the role that topological equivalence plays, we first explain how it allows us to obtain density of topologically stable mappings in all dimensions. At the heart of this is the characterization of C^∞-stability by Mather in terms of multitransversality. Recall in §3, we related the stability of a germ (for $\mathcal{A}$-equivalence), via versality, with transversality to "bundles of $\mathcal{K}$-orbits" in $J^k(\mathbb{R}^n, \mathbb{R}^p)$ (or more simply "$\mathcal{K}$-orbits" in $J^k(\mathbb{R}^n, \mathbb{R}^p)$).

The construction of the jet bundle can be made for any pair of manifolds N and P. It is a fiber bundle $J^k(N, P)$ with fiber $J^k(n, p)$ and base $N \times P$; and any $f: N \longrightarrow P$ induces a jet extension $j^k(f) : N \longrightarrow J^k(N, P)$ sending $x \mapsto j^k(f)(x)$ over the point $(x, f(x))$. Furthermore, this can be extended to simultaneously keep track of k–jets at r distinct points by forming $J_r{}^k(N, P)$ which equals the restriction of $(J^k(N, P))^r$ (r copies) to $N^{(r)} \times P^r$ with $N^{(r)}$ denoting $\{ (x_1, \ldots , x_r) \in N^r : x_i \neq x_j$ if $i \neq j \}$. There are analogous "$\mathcal{K}$-multiorbits" in $J_r{}^k(N, P)$. Again f induces $j_r{}^k(f) : N^{(r)} \longrightarrow J_r{}^k(N, P)$ by $j_r{}^k(f)(x) = (j^k(f)(x_1) , \ldots , j^k(f)(x_r))$ where $x = (x_1, \ldots , x_r)$.

Now the Thom transversality theorem (see e.g. [GG,chap2]) asserts

Theorem 7.2 : *Given a submanifold* $W \subset J^k(N, P)$ (*or even a countable collection* $\{W_i\}$) *set of* $f \in C^\infty(N, P)$ *for which* $j^k(f)$ *is transverse to* W (*or all* W_i) *is a residual set, i.e. a countable intersection of open dense sets.*

There is an analogous multitransversality theorem for $J_r{}^k(N, P)$ and $j_r{}^k(f)$ (see [Ma1–V] or [GG]). Multitransversality refers to the transversality of $j_r{}^k(f)$ to submanifolds of $J_r{}^k(N, P)$.

In applications of the transversality theorem, the important submanifolds $W \subset J^k(N, P)$ are those consisting of jets sharing some property, e.g. belonging to the same orbit under one of the natural group actions. Similarly, interesting submanifolds of the multijet bundle keep track of r –distinct jets with each individual jet having some property; the importance is for the case when some of the germs of f at the x_i map to a common target and the submanifold captures a property of the interaction of the images of these distinct germs. Since (multi)transversality to a submanifold W of codimension $> n$ means that the image of $j^k(f)$ (or $j_r{}^k(f)$) misses W, it follows from transversality that "undesirable properties" defined by submanifolds of codimension $> n$ won't occur and other properties occur in a geometrically nice way.

Since residual sets are dense in $C^\infty{}_{pr}(N, P)$, if it is possible to characterize stability in terms of transversality or multitransversality to at most a countable number of submanifolds then density would follow. Mather does exactly that.

Theorem 7.3 [Ma1-V] : *A proper mapping* $f: N \longrightarrow P$ *is* C^∞*-stable iff it is multitransverse to all $\mathcal{K}$-multiorbits of* $J_r^k(N, P)$.

For example, if germs of f at individual points x_i are stable and $f(x_i) = y$ then multitransversality to the $\mathcal{K}$-multiorbits means that the images of the singular sets of f near the x_i intersect transversely near y. Thus, the $\mathcal{K}$-multiorbit at such a point $x = (x_1, \dots, x_r)$ is determined by the $\mathcal{K}$-orbits of the individual stable germs of f at x_i. Thus, if there are a countable number $\mathcal{K}$-orbits of codimension $\leq n$ then there will likewise only be a countable number $\mathcal{K}$-multiorbits of codimension $\leq n$. By theorem 7.3, the stable mappings will be dense.

Conversely if there are an uncountable number of orbits, then there are moduli present and there are submanifolds $W \subset J^k(N, P)$ (for appropriate k) of codimension $\leq n$ which are foliated by $\mathcal{K}$-orbits corresponding to the modulus. Then, Mather shows under

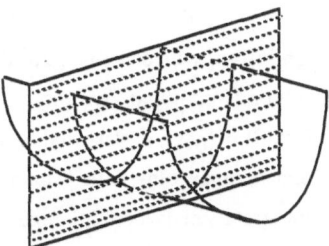

fig. 7.4

such circumstances there exist mappings f whose k–jet at, say, x is transverse to W but not any $\mathcal{K}$-orbit in W; nor can f be perturbed to one which does. We can visualize this via fig. 7.4. The paraboloid is transverse to the plane but <u>not</u> to all horizontal lines; also, any small perturbation will still have a local minimum for its intersection with the plane and so fail to be transverse to the line through that point.

Using this criteria, Mather completely determined the codimension $\sigma(n, p)$ of the strata where moduli first appear for the pairs of dimensions (n, p). He deduces those dimensions $n < \sigma(n, p)$ in which the stable mappings are dense [Ma1-VI]. These are called the "nice dimensions" and are the pairs indicated in fig. 7.5

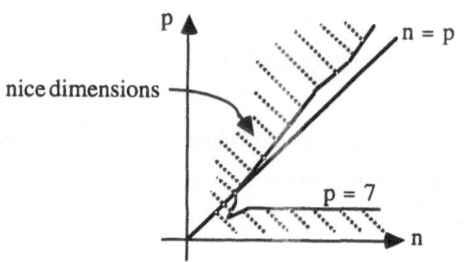

fig. 7.5

Topological Stability

The way to obtain density as described by Thom [Th4], [Th5] and refined by Mather [Ma3] and [Ma4] is to replace the collection of $\mathcal{K}$-orbits described by moduli by single submanifolds which are a union of such orbits. Then, we will obtain a decomposition by a finite number of submanifolds of codimension $\leq n$ and a complement of codimension $> n$. The main point is that the submanifolds so obtained have to describe common topological properties of germs.

Here Thom's second isotopy theorem is used; this is the analogue for topological equivalence of Theorem 2.7 on Homotopic Triviality. A homotopy $F: N \times I \longrightarrow P \times I$, with $F(x, t) = (\bar{F}(x, t), t)$ and $\bar{F}(x, 0) = f_0(x)$, is stratified by decomposing both $N \times I$ and $P \times I$ into a finite union of disjoint submanifolds, called strata. This is a "prestratification" of $N \times I$ and $P \times I$ and if different strata satisfy certain tangential conditions called the Whitney conditions, they form a Whitney stratification (see [Th4], [Ma3], or [Gi]). Finally the mapping f must be well-behaved relative to the strata; i.e. f maps each stratum of N submersively onto a stratum of P and its behavior on different stata is related by a condition called the "Thom condition a_f" (again see [Th4], [Ma3], or [Gi]). Such a mapping is said to be *Thom Stratified.*

Theorem 7.6 (Thom's second isotopy theorem) : *Suppose the proper mapping* $F: N \times I \longrightarrow P \times I$ *is Thom stratified and each stratum maps submersively onto I, then F is topologically trivial i.e. there are homeomorphisms of the form* $\varphi : N \times I \longrightarrow N \times I$ *with*

$\varphi(x, t) = (\bar{\varphi}(x, t), t)$ *and* $\psi : P \times I \longrightarrow P \times I$ *with* $\psi(y, t) = (\bar{\psi}(y, t), t)$ *so that*

$$F = \psi \circ (f_0 \times id) \circ \varphi^{-1}$$

Mather [Ma3] and [Ma4] uses the stable unfoldings of mappings to construct the desired stratification of the jet space, called the *canonical stratification*. If a family of mappings is multitransverse to this stratification then the family can be Thom stratified and hence is topologically trivial. The density of topologically stable mappings then follows as in the smooth case. A proper mapping multitransverse to the canonical stratification has a neighborhood of mappings which is locally pathwise contractible and such that each family in the neighborhood remain multitranverse. Applying Thom's isotopy theorem (instead of (2.7) homotopic triviality) gives the result.

Theorem 7.7 [Th4], [Ma3], [Ma4] (see also [Gi]) : *The topologically stable mappings are dense in* $C^{\infty}{}_{pr}(N, P)$.

We refer to the mappings which are multitransverse to the canonical stratification as *mt-stable mappings*. Just as in the case of smooth stability, the conclusion of the topological stability theorem is stonger than just the density of the mt–stable mappings. In addition they have the following properties:

(7.8) 1) (stability under suspension) if f is mt-stable then $f \times id : N \times M \longrightarrow P \times M$ is also mt–stable and

2) (continuous dependence of homeomorphisms) if f is mt-stable and $F : N \times U \longrightarrow P \times U$ is an unfolding of f with U a neighborhood of $0 \in \mathbb{R}^q$, then there is an open $U' \subseteq U$ and unfoldings of homeomorphisms $\varphi : N \times U' \longrightarrow P \times U'$ and $\psi : P \times U' \longrightarrow P \times U'$ so that $F | N \times U' = \psi \circ (f \times id_{U'}) \circ \varphi^{-1}$

Remark 7.9: A topologically stable mapping as defined in (2.1) does not a priori have any additional properties such as (7.8). However, property 2), that of homeomorphisms depending continuously on the mappings in finite parameter families, is a basic property exhibited by all of the usual constructions of topologically stable mappings. Also, it is not too hard to show that if N is compact and property 2) holds for a topologically stable mapping, then property 1) is also satisfied. However, May has given examples of topologically stable mappings for which 2) fails and which are not mt–stable (see [My] or e.g. [D11]) so the Thom–Mather method does not capture all topologically stable germs.

Questions : How does one explicitly construct the mt–stable mappings? Which other topologically stable mappings are missed? How do the topologically stable mappings compare with C^∞–stable ones, especially in the nice dimensions where both are dense?

Relating C^∞ and Topological Stability:

In the nice dimensions it turns out that the stable germs behave very well with respect to topological equivalence. For stable germs $f : \mathbb{R}^n,0 \longrightarrow \mathbb{R}^p,0$ we have topological invariants: the local multiplicity $m(f)$ for $n \leq p$; and the local topological type of $f^{-1}(0)$ at 0 for $n > p$. Locally for all x near 0 we can consider the germ of f at x, denoted (f,x). Decompose $\mathbb{R}^n$ into subsets $\{T_i\}$ on which the local invariants $m(f, x)$ or $(f^{-1}(f(x)), x)$ are constant. For example, if $n \leq p$, we can let $T_k = \{x : m(f, x) = k\}$ for $k \leq m(f) (= m(f,0))$. For $n \geq p$, as f is stable there are only a finite number of possibilities for the local topological type of the set $f^{-1}(f(x))$ near x; so the decomposition is again finite. These T_i are not manifolds; however, they uniquely determine stable germs (see e.g. [D10] or [D11] and the references there)

Theorem 7.10 (topological classification)   : *Let f and g: $\mathbb{R}^n,0 \longrightarrow \mathbb{R}^p,0$ be stable germs with (n, p) in the nice dimensions. If there is a germ of a homeomorphism $\varphi :\mathbb{R}^n,0 \longrightarrow \mathbb{R}^n,0$ such that $\varphi(T_i(f)) = T_i(g)$ for all i then f and g are C^∞–equivalent.*

For example, the real germs x^2-y^2 and x^2-y^4 are topologically equivalent; however, the stable unfoldings $\mathbb{R}^4,0 \longrightarrow \mathbb{R}^3,0$ defined by

$\quad$ $(x, y, u, v) \mapsto (x^2-y^2, u, v)$ and $(x^2-y^4+uy^2+vy, u, v)$

are topologically distinct and have distinct decompositions. The set of points where the local topological type is that of a curve with two branches (i.e. topologically a pair of crossed lines) is a plane for the first but a cut plane for the second.

This topological uniqueness implies that in the nice dimensions the $\mathcal{K}$–orbits of codimension $\leq n$ are local topological invariants in the following sense.

Definition 7.11: Let $\Gamma \subset J^k(n, p)$ be a submanifold invariant under the action of $\mathcal{K}$. We say Γ is a *local topological invariant* if there is a neighborhood U of Γ such that a stable germ f with $j^k(f) \in \Gamma$ is topologically distinct from a stable g with $j^k(g) \in U \backslash \Gamma$.

This provides a mechanism for comparing topologically stable and C^∞-stable mappings. We consider mt–stable mappings, or more generally C^0-stable mappings with continuous dependence of homeomorphisms (2 of (7.8)). Either of these satisfy the condition of stability under suspensions (1 of (7.8)), which we refer to as *S–stability*. Then, mutitransversality for S–stable mappings is provided by the following proposition *in the case N is compact.*

Proposition 7.12: *Let* $\{\Gamma_i\} \subset J^k(n, p)$ *be a collection of local topological invariants. Then, S–stable mappings are multitransverse to the* $\{\Gamma_i\}$.

Remark : There are certain additional technical points which should be mentioned regarding these local topological invariants; however, they are satisfied for all interesting Γ (see [D10,§4]).

This proposition asserts that: *S–stable mappings will be multitransverse to any natural stratification consisting of local topological invariants.* In particular, in the nice dimensions, this means the $\mathcal{K}$-orbits. So together with theorem 7.3 we conclude

Theorem 7.13: *In the nice dimensions with N compact, S–stable mappings are C^∞-stable.*

Remark : For most circumstances, the topologically stable mappings that are constructed will satisfy the condition of continuous dependence of homeomorphisms, and hence be S–stable; thus, these results apply. However, for the "plain vanilla" definition of topological stability, life becomes more complicated. There are no longer clean statements about multitransversality in term of local topological invariants as in (7.12) and the statement that follows it. It is necessary to introduce more technical notions to take care of various cases. This is being worked out by C.T.C. Wall and A. DuPlessis in a forthcoming book on topological stability.

The preceding implies that the canonical strata are exactly the $\mathcal{K}$-orbits in codimension $< \sigma(n, p)$. Hence, if e.g. $n \leq p$, then even outside the nice dimensions S-stable mappings will be C^{∞}-stable off a set of codimension $\sigma(n, p)$. Also, this suggests an approach to identifying the canonical strata by finding local topological invariants. This has been partially carried out for a larger class of singularities, those of "discrete algebra type" (see [DGal] and [D11]). It also provides for the possibility that topologically stable mappings which are not mt–stable can exist if the canonical strata are not local topological invariants. Although this may seem far fetched, this not only can happen, but does so with surprising frequency. It was discovered first by Looijenga [Lo], but later has appeared in many settings [Wi] and [D13-II].

Looijenga's result concerns the stable unfoldings of the families of germs $f_t(x_1, \ldots , x_n)$ in (7.14) which are the first families to exhibit moduli (and include (7.1) considered earlier).

$$
\begin{array}{lll}
\tilde{E}_6 & & x_2(x_1-x_2)(x_1-tx_2) + x_2x_3{}^2 + q(x_4, \ldots , x_n) \\
\textbf{(7.14)} \quad \tilde{E}_7 & & x_1x_2(x_1-x_2)(x_1-tx_2) + q(x_3, \ldots , x_n) \\
\tilde{E}_8 & & x_1(x_1-x_2{}^2)(x_1-tx_2{}^2) + q(x_3, \ldots , x_n)
\end{array}
$$

(here q denotes a nonsingular quadratic form and t is a modulus in each case, and must avoid certain values producing degeneracies such as $t = 0, 1$ for $\tilde{E}_7$).

Theorem 7.15 [Lo]: *The stable unfoldings of the families in (7.14) are topological trivial along the t– axis at all values (excluding the degeneracies).*

Furthermore, as Looijenga points out, if one takes the unfolding by all terms except the modulus term (i.e. the modulus parameter is fixed) then the resulting germ is topologically stable (a proof which applies more generally is given in [D13-II]). However, combined work of Bruce, Giblin, Greuel, and Wall have identified additional values of the parameters t where the Whitney stratification condition fails. Hence, these values identify topologically stable (in fact S-stable) germs which are not mt-stable (for a more detailed discussion see e.g. [D12,§3]). Thus, Looijenga's results suggest a different approach to topological equivalence which provides topological information undetected by stratification methods. In further pursuing this approach, two key elements of Looijenga's work are: the role of weighted homogeneity (for the stable unfoldings) and the explicit construction of "stratified vector fields" from algebraic data.

Topological Equivalence:

Recall that Bezout's theorem and Palomodov's theorem were the key results for complex germs against which we measured the corresponding results for the real case. In the same way, results on topological equivalence of mappings are measured against the incredibly successful results for holomorphic germs $f : \mathbb{C}^n, 0 \to \mathbb{C}, 0$ with isolated singularities at 0. In concentrating on this case we ignore a considerable body of work on more general complex singularities (see e.g. [AGV, vol II], [Lo2], and [LT]).

The reason for the success is the rich geometric structure of such a germ. For sufficiently small balls $B_\varepsilon \subset \mathbb{C}^n$ and $B_\delta \subset \mathbb{C}$, let $B_\delta{}^* = B_\delta \setminus \{0\}$. The restriction of f mapping $f^{-1}(B_\delta{}^*) \cap B_\varepsilon \to B_\delta{}^*$ is a smooth fibration over the punctured ball $B_\delta{}^*$ (see fig. 7.16).

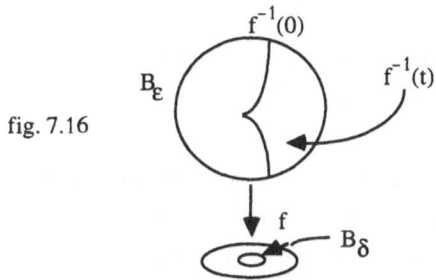

fig. 7.16

The fiber, called the Milnor fiber, is a smooth manifold with especially simple structure– it is homotopy equivalent to a bouquet of spheres of dimension $n-1$ [Mi2]. The number of such spheres, called the Milnor number and denoted by $\mu(f)$, is given by the remarkable formula

$$(7.17) \qquad \mu(f) = \mathcal{R}_e\text{-codim}(f) \quad (= \dim_{\mathbb{C}} \mathcal{O}_n / (\frac{\partial f}{\partial z_1}, \dots, \frac{\partial f}{\partial z_n}).$$

Because of the remarkably simple structure, methods of geometric topology can be applied.

Let $f_t : \mathbb{C}^n, 0 \to \mathbb{C}, 0$ be a germ of a deformation with 0 the isolated critical point for each t. Then, topological triviality of f_t is determined by the behavior of this single number. We say f_t is a μ-constant deformation if $\mu(f_t)$ is constant for each fixed t.

Theorem 7.18 : *Let f_t be a μ-constant deformation with n $\neq$ 3. Then,*

i)   (Lê-Ramanujam [LêR])   *the Milnor fibration of f_t is trivial (i.e. topologically equivalent to a product fibration along t-direction);*

ii)   (Timourian [Ti])   *The deformation f_t is topologically trivial for right equivalence (i.e. there is a parametrized family of germ sof a homeomorphisms $\varphi_t : \mathbb{C}^n, 0 \to \mathbb{C}^n, 0$ so that $f_0 = f_t \circ \varphi_t$).*

Morover, Teissier is able to give an equally simple criterion for triviality in a Whitney stratified sense. He constructs sequence of numbers $\mu_* = (\mu_1, \dots, \mu_n)$ with $\mu_k = \mu(f| \Pi_k)$ for Π_k a sufficiently general k-dim subspace of $\mathbb{C}^n$.

Theorem 7.19 [Te]: *Let f_t be a μ_*-constant deformation then f_t is trivial in a Whitney stratified sense.*

Remark 7.20 :   The exceptional value n = 3 in (7.18) (but not (7.19)) occurs because the geometric methods used (which involve the h-cobordism theorem) can not be applied to the 4-dimensional Milnor fiber. There are also examples such as $f_t = x^5 + z^{15} + zy^7 + txy^6$, due to Briancon-Speder, which are μ-constant but not μ_*-constant, so topological triviality holds even when Whitney conditions fail.

There have been results to compute μ under certain restrictions so that the μ-constant condition can be verified. In [A5], Arnold shows that if we deform a weighted homogeneous germ by terms of higher weight then μ remains constant; and Milnor-Orlik explicitly compute μ in this case [MOr]. Also, Kouchnirenko [K] has found a formula for μ in terms of the associated Newton diagram provided the function is *nondegenerate* in a well-specified sense. These results allow one to determine topological triviality for such cases provided n $\neq$ 3.   Moreover, Oka [O] has shown that deformations of such a nondegenerate germ by terms on or above the Newton diagram are Whitney stratified.

To also include real germs, different methods have to be employed which do not rely on the geometry in such a crucial way.

Finite Determinacy and Order of Vanishing:

To consider what might be possible in the real case, we recall the theorem on topological determinacy mentioned earlier. We state a simplified version of this result for a single jet.

Theorem 7.21 : *Given an ℓ-jet $g \in J^\ell(n, p)$ and an integer $k > \ell$, let $\pi_{k\ell} : J^k(n, p) \longrightarrow J^\ell(n, p)$ denote projection (i.e. ignoring terms $> \ell$). There exists a proper semialgebraic subset $\Sigma \subset \pi_{k\ell}^{-1}(g)$ such that if f, h : $\mathbb{R}^n, 0 \longrightarrow \mathbb{R}^p, 0$ with $j^k(f)$ and $j^k(h)$ belonging to the same connected component of $\pi_{k\ell}^{-1}(g) \backslash \Sigma$ then f and h are topologically equivalent (for $\mathcal{A}$-equivalence). In particular, if $j^k(f) \in \pi_{k\ell}^{-1}(g) \backslash \Sigma$ then f is k-topologically determined.*

As already mentioned, this was originally stated by Thom [Th3], and first proven by Varchenko [Va]. Later DuPlessis [DP] gave another proof using the same stratification techniques that Thom–Mather had used for topological stability. Neither of the proofs provide an effective mechanism for explicitly identifying Σ to determine when a given f with $j^\ell(f) = g$ will be k–determined.

Even before this theorem was proven, Kuo [Ku1] [Ku2] had begun proving versions of the topological determinacy theorem for germs f : $\mathbb{R}^n, 0 \longrightarrow \mathbb{R}, 0$ and $f^{-1}(0)$ for germs f : $\mathbb{R}^n, 0 \longrightarrow \mathbb{R}^p, 0$. The form of his results call to mind the remark we made in §2 concerning order of vanishing; namely,

$$(7.22) \qquad\qquad |h(x)| = O(\|x\|^k) \quad\Longleftrightarrow\quad h \in m_x^k$$

Kuo expresses his conditions in terms of such inequalities.

Theorem 7.23 [Ku1]: *Given a germ f : $\mathbb{R}^n, 0 \longrightarrow \mathbb{R}, 0$ with $j^k(f) = j^k(h)$ for a polynomial h of degree k. Then f is k-topologically determined for right equivalence if the is an $\varepsilon > 0$ so that in a neighborhood of 0*

$$(7.23) \qquad\qquad \|grad(h)\| \geq \varepsilon \|x\|^{k-1}$$

Secondly, consider $f : \mathbb{R}^n, 0 \longrightarrow \mathbb{R}^p, 0$ with $f = (f_1, \ldots, f_n)$. Given vectors $\{v_1, \ldots, v_n\}$ we let V_i = subspace spanned $\{v_1, \ldots, v_{i-1}, v_{i+1}, \ldots, v_n\}$ and $d(v_1, \ldots, v_n) = \min \text{dist}(v_i, V_i)$. Lastly, let $H_r(f, c) = \{x \in \mathbb{R}^n : \|f(x)\| \le c\| x\|^r\}$. Then, Kuo gives a sufficient condition for k–determinacy for topological $\mathcal{K}$–equivalence (strictly speaking he considers "V–sufficiency" where only nonambient homeomorphisms of $f^{-1}(0)$ are considered).

Theorem 7.24 [Ku2]:   *Given a germ* $f : \mathbb{R}^n, 0 \longrightarrow \mathbb{R}^p, 0$ *with* $j^k(f) = j^k(h)$ *for polynomial h of degree k with* $h = (h_1, \ldots, h_p)$. *Then f is k–topologically determined for $\mathcal{K}$–equivalence if there are c, δ and $\varepsilon > 0$ so that on $H_k(f, c)$ in a neighborhood of 0*

(7.24) $$ d(\text{grad}(h_1), \ldots, \text{grad}(h_p)) \ge \varepsilon\| x\|^{k-\delta} $$

     Just as explained in the smooth case, topological determinacy is proven by proving that deformations by higher order terms are topologically trivial. Just as for the isotopy theorems and topological stability results of Thom–Mather and the topological triviality result of Looijenga, these results are proven by integrating certain vector fields to obtain continuous flows. Then, Kuo's conditions ensure that the vector fields which he defines are integrable.

     More generally, to apply topological analogues of Mather's methods to these and other situations requires the following:

1)   substituting algebraic conditions for geometric conditions (such as (7.23) and (7.24));

2)   finding a general algebraic framework for measuring order of vanishing; and

3)   obtaining results for integrability of stratified vector fields compatible with the algebraic conditions in 1) and 2).

     We first illustrate how algebraic conditions can imply geometric conditions, as for example (7.23). Suppose

(7.25) $\qquad m_x{}^k \subseteq T\mathcal{R}_e{\cdot}f = (\frac{\partial f}{\partial x_1}, \ldots, \frac{\partial f}{\partial x_n})$ (the ideal generated by the $\frac{\partial f}{\partial x_i}$).

It follows by standard singularity theory (see e.g. [Si]) that (7.25) implies that f is k+1 determined for $\mathcal{R}$–equivalence. More generally suppose for some $\ell > 0$

(7.26) $\qquad\qquad\qquad m_x{}^{k+\ell} \subseteq m_x{}^{\ell}{\cdot}T\mathcal{R}_e{\cdot}f$

We can not recover (7.25) from (7.26). even though (7.25) implies (7.26). However, we claim that (7.26) does imply the same topological determinacy results.

Multiplying (7.26) by any power m_x^r implies the (7.26) is true for all $\ell \geq \ell_0$. If $g \in m_x^{k+\ell}$ then $g \in m_x^\ell \cdot T \, \mathcal{R}_e \cdot f$, so there is a constant $C > 0$ such that

(7.27) $|g(x)| \leq C \cdot \|x\|^\ell \cdot \|\text{grad}(f)\|$ on a neighborhood of 0.

In particular, applying this for $\ell = (2r-1)k$ for r sufficiently large implies (7.27) holds for all monomials $x^\alpha \in m_x^{2rk}$. This implies that on a neighborhood of 0

(7.28) $\|x\|^{2rk} \leq C \cdot \|x\|^{(2r-1)k} \cdot \|\text{grad}(f)\|$

Lastly, dividing (7.28) by $\|x\|^{(2r-1)k}$ implies that on a neighborhood of 0, excluding 0,

$$\|x\|^k \leq C \cdot \|\text{grad}(f)\|$$

By continuity, it also holds at 0, so (7.23) is satisfied and we conclude that f is $k+1$ determined for topological $\mathcal{R}$-equivalence.

In fact, we want to do even better than conditions such as (7.26). This is the goal of conditions 1) and 2) mentioned above. First, we indicate how we can improve measurements of order of vanishing.

Consider a germ such as $f(x, y, z) = x^5 + y^6 - xz^{12}$. If the preceding results are applied to f, we can find that it is smoothly 27–determined and topologically 15–determined for right–equivalence (provided we cleverly make the estimates). Thus, we can say nothing about adding terms such as $y^2x^3z^2$, $y^3x^2z^2$, and other such lower order terms. However, it turns out that adding these terms does not change the topological type of f. The point is that the method for measuring "order of vanishing" used by Kuo does not take into account the relative importance of the variables in the expressions for f. A different approach occurs in Looijenga's result where he makes considerable use of the weighted homogeneity of the germs, both algebraically and geometrically. We take this even further through the use of the weight filtration and its extension for more general filtrations.

A filtration on C_x is a decreasing sequence of ideals (having finite codimension in C_x) $w_x^{(0)} \supset w_x^{(1)} \supset ... \supset w_x^{(k)}$ such that $w_x^{(k)} \cdot w_x^{(\ell)} \subseteq w_x^{(k+\ell)}$. Such a filtration is said to be *convex* if there is an m and $g_1, ..., g_r \in w_x^{(m)}$ so that if $\rho = \sum |g_i|^2$ and $h \in w_x^{(k)}$ then there is a constant $C > 0$ so that

(7.29) $|h| \leq C \cdot \rho^{k/2m}$ in a neighborhood of 0

Inequalities such as (7.29) are often referred to as Lojasiewicz-type inequalities.

Convexity provides us with an alternate way of measuring the order of vanishing. Instead the norm $\| x \|$, the ideals $m_x{}^k$, and (7.22) are replaced by $\rho^{1/2m}$, the filtration $\{w_x{}^{(k)}\}$, and the estimates (7.29). Any such ρ which satisfies (7.29) is referred to as a *control function of filtration 2m*. It is highly nonunique. Given h with fil(h) = k, the initial part of h, denoted in(h), consists of those terms of h in $w_x{}^{(k)} \setminus w_x{}^{(k+1)}$. This is the "lowest order" part of h

The first example of such a filtration is the *weight filtration*. If we have assigned weights $\text{wt}(x_i) = a_i > 0$ then we let $w_x{}^{(k)}$ denote the ideal generated by monomials of weight $\geq$ k. This filtration was used by Arnold [A5] in connection with the smooth classification of germs. The g_i can be constructed by choosing b_i so that $a_i b_i = m$ for all i and using $g_i = x_i{}^{a_i}$. However, by [D14–II,lemma 1.13], given any collection $\{g_i\}$ for which each fil(g_i) = m and $\{in(g_i)\}$ generate an ideal of finite codimension, then the g_i satisfy the convexity condition.

A second example is a more refined *Newton filtration* given in [K]. From a germ f : $\mathbb{R}^n,0 \rightarrow \mathbb{R},0$ we can define the Newton diagram which is the convex hull in $\mathbb{R}_+{}^n$ of the sets $\alpha + \mathbb{R}_+{}^n$ for lattice points $\alpha \in \mathbb{Z}_+{}^n$ which correspond to nonzero monomials x^α

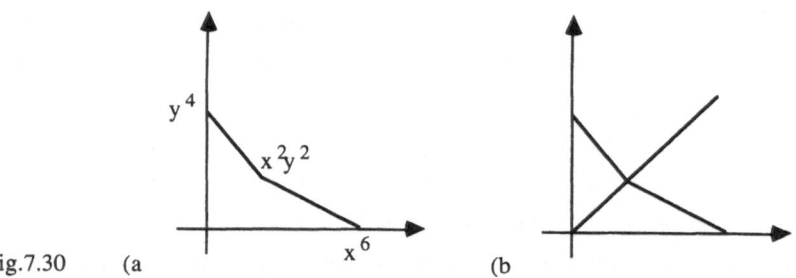

fig.7.30   (a   (b

appearing in f. For $f(x, y) = x^6 - x^2 y^2 + y^4$ we obtain the Newton diagram in fig. 7.30a. A filtration is constructed from the diagram by choosing a piecewise linear function φ linear on the cones through the vertices of the Newton diagram (fig. 7.30b), and constant on the diagram. We define for $f = \sum a_\alpha x^\alpha$

$$\text{fil}(f) = \min \{\varphi(\alpha) : a_\alpha \neq 0\}$$

This defines a filtration [K] and it is convex using the vertices of the Newton diagram

[DGaf]. The nondegeneracy condition of Kouchnirenko can be stated in terms of the

algebraic properties of $\{ in(x_i \cdot \frac{\partial f}{\partial x_i}) \}$ relative to the Newton filtration.

Yet a third example is the *filtration by integral closures of ideals* due to Teissier [Te], with which he gives an algebraic characterization of the μ_*-constant condition.

It was discovered in [D13-I] for $\mathcal{A}$-equivalence that the key idea in the weighted homogeneous case, which also underlies the algebraic calculations of Looijenga, is the finite codimension of the germ. In general, in either the real or complex case, let $f_0 = (f_{01}$, ... , $f_{0p}) : k^n, 0 \to k^p, 0$ be weighted homogeneous and let f denote an *unfolding* of f_0 *of nondecreasing weight*, i.e. f deforms by terms in the i-th coordinate of weight $\geq$ wt(f_{0i}).

Theorem 7.31 [D13-I] : *If f_0 is a finitely ($\mathcal{A}$-)determined weighted homogeneous germ and f is an unfolding of nondecreasing weight, then f is a topologically trivial unfolding.*

Thus, to verify that a deformation does not topologically alter f_0 it is enough to check whether lower weight terms occur. A similar situation occurs for germs f_0 : $k^n, 0 \to k, 0$. If for given weights, in(f_0) still defines an isolated singularity (i.e. finite $\mathcal{R}$-codimension), then f_0 is said to be semi-weighted homogeneous (in agreement with §6). A complement to the earlier results which places the real and complex cases on an equal footing in certain situations is the following.

Theorem 7.32 [DGaf]: i) *If f_0 is semi-weighted homogeneous then any deformation of nondecreasing weight is topologically trivial; hence, f_0 is topologically determined by* in(f_0).
ii) *If f_0 is nondegenerate (in the sense of Kouchnirenko), then any deformation of nondecreasing Newton filtration (i.e. by terms on or above the Newton diagram) is topologically trivial; hence, f_0 is topologically determined by* in(f_0) *(for the Newton filtration), and in particular by the Newton diagram if* $k = \mathbb{C}$.

Remark : We note that even the annoying exceptional case n = 3 is allowed for these two classes of singularites. Moreover, the results in [DGaf] give criteria for topological triviality even in if the germs are degenerate in Kouchnirenko's sense.

If we return to the example $f(x, y, z) = x^5 + y^6 - xz^{12}$ we now see it is weighted homogeneous for weights $wt(x, y, z) = (6, 5, 2)$. Thus, we see by the theorem that it is topologically determined at weight 30, which in particular includes ordinary degree 15. As terms such as $y^2x^3z^2$, and $y^3x^2z^2$ have weight > 30, adding them will not topologically change f.

These last two theorems illustrate how topological methods can be used to simplify the work involved in determining the structure for germs; instead of computing terms up to degree 26 it's enough to compute terms up to weight 30, i.e. at most degree 15. However, it must be stressed that the choice of an appropriate filtration is more of an art than a science. It depends on a choice of coordinates and a germ may not satisfy the conditions if coordinates are changed. For example, $x^6 - x^2y^2 + y^4$ will not be semi-weighted homogeneous for any choice of weights or coordinates; however, it is nondegenerate in Kouchnirenko's sense, and so topologically determined by its Newton diagram 7.30a.

G-topological   Triviality

Lastly, we describe how to obtain topological analogues of the main theorems proven for G-equivalence in §4. These topological results will follow the original Thom–Mather approach for solving problems by reducing them to infinitesimal problems which can be solved algebraically. We will apply them in the weighted homogeneous case to the examples mentioned at the beginning of this section.

To explain how such an approach may work for topological equivalence, we must replace G by its topological analogue G^{top}. For example, in place of $\mathcal{A}$ we would use $\mathcal{A}^{top}$, by replacing germs of diffeomorphisms by germs of homeomorphisms. However, there are inherent difficulties with any straightforward approach. To begin with, there is no orbit map $G^{top} \longrightarrow \mathcal{F}$. Also, TG^{top} will be a module of germs of continuous vector fields, which need not even be locally integrable. In addition, we must individually specify for each G what G^{top} should be.

The arguments of §4 could be applied in the topological setting to questions of topological equivalence, determinacy, etc. provided a topological analogue of the infinitesimal reduction lemma (4.4) could be proven and applied to appropriate unfoldings. For us applicability means that the conditions can be expressed by algebraic conditions in terms of the infinitesimal orbit map.

To explain the nature of these algebraic conditions, we observe that virtually any result for topological equivalence requires the integration of nondifferentiable vector fields. The algebraically stratified vector fields which we use are variants of the stratified vector fields used by Thom–Mather for topological stability, replacing their tubular neighborhoods and control conditions by algebraically defined analogues.

These vector fields are constructed, via partitions of unity, from pieces of vector fields in $T\mathcal{G}_{un,e}$ allowing division by functions in the rings and continuous extensions across certain natural strata defined by the algebraic structure of $T\mathcal{G}_{un,e}$. For these vector fields to be locally integrable, we must control their behavior relative to strata. This is achieved by controlling their order of vanishing along strata. Not surprisingly, if we construct the pieces using the original infinitesimal orbit map, we must control how it behaves relative to orders of vanishing.

We have seen how order of vanishing can be measured by a filtration on an algebra of germs. This idea extends to the modules of vector fields, just as we defined higher order terms for vector fields. We use filtrations by submodules $T\mathcal{G}^{(\ell)}_{un,e}$ and $T\mathcal{F}^{(\ell)}_{un,e}$ so that

$$d\alpha_f : T\mathcal{G}^{(\ell)}_{un,e} \longrightarrow T\mathcal{F}^{(\ell)}_{un,e}.$$

Example (7.32) : The easiest example is the weight filtration for $f_0 = (f_{01}, \dots, f_{0p})$

$: k^n,0 \longrightarrow k^p,0$. If $wt(x_i) = a_i$, then we assign $wt(y_i) = wt(f_{0i})$ (say $= d_i$) and $wt(\frac{\partial}{\partial x_i}) =$

$-a_i$, $wt(\frac{\partial}{\partial y_i}) = -d_i$, with $wt(g_i \cdot \frac{\partial}{\partial x_i}) = wt(g_i) - a_i$, etc. For most $\mathcal{G}$ we have listed, $d\alpha_{f_0}$

will preserve the weight filtration; and if f is an unfolding of f_0 by terms of nondecreasing weight then so does $d\alpha_f$.

Then, the analogue of finite codimension for topological equivalence is: a germ f_0

has *finite graded codimension* relative to the given filtrations if there is an ℓ_0 so that

(7.33) $T\mathcal{F}^{(\ell)}{}_e \;=\; d\alpha_{f_0}(T\mathcal{G}^{(\ell)}{}_e) + T\mathcal{F}^{(\ell+1)}{}_e$ for all $\ell \geq \ell_0$.

(as usual, there is an ℓ_1 so that if (7.33) holds for $\ell_0 \leq \ell \leq \ell_1$ then it holds for all $\ell \geq \ell_0$). For the weight filtration, we can define the notion of semi–weighted homogeneous for any group $\mathcal{G}$.

Definition 7.34 : A germ f_0 is *semi–weighted homogeneous* for $\mathcal{G}$ if there are weights so that f_0 has finite graded codimension for the weight filtration (for most $\mathcal{G}$ this is equivalent to $in(f_0)$ having finite $\mathcal{G}$-codimension).

Verifying conditions ensuring integrability requires specific information about the behavior of the vector fields. Along with the algebraic conditions, there is an important geometric condition, called the *stratification condition*, which is needed for the algebraically stratified vector fields. We just mention that it is a condition that is often universally satisfied for finite $\mathcal{G}$-codimension germs, and it dictates on which strata the topological equivalence will be smooth. For example, this is the case for all of the preceding examples: for $\mathcal{R}$ and $\mathcal{K}$ the strata in $\mathbb{R}^n$ will be $\{\mathbb{R}^n \backslash \{0\}, \{0\}\}$; for imperfect bifurcation groups $\mathcal{K}^*$ or $\mathcal{K}^{G*}$ the strata in $\mathbb{R}^{n+1}$ will be $\{\mathbb{R}^{n+1} \backslash \mathbb{R}^n, \mathbb{R}^n \backslash \{0\}, \{0\}\}$, etc.

Because of these considerations, the class of allowable groups $\mathcal{G}$ has to be specified more precisely, and leads to a Special Class of Geometric Subgroups (see [D13§8]); however, additional requirements are of a sufficiently general nature that all of the examples from §4 are still included; thus, we do not dwell further on this point.

Also, by the method of construction, the geometric properties of $T\mathcal{G}_{un,e}$ are retained at the infinitesimal level by these stratified vector fields, in turn, these translate into the corresponding properties for the topological version of $\mathcal{G}$, overcoming the problem of individually defining the topological equivalence for each $\mathcal{G}$.

Then, just as for the case of equivalence, the topological analogue of the infinitesimal reduction lemma (4.4) can be proven [D14-I, §10]. Then the topological versions of the theorems hold. Rather than state each of the theorems for the topological case, we compare the topological results with the earlier results for $\mathcal{G}$-equivalence. The results will be stated in terms of algebraic conditions relative to the filtrations. Then we will concentrate on the consequences of the theorems for semi–weighted homogeneous germs.

Comparison (7.35) (see [D14–I,II])

G-Equivalence	Topological G-equivalence

Conditions

For f_0 having finite G– codimension

For f_0 having finite graded G-codimension

Triviality of Unfoldings

1) For the case of a few G's, there has been identified a subspace $M \subseteq T G_e \cdot f_0$ such that if $\dfrac{\partial f}{\partial u_i} \in M$ $\quad 1 \le i \le q$ then f is a G-trivial unfolding of f_0

1) For all G, there is a number $L(f_0)$– determined by f_0 and G such that if $\dfrac{\partial f}{\partial u_i}$ has algebraic filtration $\ge L(f_0)$, $1 \le i \le q$ then, f is a topologically G-trivial unfolding of f_0

Order of Determinacy

2) there is a large K determinedby f_0 and G such that if $j^K(g_0) = j^K(f_0)$ then g_0 and f_0 are G–equivalent For a few G, an explicit small K can be given.

2) For each G, there is an explicit K determined by f_0,and G such that if $g_0 = f_0$ mod terms of filtration $\ge$ K+1 then g_0 is topologically G–equivalent to f_0. This K is virtually always smaller than the one predicted for equivalence.

Versality

3) If $d\alpha_f \colon T G_{un,e} \longrightarrow T \mathcal{F}_{un,e}$ is surjective then f is G-versal.

3) If $d\alpha_f \colon T G_{un,e} \longrightarrow T \mathcal{F}_{un,e}$ has finite graded–codimension, then f is topologically G–versal.

In the examples, we shall concentrate on the applications for topological classification and the order of topological determinacy. The topological versality theorem is considerably more difficult to apply because of the algebra involved (see e.g. [D14–II] and references therein). We do make the important remark that unlike the smooth case, if two germs are only G-topologically equivalent then they need not have topologically equivalent G-versal unfoldings; this was discovered quite a while ago by Pham.

If $f_0 = (f_{01}, \ldots, f_{0p}) : k^n, 0 \rightarrow k^p, 0$ is semi-weighted homogeneous for any group G then the number $L(f_0) = 0$ in condition 2) of (7.36). Thus, we obtain as a corollary one form of the *topological determinacy theorem* (see [D14–II,§2]).

Corollary 7.36 : *If f_0 is semi-weighted homogeneous for G then*

i)   *An unfolding f of f_0 by terms of nondecreasing weight is G-topologically trivial;*

ii)   *f_0 is G-topologically determined by* in(f_0).

Examples :

(7.37)   First consider G–equivariant germs $F : \mathbb{R}^n, 0 \longrightarrow \mathbb{R}^p, 0$ where G acts linearly on $\mathbb{R}^n$ and $\mathbb{R}^p$.   Suppose F is semi–weighted homogeneous for equivariant $\mathcal{K}$-equivalence. Then, by the topological determinacy theorem in the form of Cor. 7.36, if we denote in(F) = F_0 then there is a G–equivariant homeomorphism $\varphi : \mathbb{R}^n, 0 \longrightarrow \mathbb{R}^n, 0$ such that $\varphi(F_0^{-1}(0)) = F^{-1}(0)$. In particular, in the case p = n–1, with G finite, these curve singularities have the same representations of G on the (half)branches. Thus, this justifies the use of F_0 = in(F) in (6.21). The homeomorphism is actually smooth off 0 (since the strata are $\{\mathbb{R}^n \backslash \{0\}, \{0\}\}$). If there is no group, then the result also follows for homogeneous germs using (7.24) of Kuo or a result of Buchner–Marsden–Schecter [BMS].

(7.38)   For equivariant Hamiltonian systems in§4, Montaldi–Roberts–Stewart consider the $\Gamma \times S^1$ -equivariant map (4.27) under $\mathcal{K}^{\Gamma \times S^1}$-equivalence.

$$\Psi : \mathbb{R}^{2n} \times \mathbb{R} \times \mathbb{R}, 0 \longrightarrow \mathbb{R}^{2n}, 0$$

$$((p, q), \lambda, \tau) \longmapsto J \cdot d_x(-\tau H_2 + H_k) + \lambda d_x H_2$$

If Ψ' denotes Ψ when we replace H_k by $H' = H - H_2$. Then, as in (4.29), $H_2 = N$, k = 4, and

$$H_4 = \alpha_1 N^2 + \alpha_2 P + \beta_1 NQ + \beta_2 Q^2$$

If we assign weights wt(p_i) = wt(q_i) = 1 and wt(λ) = wt(τ) = 2, then in(Ψ') = Ψ. Finally, Montaldi–Roberts–Stewart show that Ψ is finitely $\mathcal{K}^{\Gamma \times S^1}$-determined. Thus, Ψ' is semi–weighted homogeneous. Thus, by corollary 7.36 (or [D3]) the zero sets $\Psi'^{-1}(0)$ and

$\Psi^{-1}(0)$ are $\Gamma \times S^1$ -equivariantly homeomorphic. Then [MRS,§4] identifies these with the periodic orbits of the Hamiltonian systems.

(7.39) Next, let's consider a more specialized equivalence of G-equivariant bifurcation for germs $f(x, \lambda) : \mathbb{R}^{n+1}, 0 \longrightarrow \mathbb{R}^n, 0$ where now there is a trivial action of G on the last factor in $\mathbb{R}^{n+1}$. If now f is semi-weighted homogeneous for equivariant bifurcation equivalence, then again f is topologically equivalent to $f_0 = \text{in}(f)$ for this equivalence. In particular, by the remarks preceding (7.36), the topological equivalence will be smooth on $\mathbb{R}^{n+1} \setminus \mathbb{R}^n$. Thus, differentiable information about the function on $f^{-1}(0)$, such as any stability information for the bifurcation problem which f models, will be preserved when $\lambda \neq 0$. Also, information such as whether a branch has $\lambda > 0$ or < 0 will be preserved. Again in the absence of a group, a version of determinacy has been obtained by Percell–Brown [PB].

(7.40) The first example in the classification of Golubitsky–Roberts in §4 is given by
$$f_0 = (\alpha N + \varepsilon' \lambda) \cdot (x, y) + \varepsilon \delta \cdot (x, -y) \qquad \alpha \neq 0, \varepsilon$$
where $N = x^2 + y^2$, and $\delta = x^2 - y^2$. Thus, we see f_0 is weighted homogeneous with $\text{wt}(x, y, \lambda) = (1, 1, 2)$. Thus, by the above, f_0 is topologically constant for equivariant bifurcation (i.e. $\mathcal{K}^{D_4}*$)-equivalence on the intervals where $\alpha \neq 0, \varepsilon$.

(7.41) The stability results of Field–Richardson (§6) for equivariant bifurcation can be partially obtained as consequences of the topological determinacy results. For this the key observation is that the topological equivalences are actually smooth on $\mathbb{R}^{n+1} \setminus \mathbb{R}^n$ so if the branches remain off $\mathbb{R}^n$ then their smooth properties remain. The standard form is given by

(6.18) $\qquad j^d(F) \ (= \ F_0) \ = \ R + Q + \lambda \cdot \text{id} \ :$

In (6.20) we pointed out that if $F_0 = Q + \lambda \cdot \text{id}$ is finitely $\mathcal{K}$-determined then F is $\mathcal{K}$-equivalent to a semi-weighted homogeneous germ with initial part $= F_0$. Thus, the set of half-branches with its G-action is stable under perturbation of F. If, in addition, F_0 is finitely determined for equivariant bifurcation equivalence, then the equivariant branching pattern including $\text{sign}(\lambda)$, is stable for the case $R = 0$. The only time we can assure that the index for hyperbolic branches (number of eigenvalues for $d_x F$ with negative real part) will be preserved is when F is gradient so the eigenvalues are real. Then the number cannot change because the germ remains finitely determined for bifurcation equivalence.

Finally, to allow nonzero R and obtain stronger stability, it would be sufficient to know that $R + Q : \mathbb{R}^n, 0 \longrightarrow \mathbb{R}^n, 0$ has an isolated singularity for any $R = h \cdot \mathrm{id}$ of lower degree; for then the only way $\mathrm{sign}(\lambda)$ or the hyperbolic index (for gradient F) could change would be for a branch to pass through $\lambda = 0$, which would be impossible.

The stability for the general hyperbolic case can not be treated by these singularity methods; however, perhaps considering an analogue of the energy–momentum map (as in §4) might work.

(7.42)   The last example we mention is the energy–momentum map considered by Duistermaat for Hamiltonian systems. It is an S^1-equivariant map

$$f = (H_2, H) : \mathbb{R}^{2n}, 0 \longrightarrow \mathbb{R}^2, 0$$

which is S^1-equivariant $\mathcal{A}$-equivalent to a germ with H replaced by (4.23), with b a modulus.

$$\hat{H} = k\rho_1 + \ell\rho_2 + (a/2)\rho_1^2 + b\rho_1\rho_2 + (c/2)\rho_2^2 + \psi$$

Referring to §4, we see that H and f will not be semi–weighted homogeneous unless $(k, \ell) = (\pm 4, 1)$. For the other cases, it is necessary to use a Newton–type filtration on the ring of invariants; we do not discuss this further here (see examples in [D3], [D4], and [D14–II]).

Concluding Remarks :   There are other related results for obtaining topological equivalence such as, e.g. "modified analytic triviality" intorduced by Kuo. When such methods are applied to examples such as nondegenerate germs, the arguments closely resemble the arguments already used in [DGaf], so these results are not really so different. Given that further applications of singularity theory to nonlinear problems will make increased use of topological methods, there is an ongoing need for continued development of the many different facets of these methods.

Bibliography

The references are listed in the section in which they were first made, which is usually the main section of reference. However, a short list of books which are general references are given at the beginning.

General References

AGV   Arnol'd, V.I., Gusein-Zade, S.M., Varchenko, A.G. Singularities of Differentiable Maps , Birkhauser, Boston-Basel-Berlin (1988)

BG1   Bruce, J. W. and Giblin, P. J. Curves and Singularities, Cambridge Univ. Press, Cambridge-NewYork (1984)

CH   Chow, S.N. and Hale, J.K. Methods of Bifurcation Theory, Grundlehren Math. Wissenschaften 251, Springer-Verlag, Heidelberg-New York (1982)

GG   Golubitsky, M. and Guillemin, V. Stable Mappings and their Singularities, Springer-Verlag, New York (1973).

GS1   Golubitsky, M. and Schaeffer, D. Singularities and Groups in Bifurcation Theory, Springer-Verlag, New York (1985).

GSS   Golubitsky, M., Stewart, I., and Schaeffer, D. Singularities and Groups in Bifurcation Theory Vol. II, Springer-Verlag, New York (1988).

Mar1   Martinet, J. Singularities of Smooth Functions and Maps. London Math. Soc. Lect. Notes 58, Cambridge Univ. Press (1982)

PS   Poston, T. and Stewart, I. Catastrophe Theory and its Applications, Putnam, London-New York (1978)

Sa1   Sattinger, D.   <u>Group Theoretic Methods in Bifurcation Theory</u>, Springer Lect.
        Notes in Math. 762 (1979)

Th1   Thom, R.   <u>Structural Stability and Morphogenesis</u>, (1972) W.A.
        Benjamin-Addison-Weseley

§1   Lyapunov-Schmidt Reduction

Fr   Fraenkel, L.   *Formulae for Higher Derivatives of Composite Functions*,
        Proc. Camb. Phil. Soc. 83 (1978), 159–165

N   Nirenberg, L.   <u>Topics in Nonlinear Functional Analysis</u>, Courant Lecture
        Notes, New York (1974)

Rg   Ronga, F.   *A new look at Faá de Bruno's Formula for Higher Derivatives*
        of Composite Functions and the Expression of some Intrinsic Derivatives, Proc.
        Sym. Pure Math. 40 Pt 2 (1983) 423–432

§2   Stability of Mappings

Ha   Hamilton, R. S.   *The Inverse Function Theorem of Nash and Moser*, Bull.
        Amer. Math. Soc. 7 (1982) 65–222

Mg   Malgrange, B.   <u>Ideals of differentiable functions</u>, Oxford Univ.Press (1966).

Ma1   Mather, J.   *Stability of C^∞-mappings:*
        I. *The Division Theorem*, Ann. of Math. 89 (1969), 89–104.
        II. *Infinitesimal stability implies stability*, Ann. of Math. (2) 89
        (1969), 254–291; III. *Finitely determined map germs*, Inst. Hautes Etudes Sci.
        Publ. Math. 36 (1968), 127–156;

IV. *Classification of stable germs by* **R**-*algebras*, Inst.
Hautes Etudes Sci. Publ. Math. 37 (1969), 223–248.
V. *Transversality*, Advances in Math. 4 (1970), 301–336.
VI. *The nice dimensions*, Liverpool Singularities Symposium
I, Springer Lecture Notes in Math. 192 (1970), 207–253.

Mi   Milnor, J.   <u>Morse Theory</u>, Annals Math. Studies 51 (1963) Princeton Univ.
        Press

Th2   Thom, R.   *Les Singularités des applications differentiables*, I; Annales
        Inst. Fourier (Grenoble) 6 (1955), 43–87.

Th3   -----   *Local Topological Properties of Differentiable Mappings*,
        Colloquium on Differential Analysis, Tata Institute Bombay, Oxford Univ. Press
        (1964), 191–202

ThL   Thom, R. and Levine, H.   *Singularities of Differentiable Mappings*, Proceedings
        Liverpool Singularities Symposium, Springer Lecture Notes 192 (1971) 1–89

Wa1   Wall, C.T.C.   *Lectures on C^∞-Stability and Classification*, Proceedings of
        Liverpool Singuarities Symposium, Springer Lecture Notes 192 (1971), 178–206.

Wh1   Whitney, H.   *Analytic Extensions of Differentiable Functions defined in*
        Closed Sets, Trans. Amer. Math. Soc. 36 (1934) 63–89

Wh2   -------   *On the Singularities of Mappings of Euclidean Spaces I,*
        Mappings of the plane to the Plane, Ann. Math. 62 (1955) 374–410.

Wh3   ------   *On the Singularities of Smooth n-Manifolds into (2n-1)-*
        space, Ann. Math. 45 (1944) 247–293

§3   Local Theory as Paradigm

BDW   Bruce, J. W., DuPlessis, A., and Wall, C.T.C.   *Unipotency and Determinacy*, Invent. Math. 88 (1987) 521–554

DuG   DuPlessis, A. and Gaffney, T.   *More on the determinacy of smooth map germs*, Invent. Math. 66 (1982), 137–163.

Gaf1   Gaffney, T.   *A Note on the Order of Determination of a Finitely Determined Germ*, Invent. Math. 52 (1979), 127–135.

Ma2   Mather, J.   Notes on right equivalence, 1970, unpublished notes

Mar2   Martinet, J.   *Deploiements versels des applications differentiables et classification des applications stables*, Singularites d'Applications Differentiables, Plans–Sur– Bex, Springer Lecture Notes 535 (1975) 1–44.

Si   Siersma, D.   Classification and Deformation of Singularities, Thesis, University of Amsterdam, 1974.

Tg   Tougeron, J.-Cl.   *Ideaux de fonctions differentiables*, I; Annales Inst. Fourier (Grenoble) 18 (1968), 177–240.

Ws1   Wasserman, G.   Stability of Unfoldings, Springer Lecture Notes 393 (1974)

Wa2   Wall, C.T.C.   *Finite Determinacy of Smooth Map Germs*, Bull. London Math. Soc. 13 (1981) 481–539

§4   Singularity Theory with Special Conditions

A1   Arnold, V. I.   *Wave front evolution and equivariant Morse lemma*, Comm. Pure App. Math. 29 (1976), 557–582.

A2   -------   *Normal forms for functions, nondegenerate critical points, the Weyl Groups A_n, D_n, E_n and Lagrangian singularities,* Funct. Anal. and Appl.6 (1972) , 254–272.

A3   -------   *On Matrices Depending on Parameters,* Russian Math. Surveys 26 (1971) 29–43

A4   -------   *Singularities of Systems of Rays,* Russian Math. Surveys 38, (1983) 77–147

Ba   Baas, N.   *Structural stability of composed mappings,* preprint, 1974

Bi   Bierstone, E.   <u>The structure of orbit spaces and the singularities of equivariant mappings</u>, I.M.P.A. Monograph 35 (1980).

BG2   Bruce, J. W. and Giblin, P. J.   *Smooth maps of discriminant varieties,* Proc. London Math. Soc. (3), 50 (1985), 535–551.

BG3   ------   *Projections of Surfaces with Boundaries,* Proc. London Math. Soc. 60 (1990) 392–416

BR   Bruce, J. W. and Roberts, M.   *Critical points of functions on analytic varieties,* Topology 27 (1988) 57– 90.

D1   Damon, J.   a) *The unfolding and determinacy theorems for subgroups of $\mathcal{A}$ and $\mathcal{K}$,* Proc. Sym. Pure Math. 40 (1983) 233–254. b) *The unfolding and d eterminacy theorems for subgroups of $\mathcal{A}$ and $\mathcal{K}$,* Memoirs A.M.S. 50, no. 306 (1984).

D2   ------   *Deformations of sections of singularities and Gorenstein surface singularities,* Amer. J. Math. 109 (1987) 695-722.

D3   ------   *Topological equivalence of bifurcation problems,* Nonlinearity 1 (1988), 311-331

D4   ------   *Topological invariants of* μ-*constant deformations of complete intersection singularities,* Quart. J. Math.. 40 (1989), 139- 160

D5   ------   *Generic Properties of Solutions to Partial Differential Equations,* preprint

DS   Dangelmayr, G. and Stewart, I.   *Classification and unfolding sequential bifurcations,* SIAM J. Math. Anal. 15 (1984), 423-445.

DD   Dubois, J. and Dufour, J. P.   *Singularités de Solutions d''equations aux Derivées Partielles,* Jour. Diff. Eqtns 60, (1985) 174-200

Du1   Dufour, J. P.   *Deploiements de cascades d'applications differentiables,* C. R. Acad. Sci. Paris Ser. A281 (1975), 31-34.

Du2   ------   *Sur la stabilité des diagrammes d'applications différentiables,* Ann. Sci. Ecole Norm. Sup. (4) 10 (1977), 153-174.

Du3   ------   *Famille de courbes planes différentiables,* Topology 22 (1983), 449-474

Du4   ------   *Stabilité simultanée de deux fonctions,* Ann. Inst. Fourier 29 (1979), 263-282.

Dm   Duistermaat, J.J.   *Bifurcations of Periodic Solutions near Equilibrium Points of Hamiltonian Systems*, Bifurcation Theory and Applications, Springer Lecture Notes 1057 (1984) 57–105

FR   Fadell, E. and Rabinowitz, P.   *Generalized Cohomological Index Theories and Lie Group Actions with an Application to Bifurcation Questions for Hamiltonian Systems*, Invent. Math. 45 (1978) 139–174

FM   Favaro, L. and Mendes, C. M.   *Global stability of diagrams of differentiable applications*, Annales Inst. Fourier 36 (1986) 133– 154

Gaf2   Gaffney, T.   *New Methods in the Classification Theory of Bifurcation Problems*, Contemp. Math. 56, Amer. Math. Soc. (1986) 97–116

Gal2   Galligo, A.   *Théorème de division et stabilité en géométrie analytique locale*, Annales Inst. Fourier 29 (1979),107–184.

Ge1   Gervais, J. J.   *Germes de G–determination finie*, C. R. Acad. Sci. Paris Ser. A–B 284A (1977), 291–293.

Ge2   Gervais, J. J.   *Criteres de G–stabilité en terms de transversalité* , Canad. J. Math. 31 (1979), 264–273.

GR   Golubitsky, M. and Roberts, M.   *A Classification of Degenerate Hopf Bifurcations with O(2) Symmetry*, Jour. Diff. Eqtns. 69 (1987) 216–264.

GS2   Golubitsky, M. and Schaeffer, D.   *A theory for imperfect bifurcation via singularity theory*, Comm. Pure. Appl. Math. 32 (1979), 21–98.

GS3   Golubitsky, M. and Schaeffer, D.   *Imperfect bifurcation in the presence of symmetry*, Comm. Math. Phys 67 (1979), 203–223.

GS4   Golubitsky, M. and Schaeffer, D.   *A discussion of symmetry and symmetry breaking*, Proc. Sym. Pure Math. 40 (1983), 499–516.

GSt   Golubitsky, M. and Stewart, I.N.   *Symmetry and Stability in Taylor–Couette flow*, SIAM J. Math. Anal. 17 (1986) 249–288

G   Goryunov, V. V.   *Projections of generic surfaces with boundaries*, Adv. Soviet Math. 1 (1990), 157– 200

Iz1   Izumiya, S.   *Generic bifurcations of varieties*, Manuscripta Math. 46 (1984) 137– 164.

Iz2   ------   *Stability of G–unfoldings*, Hokkaido Math. J. IX 1 (1980) 31–45.

Iz2   ------   *First Order Partial Differential Equations and Singularities*, preprint

LL   Lari–Lavasani, A.   *Multiparameter Bifurcation with Symmetry via Singularity Theory*, Thesis Ohio State Univ. 1990

La   Latour, F.   *Stabilité des Champs d'applications Différentiables: Generalization d'un Théorème de J. Mather*, C.R. Acad. Sci. Paris Sér. A–B 268A (1969), 1331–1334.

Lê   Lê, D.T.   *Calcul du nombre de cycles evanouissants d'une hypersurface complex*, Annales Inst. Fourier 23 (1973), 261–270.

Ly   Lyashko, O.   *Classification of critical points of functions on a manifold with singular boundary*, Funct. Anal. and Appl.(1984), 187–193.

MRS1 Montaldi, J., Roberts, M., and Stewart, I. *Periodic Solutions near Equilibria of symmetric Hamiltonian systems*, Phil. Trans. Roy. Soc. London A 325 (1988) 237–293.

MRS2 ------- *Existence of nonlinear normal modes of symmetric Hamiltonian systems*, Nonlinearity 3 (1990) 695–730

Mo Moser, J. *Periodic Solutions near an Equilibrium and a Theorem of Alan Weinstein*, Comm. Pure Appl. Math. 29 (1976) 727–747.

Pk Pellikan, R. *Finite Determinacy of Functions with Nonisolated Singularities*, Proc. London Math. Soc. 57, (1988), 357–382.

Pr Peters, M. *Classification of Two Parameter Bifurcations*, Singularity Theory and its Applications: Warwick 1989 Pt. 2, Springer Lecture Notes 1463 (1991) 294–300.

P Poenaru, V. Singularités C^∞ en présence de symétrie, Lecture Notes in Math., vol. 510, Springer–Verlag, Berlin and New York, 1976.

Ra Rabinowitz, P. *Periodic Solutions of Hamiltonian Systems: a Survey*, SIAM J. Math. Anal. 13 (1982) 343–352

Rb Roberts, M. *Characterizations of finitely determined equivariant map germs*, Math. Annalen 275 (1986) 583–597

Sa2 Sattinger, D. *Bifurcation and Symmetry Breaking in Applied Mathematics*, Bull. Amer. Math. Soc. 3 (new series) (1980) 779–819

Si2 Siersma, D. *Singularities of functions on boundaries, corners, etc.*, Quart. J. Math. 32 (1981), 119–127.

Sz   Schwarz, G.   *Smooth functions invariant under the action of a compact Lie group*, Topology 14 (1975), 63–68.

Sw   Swift, J.W.   Bifurcation and Symmetry in Convection, Thesis Univ. of California, Berkeley 1984

Ta   Tari, F.   *Projections of Piecewise Smooth Surfaces*, J. London Math. Soc. 44 (1991), 155–172≥≥

VM   Van der Meer, J.C.   The Hamiltonian–Hopf Bifurcation, Springer Lecture Notes 1160 (1985)

Ws2   Wasserman, G.   *Stability of unfoldings in space and time*, Acta. Math. 135 (1975), 57– 128.

We   Weinstein, A.   *Normal Modes for Nonlinear Hamiltonian Systems*, Invent. Math. 20 (1973) 47–57

Za   Zakalyukin, V.   *Lagrangian and Legendrian Singularities*, Funct. Anal. Appl. 10, (1976) 23–31

Ze   Zeeman, E.C.   *Differential Equations for the Heartbeat and Nerve Impulse*, in Dynamical Systems, M. Peixoto, Editor, Academic Press, New York, 1973, 683–741.

§5   Structure of Nonlinear Operators

AP   Ambrosetti, A. and Prodi, G.   *On the inversion of some differentiable maps with singularities*, Annali. di Math. 93 (1972), 231–246.

BC   Berger, M. and Church, P.   *Complete integrability and perturbation of a
      nonlinear Dirichlet problem I.*, Indiana Univ. Math. J. 28
      (1979), 935–952.

BCT   Berger, M., Church, P., and Timourian, J.   *An application of singularity theory to
      nonlinear elliptic partial differential equations*, Proc. Sym. Pure Math. 40 (1983),
      119–126.

BCT2 Berger, M., Church, P., and Timourian, J.   *Folds and Cusps in Banach Spaces:
      With applications to partial differential equations I*, Indiana Univ. Math. J. 34
      (1985), 1–19.

CT   Church, P., and Timourian, J. *Global Fold Maps in Differential and Integral
      Equations*, Nonlinear Anal. 18, (1992) 743–758

C   Cafagna, V.   *Global Invertibility and Finite Solvability*, Nonlinear Functional
      Analysis, Ed. P.S. Milojević, Lect. Notes Pure and Appl. Math. 121 (1990),
      Dekker , New York, 1–30

D6   Damon, J.   *A theorem of Mather and the Local Structure of Nonlinear
      Fredholm Maps*, Proc. Sym. Pure Math. vol. 45 pt. 1 (1985)
      339–352

D7   -----   *Time Dependent Nonlinear Oscillations with many Periodic
      Solutions*, SIAM Jour. Math. Anal. 18 (1987) 1294–1316

HR   Hale, J.K. and Rodriguez, H.M.   *Bifurcation in the Duffing Equation with
      Independent Parameters I*, Proc. Roy. Soc. Edinburgh 77A,
      (1977) 57–65; *II*, 79A (1978) 317–326

MD   Mather, J. and Damon, J.   Preliminary Notes on C^∞-Stability of Mappings, unpublished

Mc   McKean, H.P.   *Singularities of a Simple Elliptic Operator*, Jour. Diff. Geom. 25 (1987) 157-165

MSc   McKean, H.P. and Scovel, J.C.   *Geometry of Some Nonlinear Differential Operators*, Annali Scu. Norm. Sup. Pisa 13 (1986) 299-346

Te   Teissier, B.   *Cycles évanescents, sections planes, et conditions de Whitney, Singularités à Cargèse*, Astérisque 7,8 (1973), 285-362.

§6   Multiplicities of Solutions

AFN   Aoki,K., Fukuda, T. and Nishimura T.   *On the number of branches of the zero l ocus of a map germ* $(\mathbb{R}^n,0) \longrightarrow (\mathbb{R}^{n-1},0)$, Topology and Computer Sciences, S. Suzuki ed. Kinokuniya Co. Ltd. Tokyo (1987) 347-367.

AFN2 - - - - -   *An algebraic formula for the topological types of one parameter bifurcation diagrams*, Arch. Rat'l. Mech. and Anal. 108 (1989) 247-266.

Brc   Briancon, J.   *Description de Hilb$^n\{x, y\}$*, Invent. Math. 41 (1977) 45-89

BGal Briancon, J. and Galligo, A.   *Déformations de points de* $\mathbb{R}^2$ *ou* $\mathbb{C}^2$, Astérisque 7 & 8 (1973) 129-138

Bu   Buchberger, B.   *An Algorithmic Method in Polynomial Ideal Theory*, in Multidimensional Systems Theory, N. K. Bose Ed., D. Reidel Publ., Dordrecht (1985), 184-232

D8   Damon, J.   *G-signature, G-degree, and the symmetries of branches of curve singularities*, Topology 30 (1991) 565–590

D9   ------   *Equivariant bifurcations and Morsifications for finite groups*, Singularity Theory and its Applications: Warwick 1989 Pt. 2, Springer Lecture Notes 1463 (1991) 80–106

DGal Damon, J. and Galligo, A. *A Topological Invariant for Stable Map Germs*, Invent. Math. 32 (1976) 103–132

DGal2   ------   *On the Hilbert–Samuel Partition of Stable Map Germs*, Bull. Soc. Math. France 111, (1983) 327–358

EL   Eisenbud, D. and Levine, H. *An algebraic formula for the degree of a C^∞-map germ*, Annals of Math. 106 (1977) 19–44

FRc1 Field, M. and Richardson, R.W. *Symmetry Breaking and the maximal isotropy subgroup conjecture for reflection groups*, Arch. Rat'l. Mech. and Anal. 105 (1989) 61–94

FRc2 ------   *Symmetry breaking and Branching Patterns in Equivariant Bfurcation Theory*, I and II, Arch. Rat. Mech. Anal. 118 (1992) 297–348 and 120 (1992) 147–190

Gal2 Galligo, A.   *A propos du Théorèm de préparation de Weierstrass*, Springer Lecture Notes in Math. 409 (1974), 543–579.

GH   Griffiths, P. and Harris, J. Principles of Algebraic Geometry, John Wiley, New York (1978)

GZ   Gusein–Zade, S. M.   *On the Degree of an Equivariant Map*, Singularity Theory and its Applications: Warwick 1989 Pt. 2, Springer Lecture Notes 1463 (1991) 185–193.

Hr   Hartshorne, R.   <u>Residues and Duality</u>, Springer Lecture Notes 20 (1966).

Hi   Himshiashvili, G. *The Local Degree of a Smooth Mapping*, Comm. Acad. Sci. Georgian SSR 85 (1977) 309–311

Ia1   Iarrobino, A.   *Reducibility of the Families of 0-Dimensional Schemes on a Variety*, Invent. Math. 15 (1972) 72–77

Ia2   ------   *Punctual Hilbert Schemes*, Memoirs Amer. Math. Soc. 10 (1977)

IMV   IzeJ., Massabo, I., and Vignoli, A.   *Degree Theory for Equivariant Maps I*, Trans. Amer. Math. Soc. 315, (1989) 433–510

IMV2   -------   *Degree Theory for Equivariant Maps, the General S^1-action*, Memoirs Amer. Math.Soc. to appear

Kh   Khovanski, A.G.   *Fewnomials and Pfaff Manifolds*, Proc. Int. Cong. Math. (1983) 549–564

MvS   Montaldi, J. and van Straten, D.   *One-forms on singular curves and the topology of real curves singularities*, Topology 29, (1990) 501–510

Pa   Palamodov, V.P.   *Multiplicity of Holomorphic Mappings*, Funct. Anal. Appl. 1 (1967) 218–226

§7   <u>Topological Equivalence</u>

A5   Arnold, V. I.   *Normal Forms of Functons in Neighborhoods of Degenerate Critical Points*, Russian Math. Surveys 29, (1974) 10–50

BMS Buchner, M., Marsden, J., and Schecter, S.   *Applications of the blowing–up construction and algebraic geometry to bifurcation problems.* Jour. Diff. Eqtns. 48 (1983) 404 –433.

D10   Damon, J.   *Topological Stability in the Nice Dimensions*, Topology 18 (1979) 129–142

D11   ------   *The Relation between C^∞ and Topological Stability*, Bol. Soc. Bras. Mat. 8, (1977) 1–38

D12   ------   *Topological Triviality of Versal Unfoldings* ,Proc. Sym. Pure Math. 40 (1983) 255–266.

D13   ------   *Finite determinacy and topological triviality* I.,Invent. Math. 62 (1980), 299–324.
II. *Sufficient conditions and topological stability*, Compositio Math. 47 (1982) 101–132.

D14   --------   *Topological triviality and versality for subgroups of $\mathcal{A}$ and $\mathcal{K}$* Memoirs Amer. Math. Soc. 389 (1988)
II: *Sufficient Conditions and Applications*, Nonlinearity 5 (1992), 373–412

DGaf Damon, J. and Gaffney, T.   *Topological triviality of deformations of functions and Newton filtrations*, Invent. Math. 72 (1983) 335–358.

DP   DuPlessis, A.   *On the genericity of topologically finitely determined map germs*, Topology 21 (1982), 131– 156

Gi   Gibson, C. G. et al   <u>Topological Stability of Smooth Mappings</u>, Springer Lect.
                               Notes 552, (1976)

K   Kouchnirenko, A. G.   *Polyedres de Newton et nombres de Milnor*, Invent.
                               Math. 32 (1976) 1–31.

Ku1   Kuo, T. C.   *On C^0-sufficiency of jets of potential functions*, Topology 8
                               (1969), 167–171.

Ku2   Kuo, T. C.   *Characterization of v-sufficiency of jets*, Topology 11 (1972),
                               115– 131.

LR   Lê, D. T. and Ramanujam, C. P.   *Invariance of Milnor's number implies*
                               the invariance of topological type, Amer. J. Math. 98 (1976),
                               67–78.

LT   Lê, D. T. and Teissier, B.   *Cycles évanescents, sections planes, et conditions de*
                               Whitney II   Proc. Sym. Pure Math. 44 Part II (1983) 65–
                               103

Lo   Looijenga, E.   *Semi-universal deformation of a simple elliptic hypersurface*
                               singularity: I. Unimodularity, Topology 16 (1977) 257–262.

Lo2   ------   <u>Isolated Singular Points on Complete Intersections</u>, London
                               Math. Soc. Lect. Notes 77, Camb. Univ. Press (1984)

Ma3   Mather, J.   *Stratifications and mappings*, in Dynamical Systems, M.
                               Peixoto, Editor, Academic Press, New York, 1973, 195–232.

Ma4   ------   *How to stratify mappings and jet spaces,* in Singularites

d'Applications Differentiables, Plans–sur–Bex, Springer
Lectures Notes in Math., vol. 535, (1975), 128–176

Ma5   ------   *Generic Projections*, Ann. Math. 98, (1973) 226–245

My   May, R.   *Stability and Transversality*, Bull. Amer. Math. Soc. 80, (1974)
85–89

Mi2   Milnor, J.   Singular Points of Complex Hypersurfaces, Annals Math.
Studies 61 (1968) Princeton Univ. Press

MOr   Milnor, J. and Orlik, P.   *Isolated Singularities defined by Weighted
Homogeneous Polynomials*, Topology 9 (1970), 385–393

O   Oka, M.   *On the Bifurcation of the Multiplicity and Topology of the
Newton Boundary* , Jour. Math. Soc. Japan 31 (1979), 435–
450

PB   Percell, P. B. and Brown, P. N.   *Finite Determination of Bifurcation Problems*,
SIAM J. Math. Anal. (1985), 28–46.

Th4   Thom, R.   *Ensembles et Morphismes Stratifiés*, Bull. Amer. Math. Soc. 75
(1969) 240–284

Th5   -----   *La Stabilité Topologique des applications Polynomiales*,
Enseignment Math. 8 (1962) 24–33

Ti   Timourian, J. G.   *Invariance of Milnor's number implies topological triviality*,
Amer.J. Math. 99 (1979), 437–446.

Va   Varchenko, A. N.   *Local Topological Properties of Differentiable Mappings*,

USSR-Izv. 8 (1974), 1033-1082.

W   Wirthmuller, K.   <u>Universell Topologische Triviale Deformationen</u>, Thesis,
                         Univ. of Regensberg.

Positivity of Maps and Applications

E. N. Dancer

To the Memory of Peter Hess

Department of Mathematics, Statistics and Computing Science
The University of New England
Armidale, NSW 2350, Australia

In this survey, we discuss some recent work on maps in cones on Banach spaces. We emphasize developments more recent than Amann's survey article [2], and try to limit the intersection with Nussbaum's survey [38]. We emphasize results which depend upon fixed point arguments. We also, at times, discuss applications to ordinary and partial differential equations. We do not discuss applications to delay equations. These are important but they tend to be outside the author's expertise. We do, however, discuss applications to dynamical systems. We will use the fixed point index as a tool and we summarize its properties. A survey of the fixed point index for more general maps appears in Nussbaum [37].

This survey really concentrates on two topics. First, we discuss fixed point index calculations for maps on convex sets. More precisely, assume that C is a closed bounded convex set in a Banach space and $U \subseteq C$ is open and $f : \overline{U} \to C$ is completely continuous. If f has no fixed points on ∂U, the fixed point index $i_C(f, U)$ is defined. We are interested in the contribution of an isolated fixed point x_0 of f in C to this fixed point index. (Sometimes, we replace x_0 by a set of fixed points.) In the interesting cases, x_0 is on the boundary of C. This situation might seem rather restrictive but we will give a number of applications which shows that this situation occurs much more frequently than one might at first suspect.

The second main part of our survey consists of monotone dynamical systems both continuous and discrete. Here we mean monotone in the sense of order. We discuss results on the behaviour of typical trajectories, the

existence of connecting orbits and related properties and the differences
between continuous and discrete dynamical systems. Here we emphasize
results which involve fixed point arguments.

Part 1. Fixed Point Index Calculations and Applications

In this part, we discuss various fixed point index calculations on convex
sets and give a number of applications.

In §1, we discuss the fixed point index and in §2, we present some
preliminary material on a closed convex set. In §3, we present a basic
fixed point index calculation and in §4 we discuss some variants. In §5 and
§6, we consider various applications of the results in §3 and §4 to partial
differential equations. Finally in §7, we consider some results which say
that certain branches of positive solutions exist globally.

We begin by first reviewing the fixed point index.

§1. The Fixed Point Index

We will not consider the most general situation. We assume that C is a
closed convex set in a Banach space, U is a relatively open bounded subset
of C, and $f : \overline{U} \to C$ is completely continuous (that is f is continuous
and $f(\overline{U})$ has compact closure) and $x \neq f(x)$ on ∂U (where the boundary
is with respect to C). Then (see [37]) the fixed point index $i_C(f, U)$ is
defined and is an integer and provides a "count" of the number of solutions
of $x = f(x)$ in U. Note that a wide variety of notations are used for this.
It has the following basic properties (see [37], §1)

   (i) (additivity) If U_1 and U_2 are disjoint bounded open subsets of U
       such that $x \neq f(x)$ on $\overline{U} \setminus (U_1 \cup U_2)$, then

$$i_C(f, U) = i_C(f, U_1) + i_C(f, U_2)$$

   where $f : \overline{U} \to C$ is assumed to be completely continuous. Moreover, if $x \neq f(x)$ on $\overline{U}$, $i_C(f, U) = 0$.
  (ii) (homotopy invariance) If $F : \overline{U} \times [0, 1] \to C$ is completely con-
       tinuous and $x \neq F(x, t)$ on $\partial U \times [0, 1]$, then $i_C(F(\,, t), U)$ is inde-
       pendent of t for $t \in [0, 1]$.
 (iii) (normalization) If C is bounded and $f : C \to C$ is completely
       continuous, $i_C(f, C) = 1$
  (iv) (commutativity) Assume that C_1 and C_2 are closed convex sets
       and w_i is an open subset of C_i $i = 1, 2$. Assume $f_1 : \overline{w}_1 \to C_2$ and

$f_2 : \overline{w}_2 \to C_1$ are continuous. Assume that $S = \{x \in f_1^{-1}(\overline{w}_2) : f_2(f_1(x)) = x\}$ is compact, is contained in $f_1^{-1}(w_2)$ and that f_1 is completely continuous on some open neighbourhood of S. Then one has $i_{C_1}(f_2f_1, f_1^{-1}(w_2)) = i_{C_2}(f_1f_2, f_2^{-1}(w_1))$

A few points need to be made here. The fixed point index can be defined for much more general sets than closed convex sets (in particular for metric absolute neighbourhood retracts in the sense of Hu [32]). We do not do this here because we do not need it. However, we will discuss this a little more later. (In this case, (iii) would need to be formulated differently.) It is easy to deduce that we can delete the condition that U is bounded if we assume that $T = \{x \in U : x = f(x)\}$ is compact and f is completely continuous in some neighbourhood of T. In this case (ii) and (iii) need to be formulated differently. Indeed, we have implicitly done this to obtain a convenient formulation for (iv). Note that (i) is frequently stated as two separate properties. The commutativity property is a very useful property but seems to be poorly understood. For example, it can frequently be used to prove that a fixed point index is independent of the particular Banach space X lies in. An important special case is that if $f(\overline{U}) \subseteq C_2$ where C_2 is a closed convex subset of C, then $i_C(f, U) = i_{C_2}(f|_{C_2}, U \cap C_2)$. Finally, note that in (ii) we could easily deduce a variant where θ is a bounded open subset of $C \times [0,1]$ such that $x \neq F(x,t)$ on $\partial\theta_t$ for $t \in [0,1]$ (where $\theta_t = \{x \in C : (x,t) \in \theta\}$) and F is completely continuous on $\overline{\theta}$. The conclusion is $i_C(F(\,,t), \theta_t)$ is independent of t. This is useful later.

It follows immediately from property (i) of the fixed point index that if $x_0 \in U$ is an isolated fixed point of f in U, then $i_C(f, B_\varepsilon(x_0) \cap C)$ is independent of ε for sufficiently small positive ε (and is independent of U). Here $B_\varepsilon(x_0)$ is the open ball centre x_0 and radius ε. We denote this by $i_C(f, x_0)$. We are interested in obtaining formulae for $i_C(f, x_0)$. Note that this is important because it enables us to calculate $i_C(f, U)$ in many cases.

There is one case which is classical. Assume that there is an $\varepsilon > 0$ such that $B_\varepsilon(x_0) \subseteq U \subseteq C$. Then, by the commutativity theory for the degree, $i_C(f, x_0) = i_X(f, x_0)$ and we are back to the classical Leray-Schauder case. Note that even in this case, the calculation of the fixed point index is not at all easy and usually depends on smoothness properties of f (and even then is not known in full generality). See [33], [35] and [49] where further references can be found. Thus we do not expect it to be easy to obtain formula for $i_C(f, x_0)$ and we expect results to depend on smoothness assumptions of f. Note that the conditions that $B_\varepsilon(x_0) \subseteq X$ is frequently not satisfied for two reasons. Frequently in applications, the smallest closed subspace containing C (denoted $\overline{\lim}\, C$) is X but C has non-empty interior in X. Secondly, even if C has empty interior, x_0 is frequently on the boundary of C in applications.

We will see in the next section that in the case where f is smooth on X, the formula for $i_C(f, x_0)$ is different in cases where x_0 is in the interior of C or the boundary of C. On the other hand, at the topological level (that is up to homeomorphism) what is on the boundary of C and what is in the interior of C is unclear in the infinite dimensional case. (More precisely, if C is closed and convex with non-empty interior in an infinite dimensional Banach space and if $x_0, x_1 \in C$, then there is a homeomorphism $h : C \to C$ with $h(x_0) = x_1$. For example, this follows easily from Corollary VI.6.1 in Bessaga and Pelczynskii [5].) These two results together strongly suggest that it is difficult to obtain good results for the index of an isolated solution of a non-differentiable map on a convex set. We will meet another example in the next section. The point seems to be that the differentiability builds in some uniformity in the directions in the set and many directions close to the set.

Lastly for this section, we discuss in a little more detail on which sets the fixed index can be defined. It turns out that it can be defined for the class of ANR(M) that is absolute neighbourhood retracts for metric spaces in the sense of [32] . We do not define this class explicitly here except to say that it is defined by extension properties and is discussed in [32]. Note that by a result in [37] a finite union of closed convex sets in a Banach space is an ANR(M). To a large extent an ANR(M) is a space with good local properties (for example a sufficient condition to ensure that it is well behaved locally is to assume that it is locally homeomorphic to a convex set). Unfortunately, in the infinite dimensional case, examples in [32] show that the simplest local conditions (local contractibility) does not determine the class ANR(M). On the other hand there is a result of Hammer (see [38]) which essentially says that a space which is locally an ANR(M) is an ANR(M). (A more precise statement appears in [37].)

§2. Some Remarks on Convex Sets

We briefly discuss some conditions on convex sets that we need for the index formula of §3. Most of this material comes from [11].

Assume that C is a closed bounded convex set in X and $y \in C$. Let $C_y = \{z \in X : ty + (1-t)z \in C$ for some small positive $t\}$. By the convexity, if $z \in C_y$, then $ty + (1-t)z \in C$ for all sufficiently small positive t. It is easy to check that C_y is convex, $0 \in C_y$, $\alpha C_y \subseteq C_y$ for $\alpha \geq 0$. Hence $\overline{C}_y$ has the same properties. (A closed set with these properties is usually called a **wedge**. A wedge W is said to be a **cone** if $x \in W \setminus \{0\}$ implies $-x \notin W$.) Let $S_y = \{u \in \overline{C}_y : -u \in \overline{C}_y\}$. It is easy to check this is a closed subspace of X. Note that $\overline{C}_y$ is some form of "linear" approximation to C near y. If y is in the interior of C, $C_y = X$ while if y is in the boundary of

∂C of C, if C has non-empty interior and if ∂C is a smooth manifold near y, then it is easy to proof that $\overline{C}_y$ is a half space bounded by the tangent hyperplane to C at y (translated to zero).

Usually it is easiest to calculate $\overline{C}_y$ directly but it is also not difficult to use the Hahn-Banach theorem to give a dual characterization of $\overline{C}_y$. Let

$$C_y^* = \{f \in X^* : f(x) \geq f(y) \text{ for } x \in C\}.$$

Then it is easy to prove that

$$\overline{C}_y = \{x \in X : f(x) \geq 0, \ \forall f \in C_y^*\}$$

and hence

$$S_y = \{x \in X : f(x) = 0, \ \forall f \in C_y^*\}.$$

Note that there has been some change of notation from [11]. This also provides a convenient method for calculating $\overline{C}_y$ and S_y.

The following useful result can be easily proved by either characterization. Here if Ω is a bounded open set in R^m. Let $C_0(\Omega)$ denote the set of continuous functions on $\overline{\Omega}$ which are zero on $\partial\Omega$ and let K denote the set of non-negative functions in $C_0(\Omega)$.

Lemma 1. *If $y \in C_0(\Omega)$ and $y(x) > 0$ for all $x \in \Omega$, then $\overline{K}_y = C_0(\Omega)$.*

Remarks

1. This illustrates a general point that $\overline{C}_y$ may be all of X even when C has empty interior.

2. If y is non-negative and $A = \{x \in \Omega : y(x) = 0\}$, it is not difficult to prove that $\overline{K}_y = \{z \in C_0(\Omega) : z(x) = 0 \text{ on } A\}$.

3. There is a variant of the lemma with a similar proof. If $1 \leq p < \infty$, $y \in L^p(\Omega)$ and $y(x) > 0$ a.e. in Ω, then $\overline{K}_y = L^p(\Omega)$ where K now denotes the set of non-negative functions in $L^p(\Omega)$. (The dual characterization is most convenient for proving this.)

An element y of C is said to be demi-interior to C if $\overline{C}_y = X$. If C has non-empty interior, the Hahn Banach theorem implies that the demi-interior points are simply the interior points. On the other hand Lemma 1 shows that a closed convex set with empty interior may have demi-interior points. In fact if X is separable and if the smallest closed subspace of X containing C is all of X (usually written as $X = \overline{\text{lin}}\ C$) then the set of demi-interior points of C are dense in C. (This is a simple variant of Theorem 5.7.6 in Schaeffer [48].) In the non-separable case, there may be no demi-interior points at all. The simplest example is the set C of points with non-negative coefficients in a non-separable Hilbert space H (for some orthonormal basis of H). Note that problems with compact maps usually effectively reduce to problems on separable spaces.

In the case of a Banach lattice, (as defined in [48]) it can be shown (see [11]) that, if K is the natural cone and $y \in K \setminus \{0\}$, then S_y is the closure of

$$\{u \in X, \ -\gamma y \le u \le \gamma y \text{ for some } \gamma > 0\}.$$

Alternatively, in the dual characterization of S_y, we need only consider positive functionals. This is useful because it implies that if X is a Banach lattice with natural cone K, if $a \le b$ in X (that is $b - a \in K$) and if $a \le y \le b$, then y is demi-interior to the order interval $[a, b] \equiv \{x \in X : a \le x \le b\}$ provided that $y - a$ and $b - y$ are both demi-interior to K. This is useful because order intervals occur naturally in work with monotone maps.

Lastly for this section, assume W is a wedge in X and L is a continuous linear map of X into itself such that $L(W) \subseteq W$. Then it is easy to see (see [11]) that L maps $\tilde{W} \equiv \{x \in W : -x \in W\}$ into itself (by linearity) and hence L induces a continuous linear mapping $\tilde{L}$ of $X/\tilde{W}$ into itself. In Section 3, we will be interested in whether the spectral radius $r(\tilde{L})$ is less than or larger than 1. In nearly all applications, $\tilde{W}$ has a closed complement Y in X. Since $L(\tilde{W}) \subseteq \tilde{W}$,

$$\tilde{L}(w, y) = (L_1 w, L_2 w + L_3 y) \quad \text{for} \quad w \in \tilde{W}, \quad y \in Y.$$

Then $X/\tilde{W}$ is isomorphic to Y and $\tilde{L}$ is isomorphic to $L_2 \equiv PLP|Y$ (where P is the natural projection of X onto Y parallel to $\tilde{W}$). Thus $r(\tilde{L}) = r(L_2)$. This is usually the most convenient method for calculating $r(\tilde{L})$.

§3. A Basic Index Calculation

Generally, if $A : X \to X$ is completely continuous, x_0 is an isolated fixed point of A and A is differentiable at x_0 then $\text{index}_X(A, x_0) = \pm 1$ (with a well-known formula for when it is 1 or -1). By generally, we mean that this holds when $I - A'(x_0)$ is invertible. Krasnosel'ski first realized that in the case of convex sets (in particular in the case of a cone), generally the index (relative to C) may also take the value zero.

We assume that Z is a neighbourhood of the closed convex set C in X and $A : \overline{Z} \to X$ is completely continuous, differentiable at $y \in C$ and $A(C) \subseteq C$. In addition we assume that $\overline{\text{lin}}\, C = X$. (This could be avoided though it nearly always seems to be true in applications.)

Theorem 1. ([11], Theorem 1 and [14], Proposition 1) *Assume that y is a fixed point of A in C and the above conditions hold. Finally assume that $z \ne A'(y)z$ if $z \in \overline{C}_y \setminus \{0\}$. Then y is an isolated fixed point of A in C and*

(i) $i_C(A, y) = i_X(A, y) = i_X(A'(y), 0) = i_{S_y}(A'(y), 0)$ if $z \neq A'(y)z$ on X and $r(\tilde{A}'(y)) < 1$.

(ii) $i_C(A, y) = 0$ if $r(\tilde{A}'(y)) > 1$.

(iii) $i_C(A, y) = 0$ if $N(I - A'(y)) \cap (X \setminus \overline{C}_y)$ is non-empty.

This is the basic index formula. Note that since $A(y) = y$ and $A(C) \subseteq C$, it is easy to prove (as in [11]) that $A'(y)$ maps $\overline{C}_y$ into itself and hence maps S_y into itself. As in the previous section $\tilde{A}'(y)$ denotes the corresponding induced map of X/S_y into itself. Note that by Krasnosel'skii and Zabrieko [33],) $A'(y)$ is a compact linear map and hence $i_X(A'(y), 0)$ is defined and we can use Theorem 2.4.5 in Lloyd [35] to evaluate it. (The result depends on the eigenvalues of $A'(y)$ and their multiplicity.) If $r(\tilde{A}'(y)) = 1$, we can argue as in §1 of [11] to prove that $N(I - A'(y)) \cap \overline{C}_y \neq \{0\}$ (by using the Krein-Rutman theorem). Hence our other assumptions ensure that $r(\tilde{A}'(y)) \neq 1$. Thus the above theorem gives a formula for the index in all cases where $N(I - A'(y)) \cap \overline{C}_y = \{0\}$. In applications the theorem tends to be relatively easy to apply. We will meet examples later.

The key point in the proof is to use a careful approximation and homotopy argument to prove that $i_C(A, y) = i_Z(A'(y), 0)$ where $Z = \overline{C}_y$. Then homotopies and linear operator theory are used to calculate this last index.

There is one curious point in this argument. For most of the argument the linearity of $A'(y)$ is not needed. (One needs to assume that $A'(y)$ is completely continuous and positive homogeneous on $\overline{C}_y$ and that $\|A(u) - A(y) - A'(y)(u-y)\| = o(\|u-y\|)$ as u tends to y in C (where the limit holds uniformly on C). More precisely, if we assume the above differentiability condition on A, and the other assumptions of the theorem hold, then it is still true that $i_C(A, y) = i_Z(A'(y), 0)$ where $Z = \overline{C}_y$. However, it is not clear if there is a simple formula for this last index in this more general case. However, the whole theory (not (iii) of Theorem 1) works if $A'(y)$ induces a well-defined monotone map $\tilde{A}'(y)$ on $\overline{C}_y/S_y$ (monotone for the induced order). Here the spectral radius is replaced by the supremum of positive λ for which $\tilde{A}'(y)$ has a non-trivial eigenvector in $\overline{C}_y/S_y$. More generally, one can at least prove that $i_Z(A'(y), 0) = i_{S_y}(A'(y)|_{S_y}, 0)$ if $A'(y)(S_y) \subseteq S_y$ and if $x - tA'(y)x \in S_y$ $x \in \overline{C}_y$ and $0 \leq t < 1$ implies that $x \in S_y$.

This nonlinear extension might look quite artificial but there are some examples in [41] coming from demography where $A'(y)$ is not linear but the above conditions hold. (Indeed, in this case, it turns out that $A'(y)$ is not linear but has many of the properties of a linear map.) The particular model in [41] can be handled more easily directly and stronger results obtained but the index calculations above could be used to prove existence of solutions $x \in R^n$ with each $x_i > 0$ in many variants of this model. (These comments are joint work with Professor Nussbaum.)

Note that in the above theorem, the requirement that y be demi-interior C seems to be exactly the correct condition to ensure that $i_C(A, y) = i_X(A, y)$ without making extra assumptions on $A'(y)$ (other than $N(I - A'(y)) \cap \overline{C}_y = \{0\}$). It would be interesting to know if the equality of the above indices is still true if we delete the condition on $N(I - A'(y))$ but assume that y is an isolated fixed point of A in X (and A is C^1). This is trivial if C has non-empty interior. It is proved in some other cases, in particular, C and very well behaved mappings A in [14, Proposition 3], but the general case is open. The problem is that, if we perturb A, it is likely that some of the fixed points of the perturbed map near y may not belong to C. An example below shows that the result is frequently not true if A is continuous rather than C^1.

We prove that, given any closed bounded convex set C in X such that $C - C \equiv \{c_1 - c_2 : c_1, c_2 \in C\}$ has empty interior and is closed, given $y \in C$ and given any completely continuous map $f : C \to C$ with fixed point y isolated in C, f can be extended to a completely continuous mapping $\tilde{f} : X \to X$ such that y is an isolated fixed point of $\tilde{f}$ and $i_X(\tilde{f}, y) = 0$. This gives the required example if we choose f to be the mapping which maps all of C to y. (In this case, $i_C(f, y) = i_C(f, C) = 1$.) To construct this example, we choose $z \in X$ so that $\{\alpha z : \alpha \neq 0\}$ does not intersect $C - C$. (Here we use the convexity of C and the lemma on p.139 of Holmes [31].) Let $\tilde{C} = \{m + \alpha z : m \in C, \alpha \in R\}$. The choice of z ensures that each element of $\tilde{C}$ can be uniquely written in this way. Define $H : C \oplus R \to \tilde{C}$ by $H(m, \alpha) = m + \alpha z$. This is clearly continuous $1 - 1$ and onto and H^{-1} is also continuous. (To see the latter, note that if $m_n + \alpha_n z \to u$ as $n \to \infty$, the boundedness of C ensures that $\{\alpha_n\}$ is bounded. Hence we can choose a subsequence so $\{\alpha_{n(i)}\}$ converges and hence $\{m_{n(i)}\}$ converges. The continuity of H^{-1} follows easily from this and the one to oneness of H.) Thus H is a homeomorphism between $\tilde{C}$ and $C \oplus R$. A similar argument shows that $\tilde{C}$ is closed in X. Choose $s : R \to R$ such that s is continuous, s has zero as its only fixed point, and $i_R(s, 0) = 0$. Define $\tilde{f} : \tilde{C} \to \tilde{C}$ by $\tilde{f}(m + \alpha z) = f(m) + s(\alpha)z$. It is easy to prove that $\tilde{f}$ is completely continuous. (Note, if $Z \subseteq \tilde{C}$ is bounded, the set of corresponding α's is bounded since C is bounded.) Moreover y is an isolated fixed point of $\tilde{f}$. Define $\hat{f} : C \oplus R \to C \oplus R$ by $\hat{f}(m, t) = (f(m), s(t))$. Note that $\tilde{f} = H\hat{f}H^{-1}$

$$i_{\tilde{C}}(\tilde{f}, y) = i_{C \times R}(\hat{f}, (y, 0))$$

(by the commutativity theorem for the degree)

$$= i_C(f, y)i_R(s, 0)$$

(by the product theorem)

$$= 0.$$

Since $\tilde{C}$ is closed in X, Schwartz [49, Lemma 3.58] implies that we can extend $\tilde{f}$ to a completely continuous map on X into $\tilde{C}$. Thus y is still an isolated fixed point of $\tilde{f}$ on X and

$$\text{index}_X(\tilde{f}, y) = \text{index}_{\tilde{C}}(\tilde{f}, y) \quad (\text{since } \tilde{f}(X) \subseteq \tilde{C})$$
$$= 0.$$

This gives the claimed example. This corrects an example on p. 55 of [14]. Note that $C - C$ is always closed if X is reflexive (by using the weak topology). Note that $C - C$ does contain a neighbourhood of zero if $C \supseteq B_\varepsilon(y) \cap K$ where $y \in K$ and K is a reproducing cone in X. (Here a cone K is said to be reproducing if $X = K - K$.) We suspect that similar examples still exist if C has empty interior but $C - C$ contains a neighbourhood of zero.

Note that Theorem 1 readily generalizes to any set locally C^1 diffe-ormorphic to a closed convex set. V.I. Arnold [4] independently obtained related results in the case of a smooth finite-dimensional manifold with boundary.

R. Nussbaum [40] has some fixed point index calculations for what he calls ejective fixed points. He does not require any differentiability. It wold be interesting to better understand the relationship of his work to ours.

§4. Index Calculations in Product Cones

In this section, we consider in more detail the special case where C_i is a cone in E_i $(i = 1, 2)$. Let $C = C_1 \oplus C_2$ and $E = E_1 \oplus E_2$. Then C is a cone in E. In a number of applications, we are interested in the index of points of the form $(u_1, 0)$ or $(0, u_2))$ where $(u_1, 0)$ is an isolated fixed point of a completely continuous map $A : C \to C$ (or certain isolated sets of fixed points). This case occurs frequently in applications.

Theorem 2. ([21] , Theorem 2.1) *Assume that $A : C \to C$ is completely continuous $C_2 - C_2$ is dense in E_2, $u_1 \in C_1$, and $A(\overline{u}_1, 0) = (\overline{u}_1, 0)$. Write $A = (A_1, A_2)$ where $R(A_i) \subseteq C_i$. Assume that $A_2(u_1, 0) = 0$ on C_1 and that A_2 extends to a continuously differentiable mapping of a neighbourhood of $(u_1, 0)$ in E into E_2. Finally, assume that $\overline{u}_1$ is isolated as a fixed point of $A|_{C_1 \oplus \{0\}}$.*

(i) *Then $i_C(A, (\overline{u}_1, 0))$ is defined and equals $i_{C_1}(A_1|_{C_1}, \overline{u}_1)$ if $r(A_2'(\overline{u}_1, 0)|_{E_2}) < 1$, and*

(ii) *$i_C(A, (\overline{u}_1, 0))$ is defined and equals zero if $r(A_2'(\overline{u}_1, 0)|_{E_2}) > 1$ and 1 is not an eigenvalue of $A_2'(\overline{u}_1, 0)|_{E_2}$ corresponding to an eigen-vector in C_2.*

The difference from Theorem 1 is that we do not assume non-degeneracy (where by non-degeneracy we mean that $u \neq A'(\overline{u}, 0)u$ for $u \in E \setminus \{0\}$). Indeed, in the non-degenerate case Theorem 2 follows easily from Theorem 1. (In this case, $\overline{C}_y$ (where $y = (\overline{u}_1, 0)$ contains $\{0\} \oplus K_2$ as a direct summand.) The proof is by constructing suitable homotopies. Note, since A_2 is completely continuous on a neighbourhood of $(\overline{u}, 0)$ in E, $A_2'(\overline{u}_1, 0)$ is a compact linear operator and hence $A_2'(\overline{u}_1, 0)|_{E_2}$ is compact. As in [21], Theorem 2 can be generalized to the case of a compact set of fixed point T of A being in $C_1 \oplus \{0\}$ provided that no point of T is the limit of fixed points of A in $C_1 \oplus \{0\}$ and provided that (i) holds for **every** $(\overline{u}_1, 0) \in T$ or (ii) holds for every $(u_1, 0) \in T$. It is clear that the situation is more complicated when (i) holds for some points of T and (ii) holds for some points of T. We will give an example later where this occurs. The great advantage of Theorem 2 is that it does not need non-degeneracy and non-degeneracy can be difficult to verify. We conjecture that a variant of Theorem 2 (ii) holds here we delete the assumption that 1 is not an eigenvalue of $A_2'(\overline{u}_1, 0)|_{E_2}$ corresponding to an eigenvector in C_2 but assume that $(\overline{u}_1, 0)$ is an isolated fixed point of A in C. A partial result of this type is discussed in [14]. Note that while the assumption that $A_2(u_1, 0) = 0$ on C_1 might appear unnatural, it seems to be true in applications whenever we have mappings $A : C \rightarrow C$ which have fixed points on $C_1 \oplus \{0\}$. There is an obvious variant where $A_1(0, u_2) = 0$ on C_2 and $(0, \overline{u}_2)$ is a fixed point of A.

We now want to consider briefly the simplest situation where $r(A_2'(\overline{u}_1, 0)|_{E_2}) = 1$. We assume the basic conditions of Theorem 2 hold except that we assume the continuous differentiability condition holds for A. Since $A_2(u, 0) = 0$ for all $u \in C_1$, we easily see that $A_2'(\overline{u}_1, 0)|_{C_1} = 0$. Hence $A'(\overline{u}_1, 0)$ is of the form $\begin{pmatrix} B & C \\ O & D \end{pmatrix}$ (where $B = A_1'(\overline{u}_1, 0)|_{E_1}$ etc.). Assume that $I - B$ is invertible on E_1 and $A_2'(\overline{u}_1, 0)|_{E_2}$ has 1 as a simple eigenvalue with kernel spanned by h. (Thus we are looking at the simplest situation in which $r(A_2'(\overline{u}_1, 0)|_{E_2}) = 1$ but one which is likely to occur frequently.) In fact it can be proved that the simplicity implies the invertibility of $I - B$. By the Krein Rutman theorem, $h \in C_2$. It is easy to see that 1 is a simple eigenvalue of $A'(\overline{u}_1, 0)$ with eigenvector of the form $z = (k, h)$. By a standard bifurcation analysis (that is, a Liapounov-Schmidt reduction) the analysis of small solutions of $x = A(x)$ near $(\overline{u}_1, 0)$ reduces to the analysis of a bifurcation equation $\phi(\alpha) = 0$ where $\phi : R \rightarrow R$. We need to be a little more precise how this is done since the sign of ϕ is crucial. By the simplicity of 1 as an eigenvalue of $A'(\overline{u}, 0)$, $E = N(I - A'(\overline{u}_1, 0)) \oplus R(I - A'(\overline{u}_1, 0))$ where R denotes the range. We write elements of E in the form $\alpha z + w$ where $w \in R(I - A'(\overline{u}_1, 0)) \equiv \mathcal{R}$ and let P be the corresponding projection onto $\mathcal{R}$. We then use the implicit function theorem to solve the equation $P(z(\alpha) - A(z(\alpha))) = 0$ for z

as a function of α where $z(\alpha) = (\overline{u}_1, 0) + \alpha z + p(\alpha)$ with $p(\alpha) \in \mathcal{R}$. Then $\phi(\alpha)z = (I - P)(z(\alpha) - A(z(\alpha)))$.

Theorem 3. *Assume that the above conditions hold and $\phi(\alpha) \neq 0$ for small positive α. Then*

   (i) $i_C(A, (\overline{u}_1, 0)$ *is defined and equals zero if $\phi(\alpha) < 0$ for small positive α,*

   (ii) $i_C(A, (\overline{u}_1, 0))$ *is defined and equals $index_{C_1}(A_1|_{C_1}, \overline{u}_1)$ if $\phi(\alpha) > 0$ for small positive α.*

Remarks. If there is a sequence $\{\alpha_i\}_{i=1}^{\infty}$ decreasing to zero where $\phi(\alpha_i) = 0$, then $z(\alpha_i)$ is a fixed point of A and $z(\alpha_i) = (\overline{u}, 0) + \alpha_i z + o(\alpha_i)$. Thus $z(\alpha_i) \in C$ or is very close to C. In fact, if $\overline{u}_1$ is demi-interior to C_1 and if $A'_2(u, v)|_{C_2}$ is a demi-interior operator on C_2 in the sense of [14] for all (u, v) in C near $(\overline{u}_1, 0)$ it is possible to modify arguments in §1 and §2 of [14] to prove that $\phi(\alpha) \neq 0$ for small positive α whenever we assume $(\overline{u}_1, 0)$ is an isolated fixed point of A in C. Note that we can use Theorem 1 to evaluate the index which appears in Theorem 3(ii).

Proof of Theorem 3. (sketch) The proof is a modification of part of the proof of Theorem 1 in [14]. We leave out a number of tedious details which are not hard to fill in. We consider the maps $A_\lambda(u_1, u_2) = (A_1(u_1, u_2), \lambda A_2(u_1, u_2))$ for λ near 1. Note that $A_1 = A$, $A_\lambda(C) \subseteq C$ and that $A_\lambda(\overline{u}_1, 0) = (\overline{u}_1, 0)$. It is easy to check that this map satisfies the assumption of the Crandall and Rabinowitz theorem [8] for bifurcation at a simple eigenvalue and there is a curve of solutions bifurcating from 0 for $\lambda = 1$ of the form $\lambda = \lambda(\alpha)$, $(u_1, u_2) = (\overline{u}_1, 0) + \alpha z + o(\alpha)$ which is defined for all small α. Clearly, by considering the second component, these can only belong to C for $\alpha \geq 0$ (when α is small). These solutions are determined by a bifurcation equation $\phi(\alpha, \lambda) = 0$ where $\phi(\alpha, \lambda) < 0$ if $\lambda > 1$ and if α is small and positive and $\phi(\alpha, \lambda) > 0$ if $\lambda < 1$ and if α is small and positive. (This is really a calculation on linear terms.) It is useful to note that $I - A'(\overline{u}_1, 0)^*$ is spanned by $(0, \tilde{h})$ where the linear functional $\tilde{h}$ is positive on C_2. (To prove this, we use the Krein-Rutman theorem and our matrix formula for $A'(\overline{u}_1, 0)$). We consider the case where $\phi(\alpha) > 0$ for small positive α. The other case is similar. By our earlier comments on the bifurcation equation, this implies that the only fixed points of A near $(\overline{u}_1, 0)$ are of the form $(\overline{u}_1, 0) + \alpha z + o(\alpha)$ where α is small and negative. It is easy to see that these do not belong to C. Hence $index_C(A, (\overline{u}_1, 0))$ is defined. Since $\phi(\alpha, 1) > 0$ for small positive α and $\phi(\alpha, \lambda) < 0$ if $\lambda > 1$ and α is small and positive, we see that there must be zeros (α, λ) of ϕ with $\alpha > 0$ and $\lambda > 1$ arbitrarily close to $(0, 1)$. Hence $\lambda(\alpha) > 1$ for α small and positive. (Remember that the bifurcating solutions form a continuous curve and our other assumptions ensure that $\lambda(\alpha) \neq 1$ for small positive α.) Hence the only solutions of $x = A_\lambda(x)$ in $C \times R$ near $(\overline{u}_1, 0, 1)$ have $\lambda > 1$. Thus, by the homotopy invariance of the fixed point index,

$$
\begin{aligned}
i_C(A, (\overline{u}_1, 0)) &= i_C(A_1, (\overline{u}_1, 0)) \\
&= i_C(A_\lambda, (\overline{u}_1, 0)) \text{ (for } \lambda < 1 \text{ but close to 1)} \\
&= i_{C_1}(A_\lambda|_{C_1}, \overline{u}_1) \text{ (by Theorem 2)} \\
&= i_{C_1}(A|_{C_1}, \overline{u}_1).
\end{aligned}
$$

Remark. Theorem 3 appears to be new. It admits many variants. A related result appears in [14].

§5. Applications of Index Formulae - I

In this section, we discuss rather briefly two applications of the index formula of §3. The first comes from [17] and the second from [18].

Firstly, assume that D is an annulus in R^2. We consider the solutions of the Gelfand problem

$$
\begin{aligned}
-\Delta u &= \lambda e^u \text{ in } D \\
u &= 0 \text{ on } \partial D
\end{aligned}
\tag{1}
$$

for $\lambda \geq 0$. (The case where $\lambda \leq 0$ is simple because it is not difficult to show that the solution is unique.) Moreover if $\lambda > 0$, the maximum principle implies that any solution is positive in D. We first consider the radially symmetric solutions though our main interest is in non-radially symmetric solutions. It is proved in [34] that there exist $\alpha > 0$ and continuous maps u_1, u_2 of $(0, \alpha]$ into $C_0^r(D) = \{u \in C_0(D) : u \text{ is radially symmetric}\}$ such that $u_1(t) \to 0$ at $t \to 0^+$, $\|u_2(t)\|_\infty \to \infty$ as $t \to 0^+$, $u_1(\alpha) = u_2(\alpha)$, $u_1(t) < u_2(t)$ for $0 < t < \alpha$ and $\{(u_1(t), t) : 0 < t \leq \alpha\} \cup \{(u_2(t), t) : 0 < t \leq \alpha\}$ are the radially symmetric solutions of (1) for $\lambda > 0$. $u_1(\lambda)$ is known as the minimal solution and general theory for convex mappings (as in [2]) ensures that there are no solutions at all of (1) for $\lambda > \alpha$, no other solutions for $\lambda = \alpha$ and that $u_1(\lambda)$ is a non-degenerate solution (and stable) for $0 < \lambda < \alpha$. It follows easily from this and the remark after the lemma in Dancer [10] that non-radial solutions cannot bifurcate from $u_1(\lambda)$ for any λ or from $u_2(\lambda)$ for λ near α. It is also proved in [17] that $u_2(\lambda)$ is non-degenerate in the space of radial functions for λ near α. Fix $n \geq 1$. It is also proved in [34] that there exists $\tau_n \in (0, \alpha)$

such that the smallest eigenvalue $\gamma = \gamma_n(\lambda)$ of the linearized equation

$$-\Delta h - \lambda \exp(u_2(\lambda))h = \gamma h \text{ on } D$$
$$h = 0 \text{ on } \partial D \tag{2}$$

with an eigenfunction of the form $\bar{h}(r)\cos n\theta$ satisfies $\gamma_n(\lambda) > 0$ if $\tau_n < \lambda < \alpha$ while $\gamma_n(\lambda) < 0$ if $0 < \lambda < \tau_n$. Moreover τ_n is the only point where (2) has a solution of the form $\bar{h}(r)\cos n\theta$ and $\bar{h}$ is non-negative. It is easy to show (by using the equation for the radial part of an eigenfunction) that $\tau_{n+1} < \tau_n$ for all $n \geq 1$ and $\tau_n \to 0$ as $n \to \infty$. Note that this can easily be done in this case because there are explicit formulae for $u_1(\lambda)$, $u_2(\lambda)$ and $\gamma_n(\lambda)$.

We now define a suitable narrow cone. We define

$$C_0^n(D) = \left\{ u \in C_0(D) : u\left(r, \theta + \frac{2\pi}{n}\right) = u(r,\theta) \right\},$$

$K^n = \{u \in C_0^n(D) : u \geq 0 \text{ in } D,\ u \text{ is even in } \theta,\ u(r,\theta) \text{ is decreasing in } \theta \text{ for}$

$$0 \leq \theta \leq \frac{\pi}{n}, \ a \leq r \leq 1$$

and $K_0^n = K^n \cap C_0^n(D)$. Here $D = \{\underline{x} \in R^2 : a < r < 1\}$ and we are using polar coordinates. It is easy to see that $C_0^n(D)$ is a closed subspace of $C_0(D), K_0^n$ is a cone in $C_0^n(D)$ and $K_0^n - K_0^n$ is dense in $C_0^n(D)$. Moreover, as in [17, Lemma 1] it is not difficult to prove that if $f \in K^n$, the solution of

$$-\Delta u = f \text{ in } D$$
$$u = 0 \text{ on } \partial D$$

belongs to K_0^n. Moreover, it is easy to see that the mapping $u \to \exp u$ maps K_0^n into K^n. Hence we see that the map A defined by $A(u) = (-\Delta)^{-1}(\exp u)$ maps K_0^n into itself (and $C_0^n(D)$ into itself).

Theorem 4. *There is an unbounded connected subset T_n of $K^n \times (0,\infty)$ consisting of non-radially symmetric positive solutions of (1) with $(u_2(\tau_n), \tau_n) \in \bar{T}_n$. Moreover, $T_n \cap T_m = \emptyset$ if $m \neq n$. In particular, for each $\lambda \in (0, \tau_n)$ there is a non-radially symmetric solution of (1) in K^n and there are at least n distinct non-radially symmetric solutions. (Here distinct means that they can not be obtained from each other by the symmetries.)*

Sketch of Proof. Let $C = K_0^n$. We prove that $i_C(A, u_2(\lambda))$ is defined for $\lambda \neq \tau_n$ and that $i_C(A, u_0(\lambda)) = \pm 1$ if $\lambda > \tau_n$ (but is independent of λ for $\lambda > \tau_n$) and is 0 if $\lambda < \tau_n$. (In fact, the first index is -1 but we do not need this.) Since there is a change of the fixed point index at $\lambda = \tau_n$, there must be bifurcation in C at $\lambda = \tau_n$. In fact by rather standard degree and connected set arguments (which go back to Rabinowitz [47]) the existence of the global branch T_n follows. The details appear in [17]. That T_n does not intersect T_m follows easily because the maximum principle implies that solution in T_n are strictly decreasing in θ for $a < r < 1, 0 < \theta < n^{-1}\pi$.

Hence it suffices to establish the index formula. Let $y = u_2(\lambda)$. It is easy to prove that

$$\overline{C}_y = \left\{\phi \in C_0^n(D) : \phi \text{ is decreasing in } \theta \text{ for } 0 \leq \theta \leq \frac{\pi}{n}\right\}$$

and hence that $S_y = \{\phi \in C_0(D) : \phi \text{ is radially symmetric}\}$.

We choose a complement to S_y in $C_0^n(D)$ to be those functions in $C_0(D)$ which have a Fourier series expansion $\sum_{j=1}^{\infty} c_j h_j(r) \cos jn\theta$. It is easy to see that $A'(u_2(\lambda))$ maps S_y and W into themselves (which is basically due to the symmetries or to self-adjointness). By our earlier remarks above and by using Fourier expansions (as above) one finds that

$$h \neq A'(u_2(\lambda))h \text{ if } h \in \overline{C}_y \setminus \{0\} \text{ and if } \lambda \neq \tau_n \text{ and that}$$

$$r(A'(u_2(\lambda))|_W) \begin{cases} < 1 & \text{if } \lambda > \tau_n \\ > 1 & \text{if } \lambda < \tau_n. \end{cases}$$

(In the proof, one needs to use results on the spectrum of ordinary differential operators.) The details appear in [17]. Hence our index results follow from Theorem 1. This completes our sketch of the proof.

Remark. There are a number of variants. One can vary the nonlinearity or use the inner radius of the annulus as a bifurcation parameter. (This is discussed more in [17].)

Lastly, for this section, we sketch a second example which arose in [18]. We consider Ω a smooth domain in R^k and look for positive solutions of the following problem which depend on x_{k+1}

$$\begin{aligned} -\Delta u &= f(u) \text{ in } \hat{D}_n = \Omega \times [0, n] \\ u &= 0 \text{ on } \partial\Omega \times [0, n] \\ \frac{\partial u}{\partial x_{k+1}} &= 0 \text{ if } x_{k+1} = 0 \text{ or } n. \end{aligned} \qquad (3)$$

Here f is C^1 $f(0) \geq 0$, f is increasing and Ω is a smooth bounded domain in R^k. (The increasing assumption is purely for simplicity.)

We also have the reduced problem

$$-\Delta_k u = f(u) \text{ in } \Omega$$
$$u = 0 \text{ on } \partial\Omega. \tag{4}$$

Here Δ_k is the Laplacian on R^k. We assume that (4) has two solutions v_1, v_2 on Ω such that $v_1(x) < v_2(x)$ on Ω and both v_1 and v_2 are stable or neutrally stable for the natural parabolic corresponding to (4) and that there are no neutrally stable solutions of (4) in the order interval $[v_1, v_2]$ between v_1 and v_2 in $C_0(\Omega)$. Then we prove that there is a solution of (3) which strictly decreasing in x_{k+1} if n is large. To prove this, we use the space

$$X = \{u \in C(\Omega \times [0, n]) : u(x', x_{k+1}) = 0 \text{ if } x' \in \partial\Omega\}$$

and $\tilde{K} = \{u \in X : u \geq 0 \text{ on } \Omega \times [0, n] : \text{ is non-increasing in } x_{k+1}\}$.

Here we are writing elements of R^{k+1} as (x', x_{k+1}) with $x' \in R^k$, $x_{k+1} \in R$. It is easy to see that $\tilde{K}$ is a cone in X and $A(u) = (-\Delta)^{-1} f(u)$ maps $\tilde{K}$ into itself (where $(-\Delta)^{-1}$ denotes the inverse of the Laplacian on $\Omega \times [0, n]$ under the boundary conditions in (3)). We consider the closed convex set

$$C = \{p \in \tilde{K} : v_1(x') \leq p(x', x_{k+1}) \leq v_2(x') \text{ on } \Omega \times (0, n)\}.$$

Since v_1 and v_2 are fixed points of A and A is monotone for the usual order on $C(\Omega \times [0, n])$, it is easy to see that A maps C into itself and is completely continuous. Note that we obtain regularity at the corners on $t = n$ by extending solutions to be even in t about $t = n$ and by using a similar device on $t = 0$. Hence $i_C(A, C) = 1$ by the normalization property of the degree. For **simplicity**, we assume that any unstable solution of (3) in $[v_1, v_2]$ is non-degenerate. As in [18], this can be avoided by a careful approximation argument.) We will prove that $i_C(A, v_i) = 1$ for $i = 1, 2$, $i_C(A, v) = 1$ if v is a stable solution of (4) in C and $i_C(A, v) = 0$ if v is an unstable solution of (4) in $[v_1, v_2]$ and n is large. Hence the sum fixed point indices of the solutions (relative to C) of (3) in $[v_1, v_2]$ is $2 + s$ where s is the number of stable solutions of (3) in (v_1, v_2). Since $2 + s > 1$, it follows that there is a solution w of (3) in C which is not a solution of (4). Thus w depends on x_{k+1} and is decreasing in x_{k+1}. Since $\frac{\partial w}{\partial x_{k+1}}$ satisfies a linear elliptic equation, the maximum principle implies that $\frac{\partial w}{\partial x_{k+1}} < 0$ on $\Omega \times (0, n)$ and our claim follows.

Thus it remains to check our index calculations. We only calculate $i_C(A, v)$ where v is a solution of (3) in $[v_1, v_2]$. The other cases are similar, but rather easier. It is easy to see that

$$\overline{C}_v = \{u \in X : u \text{ is decreasing in } x_{k+1}\}$$

and $\overline{S}_v = \{u \in X : u \text{ is independent of } x_{k+1}\}.$

This is similar to the previous example. Moreover, we choose T to be the orthogonal complement in the L^2 sense to S_v in X. As in the previous example, we find that

$$i_C(A, v) = 0$$

if $\alpha_v + n^{-2}\pi^2 < 0$ and $i_C(A, v) = i_{S_v}(A'(v), 0)$ if $\alpha_v + n^{-2}\pi^2 > 0$. Here α_v is the smallest eigenvalue of $-\Delta - f'(v)I$ on Ω (for Dirichlet boundary conditions). Note that we can calculate spectral radii by separating variables and that $\alpha_v + n^{-2}\pi^2 < 0$ if and only if $r(A'(v)|_T) > 1$. Our assumptions ensure that $\alpha_v \neq 0$ for v a solution of (3) in (v_1, v_2). Hence by compactness, α_v is bounded away from zero for solutions of (3) in (v_1, v_2). Note that, by our assumptions, v_1 and v_2 are isolated solutions of (3). Hence we see that if n is large $\alpha_v + n^{-2}\pi^2 > 0$ for every stable solution of (3) and $\alpha_v + n^{-2}\pi^2 < 0$ for every unstable solution of (3) (in (v_1, v_2)). Our index calculations follow if we note that, if v is stable, $r(A'(v)) < 1$, hence $r(A'(v)|_{S_v}) < 1$ and thus $i_{S_v}(A'(v), 0) = 1$.

Remarks

1. This is used in [18] as a preliminary step to proving the existence of solutions of our equation on an infinite strip $\Omega \times R$ which are even in x_{k+1} and strictly decreasing (or increasing) in x_{k+1} for $x_{k+1} > 0$. There are a number of variants of this result in [18].

2. We need n large enough to ensure that $\alpha_v + n^{-2}\pi^2 < 0$ for every unstable solution v of (3) in (v_1, v_2).

3. A variant appears in [18], §4.

§6. Applications of Index Formulae – II

We discuss the application of our index calculations to proving the existence of positive solutions of population models. Many people have studied these problems. Further references can be found in [15] and [21].

We consider the problem

$$
\begin{aligned}
-\Delta u &= u(a - u - cv) &\quad \text{in } \Omega \\
-\Delta v &= v(d - v - eu) &\quad \text{in } \Omega \\
u = v &= 0 &\quad \text{on } \partial\Omega.
\end{aligned}
\tag{5}
$$

This is a Lotka-Volterra model for competing species where there is diffusion. Thus we assume $c, e > 0$. We could consider other boundary conditions and allow the coefficients to have space dependence, but we restrict ourselves to the coonstant coefficient Dirichlet case. Because u and v represent populations it is natural to only look at solutions with u, v nonnegative on Ω. A particular point of interest is the existence of solutions

(u, v) with $u(x) > 0$ in Ω, $v(x) > 0$ in Ω. These are usually called positive solutions.

It follows easily from the maximum principle that the non-negative solutions consists of $(0,0)$ (usually called the trivial solution), $(u, 0)$ with $u(x) > 0$ in Ω or $(0, v)$ with $v(x) > 0$ in Ω (usually called semitrivial solutions) and positive solutions.

If we look for solutions of the form $(u, 0)$, then u must satisfy

$$-\Delta u = u(a - u) \text{ in } \Omega$$
$$u = 0 \text{ on } \partial\Omega. \tag{6}$$

It is well known (see [12]) that this equation has a non-trivial non-negative solution if and only if $a > \lambda_1$. If this holds, there is a unique non-trivial non-negative solution $\bar{u}$ which is positive on Ω. Here λ_1 denotes the first eigenvalue of $-\Delta$ for Dirichlet boundary conditions.

Similarly there is a semitrivial solution of the form $(0, v)$ if and only if $d > \lambda_1$ in which case there is a unique one $(0, \bar{v})$.

If (u, v) is a positive solution of (5) , then

$$-\Delta u \leq u(a - u) \text{ in } \Omega,$$
$$u = 0 \text{ on } \partial\Omega,$$

and hence u is a subsolution of (6). Since a large positive constant is a supersolution of (6), it follows from the method of sub and supersolutions (see Amann [2]) that (6) has a solution u_1 with $u_1 \geq u$. Thus u_1 is positive and, by our comments above, it follows that $a > \lambda_1$ and $u_1 = \bar{u}$. Hence we see that $a > \lambda_1$ is a necessary condition for the existence of a positive solution of (5) and that if (u, v) is a positive solution, then $u \leq \bar{u} \leq a$. (The last inequality is simply proved by looking at where $\bar{u}$ has its maximum and using that $\Delta\bar{u}(\bar{x}) \leq 0$ if $\bar{u}$ has its maximum at $\bar{x}$.) Similarly, $d > \lambda$ is a necessary condition for the existence of a positive solution and, if (u, v) is a positive solution $v \leq \bar{v} \leq d$.

Henceforth we assume that $a, d > \lambda_1$. Let $\lambda_1(\phi)$ denote the smallest eigenvalue of

$$-\Delta w - \phi w = \lambda w \text{ in } \Omega$$
$$w = 0 \text{ on } \partial\Omega.$$

By our assumption, $\lambda_1(a) = \lambda_1 - a < 0$. It is not difficult to prove that there is a $\bar{c} > 0$ such that

$$\lambda_1(a - c\bar{v}) \begin{cases} < 0 & \text{if } c < \bar{c} \\ = 0 & \text{if } c = \bar{c} \\ > 0 & \text{if } c > \bar{c}. \end{cases}$$

One uses that $a - c\bar{v}$ is strictly decreasing on Ω for increasing c and $a - c\bar{v}$ is large negative on most of Ω if c is large. One can either use the variational characterization of eigenvalues or positive operator theory. Similarly there is an $\bar{e} > 0$ such that

$$\lambda_1(d - e\bar{u}) \begin{cases} < 0 & \text{if } e < \bar{e} \\ = 0 & \text{if } e = \bar{e} \\ > 0 & \text{if } e > \bar{e}. \end{cases}$$

Note that it is not difficult to show that $(\bar{u}, 0)$ is stable as a solution of the natural parabolic system corresponding to (5) if $e > \bar{e}$ and is unstable if $e < \bar{e}$. This provides a natural interpretation of $\bar{e}$. An analogous result holds for $(0, \bar{v})$.

We let C denote the cone $K \oplus K$ in $E = C_0(\Omega) \oplus C_0(\Omega)$ where K is the usual cone in $C_0(\Omega)$ and choose $S > 1 + d$ and then choose $M > 0$ such that $a + M - u - cv \geq 0$ and $d + M - v - eu \geq 0$ if $0 \leq u, v \leq S$. Define $A : E \to E$ by

$$A(u, v) = (-\Delta + MI)^{-1}((u(a + M - u - cv),\ v(d + M - v - eu)),$$

where $(-A + MI)^{-1}$ denotes the inverse for Dirichlet boundary conditions and

$$(-\Delta + MI)^{-1}$$

acts on each component. It is easy to see that A is completely continuous, A maps $\overline{\Omega}$ into C where $\Omega = \{(u, v) \in C : \|u\|_\infty, \|u\|_\infty < S\}$, Ω is open in C and the fixed points of A in $\overline{\Omega}$ are then non-negative solutions of (5) (and hence do not belong to $\partial\Omega$, where the boundary is with respect to C). Hence $i_C(A, \Omega)$ is defined. By homotopy invariance of the fixed point index, we see that $i_C(A, \Omega)$ is unchanged if we decrease a and d. In particular, if we choose $a, d < \lambda_1$, our earlier remarks show that the only fixed point of A in $\overline{\Omega}$ is $(0, 0)$. An easy computation shows that

$$A'(0)(h, k) = (-\Delta + MI)^{-1}((a + M)h,\ (d + M)k),$$

and thus $r(A'(0)) < 1$. Hence, by Theorem 1 of §3, $i_C(A, 0) = 1$ if $a, d < \lambda_1$. Hence, by our above comments $i_C(A, \Omega) = 1$ if $a, d > \lambda_1$.

By using our computations above, and similar arguments we see that, if $a, d > \lambda_1$, $r(A'(0)) > 1$, $i_C(A, 0)$ is defined and equals zero. (Here we have to check that

$$A'(0)(h, k) \neq (h, k)$$

if $(h, k) \in C$.)

We can apply either Theorem 1 of §3 or Theorem 2 of §4 (the latter is slightly easier) to prove that

$$i_C(A, (\overline{u}, 0)) = \begin{cases} 1 & \text{if } e > \overline{e} \\ 0 & \text{if } e < \overline{e}, \end{cases}$$

where part of the result is that the index is defined. We use here that $\overline{u}$ has index 1 as a solution of

$$u = (-\Delta + MI)^{-1}(au + Mu - u^2)$$

on K. This is easily proved by Theorem 1 (or see [15]).

Similarly,

$$i_C(A, (0, \overline{v})) = \begin{cases} 1 & \text{if } c > \overline{c} \\ 0 & \text{if } c < \overline{c}. \end{cases}$$

Hence we see that the sum of the indices of the trivial and semitrivial solutions is 0 if $e < \overline{e}$ and $c < \overline{c}$ and it is 2 if $e > \overline{e}$ and $c > \overline{c}$. Since $i_C(A, \Omega) = 1$, we have established the following result.

Theorem 5. *Assume that $a, d > \lambda_1$. Then (5) has a positive solution if $e < \overline{e}$ and $c < \overline{c}$ or if $e > \overline{e}$ and $c > \overline{c}$.*

Remarks.

1. If $e > \overline{e}$ and $c < \overline{c}$ or if $e < \overline{e}$ and $c > \overline{c}$, the method does not prove the existence of a positive solution and indeed there may or may not be a positive solution in this case.

2. The method can also be used for other population problems (for example predator-prey problems as in [13]).

3. The above problem has additional features. If $(u_1, v_1), (u_2, v_2) \in C$, we write $(u_1, v_1) \geq_s (u_2, v_2)$ if $u_1 - u_2, v_2 - v_1 \in K$. Then it is not difficult to prove that A is monotone for this order in the sense that $A(u_1, v_1) \geq_s A(u_2, v_2)$ if $(u_1, v_1) \geq_s (u_2, v_2)$,

$$(u_1, v_1) \in \overline{\Omega} \cap C,$$

$(u_2, v_2) \in \overline{\Omega} \cap C$. Using this, one can prove that there is always a stable positive solution of (1) if $e < \overline{e}, c < \overline{c}$ while there is always an unstable solution if $e > \overline{e}, c > \overline{c}$. This trick does not seem to extend to more than two equations.

We now consider the case of three competing species equations. Thus we consider the system

$$-\Delta u = u(a_1 - u - c_{12}v - c_{13}v) \quad \text{in } \Omega$$
$$-\Delta v = v(a_2 - v - c_{21}u - c_{23}w) \quad \text{in } \Omega$$
$$-\Delta w = w(a_3 - w - c_{31}u - c_{32}v) \quad \text{in } \Omega$$
$$u = v = w = 0 \quad \text{on } \partial\Omega.$$

Here $c_{ij} > 0$. The interest here is again the existence of positive solutions
(that is solutions with $u(x), v(x), w(x) > 0$ on Ω). It is easy to show as
before that $a_i > \lambda_1$, $i = 1, 2, 3$ is a necessary condition for the existence
of a positive solution. This is a much more complex problem. We explain
why. There are now two types of semi trivial solutions, solutions with two
components zero such as $(\overline{u}, 0, 0)$ or solutions with only one component
zero such as $(u, v, 0)$ (where $u(x) > 0$, $v(x) > 0$). The problem is that
the number of solutions of this latter type can vary (as we vary the coefficients). Suppose that there are two positive solutions of the type $(u_1, v_1, 0)$,
$(u_2, v_2, 0)$. Now, by Theorem 2, the contribution of these to the natural
analogue of $i_C(A, \Omega)$ is greatly affected by whether a certain spectral radius is less than or larger than 1. Unfortunately, it may happen that this
spectral radius is larger than 1 for one of the these solutions and less than
1 for the other. Thus these two solutions will tend to contribute very dif-
ferently to the sum of the fixed point indices of the semi-trivial solutions.
This makes the problem rather complicated. This also suggests that the
sum of the indices of the positive solutions need not only be ± 1 or 0 (as
in the two-species case) but could also take large values. This is shown to
be the case in [21]. Another difficulty is that if we try to apply Theorem
2 we frequently have conditions coming from different types of semi-trivial
solutions which are not independent.

Some simple conditions are known for the existence of positive solutions. For example, it is proved in [21] and [25] that there is a positive
solution if

$$\lambda_1(a_1 - c_{12}\overline{v} - c_{13}\overline{w}) < 0$$
$$\lambda_1(a_2 - c_{21}\overline{u} - c_{23}\overline{w}) < 0$$
$$\text{and} \quad \lambda_1(a_3 - c_{31}\overline{u} - c_{32}\overline{v}) < 0.$$

Here $(\overline{u}, 0, 0)$, $(0, \overline{v}, 0)$, $(0, 0, \overline{w})$ denote the three semi-trivial solutions with
two components vanishing identically. This problem of the existence of
positive solutions is investigated much further in [21] but it seems very
difficult to say exactly when there is a positive solution. The particular
result above is generalized to systems of n equations in [20]. This last
paper also considers the case where the coefficients depend periodically on
time.

§7. Some Global Branches

In this section we discuss some global results for positive mappings which
depend on monotonicity. These illustrate that monotonicity and the conse-
quent invariance of certain order intervals have more effect on the structure
of solutions than is at first apparent. This also relates to §2 of Part II.

We assume that K is a cone in X and $A : X \times [0, \infty) \to X$ is completely continuous and monotone in the sense that $A(x, \lambda) \geq A(y, \mu)$ if $x \geq y$ and $\lambda \geq \mu$. We strengthen the complete continuity by requiring that $A([x_1, x_2] \times [0, r])$ has compact closure for each $r > 0$. {There is no additional restriction if the cone K is **normal** in the sense that there is an $M > 0$ such that $\|x\| \leq M \|y\|$ if $0 \leq x \leq y$. Many of the most used cones are normal.}

We first consider two very simple results. Results of this type seem to have been known to a number of people (see [14], [24]). Let

$$D = \{(x, \lambda) : x \in X, \ \lambda \geq 0, \ x = A(x, \lambda)\}$$

Proposition 1. *Assume that the basic conditions above hold.*

(i) *Assume that (x_1, λ_1) and (x_2, λ_2) are solutions of $x = A(x, \lambda)$ with $x_2 \geq x_1$, $\lambda_2 \geq \lambda_1$ and*

$$i_{[x_1, x_2]}(A(\ , \lambda_1), x_1) \neq 0$$

(where part of the assumption is that this index is defined). Then the component of

$$\{(x, \lambda) : x = A(x, \lambda) : x_1 \leq x \leq x_2, \ \lambda_1 \leq \lambda \leq \lambda_2\}$$

containing (x_1, λ_1) contains a point (x, λ_2) with $x_1 \leq x \leq x_2$ or a point (x_3, λ_1) where $x_1 < x_3 \leq x_2$.

(ii) *Assume that (x_1, λ_1) is a solution of $x = A(x, \lambda)$ where $0 \leq x_1$ and $\lambda_1 > 0$, such that $i_{[x_1, \infty)}(A(\ , \lambda_1), x_1) \neq 0$. Here $[x_1, \infty) = \{x \in X : x \geq x_1\}$. Then the component of $\{(x, \lambda) : x = A(x, \lambda) : x \geq x_1, \ \lambda \geq \lambda_1\}$ containing (x_1, λ_1) is unbounded or contains (x_2, λ_1) where $x_2 > x_1$.*

Proof. (i) By the monotonicity, one easily sees that $A(\ , \lambda)$ maps $[x_1, x_2]$ into itself for each $\lambda \in [\lambda_1, \lambda_2]$. The result then follows by a standard Rabinowitz type continuation argument applied on the space $[x_1, x_2]$ rather than X. We sketch the idea rather briefly. If the result were false, we could use the Whyburn separation lemma [56] to write the compact set

$$T = \{(x, \lambda) : x = A(x, \lambda) : x_1 \leq x \leq x_2, \ \lambda_1 \leq \lambda \leq \lambda_2\} = D_1 \cup D_2,$$

where D_1 and D_2 are disjoint $(x_1, \lambda_1) \in D_1$ and all solutions (x, λ) in T with $\lambda = \lambda_2$ or $\lambda = \lambda_1$ and $x > x_1$ lie in D_2. We then choose an open subset θ of $[x_1, x_2] \times [\lambda_1, \lambda_2]$ such that $D_1 \subseteq \theta$ and $\bar{\theta}_2 \cap D_2 = \emptyset$. By homotopy invariance, $i_{[x_1, x_2]}(A(\ , \lambda), \theta_\lambda)$ is independent of λ for $\lambda \in [\lambda_1, \lambda_2]$ where $\theta_\lambda = \{x \in [x_1, x_2] : (x, \lambda) \in \theta\}$. Note that θ_λ is relatively open in $[x_1, x_2]$ and that our choice of θ ensures that $x \neq A(x, \lambda)$ if $x \in \partial \theta_\lambda$. Here we use the more general version of homotopy invariance mentioned in the remarks in §1. If $\lambda = \lambda_2$, this fixed point index is zero since the equation $x = A(x, \lambda_2)$ has no solutions in θ_{λ_2}. On the other hand the only solution of $x = A(x, \lambda_1)$ in θ_{λ_1} is x_1 (by the choice of θ) and hence,

$$i_{[x_1, x_2]}(A(\ , \lambda_1), \theta_{\lambda_1}) = i_{[x_1, x_2]}(A(\ , \lambda_1), x_1) \neq 0,$$

and we have a contradiction. Thus the result is proved.

(ii) The proof of this is similar except we use $C = [x_1, \infty)$.

Remarks.

1. If x_1 is the only solution of $x = A(x, \lambda_1)$ in $[x_1, x_2]$, then the index criteria is automatically satisfied because $A(\ , \lambda_1)$ maps $[x_1, x_2]$ into itself (as seen above) and

$$i_{[x_1, x_2]}(A(\ , \lambda_1), x_1) = i_{[x_1, x_2]}(A(\ , \lambda_1), [x_1, x_2])$$
$$\text{(since } x_1 \text{ is the only solution for } \lambda = \lambda_1)$$
$$= 1 \quad \text{(by the normalization property.)}$$

2. We do not really need the monotonicity of $A(\ , \lambda)$ in (1) but rather that it maps $[x_1, x_2]$ into itself for $\lambda \in [\lambda_1, \lambda_2]$. (For example, it suffices to assume $A(\ , \lambda)$ is monotone on $[x_1, \infty)$.)

3. Note that the component in (ii) is always unbounded if x_1 is a maximal solution of $x = A(x, \lambda_1)$ (and the index criteria is satisfied.) In this case, if $x_2 > x_1$ is a fixed point of $A(\ , \lambda_2)$ for $\lambda_2 > \lambda_1$, then, as above $i_{[x_1, x_2]}(A(\ , \lambda_1), x_1) = 1$. It seems to be very frequently the case (and is trivially true if int $K \neq \emptyset$ and $x_2 - x_1 \in$ int K by the commutativity theorem) that $i_{[x_1, x_2]}(A(\ , \lambda_1), x_1) = i_{[x_1, \infty)}(A(\ , \lambda_1), x_1)$. Hence we see that the condition in (ii) holds frequently. Some other conditions guaranteeing that this condition holds can be found in [16].

4. The stronger compactness condition is not needed for (ii) of the proposition.

5. Many conditions implying that the indices in the proposition are non-zero are discussed in [14] and [16].

If we make stronger assumptions on A, we can obtain more information on the solutions. Assume that int $K \neq \emptyset$, that A is strongly increasing in λ that is

$$A(x, \lambda) - A(x, \mu) \in \text{int } K,$$

if $\lambda > \mu \geq 0$ and $x \in [x_1, \infty))$ and A is strongly increasing in x for $\lambda > \lambda_1$ (that is $A(y, \lambda) - A(x, \lambda) \in$ int K if $y - x \in K \setminus \{0\}$, $x \in [x_1, \infty)$ and $\lambda > \lambda_1$. We also assume that x_2 is the minimal solution of $x = A(x, \lambda_2)$ in $[x_1, \infty)$. In this case, it is easy to show under the assumptions of (ii), that any connected subset C of $\tilde{T} = \{(x, \lambda) \in [x_1, \infty) \times [\lambda_1, \lambda_2] : x = A(x, \lambda)\}$ containing (x_1, λ_1) must lie in $[x_1, x_2] \times [\lambda_1, \lambda_2]$. This follows because the strong positivity implies that, if $(x, \lambda) \in \tilde{T}$, $x \leq x_2$ and $\lambda < \lambda_2$ then $x \ll x_2$ (where $a \ll b$ means $b - a \in \text{int } K$). It follows easily that $\overline{C} \cap ([x_1, x_2]) \times [\lambda_1, \lambda_2])$ is open and closed in $\overline{C}$ and hence must equal $\overline{C}$. (If $\text{int } K = \phi$, a similar result for certain differentiable demi-interior operators can be proved by the methods of [14]. Here a demi-interior operator is defined in [14]). It follows easily by connectedness that, if $\lambda_1 < \lambda_3 < \lambda_2$, then any connected subset of $\tilde{T}$ containing (x_1, λ_1) and containing points with $\lambda \geq \lambda_3$ must contain $(\hat{x}(\lambda_3), \lambda_3)$ where $\hat{x}(\lambda_3)$ is the minimal solution for $\lambda = \lambda_3$ in $[x_1, x_2]$. (A standard iteration as in [2] shows that minimal solutions exist.) Thus any connected subset of $\tilde{T}$ containing (x_1, λ_1) and a point with $\lambda = \lambda_2$ must include all the minimal solutions $\{(x_1(\lambda), \lambda) : \lambda_1 \leq \lambda \leq \lambda_2\}$. This gives extra information on the structure of the solutions. This includes many of the results in Clément [6] and Yihong Du [24] . Most of the other results in these two papers can be easily deduced from these ideas.

We consider another application of these ideas. **We assume the conditions on A of the previous paragraph (except that we require the strong positivity for $x > 0$ and $\lambda > 0$), assume that $A(0,0) \geq 0$ and assume that A is C^1 on $\text{int } D \times (0, \infty)$. We assume that $D \cap (K \times [0, \infty))$ contains a bounded component $\hat{D}$. Let $\hat{\lambda} = \inf\{\lambda : (x, \lambda) \in \hat{D}\}$ and then choose $\hat{x}$ maximal among $\{x : (x, \hat{\lambda}) \in \hat{D}\}$. We also assume $x \neq A(x, 0)$ if $x \in K \setminus \{0\}$. Thus if $\hat{\lambda} = 0$, $(0, 0) \in \hat{D}$** and we would contradict Theorem 1 of Dancer [9] . Hence $\hat{\lambda} > 0$. Thus A is differentiable at $(\hat{x}, \hat{\lambda})$. (By the strong positivity, $\hat{x} \gg 0$.) By the implicit function theorem and the strong positivity, $I - A_1'(\hat{x}, \hat{\lambda})$ is not invertible. We also assume **that whenever $r(A_1'(\hat{x}, \hat{\lambda})) = 1$, $I - A_1'(\hat{x}, \hat{\lambda})$ has a one-dimensional kernel spanned by $h \in \text{int } K$ and $A_2'(\hat{x}, \hat{\lambda}) \notin R(I - A_1'(\hat{x}, \hat{\lambda}))$. (In practice, this is the key assumption to verify.) Then the conclusion is that $r(A_1'(\hat{x}, \hat{\lambda})) > 1$.** Before proving this result, we note that these assumptions frequently hold when $\text{int } K \neq \phi$ and A is strongly increasing in x. For example, variants of the Krein Rutman theorem in [48] easily imply that the last condition holds if $A_1'(\hat{x}, \hat{\lambda})$ is strongly positive (that is, $A_1'(\hat{x}, \hat{\lambda})(K \setminus \{0\}) \subseteq \text{int } K$ and if $A_2'(\hat{x}, \hat{\lambda}) \neq 0$.)

To prove the main result in the previous paragraph, we first note that $r(A_1'(\hat{x}, \hat{\lambda})) < 1$ is impossible since this implies $I - A_1'(\hat{x}, \hat{\lambda})$ invertible. Hence it suffices to prove the result when $r(A_1'(\hat{x}, \hat{\lambda})) = 1$. In this case, our assumptions ensure that the conditions of a result of Amann [3, Theorem 2.1] are satisfied. This ensures that the solutions near $(\hat{x}, \hat{\lambda})$ form a smooth curve parametrized by α (necessarily part of $\hat{D}$ by connectedness) and are of the form $(w(\alpha), \phi(\alpha))$ where $\phi(0) = \hat{\lambda}$. and $w(\alpha) = \hat{x} + \alpha \hat{h} + o(\alpha)$. By the definition of $\hat{\lambda}$, $\phi(\alpha) \geq \hat{\lambda}$ for small α. Since $\hat{h} \in \text{int } K$ (by assumption), we see that $\hat{x} + \alpha h + o(\alpha) \geq \hat{x}$ for small positive α. Since $\hat{x}$ is a maximal solution of $x = A(x, \hat{\lambda})$ in $\hat{D}$, it follows that $\phi(\alpha) > \hat{\lambda}$ for small positive α. Since $\hat{x} + \alpha h + o(\alpha) \notin [\hat{x}, \infty)$ for small negative α (since $-\hat{h} \notin K$), we see that $i_{[\hat{x}, \infty)}(A(\ , \hat{\lambda}), \hat{x})$ is defined. By choosing a small positive α, we find $x_2 > \hat{x}$ and $\lambda_2 > \hat{\lambda}$ such that $(x_2, \lambda_2) \in D$. If we know the equation $x = A(x, \hat{\lambda})$ has $\hat{x}$ as its only solution in $[\hat{x}, x_2]$, we can argue as before to prove that $i_{[\hat{x}, x_2]}(A(\ , \hat{\lambda}), \hat{x}) = 1$ and hence, since $\text{int } K \neq \emptyset$, $i_{[\hat{x}, \infty)}(A(\ , \hat{\lambda}), \hat{x}) = 1$. We can then apply Proposition 1(ii) to deduce there is a connected set W in

$$\{(x, \lambda) : x = A(x, \lambda) : x \geq \hat{x}, \ \lambda \geq \hat{\lambda}\}$$

containing $(\hat{x}, \hat{\lambda})$ which is unbounded or contains another point $(\tilde{x}, \hat{\lambda})$ with $\tilde{x} > \hat{x}$. Since W is connected, $W \subseteq \hat{D}$. Since $\hat{D}$ is bounded, the second alternative must hold. However, this contradicts the maximality of $\hat{x}$ as a solution of $x = A(x, \hat{\lambda})$ in $\hat{D}$. Hence the proof of our result reduces to proving that if we choose (x_2, λ_2) carefully, then $\hat{x}$ is the only solution of $x = A(x, \hat{\lambda})$ in $[\hat{x}, x_2]$. Since $\hat{h} \in \text{int } K$, it follows easily from the formula for $w(\alpha)$ that we can choose a decreasing sequence $\{\alpha_i\}_{i=1}^{\infty}$ in R such that $\{w(\alpha_i)\}_{i=1}^{\infty}$ is decreasing in K and $\alpha_i \to 0$ as $i \to \infty$ (and thus $w(\alpha_i) \to \hat{x}$ as $\alpha_i \to 0$). By our assumption,

$$Z = \{(x, \lambda) : x = A(x, \lambda) : \hat{x} \leq x \leq w(\alpha_1), \ \hat{\lambda} \leq \lambda \leq \phi(\alpha_1)\}$$

is compact. It the result is false, there exist z_i such that $\hat{x} < z_i \leq w(\alpha_i)$ and $(z_i, \hat{\lambda}) \in D$. Since $w(\alpha_i) \leq w(\alpha_1)$, $z_i \in Z$ and hence, by the compactness of Z, $\{z_i\}$ has a convergent subsequence converging to z. Now if $j > i$, $z_j \leq w(\alpha_j) \leq w(\alpha_i)$. Thus, since order intervals are closed, $z \in [\hat{x}, w(\alpha_i)]$. Since $w(\alpha_i) \to 0$ as $i \to \infty$, this implies that $z = \hat{x}$, that is $z_i \to \hat{x}$ as $i \to \infty$. However, this is impossible because every solution close to $(\hat{x}, \hat{\lambda})$ of $x = A(x, \hat{\lambda})$ lies in $\hat{D}$ and $\hat{x}$ is a maximal solution of $x = A(x, \hat{\lambda})$. Hence our claim is true and the result is true. (The proof becomes much simpler if the cone is normal.)

In many applications (for example if $A_1'(\hat{x}, \hat{\lambda})$ is strongly positive) the conditions that $r(A_1'(\hat{x}, \hat{\lambda})) > 1$ and $I - A_1'(\hat{x}, \hat{\lambda})$ is not invertible imply that there is a solution h of $h = A_1'(\hat{x}, \hat{\lambda})h$ which is not in $K \cup (-K)$. Note that one can use the techniques of [14] and [16] to prove some results of this type if $\text{int } K = \emptyset$ (for restricted classes of mappings).

Summarizing the last result, we see that, under reasonable hypotheses, if D has bounded component $\hat{D}$ and if $\hat{\lambda} = \inf\{\lambda : (x, \lambda) \in \hat{D}\}$ then there is an $\hat{x} \in K$ and $h \notin K \cup (-K)$ such that $\hat{x} = A(\hat{x}, \hat{\lambda})$ and $h = A_1'(\hat{x}, \hat{\lambda})h$.

This provides a useful tool for precluding bounded components. It appears to be new. For example, if the nonlinear terms are not too large, one may be able to use comparison results for linear operators to show that if $x = A(x, \lambda)$ if $x \in K$ and if $h = A_1'(x, \lambda)h$, then $h \in K$. It is easy to give examples for differential equations. Secondly, one can sometimes establish results of these types for perturbed problems, even if the perturbations are not regular. For example (and we omit the tedious details) one can prove the connectedness claim in the third example in §5 of [18] that on long domains (that is, domains $\Omega \times (-L, L)$ where Ω fixed in R^{n-1} and L is large). It is also sometimes useful to help prevent bounded components in some domain variation problems as in some earlier work of the author.

Note that the above techniques give some information even about un-bounded components of D. One can frequently obtain some information about unbounded components because it is frequently easier to obtain in-formation about the behaviour of solutions when λ is large or $\|x\|$ is large (cp. [19]). Note that one can use similar techniques if $\hat{D}$ is a bounded component of $D \cap (K \times [0, \infty))$ and if we let $\tilde{\lambda} = \sup\{\lambda : (x, \lambda) \in \hat{D}\}$. We then define $\tilde{x}$ to be a minimal element of $\{x : (x, \lambda) \in \hat{D}\}$. (Note that this frequently exists in many cases even when $\hat{D}$ is not bounded.) Under similar assumptions on A at $(\tilde{x}, \tilde{\lambda})$ to before, we can use $[0, \tilde{x}] \times [0, \tilde{\lambda}]$ to prove that $\hat{D}$ contains $(0, 0)$ (and hence is unbounded by Theorem 1 in [9]) if $r(A_1'(\tilde{x}, \tilde{\lambda})) = 1$. This variant is more likely to give information on unbounded components of $D \cap (K \times [0, \infty))$.

Part II. Monotone Dynamical Systems

In this part we discuss monotone dynamical systems. We consider both the continuous and the discrete cases, with the main emphasis on the latter. Indeed, one of the main points of interest is the similarities and differences between the continuous and discrete cases (and also how they differ from general dynamical systems).

In §1, we introduce monotone dynamical systems while, in §2, we dis-cuss some results on connecting orbits and related results provable by degree theory. Finally, in §3, we discuss the asymptotic behaviour of monotone dynamical systems (especially generic behaviour).

§1. Preliminaries

In this section, we introduce the basic definition and give some examples.

Assume, as before, that X is a Banach space, K is a cone in X and $T : X \to X$ is monotone and continuous. We think of T (or more precisely

its iterates) as a discrete dynamical system where here what tends to be of
interest is the behaviour of large iterates of T. One of the main reasons for
the interest in monotone dynamical systems is that they arise as fixed time
maps of differential equations where the coefficients depend periodically
on time. We explain this more below. (They also arise as Poincaré maps
for autonomous differential equations but the ones which arise in this way
seldom seem to be monotone.)

A continuous monotone dynamical system is a map $T : X \times [0, \infty) \to X$
and a cone K in X such that T is continuous, $T(\ ,0) = I$, $T(x, t + s) =$
$T(T(x, s), t)$ for $t, s \geq 0$, $x \in X$ and $T(x, s) \geq T(y, s)$ if $x, y \in X$, $x \geq$
y, $s \geq 0$. (Note that the cone K occurs implicitly because $x \geq y$ means
that $x - y \in K$.) We could often allow more general maps T which are not
defined for all $t \geq 0$ but we restrict ourselves to this case. An autonomous
ordinary differential equation (or a parabolic partial differential equation)
nearly always generates a continuous dynamical system provided solutions
do not blow up in finite time. Thus the only question to really worry about
is the monotonicity.

Note also that, if $T(\ )$ is a monotone continuous dynamical system and
if $r > 0$, then $T(\ ,r)$ is a monotone discrete dynamical system. Note that
$(T(\ ,r))^k = T(\ ,kr)$. This is a very useful observation for deducing properties of continuous dynamical systems from the corresponding properties
of discrete dynamical systems (by letting r tend to zero).

To complete this section, we briefly discuss some examples. We assume
that $f : R^m \to R^m$ is C^1 and has only linear growth as $\|x\| \to \infty$. Define
$T(x, t)$ to be the value of the solution of

$$x'(t) = f(x(t))$$
$$x(0) = x$$

at time t. (Note that our growth assumption on f ensures that $T(x, t)$
is defined.) Then it is easy to see that $T(x, t)$ is a continuous dynamical
system. If, in addition,

$$\frac{\partial f_i}{\partial x_j}(x) \geq 0 \text{ whenever } i \neq j \text{ and } x \in R^m \tag{7}$$

(where f_i denotes the components of f), then it follows easily from a theorem of Kamke (see Coppel [7]) that $T(x, t)$ is a monotone continuous
dynamical system for the usual cone in R^m. If the condition (7) only holds
on part of R^m, the system may be monotone on part of R^m.

We naturally obtain discrete monotone dynamical systems from a variant of the above construction. We assume that $f : R^m \times R \to R^m$ is C^1 in
x (with the derivative depending continuously on x and t), f is continuous

and is ω-periodic in t and f grows at most linearly in x. As before, we define $T(x,t)$ to be the solution of the differential equation

$$x'(t) = f(x(t), t)$$
$$x(0) = x.$$

Then $\tilde{T} \equiv T(\ , \omega)$ define a discrete dynamical system. The interest in this discrete dynamical system comes from the easily proved result that $(\tilde{T})^n = T(\ , n\omega)$. In fact, it turns out that the behaviour of $\tilde{T}$ largely determines the dynamics of the original system of equations. This is a standard way for discrete dynamical systems to arise. Once again, it turns out much as before that the discrete dynamical system is monotone if $\frac{\partial f_i}{\partial x_j}(x, t) \geq 0$ on $R^n \times [0, \omega]$ for $i \neq j$.

We also obtain continuous dynamical systems from parabolic equations. We assume $\Omega \subseteq R^k$ is a bounded domain with smooth boundary and $f : \overline{\Omega} \times R \to R$ is smooth and grows at most linearly in the last variable. Then the time map for the system

$$\frac{\partial u}{\partial t} = \Delta u + f(x, u) \ \text{ on } \Omega$$
$$u(x, t) = 0 \ \text{ if } \ x \in \partial\Omega$$
$$u(x, 0) = u_0(x)$$

define a continuous dynamical system on a number of spaces (for example L^p spaces or the corresponding fractional power spaces in the sense of Henry [27], Ch. 1.4). See [27]. It follows easily from the parabolic maximum principle [46] that it defines a monotone continuous dynamical system. Once again, if f also depends ω periodically on t, we can obtain a discrete dynamical system (which is also monotone) by a similar construction to before. Here we have to be careful of the choice of spaces if we want a cone with non-empty interior or a strongly increasing map. This is discussed in [28]. (Sometimes it is convenient to use Holder spaces rather than L^p spaces.)

We could generalize this approach to systems of parabolic equations. Unfortunately, most of the naturally occurring systems appear not be monotone. However the natural parabolic system for two competing species (but **not** for more than two species) is monotone on the subset $\{(u, v) \in C_0(\Omega) \oplus C_0(\Omega) : u, v \geq 0 \text{ on } \Omega\}$ if we use the order mentioned in §6 of Part I.

§2. Connecting Orbits and Related Results

In this section, we show that for monotone dynamical systems certain stationary points must be connected by orbits of the dynamical system. While, it only applies to certain stationary points, it gives a very quick and easy to check method when it applies. We also show degree ideas can prove a number of other results for monotone dynamical systems.

To be more precise, a stationary point of a discrete dynamical system T is a fixed point of T while a stationary point of a continuous dynamical system is a point x_0 such that $T(x_0, t) = x_0$ for all $t \geq 0$.

A connecting orbit for a discrete dynamical system T is a sequence $\{x_n\}_{n=-\infty}^{\infty}$ and two stationary points x^1, x^2 such that $x_n \to x^1$ as $n \to -\infty$, $x_n \to x^2$ as $n \to \infty$ and $x_{n+1} = T(x_n)$ for all $n \in Z$. Note that we have to be a little careful since we do not assume T is a homeomorphism (or is even $1-1$). We say the connecting orbit is monotone if $x_{n+1} \geq x_n$ for all n or if $x_{n+1} \leq x_n$ for all n. Similarly a connecting orbit for a continuous dynamical system $T(\ ,t)$ is a map $x : R \to X$ and two stationary points x^1 and x^2 of $T(\ ,t)$ such that $T(x(t), s) = x(t+s)$ for $s \geq 0$, $t \in R$, $x(t) \to x^1$ as $t \to -\infty$ and $x(t) \to x_2$ as $t \to \infty$. The connecting orbit is said to be monotone if the map $t \to x(t)$ is always increasing or always decreasing.

For a discrete monotone dynamical system, x_0 is said to be a strict subsolution if $T(x_0) > x_0$ (and hence $T^n(x_0) > x_0$ by monotonicity) and a strict supersolution if $T(x_0) < x_0$. x_0 is said to be a subsolution if it is a strict subsolution or a solution. A supersolution is defined analogously. In the continuous case, there are analogous definitions. For example, x_0 is a strict subsolution if $T(x_0, t) > x_0$ for $t > 0$.

Theorem 6. ([22]) *Assume that x_1, x_2 are stationary points of a monotone dynamical system T, $x_1 < x_2$ and there are no stationary points in (x_1, x_2). Finally, in the discrete case, assume $\overline{T([x_1, x_2])}$ is compact while, in the continuous case, assume that, for each $\varepsilon > 0$, $\overline{T([x_1, x_2] \times (\varepsilon^{-1}, \varepsilon))}$ is compact. Then there is a connecting orbit joining x_1 and x_2. Moreover, this orbit is monotone.*

Sketch of Proof. We discuss the discrete case. The key point is to prove that either there exist strict subsolutions arbitrarily close to x_1 or strict supersolutions arbitrarily close to x_2. Once we know this, we note that, if x_3 is a strict subsolution in (x_1, x_2), then $\{T^n(x_3)\}_{n=1}^{\infty}$ is increasing in n and lies in $(x_1, x_2]$ by the monotonicity. Hence, by compactness, the sequence must converge to a stationary point $z \in (x_1, x_2]$. By our assumption on stationary points, $z = x_2$. Hence we have an orbit starting close to x_1 and increasing to x_2. We then use a careful diagonalization argument.

The details of the diagonalization argument appear in [22]. The case of supersolutions is similar.

To prove the result on sub and supersolutions, we prove that either $i_{[x_1,x_2]}(T, x_1) = 1$ or there exist strict subsolutions arbitrarily close to x_1. We can use a similar argument to prove that either $i_{[x_1,x_2]}(T, x_2) = 1$ or there exist strict supersolutions arbitrarily close to x_2. Then

$$1 = i_{[x_1,x_2]}(T, [x_1, x_2])$$

(by the normalization property of the degree)

$$= i_{[x_1,x_2]}(T, x_1) + i_{[x_1,x_2]}(T, x_2)$$

(by our assumption on stationary points),

and hence it is impossible for both indices to be 1. Thus our claim follows.

To prove our claim on $i_{[x_1,x_2]}(T, x_1)$, note that we can use a change of origin to ensure $x_1 = 0$. We then use the homotopy $H(x, t) = tT(x)$ which is easily seen to map $[0, x_2]$ into itself if $0 \le t \le 1$. If $x = H(x, t)$ where $x > 0$ and $0 < t < 1$, then $x = tT(x) < T(x)$. Hence, either there exist strict subsolutions close to zero, or $i_{[0,x_2]}(tT, B_\varepsilon(0) \cap [0, x_2])$ is defined for small ε. Here, as before, $B_\varepsilon(0)$ denotes the open ball of centre 0 and radius ε. In the latter case, homotopy invariance implies that

$$i_{[0,x_2]}(T, 0) = i_{[0,x_2]}(0, 0)$$

$$= i_{[0,x_2]}(0, [0, x_2])$$

(since 0 is the only fixed point of the zero map)

$$= 1 \quad \text{(by the normalization property)}.$$

This proves our claim.

The continuous case is proved by applying the discrete case to the map $T(\ , \frac{1}{n})$ for n large and by then using a careful limit argument. The details appear in [22].

Remarks. Matano [36] first proved a slightly weaker result in the continuous case under much stronger assumptions. As in [22] or [28], we see that our compactness assumptions are satisfied for a large class of parabolic problems in suitable fractional power spaces.

Our theory, and in particular the last theorem, is still valid if K is a closed cone in a metrizable space X and $[x_1, x_2]$ is complete. (Here we use the remarks on p. 7 of [37] to ensure $[x_1, x_2]$ is an ANR(M).) In particular Golé [26] uses this remark to obtain results for twist maps. (These occur in some Hamiltonian systems.)

If T is strongly increasing, there are some elementary cases where no connecting orbit can exist. If x_1, x_2, x_3 are fixed points of T with $x_1 < x_2 <$

x_3 (and thus $x_1 << x_2 << x_3$ by strong positivity), then there can be no connecting orbit joining x_1 and x_3 (or x_3 and x_1). This follows because if a connecting orbit starts from x_1, then by strong positivity, it will contain points z less than x_2. By monotonicity, it follows that the remainder of the orbit ($T^n z$ for $n \geq 1$) lies below x_2 and hence the orbit can not approach x_3. (A similar result is valid in the continuous case.) In particular, it follows that, for strongly positive maps, the only monotone connecting orbits are the ones in Theorem 6.

Before stating a corollary, we need some notation. A stationary point x_0 is stable from above for the discrete dynamical system T if, given $\varepsilon > 0$, there is a $\delta > 0$ such that if $z \geq x_0$ and $\|z - x_0\| \leq \delta$, then $\|T^n z - x_0\| \leq \varepsilon$ for all $n \geq 1$. Otherwise, x_0 is said to be unstable from above. Similarly, we can define stability and instability from below and the corresponding notions for a continuous dynamical system. Note that, for special classes of discrete monotone maps, these conditions are discussed extensively in [16]. Similarly, we can define the notion of a stationary point being isolated from above or below. (Here we mean isolated among the stationary points.)

Corollary. *Assume that x_0, x_1 are stationary points of T (or $T(, t)$) where T is a monotone dynamical system. Assume that $x_0 < x_1$, x_0 is stable from above and isolated from above, x_1 is stable from below and isolated from below and $T[x_0, x_1]$ is compact ($\overline{T([x_0, x_1] \times [\varepsilon^{-1}, \varepsilon])}$ is compact for every $\varepsilon > 0$ in the continuous case). Then there is an unstable stationary point in (x_0, x_1).*

Proof. Suppose the result is false. Now, by compactness, the set of stationary points is compact. Since x_0 is isolated from above, it now follows from a simple Zorn's lemma argument (as in [22]) that there is a stationary point x_2 in $(x_0, x_1]$ which is minimal in the order. We now apply Theorem 6 to $[x_0, x_2]$ and find there is a connecting orbit joining x_0 and x_2. Since x_0 is stable from above, it must join x_2 to x_0. This orbit starts arbitrarily close to x_2 and eventually approaches x_0. Hence x_2 must be unstable, as required.

Thus between stable ordered solutions there is an unstable solution. Note that we need some isolatedness type condition. For example, we must exclude the possibility of an ordered connected set of stable solutions joining x_0 and x_1. An examination of the proof shows that we need only assume the isolatedness of one of x_0 and x_1.

There naturally arises the question of whether there is always a stable solution between two suitable ordered unstable solutions. Simple examples show that this is not true except under additional assumptions. However, it is not clear what is the best result in this case. However one can prove the following.

Theorem 7. *Assume that* $intK \neq \emptyset$, T *is a monotone dynamical system,* x_0 *is a stationary point of* T *or a strict subsolution,* x_1 *is a stationary point of* T *or a strict supersolution,* $x_0 < x_1$, T *is strongly increasing on* $[x_0, x_1]$ *(strongly increasing for each* $t > 0$ *in the continuous case),* x_0 *is* <u>*unstable from above*</u> *if it is a stationary point,* x_1 *is* <u>*unstable from below*</u> *if it is a stationary point and* $\overline{T([x_0, x_1])}$ *is compact* $(\overline{T([x_0, x_1] \times [\varepsilon^{-1}, \varepsilon])}$ *is compact for every* $\varepsilon > 0$ *in the continuous case). Then there is a stable stationary point in* (x_0, x_1).

Sketch of Proof. We assume **purely for simplicity** that the cone is normal. Assume x_0 and x_1 are stationary points. By a result in [16], the assumption that x_0 is unstable from above ensures that there exist strict supersolutions x_3 arbitrarily close to x_0 with $x_3 > x_0$. Similarly, there exist strict subsolutions x_4 arbitrary close to x_0 with $x_4 < x_0$. Since T is strongly increasing, $x_1 - x_0 \in \text{int } K$. Thus we can ensure $x_3 < x_4$ by choosing x_3 close to x_0 and x_4 close to x_1. Since we can use similar arguments in the other cases, we see that we can eventually reduce to the case where x_0 is a strict subsolution and x_1 is a strict supersolution.

It is easy to see that $\{T^n x_1\}$ is increasing in n and, by compactness, the sequence converges to a fixed point y_1 of T. By the strongly increasing property, one easily sees that y_1 is stable from below. By a Zorn's lemma argument and compactness (see [22]), one can prove that $z_1 = \sup\{x \in F :$ x is stable from below$\}$ exists, where F is the set of stationary points in $[x_0, x_1]$. Moreover $z_1 \in F$ and is stable from below. On the other hand, we define z_2 by $z_2 = \inf\{x \in F : x \geq z_1, x$ is stable from above$\}$. One can show, much as before, that z_2 exists, $z_2 \in F$ and z_2 is stable from above. If $z_1 = z_2$, it is easy to see that we are finished. Otherwise one easily proves that $[z_1, z_2]$ contains no strict sub and supersolutions, that $\{z \in F : z_1 \leq z \leq z_2\}$ is a totally connected set and that each element of this set is stable. (The details appear in [22], but Theorem 6 plays a key role.) This proves the required result.

As in [16], one can prove a variant for some differentiable maps when $\text{int } K = \emptyset$. It seems that Krasnosel'skii also obtained results of this type but I do not know the reference.

It is proved in [22] by using some of these ideas that, if T is strongly increasing, if $[x_1, x_2]$ is invariant if $\overline{T([x_1, x_2])}$ is compact and if all fixed points of T in $[x_1, x_2]$ are stable, then $T^n x$ converges to a stationary point as $n \to \infty$ for each x in $[x_1.x_2]$. (There is an analogue for the continuous case.) This generalizes a result of Takac [52].

Lastly, for this section, we show that the natural looking assumption that each forward orbit of monotone dynamical system is bounded is more restrictive than one might at first suppose. This assumption is used in Smith and Thieme [50] and elsewhere. For simplicity, we only consider the discrete case. We also need to assume for the proof that the underlying space X is a Banach lattice though we **suspect** that this assumption can be removed.

More formally, we assume that X is a Banach lattice and T is a monotone completely continuous dynamical system on X such that $\{T^n x_0\}_{n=1}^\infty$ is bounded for each x_0 in X. Moreover, we assume that F, the set of fixed points of T, is bounded. We prove that there exists a strict subsolution x_0 and a strict supersolution x_1 such that all the fixed points of T (or iterates of T) lie in (x_0, x_1). (We **conjecture** that for each $x \in X, T^n x$ approaches $[x_0, x_1]$ as n tends to infinity.) Thus we see that much of the interesting behaviour of T lies in the invariant order interval $[x_0, x_1]$. To prove this result, we first note that the compactness ensures that F is compact. A simple Zorn's lemma argument (see the proof on p. 130 of [22]) ensures that F has maximal elements in the order. Suppose that $\tilde{x}_1$ is a maximal element and $x_0 \in F$. Then the sup $\tilde{x}_1 \vee x_0$ exists. By monotonicity, $T(\tilde{x}_1 \vee x_0) \geq T(\tilde{x}_1) = \tilde{x}_1$. Similarly $T(\tilde{x}_1 \vee x_0) \geq x_0$. Hence $T(\tilde{x}_1 \vee x_0) \geq \tilde{x}_1 \vee x_0$, that is, $T(w) \geq w$ where $w = \tilde{x}_1 \vee x_0$. If w is a fixed point of T, $w = \tilde{x}_1$ by the maximality of $\tilde{x}_1$ and hence $x_0 \leq \tilde{x}_1$. If $T(w) > w$, then by monotonicity, $\{T^n w\}$ is increasing in n. By our assumption, $\{T^n w\}$ is bounded and hence is precompact by the complete continuity of T. Hence $\{T^n w\}$ converges to z which is a fixed point of T. Now $\tilde{x}_1 \leq w < T(w) \leq z$, which contradicts the maximality of $\tilde{x}_1$. Hence this case does not occur and $x_0 \leq \tilde{x}_1$. Thus we see that $x \leq \tilde{x}_1$ for all $x \in F$. By a similar argument, there exists $\tilde{x}_2 \leq \tilde{x}_1$ such that $\tilde{x}_2 \in F$ and $x \geq \tilde{x}_2$ for all $x \in F$. Now, by our arguments above and the maximality of $\tilde{x}_1$, there are no subsolutions of T in $(\tilde{x}_1, \infty)$. Hence by the remark at the bottom of p. 206 in [16], there exist strict supersolutions of T in $(\tilde{x}_1, \infty)$. (This is essentially a degree argument which uses part of the proof of Theorem 6 to prove that $i_{[\tilde{x}_1, \infty)}(T, \tilde{x}_1) = 1$ and one then uses a different homotopy to obtain supersolutions.) Choose one of these and denote it by x_1. Similarly, we can construct a subsolution x_2 such that $x_2 < \tilde{x}_2$. It remains to prove that any periodic point of T (that is a fixed point of an iterate of T) lies in $[\tilde{x}_2, \tilde{x}_1]$. Assume that $T^j x_3 = x_3$ where $j > 1$ and j is minimal. Let $w = x_3 \vee Tx_3 \ldots \vee T^{j-1} x_3$. Now $w \geq x_3$ and hence, by monotonicity, $Tw \geq Tx_3$. Similarly, $w \geq Tx_3$ and hence $Tw \geq T^2 x_3$. Similarly, $Tw \geq T^3 x_3, \ldots, Tw \geq T^j x_3 = x_3$. Thus $Tw \geq x_3 \vee Tx_3 \ldots \vee T^{j-1} x_3 = w$. Much as before, the result that $Tw \geq w$ implies that there is a fixed point z of T with $z \geq w$. Hence $x_3 \leq w \leq z \leq \tilde{x}_1$. Thus $x_3 \leq \tilde{x}_1$. Similarly $x_3 \geq \tilde{x}_2$. This completes the proof.

One reason why we suspect that all points eventually approach $[x_2, x_1]$ is that one can prove by a degree argument on $[x_1, \infty)$ that for each large R there is a super solution x_3 with $x_3 \geq x_1$ and $\|x_3\| = R$. It is easy to show that, if z_1 is a subsolution, z_2 is a supersolution and $z_1 \le x_2 \le x_1 \le z_2$, then for any point $x \in [z_1, z_2]$, $T^n x$ approaches $[x_1, x_2]$ as $n \to \infty$. Thus the iterates of many points approach $[x_1, x_2]$ as $n \to \infty$. In fact, if $\|T(x) - x\| \to \infty$ as $\|x\| \to \infty$, $x \in (-\infty, x_2) \cup (x_1, \infty)$, one can refine the degree argument to prove that for any $x \in X$, there is a subsolution $\hat{x}_1$ and a supersolution $\hat{x}_2$ such that $\hat{x}_1 \le x \le \hat{x}_2$. Hence, by our comments above $T^n x$ approaches $[x_1, x_2]$ as $n \to \infty$.

Note that in the above result, we do not need to assume that F is bounded but only that there does not exist an unbounded sequence $\{x_i\}_{i=1}^\infty$ in F such that $x_i < x_{i+1}$ for all i (or $x_i > x_{i+1}$ for all i). The above result appears to be new.

Lastly, note that these degree ideas can be used to prove a number of other results, for example prove the existence of multiple solutions. See [2], [14] and [16].

§3. Generic Convergence

In this section, we discuss results of the type that for 'most' x in X, the possible limit sets of $T^n x$ as $n \to \infty$ (or $t \to \infty$ in the continuous case) are rather restricted. We discuss this in less detail because it is a little aside from the main point of this survey. If $x \in X$, the orbit of $x, O(x)$ is $\{T^n x : n \ge 0\}$ (or $\{T(x, t) : t \ge 0\}$ in the continuous case). In the discrete case, define $O(G) = \bigcup_{x \in G} O(x)$ if $G \subseteq X$. We say $z \in \omega(x)$ if there is a sequence n_i tending to infinity such that $T^{n_i} x \to z$ as $i \to \infty$. (In the continuous case, the definition is the same except we look at real sequences $\{t_i\}$ tending to infinity.) $\omega(x)$ is known as the omega limit set. It is easily seen that $\omega(x)$ is closed and $T(\omega(x)) \subseteq \omega(x)$ ($T(\omega(x), t) \subseteq w(x)$ if $t \ge 0$ in the continuous case).

We start with the continuous case.

Theorem 8. (Hirsch [30]). *Assume that T is a strongly positive monotone continuous dynamical system on X where the cone K has non-empty interior and X is separable. Finally, assume that $\overline{O(x)}$ is compact for each x in X. Then there is a dense subset A of X such that, if $x \in A$, then $\omega(x)$ is contained in the set of stationary points.*

He actually proves a more general result. If T satisfies our earlier complete continuity condition, it is not difficult to use his ideas to remove the separability condition. We omit the proof of Theorem 8 which is not short. In proving this, one obtains as intermediate steps quite a lot of important information on $\omega(x)$. For example, if $x < y$ either

$$\omega(x) \ll \omega(y)$$

or $\omega(x) = w(y) \subseteq$ set of stationary points (known as the limit set dichotomy). More recently, there has been a good deal of effort on improving the theorem. For example, one would like to know if $\omega(x)$ is a single point for a dense subset of X. Some partial results on this appear in [43] and [51].

We now consider the discrete case. The following example from [22] shows that the situation here is more complicated. We choose $X = R^2$, K the usual cone in R^2 and define $T_0 : R^2 \to R^2$ by $T_0(x, y) = (2\arctan y,\allowbreak 2\arctan x)$. Then T_0 is increasing.

Let a be the unique positive number with $a = 2\arctan a$. Then T_0 has the three fixed points $(-a, -a)$, $(0, 0)$, (a, a), whereas T_0^2 has nine fixed points: $(-a, -a)$, $(-a, a)$, $(a, -a)$, (a, a), $(a, 0)$, $(0, a)$, $(-a, 0)$, $(0, -a)$, $(0, 0)$. (To prove that these are the only fixed points of T_0^2, one uses that 2 arctan composed with itself is strongly concave for $x > 0$.) All these nine fixed points are non-degenerate, (in the sense that the derivative of $I - T_\varepsilon^2$ is invertible at the fixed points) and the first four are stable. In particular, $(-a, a)$ and $(a, -a)$ are stable fixed points of T_0^2, but not fixed points of T_0. For $\varepsilon > 0$ small let

$$T_\varepsilon(x, y) = \begin{pmatrix} 2\arctan y + \varepsilon x \\ 2\arctan x + \varepsilon y \end{pmatrix}$$

Then $T_e : \mathbf{R}^2 \to \mathbf{R}^2$ is strongly order-preserving, and, by non-degeneracy, T_ε and T_ε^2 have the same number of fixed points (also having the same stability properties as T_0 and T_0^2 respectively). (A simple estimate shows that the perturbation introduces no large fixed points of T_ε or T_ε^2.) Hence again T_ε^2 has stable fixed points, which are not fixed points of T_ε.

Taking such a fixed point $p = (x_\varepsilon, y_\varepsilon)$, we can now easily show that the limit set dichotomy theorem and generic convergence to stationary states are both not true here. Clearly $\omega(p) = \{p, T_\varepsilon(p)\}$, and p and $T_\varepsilon(p)$ are not ordered. If we take $\overline{p} > p$ close to p, it follows by stability that $\omega(\overline{p}) \gg \omega(p)$. If $\omega(\overline{p}) = \omega(p)$, certainly $\omega(p) \not\subset \mathcal{E}$, the set of fixed points of T_ε. So both alternatives in the limit set dichotomy theorem fail. Further, for all $\overline{p}$ in a neighborhood of p, the iterates $T_\varepsilon^n(\overline{p})$ remain close to $\omega(p)$ by stability and hence do not converge to a fixed point.

This example shows a very significant difference between the continuous- and discrete-time cases. We note that T_ε is not the Poincaré map for a differential equation, since its derivative has negative determinant. A similar example was also obtained in [55].

At this stage, one might ask if similar behaviour can occur for the time ω map of a parabolic partial differential equation with w periodic in time coefficients. Examples of this type were constructed in [53]. In [23] and [54] examples were constructed such that the elliptic part is self-adjoint for each t. (The methods in the two papers are quite different.)

Hence the theorem for the discrete case must be quite different. Let F_1 denote the union of the stationary and periodic points of T.

Theorem 9. ([44] and [45]). *Assume that K is a cone with non-empty interior in X and T is a completely continuous discrete dynamical system on X such that T is $C^{1,\alpha}, T$ is $1-1$ and $T'(x)$ is strongly positive for $x \in X$. (Here we mean the map $x \to T'(x)$ is locally α Hölder continuous.) If $G \subseteq X$ is open such that $O(G)$ is bounded, then there is a dense set $A \subseteq G$ such that $\omega(x) \subseteq F_1$ if $x \in A$.*

Note that, as in [2], the derivative assumption ensures that T is strongly monotone.

In [44], the result is proved under an additional assumption, known as continuous separation. It is proved in [45] that the continuous separation always holds if T is C^1 and T is $1-1$. The proof uses some ideas from Takac [55]. It seems likely that the assumptions can be weakened. Once again, the proof is too technical to discuss here. Note that in the case where T is the time map for a parabolic system one can often use backwards uniqueness theorems as in [1] to prove that T is $1-1$. More recently [29], there has been work on controlling the magnitude of the minimal period of the periodic points (including under perturbations).

Finally, note that the result at the end of the last section shows that assumptions on $O(x)$ for all x are rather more restrictive than is at first apparent.

References

[1] S. Agmon, *Unicité et Convexité dans les Problems Differentieles*, University of Montreal Press, Montreal, 1966.

[2] H. Amann, *Fixed point equations and nonlinear eigenvalue problems in ordered spaces*, SIAM Review **18**(1976), 620–709.

[3] H. Amann, *Multiple fixed points of asymptotically linear maps*, J. Functional Anal. **17**(1974), 174–213.

[4] V.I. Arnold, *Indices of singular points of 1-forms on a manifold with boundary, convolution of invariants of reflection groups and singular projections of smooth surfaces*, Uspehi Mat Nauk, **34**(1979), 3–38.

[5] C. Bessaga and A. Pelczynskii, *Selected Topics in Infinite Dimensional Topology*, PWM, Warsaw, 1975.

[6] P. Clement, *Some remarks on the continuation method of Leray-Schauder-Rabinowitz and the method of monotone operations*, MRC TSR # 2454, University of Wisconsin, Madison, 1982.

[7] W.A. Coppel, *Stability and Asymptotic Behaviour of Differential Equations*, Heath, Boston, 1965.

[8] M. Crandall and P.H. Rabinowitz, *Bifurcation from simple eigenvalues*, J. Functional Anal. **8**(1971), 321–340.

[9] E.N. Dancer, *Global solution branches for positive mappings*, Archives Rational Mech. Anal. **52**(1973), 181–192.

[10] E.N. Dancer, *On radially symmetric bifurcation*, J. London Math. Soc. **20**(1979), 287–292.

[11] E.N. Dancer, *On the indices of fixed points in cones and applications*, J. Math. Anal. and Applications **91**(1983), 131–151.

[12] E.N. Dancer, *On positive solutions of some pairs of differential equations*, Trans. Amer. Math. Soc. **284**(1984), 729–743.

[13] E.N. Dancer, *On positive solutions of some pairs of differential equations II*, J. Differential Equations **59**(1985), 236–258.

[14] E.N. Dancer, *Multiple fixed points of positive maps*, J. Reine Ang. Math. **371**(1986), 46–66.

[15] E.N. Dancer, *On the existence and uniquness of positive solutions for competing species models with diffusion*, Trans. Amer. Math. Soc. **326**(1991), 829–859.

[16] E.N. Dancer, *Upper and lower stability and index theory for positive mappings and applications*, Nonlinear Analysis **17**(1991), 205–217.

[17] E.N. Dancer, *Global breaking of symmetry of positive solutions on two-dimensional annuli*, Differential and Integral Equations **5**(1992), 903–914.

[18] E.N. Dancer, *Weakly nonlinear Dirichlet problems in long or thin domains*, to appear in Memoirs Amer. Math. Soc.

[19] E.N. Dancer, *On the number of solutions of weakly nonlinear elliptic equations when a parameter is large*, Proc. London Math. Soc. **53**(1986), 429–452.

[20] E.N. Dancer and Yihong Du, *Positive solutions of an n species competing system with diffusion*, preprint.

[21] E.N. Dancer and Yihong Du, *Existence of positive solutions for a 3 species competitive system with diffusion*, submitted.

[22] E.N. Dancer and P. Hess, *Stability of fixed points for order-preserving discrete-time dynamical systems*, J. Reine Ang. Math. **419**(1991) 125–139.

[23] E.N. Dancer and P. Hess, *Stable subharmonic solutions in periodic reaction diffusion equations*, to appear in Journal of Differential Equations.

[24] Yihong Du, *The structure of the solution set of a class of nonlinear eigenvalue problems*, J. Math. Anal. and Applications **170**(1992), 567–580.

[25] W. Feng and W.H. Ruan, **Coexistence and stability in a three species competition model**, preprint.

[26] C. Golé, *Ghost circles for twist maps*, J. Diff. Eqns. **97**(1992), 140–173.

[27] D. Henry, *Geometric Theory of Semilinear Parabolic Equations*, Lecture Notes in Mathematics vol 840, Springer-Verlag, Berlin, 1981.

[28] P. Hess, *Periodic-Parabolic Boundary Value Problems and Positivity*, Longman, Harlow, 1991.

[29] P. Hess and P. Polacik, *Boundedness of prime periods of stable cycles and convergence to fixed points in discrete monotone dynamical systems*, preprint.

[30] M. Hirsch, *Stability and convergence in strongly monotone dynamical systems*, J. Reine Ang. Math. **383**(1988), 1–58.

[31] R. Holmes, *Geometric Functional Analysis and its Applications*, Springer-Verlag, Berlin, 1975.

[32] S. Hu, *Theory of Retracts*, Wayne State University Press, Detroit, 1965.

[33] M.A. Krasnosel'skii and P.P. Zabrieko, *Geometric Methods of Nonlinear Analysis*, Springer-Verlag, Berlin, 1984.

[34] S.S. Lin, *On non radially symmetric bifurcation in the annulus*, J. Diff. Eqns. **80**(1989), 348–367.

[35] N. Lloyd, *Degree Theory*, Cambridge University Press, Cambridge, 1978.

[36] H. Matano, *Existence of non-trivial unstable sets for equilibriums of strongly order preserving systems*, J. Fac. Sci. Tokyo **30**(1984), 645–673.

[37] R. Nussbaum, *The fixed point index and fixed point theorems*, preprint.

[38] R. Nussbaum, *The Fixed Point Index and Applications*, University of Montreal Press, Montreal, 1985.

[39] R. Nussbaum, *The fixed point index for locally condensing maps*, Ann. Mat. Pura Appl. **89**(1971), 217–258.

[40] R. Nussbaum, *Periodic solutions of some nonlinear autonomous functional differential equations*, Annali di Mat. Pura ed Appl. **101**(1974), 263–306.

[41] R. Nussbaum, *Iterated nonlinear maps and Hilbert's projective metric II*, Memoirs Amer. Math. Soc. **401**(1989).

[42] A. Peressini, *Ordered Topological Vector Spaces*, Von Nostrand, New York, 1967.

[43] P. Polacik, *Convergence in smooth strongly monotone flows defined by semilinear parabolic equations*, J. Diff. Eqns. **79**(1989), 89–110.

[44] P. Polacik and I. Terescak, *Convergence to cycles as a typical asymptotic behaviour in smooth strongly monotone discrete-time dynamical systems*, to appear in Archives Rational Mech. Anal.

[45] P. Polacik and I. Terescak, *Exponential separation and invariant bundles for maps in ordered Banach spaces with applications to parabolic equations*, preprint.

[46] H. Protter and H. Weinberger, *Maximum Principles in Differential Equations*, Prentice Hall, Englewood Cliffs, 1967.

[47] P.H. Rabinowitz, *Some global results for nonlinear eigenvalue problems*, J. Functional Anal. **7**(1971), 487–513.

[48] H.H. Schaefer, *Topological Vector Spaces*, MacMillan, New York, 1966.

[49] J.T. Schwartz, *Nonlinear Functional Analysis*, Gordon and Breach, New York, 1969.

[50] H. Smith and H. Theime, *Quasiconvergence and stability for strongly order preserving semiflows*, SIAM J. Math. Anal. **21**(1990), 673–692.

[51] H. Smith and H. Thieme, *Convergence for strongly order preserving semiflows*, SIAM J. Math. Anal. **22**(1991), 1081–1101.

[52] P. Takac, *Convergence to equilibrium on invariant d-hypersurfaces for strongly increasing discrete time semigroups*, J. Math. Anal. and Applications **148**(1990), 223–244.

[53] P. Takac, *Linerly stable subharmonic orbits in strongly monotone time periodic dynamical systems*, Proc. Amer. Math. Soc. **115**(1992), 691–698.

[54] P. Takac, *A construction of stable subharmonic orbits in monotone time-periodic dynamical systems*, to appear in Monatshefte fur Mathematik.

[55] P. Takac, *Domains of attraction of generic ω-limit sets for strongly monotone discrete-time semigroups*, J. Reine Ang. Math. **423**(1992), 101–173.

[56] H. Whyburn, *Topological Analysis*, Princeton University Press, Princeton, 1958.

Topological Bifurcation

Jorge Ize

Departamento de Matemáticas y Mecánica
ITMAS – UNAM, Apartado Postal 20-726, D.F. Mexico 20

Abstract

This paper is devoted to some of the results in bifurcation theory obtained by topological methods in the last 25 years. The cases of one and several parameters will be reviewed, with "necessary" and sufficient conditions for bifurcation, both local and global, and the structure of the bifurcation set will be studied. The case of equivariant bifurcation will be considered, with a special application to the case of abelian groups.

Introduction

Bifurcation theory, like the rest of Nonlinear Analysis, has known very quick growth in the last 25 years, in both theory and applications. A great number of nonlinear phenomena lead to mathematical models of the form

$$F(\lambda, x) = 0,$$

where x represents the sought solution in a Banach space B, λ belongs to a space $\wedge$ of parameters and F maps $\wedge \times B$ into the Banach space E. In bifurcation theory one assumes that the above equation has a known set of solutions, the "trivial" solutions, and one says that (λ_0, x_0) on this set is a **bifurcation point** if, at any neighborhood of this point, one has solutions that are nontrivial.

For simplicity, one supposes that the trivial solutions form a curve or surface, according to the dimension of the parameter space of the form

$(\lambda, x(\lambda))$. In the case of a **turning point** one may take a new parametrization μ and look for zeros of $(F(\lambda, x), \lambda - \lambda(\mu))$. A natural idea is then to linearize the equation near (λ_0, x_0) and to consider the problem as a nonlinear perturbation of a singular linear problem. Assume thus that $x_0 = 0$, $\lambda_0 = 0$, $x(\lambda) = 0$, and that $F(x, \lambda)$ is expanded to be

$$F(\lambda, x) = Ax - T(\lambda)x - g(\lambda, x)$$

where A and $T(\lambda)$ are linear operators from B into E, $T(0) = 0$, $g(\lambda, x) = o(\| x \|)$. One may also truncate the Taylor series for $T(\lambda)$ after $2n$ λ-derivatives and have $g(\lambda, x) = o(\| x \| + |\lambda|^{2n+1})$.

If all the mappings are continuous and A is invertible, then the equation $F(\lambda, x) = 0$ is equivalent for (λ, x) close to 0, to $x = A^{-1}T(\lambda)x + A^{-1}g(\lambda, x)$, which, for small x and λ, has no other solution than $x = 0$. Hence a necessary condition for the existence of "nontrivial" solutions is that A has to be singular.

From here on there are many techniques that one may try in order to get information on other possible solutions: classical perturbation theory, singularity theory (unfolding and so on...) with or without group theory for the case of problems with symmetries (normal form analysis), variational methods and topological methods.

Clearly these techniques are not exclusive and, for a given problem, it is wise to try any of them provided some information is obtained. Prior to 1968, one could have rounded up all theoretical contributions to bifurcation in a few pages at the end of a book on strange phenomena in analysis, with maybe some chapters on examples. At present, it seems very difficult to make a complete report on all these new techniques in a single readable volume. Thus, one has to limit the scope to just one small part of the story.

This article is an account, more than a survey, of some of the topological ideas used in bifurcation theory. I will not try to give lists of all the contributions to the field, although I will make my best effort not to forget too many people, and I will also try to minimize the nonessential technical difficulties and to reduce them to remarks. It is also clear that this is a personal view of the field.

The plan is the following:

In a brief preliminary section, I will describe the topological results known before 1968. The first real section will be devoted to the case of one parameter bifurcation, the second section will consider the case of several parameters, the third will look at the structure of the bifurcation set and the last section will take up the problem of equivariant bifurcation. Some of the results in this section, those of subsections V and VI, are fully developed since they have not appeared in the literature.

I would like to thank Alma Rosa Rodríguez for her efficient processing of the manuscript.

0. Preliminaries

The purpose of this section is to give an idea of topological results used in bifurcation theory before 1968. Although bifurcation was given the name by Poincaré, some examples were studied using analytical methods before and after him, from Euler to Liapunov. However the first and most used "topological" result is due to Leray and Schauder in their famous paper of 1934. Let $F(\lambda, x) = Ax - T(\lambda)x - g(\lambda, x)$.

Assume $B = E, A = I - \lambda_0 K$ where K is a compact operator, $T(\lambda) = \lambda K$. Hence $A - T(\lambda)$ has a discrete number of characteristic values λ_j, including $\lambda_j = 0$, where the operator is non-invertible, corresponding to the eigenvalues $\mu_j = (\lambda_j + \lambda_0)^{-1}$ of K. Hence for λ, different from $\lambda_j, x = 0$ is an isolated zero of $A - T(\lambda)$ and has a Leray-Schauder index. By the Liapunov-Schmidt reduction, this index is $(-1)^\beta$, where β is the sum of the algebraic multiplicities of the μ_j's, for all $|\mu_j| \geq |(\lambda + \lambda_0)^{-1}|$ and μ_j of the sign of $\lambda + \lambda_0$.

In particular, if λ_0^{-1} has odd algebraic multiplicity, then there is a change in the index as λ passes through 0. In this case, if Ω is any open bounded set in $\mathbb{R} \times E$, which intersects the line $x = 0$ only in a neighborhood of λ_0 and does not contain any other characteristic value, then one must have a non-trivial zero on the boundary of Ω. This is the content of the celebrated Krasnosel'skii's bifurcation theorem [Krasnosel'skii 1964, p. 198–199]. Note that the author gives to this fact the misleading phrase of having "a continuous branch of eigenvectors which starts from 0 and goes off to infinity".

The argument here is simple: If Ω has no non-trivial zeros on its boundary, then for $|\lambda - \lambda_0|$ large, the Leray-Schauder degree of $F(\lambda, x)$, on Ω_λ the section of Ω by a constant λ, is zero since Ω is bounded. For λ close to $\lambda_0, \Omega_\lambda$ has two pieces, one which is a neighborhood of the 0 solution, where F has index $(-1)^\beta$, (a fact already proved in the Leray-Schauder paper), and another one, away from the trivial solution with zero degree (one has to make a matching argument there). Since β changes as λ crosses λ_0, this gives the contradiction.

The state of the art at this point can be seen in the books by M.A. Krasnosel'skii, J. Cronin, M. Berger and M. Berger, J. T. Schwartz, for the use of topological methods in Nonlinear Analysis, and the analytic methods (Newton diagrams and perturbations) are exposed in the Lecture Notes by Pimbley and Keller-Antman.

I. ONE PARAMETER BIFURCATION

The early thirties were marked by the fundamental papers of Leray and Schauder, Ljusternik and Schnirelmann, and Morse. But the period from 1968 to 1975 witnessed an explosion of the application of almost all the topological tools for Nonlinear Analysis. From the point of view of generalized degree theories, one of the first ingredients which was questioned in the Leray-Schauder theory was the compactness of the maps. Among these extensions one has the k-set contractions, Nussbaum 1972, the condensing maps, Sadovskii 1972, the A-proper maps, Petrsyshin 1968, Fitzpatrick 1970, the Fredholm maps, Eells 1970, with a mod 2 degree by Elworthy and Tromba, 1970, and multivalued maps, Ma 1972. In the last three classes one has already different spaces but with the same "dimension", a fact which will be used by Mawhin and his collaborators, Gaines and Laloux , to define their coincidence degree in 1977.

The case of different dimensions was considered by Nirenberg in 1971, homotopy groups of spheres, and later by Fucik-Necas, Berger-Podolak (1977), but without, at that time, a complete degree theory. Most generalized cohomology theories were used at one point or another, from cobordism, homology (K.C. Chang), cohomotopy (Ize 1974, following the ideas of Geba and Granas 1973) to K-theory (Alexander).

It is not my purpose to review all those theories in these introductory remarks. A good account of their use in Nonlinear Analysis, in the period 1968–1975, can be seen in Nirenberg's lectures notes (1974) and Berger's book (1977). From here on I will limit myself to the part of bifurcation theory closest to the application of degree theories, leaving aside (and to more qualified contributors) variational methods.

From the point of view of topological methods in bifurcation, one of the most striking results of that period is the Global Rabinowitz Alternative which, from purely local data, gives global information.

Theorem [Rabinowitz, 1971]. *Let* $F(\lambda, x) = x - \lambda Kx - g(\lambda, x)$, *from* $\mathbb{R} \times B$ *into* B, *with* K *compact and* $g(\lambda, x) = o(\| x \|)$. *Assume that* λ_0 *is a characteristic value of* K, *of odd algebraic multiplicity; then one has a continuum* C *of non-trivial zeros starting at* $(\lambda_0, 0)$, *which*

   1) *either goes to infinity in* $\mathbb{R} \times B$
   2) *or returns to a different bifurcation point* $(\lambda_1, 0)$.

The idea of the proof is the following: if the continuum (i.e. a maximum connected set) is bounded and does not return to the line of trivial zeros, then one may construct a bounded open set Ω which separates C from the rest of the non-trivial zeros of F, (the "Whyburn Lemma" which seems to be due to earlier work by people in general topology, in particular to Mazurkiewicz in 1910) and which intersects the trivial solutions only near $(\lambda_0, 0)$. Then the argument, which I have already sketched before, will give a contradiction.

The global alternative, together with new degree theories, will fuel most of the research in this period, where compactness assumptions, the linear dependence on λ, the relation between A and $T(\lambda)$, will be relaxed, while a better understanding of the role of multiplicity and of the non-linear part will be gained.

Besides the authors already mentioned, the early results were independently obtained in the PhD thesis of E.N. Dancer (Cambridge 1972), J. Ize (Courant Inst. 1974), W. Magnus (Sussex 1974) and D. Westreich (Yeshiva 1972).

1. Local Bifurcation

The simplest setting for this study will be to assume that in the equation

$$F(\lambda, x) = Ax - T(\lambda)x - g(\lambda, x) \tag{1}$$

from $\wedge \times B$ into E, A is a Fredholm operator, that is, A is continuous, ker A is of finite dimension d, Range A has finite codimension d^*, $T(\lambda)$ is continuous with $T(0) = 0$, $\| T(\lambda) \| \to 0$ as $\lambda \to 0$ and $g(\lambda, x) = o(\| x \|)$, uniformly in λ.

Let then P and Q be two projections, P from B onto ker A and Q from E onto Range A; then

$$B = \ker A \oplus B_2$$
$$E = E_2 \oplus \text{Range } A$$

with B_2 a closed subspace of B and E_2 of dimension d^*. Any x in E is written as $x = x_1 + x_2$, with $x_1 = Px$.

Since A is continuous, one-to-one from B_2 onto Range A, there is a continuous inverse K from *Range A* onto B_2, that is,

$$AKQ = Q, KA\,(I - P) = I - P.$$

One may write equation (1), for λ small, as

$$(A - QT(\lambda))(x_1 + x_2) - Qg(\lambda, x_1 + x_2) \ominus (I - Q)(T(\lambda)(x_1 + x_2) + g(\lambda, x_1 + x_2))$$

and, using the facts that $A - QT(\lambda) = A(I - KQT(\lambda))$, where for λ small, $I - KQT(\lambda)$ is an invertible mapping from B into itself, with an inverse which is given by the power series, and that $T(\lambda)(I - KQT(\lambda))^{-1}KQ = T(\lambda)KQ + (T(\lambda)KQ)^2 + \ldots = (I - T(\lambda)KQ)^{-1} - I$, as a mapping from E into E, one has that

$$\begin{aligned} F(\lambda, x) = [A - QT(\lambda)][x_2 - (I - KQT(\lambda))^{-1}KQ(T(\lambda)x_1 + g(\lambda, x))] \\ \ominus [I - Q][T(\lambda)\left((I - KQT(\lambda))^{-1}x_1\right. \\ + x_2 - (I - KQT(\lambda))^{-1}KQ(T(\lambda)x_1 + g(\lambda, x)) \\ + (I - T(\lambda)KQ)^{-1}g(\lambda, x)]. \end{aligned}$$

In order to better appreciate this formula, define

$$\begin{aligned} H(\lambda, x_1, x_2) &= x_2 - (I - KQT(\lambda))^{-1}KQ(T(\lambda)x_1 + g(\lambda, x)) \\ B(\lambda) &= -(I - Q)T(\lambda)(I - KQT(\lambda))^{-1}P \\ G(\lambda, x) &= -(I - Q)(I - T(\lambda)KQ)^{-1}g(\lambda, x). \end{aligned}$$

One then has

$$\begin{aligned} F(\lambda, x) = (A - QT(\lambda))H(\lambda, x_1, x_2) \oplus B(\lambda)x_1 + G(\lambda, x) \\ - (I - Q)T(\lambda)H(\lambda, x_1, x_2). \end{aligned} \tag{2}$$

It is clear that, if $F(\lambda, x) = 0$ and for small λ and x, then $H(\lambda, x_1, x_2) = 0$ has a unique solution $x_2 = x_2(\lambda, x_1)$ with $\| x_2 \| \le C \| x_1 \| (\| \lambda \| + 0(\| x_1 \|))$, provided $g(\lambda, x)$ is C^1 and $\| T(\lambda) \| \le C \| \lambda \|$, by using any contraction mapping argument.

Then the zeros of F coincide with those of the **bifurcation equation**

$$B(\lambda)x_1 + G(\lambda, x_1 + x_2(\lambda, x_1)) = 0 \tag{3}$$

where $B(0) = 0$, $B(\lambda)$ is a $d \times d^*$ matrix, and $G(\lambda, x_1) = o(\| x_1 \|)$.

Taking $g = 0$, one has that dim ker $(A - T(\lambda)) = $ dim ker $B(\lambda)$, while for g any element of E, Codim Range $(A - T(\lambda)) = $ Codim Range $B(\lambda)$, that is, the spectral properties of $A - T(\lambda)$ can be recovered from those of $B(\lambda)$. In particular the Range of $A - T(\lambda)$ is closed and $A - T(\lambda)$ is a Fredholm operator of index $d - d^*$.

Example 1.1. Assume A is of index 0, $B \subset E$, $\wedge = \mathbb{R}$, $T(\lambda) = \lambda I$. Then the following are equivalent:

1) $E = $ Range $A^\alpha \oplus $ ker A^α for some α
2) ker $A^{\alpha+p} = $ ker A^α, Range $A^{\alpha+p} = $ Range A^α, $\forall p \ge 0$
3) 0 is an isolated eigenvalue in the spectrum of A
4) $B(\lambda)$ is invertible for $\lambda \ne 0$.

For a proof, see Ize 1976.

If one of the above holds, then $B = B \cap \text{Range } A^\alpha \oplus \ker A^\alpha$, with $x = u \oplus v$, A sends $B \cap \text{Range } A^\alpha$ into Range A^α and ker A^α into itself and there it is nilpotent. By choosing bases, A on ker A^α is in Jordan form with d blocks of dimension k_j. On a typical block of dimension k, one has

$$
A = \begin{pmatrix} 0 & 1 & & \\ & \ddots & \ddots & \\ & & \ddots & 1 \\ 0 & & & 0 \end{pmatrix}, \quad
Q = \begin{pmatrix} 1 & & & 0 \\ & \vdots & & \\ & & 1 & \\ 0 & & & 0 \end{pmatrix},
$$

$$
K = \begin{pmatrix} 0 & & & 0 \\ 1 & & \ddots & \\ & \ddots & & \ddots \\ & & 1 & 0 \end{pmatrix},
$$

$$
I - P = \begin{pmatrix} 0 & & & 0 \\ & 1 & & \\ & & \ddots & \\ 0 & & & 1 \end{pmatrix}, \quad
(I - KQA(\lambda))^{-1} = \begin{pmatrix} 1 & & & 0 \\ \lambda & & \ddots & \\ \vdots & & & \ddots \\ \lambda^{k-1} & \ldots & & 1 \end{pmatrix}
$$

and if on the block, $x_1 = (y_1, 0, \ldots, 0)^T$ and g has components $(g_1, \ldots, g_k)$,

$$
B(\lambda)x_1 + G(\lambda, x) = \lambda^k y_1 + \lambda^{k-1} g_1 + \ldots + g_k.
$$

Thus $B(\lambda)$ is a diagonal matrix with entries $\lambda^{k_1}, \ldots, \lambda^{k_d}$, with $\Sigma k_i = m = \dim \ker A^\alpha$, the algebraic multiplicity.

Example 1.2. $B = E$, $A = I - \lambda_0 T$, $T(\lambda) = \lambda T$ with T compact. Then 0 is isolated in the spectrum of A and, from the classical Liapunov-Schmidt reduction, one has $E = \text{Range } A^\alpha \oplus \ker A^\alpha$, where α is the algebraic multiplicity of λ_0, and ker A^α is invariant under $A - T(\lambda)$. Again A is nilpotent on ker A^α; thus, in the Jordan form for A, $A - \lambda T$ will have blocks with diagonal $-\lambda/\lambda_0$ and upper diagonal $1 + \lambda/\lambda_0$. In this case $B(\lambda)$ will be diagonal with terms of the form $-\lambda^{k_j}/(\lambda_0 + \lambda)^{k_j - 1}\lambda_0$. (See also Dancer 1971).

Example 1.3. If $d = d^* = 1$, $\Lambda = \mathbb{R}$, then the bifurcation equation reduces to one equation in two variables λ and x_1. Assume that $T(\lambda) = \lambda T_1 + o(\| \lambda \|)$ and that $(I - Q)T_1 P \neq 0$; then $B(\lambda)x_1$ has a dominant term $\lambda(I - Q)T_1 x_1 + o(\| \lambda \|)x_1$ and the bifurcation equation is of the form

$$
a\lambda x_1 + \lambda b(\lambda)x_1 + h(\lambda, x_1)x_1 = 0
$$

with $b(0) = 0, h(\lambda, 0) = 0, a \neq 0$. It is then easy to put some regularity conditions on $b(\lambda)$ and $h(\lambda, x_1)$ in order to prove the existence of a unique branch of nontrivial solutions $\lambda = \lambda(x_1)$. This is the content of the Crandall-Rabinowitz theorem (1971), (a proof using Morse theory is given in Nirenberg (1974)).

Remark 1.1. Index of the zero solution. Assume $B = E$, $A = I - T_0$ with T_0 compact, $T(\lambda)$ compact. Then $d = d^*$. Suppose that for some λ_0 close to 0, $B(\lambda_0)$ is invertible. Thus, 0 is an isolated solution at λ_0 and has an index, i.e., the Leray-Schauder degree with respect to a small ball B.

By taking $g = 0$ in equation (2) and deforming λ_0 to 0 in $H(\lambda, x_1, x_2)$ and in $QT(\lambda_0)$, this index is the index of $Ax_2 \oplus B(\lambda_0)x_1$. Let L be any isomorphism from E_2 onto ker A. Then one may write this last expression as $(A \oplus L^{-1})(x_2 \oplus LB(\lambda_0)x_1)$. From the product theorem for the Leray-Schauder degree, the index of $A - T(\lambda_0)$ is the index of $A \oplus L^{-1}$ multiplied by the index of $(I_{B_2} \oplus LB(\lambda))$, that is, Index$(A - T(\lambda_0))$ = Index$(A \oplus L^{-1})$ Sign det $LB(\lambda_0)$. In particular, there is a change of index at 0 if and only if det $LB(\lambda)$ changes sign. Furthermore if $T(\lambda) = \lambda T_0$, then, from Example 1.2, det $B(\lambda)$ has the sign of $a\lambda^m$, a a constant and m the algebraic multiplicity of 1 as eigenvalue of T_0. Since the index is constant between eigenvalues and Index $I = 1$, one recovers the Leray-Schauder formula: the index is $(-1)^\beta$, β the sum of the algebraic multiplicities of the characteristic values of T_0 between 0 and λ_0.

Remark 1.2. Generalized multiplicity: d = d*. For $A = I - T_0$, but $T(\lambda)$ general, the change of index is not related to the algebraic multiplicity of 1 as an eigenvalue of T_0. Furthermore the set of singular points, i.e. where det $B(\lambda) = 0$, may be complicated, even in the case where dim $\wedge = 1$. One may assume that $B(\lambda)$ is invertible for λ close to (but different from) 0, i.e. $A - T(\lambda)$ is also invertible. This is the case if $T(\lambda)$ is analytic in λ and det $B(\lambda)$ is not identically 0. In fact, one may prove that in this case the set of λ's where $A - T(\lambda)$ is a Fredholm operator is open and that, on each connected component of this set, the index is constant and the dimension of ker$(A - T(\lambda))$ is constant except at isolated points where it can be larger. (See Ize 1976 Lemmas 6.2 and 6.3). Since $B(\lambda)$ is also analytic, det $B(\lambda) = \lambda^m a(\lambda)$ with $a(0) \neq 0$. This number m is called the *generalized algebraic multiplicity* and one may show that m and the sign of $a(0)$ are independent of the projections P and Q (Ize 1976 p. 33). By extending $A - T(\lambda)$ to be complex operators and λ complex, one may use the corresponding result for this case, or one may use Rabier's results (1989)).

There are other ways of defining this generalized multiplicity. Magnus (1976) does it with jets, Weistreich (1973) via a suspension argument,

Kielhöfer (1988) with crossing numbers. Esquinas (1988), Esquinas-López Gómez (1988) and Rabier (1989), using generalized Jordan chains, have shown that all these definitions are equivalent. In fact, from the topological point of view, what is important is the change in the index and the computation in the preceding remark. The "necessary" part, which will be given below, shows that this change of index is the only relevant aspect.

In the rest of this section, I shall suppose, except in some remarks, that $d = d^*$, $\Lambda = \mathbb{R}$ and $B(\lambda)$ is invertible for λ close to 0 ($\lambda \neq 0$). Since $H(x_1, x_2, \lambda) = 0$ can be solved uniquely for x_2 in terms if λ and x_1, one has only to look at the bifurcation equation (3).

Theorem 1.1. *If* $\det B(\lambda)$ *changes sign at* 0, *then* $(0, 0)$ *is a bifurcation point.*

Proof. One may repeat Krasnosel'skii argument, or use the following trick, which will be useful later on. Consider the mapping

$$(B(\lambda)x_1 + G(\lambda, x_1), \| x_1 \|^2 - \epsilon^2)$$

from $\mathbb{R} \times \mathbb{R}^d$ into itself. A zero of this new map will be a nontrivial ($\| x_1 \| = \epsilon$) solution of the bifurcation equation. Since $G(\lambda, x_1) = o(\| x_1 \|)$ and $B(\pm 2\rho)$ is invertible, one may choose ϵ so small that the only solution of $B(\pm 2\rho)x_1 + G(\pm 2\rho, x_1) = 0$, for $\| x_1 \| \leq 2\epsilon$, is $x_1 = 0$. Then, the above map is nonzero on the boundary of the ball $B = \{(\lambda, x_1) , |\lambda| < 2\rho,$ $\| x_1 \| < 2\epsilon\}$. Thus, the Brouwer degree of the pair with respect to B is well defined. One may deform $G(\lambda, x_1)$ to 0 and, via a linear deformation, replace $\| x_1 \|^2 - \epsilon^2$ by $\rho^2 - |\lambda|^2$. Since the map $(B(\lambda)x_1, \rho - |\lambda|^2)$ has two isolated zeros, $x_1 = 0$, $\lambda = \pm \rho$, with index equal to the sign of the Jacobian of the map, that is, $-(-1)^d \operatorname{Sign}(\lambda \det B(\lambda))$. Hence,

$$\deg(B(\lambda)x_1, \| x_1 \|^2 - \epsilon^2; B) = (-1)^d (i_- - i_+)$$

where $i(\lambda) = \operatorname{Sign} \det B(\lambda)$, $i_+(i_-)$ is this sign for $\lambda > 0$ ($\lambda < 0$). Q.E.D.

Remark 1.3 Necessary condition for bifurcation. It will be shown in the next section that if $\det B(\lambda)$ does not change its sign, then there is a $g(\lambda, x)$ such that $F(\lambda, x)$ has no nontrivial zeros near $(0, 0)$. That is the change of index is "necessary" for bifurcation unless one restricts the nonlinearities. Here I shall give a proof for the real analytic case.

Assume $B(\lambda)$ *is analytic in* λ *and* $\det B(\lambda) = a_0 \lambda^m + \ldots$, *with* m *even, then there is a nonlinearity* $G(\lambda, x)$, *real analytic in* λ *and* x, *such that* $B(\lambda)x + G(\lambda, x) = 0$, *for* $\| \lambda \| \leq \rho$, $\| x \| \leq \epsilon$, *implies* $x = 0$.

Proof [Ize 1988]. By factoring out, from each row in $B(\lambda)$ the largest possible power of λ (except possibly, in the last row), one may write $B(\lambda) = \wedge_1(\lambda)(A_0 - A_0(\lambda))$ where $\wedge_1(\lambda) = \mathrm{diag}(\lambda^{p_1}, \ldots, \lambda^{p_d})$ with Σp_i even, A_0 has a nonzero element in each row (except maybe the last one) and $A_0(0) = 0$. Hence the rank of A_0 is positive. If P_1 is the projection onto ker A_0 and Q_1 onto Range A_0, one has, as in the decomposition which lead to the bifurcation equation

$$A_0 - A_0(\lambda) = (A_0 - Q_1 A_0(\lambda)) H_1(\lambda) \oplus B_1(\lambda) - (I - Q_1) A_0(\lambda) H(\lambda),$$

where

$$H_1(\lambda) = I - P_1 - (I - K_1 Q_1 A_0(\lambda))^{-1} K_1 Q_1 A_0(\lambda) P_1$$
$$B_1(\lambda) = -(I - Q_1) A_0(\lambda)(I - K_1 Q_1 A_0(\lambda))^{-1} P_1.$$

By multiplying $Q_1 A_0(\lambda)$ by $1 - t$ in the expression for $H_1(\lambda)$ and in the first term, one has that the matrix $A_0 - A_0(\lambda)$ in deformable to $A_1 \oplus B_1(\lambda)$, where $A_1 = A_0(I - P)$.

By repeating the argument for $B_1(\lambda)$ one gets in a finite number of steps to a matrix $\wedge_1(\lambda)(A_1 \oplus \wedge_2(\lambda)(A_2 \oplus \wedge_3(\lambda)(A_3 + A_4(\ldots))))$, i.e. to $\wedge(\lambda)(\tilde{A} + \tilde{A}(\lambda))$ where $\tilde{A}$ is invertible and $\tilde{A}(0) = 0$.

Hence $B(\lambda)$ is homotopic to this last matrix through a homotopy $B(\lambda, t)$ which depends polynomially on t, and keeps fixed $\wedge(\lambda)$, and such that $B(\lambda, t)$ is invertible for $\lambda \neq 0$. If $\wedge(\lambda) = \mathrm{diag}(\lambda^{p_1}, \ldots, \lambda^{p_d})$, then $\sum p_i$ has parity of m. Note that one may also write $B(\lambda) = M(\lambda) \wedge (\lambda) N(\lambda)$ with $M(0)$ and $N(0)$ invertible using the results of Rabier (1989).

Now in $\wedge(\lambda)$ the term λ^{p_i}, with even p_i, can be deformed to a term which is never 0 by $\lambda^{p_i} + t$, while, for the submatrix $\begin{pmatrix} \lambda^{p_1} & 0 \\ 0 & \lambda^{p_2} \end{pmatrix}$ with p_1, p_2 even, one can use the deformation $\begin{pmatrix} \lambda^{p_1} & t \\ -t & \lambda^{p_2} \end{pmatrix}$. Hence one obtains a deformation $\wedge(\lambda, t)$, invertible for all λ if $t > 0$. Then if $G(\lambda, x) = \wedge(\lambda, \| x \|^2) B(\lambda, \| x \|^2) x - B(\lambda) x$, the only solution of $B(\lambda) x + G(\lambda, x) = 0$ is $x = 0$. If λ is complex and $B(\lambda)$ is complex analytic in λ with det $B(\lambda) = a\lambda^m + \ldots$, with m even, one may also construct a nonlinearity with no bifurcation. (This nonlinearity is of course not complex analytic in x. See Ize 1988 or the next section).

Remark 1.4. If the index of A is positive and ker A splits into $B_1 \oplus W$, with dim $B_1 = d^*$, such that $B(\lambda)$ restricted to B_1 is invertible for $0 < |\lambda| \leq 2\rho$ (this is always true if $B(\lambda)$ has rank d^* for $\lambda \neq 0$ and $B(\lambda)$ is analytic), then $B(\lambda)$ must have a minor which has an isolated zero at 0. Then one may complement $B(\lambda) x_1$, by $d - d^*$ equations of the form $\lambda^\epsilon w_1 - \epsilon_1, w_2 - \epsilon_2, \ldots, w_{d-d^*} - \epsilon_{d-d^*}$, with $\epsilon = 0$ if det $B(\lambda)|B_1$ changes sign and $\epsilon = 1$ if it does not. Then the augmented square matrix has a determinant which changes sign, and a local bifurcation. By varying $\epsilon_1, \ldots, \epsilon_{d-d*}$ one obtains a $(d - d*)$-family of bifurcated solutions, in fact, as we shall see in Section III, a continuum of local dimension $d - d^* + 1$ (see Ize 1976 and Ize et al. 1985).

Remark 1.5. Note that the argument in the proof of Theorem 1.1 does not need a Liapunov-Schmidt reduction, except for the computation of the indices. In fact, if the map $F(\lambda, x)$ belongs to a class for which a degree is defined, then if $F(\lambda, 0) = 0$ and there are ρ and η positive, $\rho > \eta$, such that if $F(\lambda, x) = 0$ for $\rho - \eta \leq |\lambda| \leq \rho + \eta$, and $\| x \| \leq \epsilon_0$, then $x = 0$, and $\deg(\| x \|^2 - \epsilon^2, F(\lambda, x); B)$ is defined, where $B = \{(\lambda, x) : |\lambda| \leq \rho + \eta, \| x \| \leq 2\epsilon \leq \epsilon_0\}$. The deformation $\tau(\| x \|^2 - \epsilon^2) - (1 - \tau)(\rho^2 - |\lambda|^2)$ is valid and the above degree is the sum (if the degree has this property) $\deg(\rho - \lambda, F(\rho, x); B_\rho) + \deg(\rho + \lambda, F(-\rho, x); B_{-\rho})$, where $B_{\pm\rho}$ are small balls around $(\pm\rho, 0)$. If this degree has the rather mild *suspension property*,

$$\deg(\rho \mp \lambda, F(\pm\rho, x); B_{\pm\rho}) = \mp\deg(F(\pm\rho, x); \| x \| \leq \epsilon) = \mp i(\pm\rho)$$

then one has

$$\deg(\| x \|^2 - \epsilon^2, F; B) = i(-\rho) - i(\rho)$$

and thus bifurcation if these indices are different. See Berestycki 1977, and the extensions for the different sorts of degrees.

Remark 1.6. Bifurcation from infinity. Consider again $F(\lambda, x) = Ax - T(\lambda)x - g(\lambda, x)$.

It is said that one has **bifurcation from infinity** at $\lambda = 0$, if there is a sequence (λ_n, x_n), $\lambda_n \to 0$, $\| x_n \| \to \infty$. Assume $g(\lambda, x) = o(\| x \|)$ when $\| x \| \to \infty$. For $x \neq 0$, let $u = x / \| x \|^2$, then equation (1) becomes $Au - T(\lambda)u - h(\lambda, u) = 0$, where $h(\lambda, u) = \| u \|^2 g(\lambda, u / \| u \|^2)$ if $u \neq 0$, $h(\lambda, 0) = 0$. Then, $h(\lambda, u) = o(\| u \|)$ and it is not difficult to put conditions on h so that the Liapunov-Schmidt reduction works (see Rabinowitz 1973 and later work).

2. Global Bifurcation

Assume that $F(\lambda, x)$ is such that $F(\lambda, 0) = 0$, $F^{-1}(0)$ is compact when intersected with closed bounded sets and that one has a "degree" defined for maps $G(\lambda, x) = (f_0(\lambda, x), F(\lambda, x))$ from bounded sets of $\mathbb{R} \times B$ into $\mathbb{R} \times E$ with the following properties:

    0) $\deg(G(\lambda, x); \Omega)$ is defined if $G(\lambda, x) \neq 0$ on $\partial\Omega$.
    1) If $\deg(G(\lambda, x); \Omega)$ is non-trivial then $G(\lambda, x)$ has a zero in Ω.

2) If $f(\lambda, \tau, x)$ is a continuous deformation such that the corresponding $G(\lambda, \tau, x)$ has a well defined degree, then this degree is constant for all τ.

3) If $G(\lambda, x) \neq 0$ in $\bar{\Omega}_1 \subset \Omega$ then $\deg(G(\lambda, x); \Omega) = \deg(G(\lambda, x); \Omega \backslash \bar{\Omega}_1)$, or

3') If $\Omega = \Omega_1 \cup \Omega_2$, with $\Omega_1 \cap \Omega_2 = \phi$ and $\deg(G(\lambda, x); \Omega_i)$ is defined for $i = 1, 2$, then $\deg(G(\lambda, x); \Omega)$ is the "sum" of the degrees with respect to Ω_1 and Ω_2.

Assume also that if $F(\lambda, x) = 0$ for $\| x \| \leq \epsilon_0$, $\rho - \eta \leq |\lambda| \leq \rho + \eta$ for $0 < \eta < \rho$, then $x = 0$. Let $\mathcal{S} = \{(\lambda, x)/F(\lambda, x) = 0, x \neq 0\}$ and let $\mathcal{C}$ be the connected component of $(0, 0)$ in $\bar{\mathcal{S}} \cup \{0, 0\}$ and $\mathcal{C}'$ be the connected component of $(0, 0)$ in $\bar{\mathcal{S}}$. Clearly $\mathcal{C} \subset \mathcal{C}'$. Let $B = \{(\lambda, x) : |\lambda| \leq \rho, \| x \| < 2\epsilon \leq \epsilon_0\}$.

Theorem 2.1. a) *If* $\deg(\| x \|^2 - \epsilon^2, F(\lambda, x); B)$ *is nontrivial then, if the degree theory has the properties 1-3, either* $\mathcal{C}$ *is unbounded or* $\mathcal{C}$ *returns to a different bifurcation point.*

b) *If the degree theory has property* (3') *and all bifurcation points* $(\lambda_i, 0)$ *on* $\mathcal{C}'$ *have the property that there are numbers* $\rho_i, \eta_i, \epsilon_i$ *such that if* $\| x \| \leq \epsilon_i$, $\rho_i - \eta_i \leq |\lambda - \lambda_i| \leq \rho_i + \eta_i$, *then* $x = 0$, *and if* $\mathcal{C}'$ *is bounded, one has*

$$\Sigma_i \deg(\| x \|^2 - \epsilon^2, F(\lambda, x); B_i) = 0$$

where $B_i = \{(\lambda, x)/|\lambda - \lambda_i| < \rho, \| x \| \leq 2\epsilon < \epsilon_i\}$.

Proof. If $\mathcal{C}$ is bounded and does not return to $(\lambda, 0)$, then $\mathcal{C} = \mathcal{C}'$, or if $\mathcal{C}'$ is bounded, then from "Whyburn's lemma", one may construct a neighborhood of $\mathcal{C}$ such that $\mathcal{S} \cap \partial\Omega = \phi$ and Ω intersects the line $(\lambda, 0)$ only near the bifurcation points (only $(0,0)$ in case (a)), a finite number from the hypothesis on F. Then $\deg(\| x \|^2 - r^2, F(\lambda, x); \Omega)$ is clearly well defined. By choosing r as a deformation parameter, for r large $\| x \|^2 - r^2 < 0$ in Ω, this degree is then trivial (from property (1)). From property (2), this degree in independent of r and for $r < min \{\epsilon_i\}$, from property (3) or (3') this degree is just $\deg(\| x \|^2 - \epsilon^2, F(\lambda, x); B)$ or the sum of the degrees, in case (3').   Q.E.D.

If one has a Liapunov-Schmidt reduction near λ_i then $\deg(\| x \|^2 - \epsilon^2, F(\lambda, x); B)$ is, up to an orientation factor, the jump in the local indices. For the Leray-Schauder degree, if all critical points of $A - T(\lambda)$ are isolated, one may follow these jumps and obtain explicit formulas for the orientation factor, as in Ize 1976, Dancer 1974 or Rabinowitz 1974.

In special cases one may obtain more information. For example, in a slight extension of Rabinowitz' result (1971) on bifurcation from a simple eigenvalue, assume $F(\lambda, x) = Ax - T(\lambda)x - g(\lambda, x)$, near $\lambda = 0$, as in Section 1, and A has 0 as a simple eigenvalue with a one-dimensional bifurcation equation $B(\lambda)x_1 + G(\lambda, x_1)$, such that $B(\lambda)$ changes sign at $\lambda = 0$.

Let $\mathcal{C}_\pm$ be the components of nontrivial solutions bifurcating from $\lambda = 0$ with $x_1 > 0$ (respectively $x_1 < 0$); under the hypothesis on the degree, both $\mathcal{C}_\pm$ satisfy (a) of Theorem 2.1. In fact, if $\mathcal{C}_+$, for instance, is bounded and does not meet any other bifurcation point, one would have that $\deg(\| x \|^2 - \epsilon^2, F(\lambda, x); B_+) = 0$ where $B_+ = \{\| x \| < 2\epsilon, x_1 > 0, |\lambda| < \rho\}$. This degree is equal to $\deg(x_1^2 - \epsilon^2, B(\lambda)x_1; \{|\lambda| < \rho, 0 < x_1 < 2\epsilon\}) = -\eta$, where η is 1, if $B(\lambda)$ passes from negative to positive, and -1 in the opposite case. For $\mathcal{C}_-$, the local degree is also $-\eta$ and the local degree with respect to B in -2η.

Remark 2.1. In Section III, I shall give an argument which does not depend on a formal degree theory, but only on the notion of the extension of maps.

Remark 2.2. In the recent work of Fitzpatrick and Pejsachowicz, a degree theory for Fredholm maps has been defined extending the mod-2 degree of Elworthy-Tromba. The above bifurcation result is a "local" version of what can be achieved with this degree, where interesting results are obtained which take into account the topology of the space of parameters.

Remark 2.3. One has also results for bifurcation from the edge of the continuous spectrum. This is usually handled with approximations by compact maps. See the work of Stuart 1983 and 1987. For a completely different sort of bifurcation, the reader may study the interesting paper by Chen, Jorge and Minzoni 1992; there one has bifurcation from certain points in a continuum of eigenvalues, corresponding to an oblique derivative problem. Although ker A and coker A are finite dimensional, Range A is not closed. However one may reduce to the study of the bifurcation equation, by solving for x_2 using the Nash-Moser implicit function theorem. See also Jorge and Minzoni 1993 for a simpler example.

3. Special Nonlinearities

As seen previously, in particular Remark 1.3, the change of the local index is all one needs for bifurcation if one is willing to allow any nonlinearity. If one puts more restrictions on the nonlinearity, then one may have bifurcation even in the case when there is no change in the index. For instance, assume there is no vertical bifurcation, i.e. in the bifurcation equation, $B(\lambda)x_1 + G(\lambda, x_1) = 0$, suppose that $G(0, x_1) = 0$, with $\| x_1 \| \leq \epsilon$, implies $x_1 =$

0. Then $\deg(B(\lambda)x_1 + G(\lambda_1 x_1); B_0)$ is defined for small λ, where $B_0 = \{x_1 : \| x_1 \| < \epsilon_0\}$. One may also take a neighborhood of the continuum emanating from $(0,0)$. Since, for $\lambda \neq 0$, the zero solution has an index (equal to ± 1, since the hypothesis of invertibility of $B(\lambda)$, for $\lambda \neq 0$, is still valid), then if $\deg(G(0, x_1); B_0) \neq \pm 1$, one needs another solution for $\lambda \neq 0$. Furthermore, by following the arguments of the existence of continua, it is easy to show that there is a continuum satisfying the global alternative.

There are two cases where one can be sure that this situation will happen. For simplicity, I shall assume that $F(\lambda, x)$ is of the form $x - H(\lambda, x)$, with H compact. Extensions to other cases are left to the reader.

a) *Complex analytic maps*

Suppose, again for simplicity, that the critical points of $A - T(\lambda)$ are isolated.

If $F(\lambda, x)$ is complex analytic in x, then C' is unbounded.

(The local degree computation comes from papers by Cronin and Schwartz in the early 60's. The global property is due to Dancer 1973, with a study of the structure of the branches. See also Ize, 1976, p. 82.)

In fact, in this case one knows that the degree is always positive (the real Jacobian of an analytic map is positive), and that if $\deg(F(\lambda, x); \Omega_\lambda) = 1$, then there is just one zero of $F(\lambda, x)$ and there the Frechet derivative is invertible. Hence if C' is bounded and if λ_j is a bifurcation point on C', $\deg (F(\lambda_j, x); \Omega_{\lambda_j}) \geq 2$, $\deg(F(\lambda, x); \Omega_\lambda - B_\lambda) \geq 1$, for λ close to λ_j, where B_λ is a small neighborhood of the isolated solution $x = 0$, with index 1. But if C' is bounded, this last degree should be zero for large $|\lambda|$.

If λ is complex then, under the same hypothesis, one may choose a line in the complex plane which avoids all critical points and starting at $\lambda = 0$. Then the same argument shows that C is unbounded.

b) *$G(\lambda, x)$ with homogeneous nondegenerate leading term*

Assume that $G(\lambda, x) = N(x) + H(\lambda, x)$ with $N(tx) = t^s N(x)$, for $t > 0$, s an integer, and $H(\lambda, x) = o(\| x \|^s)$.

If $N(x) = 0$ only if $x = 0$, and s is even, then one has bifurcation.

In fact in this case $\deg(G(0, x); B_0)$ is well defined if B_0 is small enough, it is the index of $N(x)$ and it is even (Cronin 1964, p. 49). See Dancer 1971, Ize 1976 p. 112.

Remark 3.1. Number of Branches. Degree theory can help count the number of bifurcated branches: assume there is a neighborhood of $(0,0)$, such that it is known that if $B(\lambda, x) \equiv B(\lambda)x + G(\lambda, x) = 0$ on $\partial\Omega$. Then $J(\lambda, x) \neq 0$ at that point, where J is the determinant of the Jacobian matrix (clearly G is supposed to be C^1 in x). Now $\deg(J, B(\lambda, x); \Omega)$ is defined. Assume that for some $\lambda_0, B(\lambda_0, x) \neq 0$ on the section $\partial\Omega_{\lambda_0}$ and that on $\Omega_{\lambda_0}, J(\lambda_0, x) \neq 0$ whenever $B(\lambda_0, x) = 0$. This implies that at λ_0, each zero belongs to a C^1 local branch of solutions parametrized by λ, and that $\deg((\lambda - \lambda_0)J, B(\lambda, x); \Omega)$ is also well defined. Let $\Omega_-(\lambda_0) = \Omega \cap \{(\lambda, x) : \lambda < \lambda_0\}, \Omega_+(\lambda_0) = \Omega \cap \{(\lambda, x) : \lambda > \lambda_0\}$. Then, for ϵ small enough one has

$$\deg((\lambda - \lambda_0)J, B; \Omega) = \deg((\lambda - \lambda_0)J, B; \Omega_-(\lambda_0 - \epsilon))$$
$$+ \deg((\lambda - \lambda_0)J, B; \Omega_+(\lambda_0 + \epsilon))$$
$$+ \deg((\lambda - \lambda_0)J, B; \Omega_+(\lambda_0 - \epsilon) \cap \Omega_-(\lambda_0 + \epsilon)).$$

The first degree on the right is $-\deg(J, B; \Omega_-(\lambda_0))$, the second $\deg(J, B; \Omega_+(\lambda_0))$ and the third is the sum of the local indices at each of the zeros of $B(\lambda_0, x)$. In a neighborhood of such a point, one may replace J by its sign and approximate $B(\lambda, x)$ by the Jacobian matrix. Hence the local index is, by the product theorem, $(\text{Sign } J(\lambda_0, x))^2 = 1$: thus, the third degree is the number $b(\lambda_0)$ of branches crossing Ω_{λ_0}.

$$b(\lambda_0) = \deg((\lambda - \lambda_0)J, B; \Omega) + \deg(J, B; \Omega_-(\lambda_0)) - \deg(J, B, \Omega_+(\lambda_0)).$$

In particular if $B(\lambda, x)$ is not zero on Ω_λ, for $|\lambda| < \rho$, and $J(\lambda_0, x) \neq 0$ if $B(\lambda_0, x) = 0$, for all (λ_0, x) in Ω with $|\lambda_0| \geq \rho$, then the first degree is $\deg(\lambda J, B; \Omega)$, the second is 0 for $\lambda_0 < -\rho$ and $\deg(J, B; \Omega)$ for $\lambda_0 > \rho$, while the third is $\deg(J, B; \Omega)$ for $\lambda_0 < \rho$ and 0 for $\lambda > \rho$. Hence

$$b(\rho) = \deg(\lambda J, B; \Omega) + \deg(J, B; \Omega)$$
$$b(-\rho) = \deg(\lambda J, B; \Omega) - \deg(J, B; \Omega).$$

One recovers a result by Nishimura et al. (1989) obtained by algebraic geometry methods. For example, in the case of a complex analytic map, $J \geq 0$, the first degree is $\deg(B(0, x); \Omega_0)$, the second is 0 and one has the same results as above.

Besides the above two cases where there is bifurcation independently of the linear part, one may also study the case where the leading part of $G(\lambda, x)$ is homogeneous. (For a study via singularity theory, see the paper by Buchner et al 1983, the results presented here are due to Dancer 1971 and Ize 1976).

Assume that $B(\lambda) = \lambda I$ and $G(\lambda, x) = N(x) + R(\lambda, x)$ with $N(tx) = t^s N(x)$ for $t > 0$, $\| R(\lambda, x) \| \leq C \| \lambda \| \| x \|^s + o(\| x \|^s)$.

Following Krasnosel'skii (1964), one says that U_0 of unit norm is a *characteristic ray* of N if $N(U_0) = \mu U_0$. If U_0 is a characteristic ray, it is easy to see that $(t^{s-1}\mu, tU_0), t \geq 0$, is a solution of $\lambda x - N(x) = 0$ and that, if U_0 is not a characteristic ray, then there are no solutions of the bifurcation equation in any small enough conical neighborhood of U_0.

If U_0 is an isolated in S^{d-1} characteristic ray, then it is not difficult to see that one may choose r and ρ small enough such that if one takes a small capped neighborhood of (μ, U_0) on the sphere $|\tilde{\lambda}|^2+ \| U \|^2 = 1 + |\mu|^2$ and one constructs a curved conical neighborhood of the form $\lambda = t^{s-1}\tilde{\lambda}, x = tu, (\lambda, \tilde{u})$ in the capped neighborhood and t such that $|\lambda|^2+ \| x \|^2 \leq \rho^2 + r^2$, then the degree of the pair $(\| x \|^2 +|\lambda|^2 - r^2, \lambda x - N(x) - R(\lambda, x))$ with respect to the curved conical neighborhood is defined and equal to the degree of $(\| x \|^2 +|\lambda|^2 - r^2, \lambda x - N(x))$.

If P is the projection on the tangent plane at U_0 to $S^{d-1}, \lambda x - N(x)$ can be written as

$$-PN(x/ \| x \|^s) \| x \|^s \oplus \| x \| (\lambda - (N(x/ \| x \|)x/ \| x \|) \| x \|^{s-1})x/ \| x \|,$$

which may be deformed to $-PN(U) \oplus \lambda - \mu \| x \|^{s-1}$. Since the index of $(\| x \|^2 +|\lambda|^2 - r^2, \lambda - \mu \| x \|^{s-1})$ with respect to $(\lambda, \| x \|)$ at its only zero is -1, the degree is $-\text{Index}(-PN(U); U_0)$.

Hence if this index is not zero, there is bifurcation along a "curve" $x(\lambda)$ with $x(0) = 0, x(\lambda)/ \| x(\lambda) \| \to U_0$ as $\lambda \to 0$. This is the case if N is a gradient (of $(N(x), x)/(s + 1)$) and U_0 is an isolated relative extremum: in this case the last index is 1 for a maximum and $(-1)^{d-1}$ for a minimum.

Note also that $(\lambda = \pm r, x = 0)$ are also isolated zeros of the pair of equations and their degree with respect to a similar cone is the degree of $(\| x \|^2 +|\lambda|^2 - r^2, \lambda x)$ and this index is 1 for $\lambda > 0$, $(-1)^{d-1}$ for $\lambda < 0$. Now if all characteristic rays are isolated, one may add all these degrees to get the degree of $(\| x \|^2 +|\lambda|^2 - r^2, \lambda x - N(x))$ with respect to the ball of radius $(r^2 + \rho^2)^{1/2}$. Since, on the boundary of this ball, the first equation is never 0, this degree is 0. Hence one recovers the Lefchetz index formula

$$\Sigma \text{ Index}(-PN(U); U_i) = 1 + (-1)^{d-1} = \chi(S^{d-1}).$$

Global results are easy to obtain, for example, for

$$(I - \lambda T)x - N(x) - R(\lambda, x) = 0$$

with T, N, R compact, N and R as above, T self-adjoint. If U_0 is an isolated characteristic ray and $\mathcal{C}$ is the connected component of nontrivial solutions bifurcating in the direction U_0, then if $\bar{\mathcal{C}}$ is compact and meets all bifurcation points $(0, \lambda_j)$ along isolated characteristics rays U_i^j, one has

$$\Sigma(-1)^{\beta_j} \text{ sign } \lambda_i \text{ Index}(PN; U_i^j) = 0$$

where N_i is the projection of N restricted to $\ker(I - \lambda_i T)$ onto that kernel and β_j is the sum of the multiplicities of $I - \mu\lambda_j T$, $0 \leq \mu \leq 1$. The proof, very similar to the previous results, is given in Ize 1976 and can be extended to other situations, as in the previous references and in Buchner et al, 1983, Hoyle 1980, Lopez-Gomez, 1985, and so on...).

Remark 3.2. Equations in cones. Since there will be other contributions on this subject, let me take the simplest possible case. Assume E has a cone K (i.e. if $x \in K$ also αx for $\alpha > 0$, if x and y are in K, also $x + y$ and if $x \in k$ then $-x \notin K$ except for $x = 0$) and assume $\tilde{F}(\lambda, x) = \lambda T_0 x + g(\lambda, x)$ is a positive map for all positive λ's, with T and g compact, i.e. it maps $K\backslash\{0\}$ into K^0, the interior of K : so here I am taking the case where $K^0 \neq \phi$. Let Ω be an open bounded subset of K. Then there is a retraction $r : E \to K$ (i.e. $rx = x$ on K). If $x - \tilde{F}(\lambda, x)$ is non-zero on $\partial\Omega$, then $\deg(x - \tilde{F}(\lambda, rx); r^{-1}(\Omega))$ is well defined, since if $x = \tilde{F}(\lambda, rx)$ then x belongs to K, $rx = x$, and the zeros of the map belong to a bounded subset of $r^{-1}(\Omega)$.

Assume that T_0 is strongly positive (i.e. $T_0 K^0 \subset K^0$); then the Krein-Rutman theorem says that the spectral radius λ_0^{-1} of T_0 is a simple eigenvalue of T_0, corresponding to a positive eigenvector x_0 and there is no other eigenvalue with positive eigenvector. Thus, the index of $x - \tilde{F}(\lambda, rx)$ is well defined, for $\lambda \neq \lambda_0$, and equal to the index of $x - \lambda T_0 rx$. Note that if $x = \lambda T_0 rx$, then, since rx belongs to K, x is in K for $\lambda \geq 0$, and x must be 0 if $\lambda \neq \lambda_0$. This implies that the index is constant for $0 \leq \lambda < \lambda_0$ and for $\lambda > \lambda_0$. In the first case the index is 1 (for $\lambda = 0$); in the second it is equal to the degree of $x - \lambda T_0 rx - \epsilon x_0$ in a small neighborhood of 0, provided ϵ is small enough. However if $x = \lambda T_0 rx + \epsilon x_0$, then, for λ and ϵ positive, x belongs to K and $\lambda T_0(x + tx_0) = x + tx_0$, for $t = \epsilon/(\lambda\lambda_0^{-1} - 1)$. In particular, $x + tx_0$ is in K if $\lambda > \lambda_0$ but then one would have another eigenvalue with a positive eigenvector. This implies that the above equation has no solution and the index is 0 for $\lambda > \lambda_0$. This change of index will initiate a branch bifurcating from $(\lambda_0, 0)$.

The alternative argument will work as well here, provided the continuum does not escape from K. This will be the case if $\tilde{F}(0, x) = 0$. Thus, one has a continuum in K which cannot return to another bifurcation point in K, since λ_0 is the only one. Hence, the continuum is unbounded. This result is due to Dancer (1973), see also Amann (1977) for further results on the type of the continuum in case $\tilde{F}$ in increasing and, at this point, I shall refer to Dancer's paper in this volume.

Note that if $\tilde{F}$ is defined on E, then the above hypothesis are so strong that one has a continuum in E, bifurcating from $(\lambda_0, 0)$. From the remarks at the end of the proof of Theorem 2.1, the continuum starts as $tx_0 + w$

where $t > 0$ and $\| w \| = o(t)$, i.e. in K. The positivity of $\tilde{F}$ will keep the continuum in K provided x does not tend to 0. If this is the case for a sequence $\{x_n\}$ then $(x_n - \tilde{F}(\lambda_n, x_n))/ \| x_n \| = 0$, λ_n converging to λ_1, and, from the compactness of T_0, $\{x_n/ \| x_n \|\}$ has a subsequence which converges to a solution of $x - \lambda_1 T_0 x = 0$, with $\| x \| = 1$. Thus, λ_1 must be λ_0, and the continuum cannot return to the trivial solutions and has to be unbounded.

Note that a similar argument will give an unbounded continuum for a nonlinear Sturm-Liouville equation with boundary conditions giving nodal properties.

Remark 3.3. Variational problems. Assume that the equation $F(\lambda, x) = Ax - T(\lambda)x - g(\lambda, x)$, of Section I, is defined on a continuously embedded subspace B of a Hilbert space E and that $F(\lambda, x)$ is the gradient of a C^2 functional $\Phi(\lambda, x)$, i.e. $\Phi_x(\lambda, x)h = (F(\lambda, x), h)$ for all h in B. Hence, A and $T(\lambda)$ are self-adjoint operators. If A is a Fredholm operator of index 0 and with an isolated eigenvalue at 0, then one may choose the projection Q to be $I - P$ and the decomposition of E as $\ker A \oplus \text{Range } A$ is orthogonal. It is easy to see that the pseudo-inverse of A and $B(\lambda)$ are symmetric.

Let $x_2(\lambda, x_1)$ be the unique solution of $H(\lambda, x_1, x_2) = 0$, with $H(\lambda, x_1, x_2) = x_2 - (I - KQT(\lambda))^{-1}KQ(T(\lambda)x_1 + g(\lambda, x_1 + x_2))$. It is clear that $x_2(\lambda, x_1)$ is C^1.

Let $\Psi(\lambda, x_1) = \Phi(\lambda, x_1 + x_2(\lambda, x_1))$; then the Frechet derivative of Ψ is such that, for h in $\ker A$

$$
\begin{aligned}
\Psi(\lambda, x_1 + h) - \Psi(\lambda, x_1) &= \Phi_x(\lambda, x_1 + x_2(\lambda, x_1))(h + x_2(\lambda, x_1 + h) \\
&\quad - x_2(\lambda, x_1)) + o(h) \\
&= \Phi_x(\lambda, x_1 + x_2(\lambda, x_1))(I - KQT(\lambda))^{-1}h + o(h) \\
&= (F(\lambda, x_1 + x_2(\lambda, x_1)), (I - KQT(\lambda))^{-1}h) + o(h) \\
&= (B(\lambda)x_1 + G(\lambda, x_1 + x_2(\lambda, x_1)), h) + o(h),
\end{aligned}
$$

where in the last equality, one uses that $F(\lambda, x_1 + x_2(\lambda, x_1))$ belongs to $\ker A$, while $KQT(\lambda)h$ belongs to Range A. Hence, $\nabla\Psi(\lambda, x_1) = B(\lambda)x_1 + G(\lambda, x_1 + x_2(\lambda, x_1))$. This result is essentially due to Rabinowitz (1977) and to Kielhöfer (1988) (see also Rabier (1989)).

Note that if $T(\lambda)$ maps $\ker A$ into itself (and hence Range A into itself), then $B(\lambda) = -PT(\lambda)P$ and $G(\lambda, x) = -Pg(\lambda, x)$. This is the case if $F(\lambda) = \lambda I$.

Let us consider first the case where $B(\lambda) = -\lambda I$ and $G(\lambda, x_1 + x_2(\lambda, x_1)) = G(x_1)$ is independent of λ. Then the classical result of Krasnoselskii (1964) says that $(0, 0)$ is always a bifurcation point. In fact the problem

$G(x_1) = \lambda x_1$, with $G(x_1) = \nabla(\Psi(0, x_1))$, may be thought of as looking for critical points of $\Psi(0, x_1)$ on the sphere $\| x_1 \| = \epsilon, \lambda$ acting as a Lagrange multiplier: hence there are at least two solutions on each sphere with $\lambda = 0(\epsilon)$.

This result does not tell the number of solutions as λ varies, or if there is a continuum bifurcating from (0,0). For the first question, one has the following result:

Theorem (Rabinowitz 1977). a) *Either* 0 *is not an isolated zero of* $G(x_1)$, *or* b) 0 *is a strict local minimum of* $\Psi(0, x_1)$ *(respectively a strict local maximum) and there is a* $\epsilon > 0$, *such that for any* λ, *with* $0 < \lambda < \epsilon$, $G(x_1) = \lambda x_1$ *has at least two nontrivial solutions (respectively for* λ, *with* $-\epsilon < \lambda < 0$), *or* c) 0 *is an isolated critical point of* $\Psi(0, x_1)$, *which is not a maximum nor a minimum and there is a* $\epsilon > 0$, *such that for* λ's, *with* $-\epsilon < \lambda < \epsilon, G(x_1) = \lambda x_1$ *has at least one nontrivial solution.*

Proof. The original proof uses Ljusternik-Schnirelman category and a primitive version of the Mountain Pass Lemma. The same technique will give at least d pairs of nontrivial solutions in (b) if $\Psi(0, x_1)$ is even (here d is the dimension of ker A). In fact if 0 is an isolated zero of $G(x_1)$, then one may choose a small ρ such that $G(x_1) \neq 0$ if $0 < \| x_1 \| \leq \rho$. Hence $G(x_1) - \lambda x_1$ is not zero for $\| x_1 \| = \rho$ and $|\lambda| \leq \epsilon_1$, for ϵ_1 small enough. Furthermore $-\Psi(0, x_1) + \lambda \| x_1 \|^2 /2$ has a strict local minimum at $x_1 = 0$ if λ is positive (the quadratic term dominates Ψ). If 0 is a strict local maximum for $-\Psi(0, x_1)$, then for $0 < \lambda < \epsilon_2$, one has that $-\Psi(0, x_1) + \lambda \| x_1 \|^2 /2 < 0$ for $\| x_1 \| = \rho$ (or for some point on that sphere in case of (c)). Hence from the Mountain Pass Lemma, there is a critical point corresponding to a positive critical value given by the infimum of all the maxima of the functional on the curves joining 0 to a point with $\| x_1 \| = \rho$ and a negative value of the functional. In case (b) one has at least one positive maximum of $-\Psi(0, x_1) + \lambda \| x_1 \|^2 /2$ in the ball B given by $\| x_1 \| \leq \rho$.

Now, if $\bar{x}$ is an isolated extremum of a functional $\phi(x)$ from $\mathbb{R}^d$ into $\mathbb{R}$, then the index of $\nabla\phi$ is defined at $\bar{x}$ and is $(-1)^d$ for a maximum and 1 for a minimum: if $\phi(\bar{x}) = 0$, take the level surfaces $\phi^{-1}(\epsilon)$, $\epsilon < 0$ for a maximum, $\epsilon > 0$ for a minimum. The component of $\bar{x}$ of the set $\phi^{-1}[\epsilon, 0]$ is a bounded neighborhood of $\bar{x}$, such that $\nabla\phi$ is the inward normal for a maximum and outward for a minimum. The index of $\nabla\phi$ at $\bar{x}$ is the Euler characteristic of $\phi^{-1}[\epsilon, 0]$ multiplied by $(-1)^d$ in the first case. Since this set is contractible, by following the gradient flow, one has the result.

Hence, if 0 is an isolated minimum of $\Psi(0, x_1)$ and if, for some $\lambda > 0$, the minimum of $\Psi(0, x_1) - \lambda \| x_1 \|^2 /2$ is the only nontrivial critical point, then its index is 1, while the index of 0 is $(-1)^d$ (it is a local maximum).

Then $deg\,(G(x_1) - \lambda x_1; B) = 1 + (-1)^d = \deg(G(x_1); B)$. But, since 0 is an isolated minimum for $\Psi(0, x_1)$, this last index is 1. This contradiction finishes the proof. For more refined results, I will refer to Rabinowitz's paper in this volume.

There is still the question of the continuum. For the equation $G(x_1) - \lambda x_1$ one may not expect a global continuum if d is even. The following example, derived from Takens (1972), shows that the set of nontrivial solutions may be quite disconnected.

In $\mathbb{R}^2$ and in polar coordinates, consider the even function

$$\phi_1(\rho, \theta) = \phi(\rho)(4^{-1} \sin 4\theta + 24(\rho - \rho_0)\rho_0^{-3} \sin 2\theta)$$

where $\phi(\rho)$ is a smooth function, positive for $1 < \rho < 2$ and 0 outside this interval, and $\rho_0^{-1} = (5 - \cos 2\theta)^{1/2}(12)^{-1/2}$ corresponds to the ellipse $x^2/3 + y^2/2 = 1$. It is easy to see that $\partial\phi_1/\partial\theta$ is zero for $\rho = \rho_0(1 + \rho_0^2/[160(\sin^4\theta - 2/5)])$ and for $\rho \leq 1$ or $\rho \geq 2$. In the annulus, $1 < \rho < 2$, $\nabla\varphi_1$ is radial on four curves: two symmetric "ears" connected to the inner circle and inside the ellipse, with a maxium on the x-axis at $x_0 = \sqrt{3}(1 - 3/64)$, and two symmetric "lips" connected to the outer circle and outside the ellipse with a minimum (for the upper curve) at $y_0 = \sqrt{2}(1 + 1/48) < x_0$. Thus any circle in the annulus intersects these curves, however, on the ellipse $\partial\phi_1/\partial\theta$ is not zero.

The radial derivative $\partial\phi_1/\partial\rho$ is 0 on the axes and, if $\phi(\rho) \equiv 1$ for $\sqrt{2} < \rho < \sqrt{3}$, if $\nabla\phi_1(x, y) = \lambda(x, y)^T$, λ has the sign of $\sin 2\theta$ in this smaller annulus. Define $\Phi(\rho, \theta) = e^{-1/\rho^2}\phi_1(2^i\rho, \theta)$ for $2^{-i} \leq \rho \leq 2^{-i+1}, i \geq 0$. It is easy to see that Φ is C^∞ and that the set of points where $\nabla\Phi$ is radial is the sequence of circles $\rho = 2^{-i}$ with scaled down external "ears" and internal "lips", disconnected by the ellipses $x^2/3 + y^2/2 = 2^{-2i}$. Other examples, but not explicit, were given by Böhme (1972) and Marino (1977).

For the general problem $\nabla\Psi(\lambda, x_1) = B(\lambda)x_1 + G(\lambda, x_1)$, where λ does not appear as a Lagrange multiplier, there is the following result due to Chow and Lauterbach (1988) using the center manifold reduction, and to Kielhöfer (1988) using the Liapunov-Schmidt approach. (See also Rabier 1989). Assume $B(\lambda)$ is invertible for $\lambda \neq 0$. Since $B(\lambda)$ is symmetric it has a well defined signature $\sigma(\lambda)$, constant for $\lambda > 0$ and for $\lambda < 0$, denoted by σ_+ and σ_-.

Theorem. *If $\sigma_+ \neq \sigma_-$, then (0,0) is a bifurcation point.*

Proof. One may mimic the proof of the previous theorem, replacing the Mountain Pass Lemma by the Saddle point theorem, or by any change in the linkings. An essentially equivalent proof is to use Conley's index. In fact if there is no vertical bifurcation then, if one consider the dynamical system

$$\frac{dx_1}{dt} = \nabla \Psi(\lambda, x_1)$$

0 is an isolated stationary point, for $\lambda = 0$, and there is an isolating neighborhood N containing it. This isolating neighborhood will retain its character for $|\lambda|$ small enough. Hence the Conley index of the invariant set S_λ contained in N is well defined and independent of λ. If S_λ reduces to 0 then its Conley index is the Morse index of 0, i.e. $(d - \sigma(\lambda))/2$. From the hypothesis, for each λ small enough, either S_λ or $S_{-\lambda}$ contains a nontrivial invariant set, in particular at least one complete orbit. If this orbit is not a stationary point, then, since $d\Psi(\lambda, x_1(t))/dt = \| \nabla \Psi(\lambda, x_1(t)) \|^2$ and the orbit is bounded, it must tend, as t goes to $+\infty$ and to $-\infty$, to two different rest points (if they are the same, then the orbital derivative would be 0 at some finite time). Hence, one of the rest points is not 0 and one has a solution of $\nabla \Psi(\lambda, x_1) = 0$ in N.   Q.E.D.

The above argument is due to Conley (1978) and can be generalized to other situations. I will refer again to the other contributions in this volume.

II. MULTIPARAMETER BIFURCATION

Many parameters may be present in a nonlinear problem and there is no a priori reason to vary just one of them. From the point of view of local bifurcation, under the hypothesis of Section I, one is reduced to the finite dimensional equation from $\mathbb{R}^k \times \mathbb{R}^d$ into $\mathbb{R}^{d*}$

$$B(\lambda)x_1 + G(\lambda, x_1) = 0,$$

where $B(0) = 0$ and $G(\lambda, x_1) = o(\| x_1 \|)$.

The purpose of this section is to give "necessary" (in a sense which will be made more precise later) and sufficient conditions for local and global bifurcation. From the topological point of view these conditions will depend only on the linear part $B(\lambda)x_1$. It is then important to have a precise knowledge of the set of points λ where $B(\lambda)$ is singular. For the most part in this section, I shall treat the zero index case, that is $d = d^*$. In this case, the singular set will be the set C_0 of point where $\det B(\lambda) = 0$. If $B(\lambda)$ is complex analytic in λ in $\mathbb{C}^k$, then this set is a union of algebraic varieties of complex dimension $k - 1$ and one could choose a complex direction λ, tranversal to C_0, and reduce the problem to one complex parameter. In the real case C_0 may be quite complicated, but one may assume that, near $\lambda = 0$, it has a "k_2-dimensional tangent space" in the sense that $\mathbb{R}^k = \mathbb{R}^{k_1} \times \mathbb{R}^{k_2}$ and for $\lambda = (\tilde{\lambda}, 0), \tilde{\lambda}$ in $\mathbb{R}^{k_1}$, then det $B(\tilde{\lambda}, 0) \neq 0$ for $\tilde{\lambda} \neq 0$ but close to 0. As we shall see in Section III, the k_2 free parameters will add to the dimension of the bifurcated surfaces if one has bifurcation in the transversal subspace $\mathbb{R}^{k_1}$.

The first results on multiparameter ($k_1 > 1$) bifurcation are given in Ize [1976], the role of the J-homomorphism was discovered independently by Alexander and Yorke (1978) and Ize. The results given in this chapter were published in Ize (1988), however the present formulation will avoid obstruction theory.

1. Sufficient Conditions for Local Bifurcation

Consider the bifurcation equation, from $\mathbb{R}^k \times \mathbb{R}^d$ into $\mathbb{R}^d$,

$$B(\lambda)x + G(\lambda, x) = 0 \tag{1}$$

with $B(0) = 0$ and $G(\lambda, x) = o(\| x \|)$. Here $B(\lambda)$ will be assumed to be invertible for λ small and nonzero. Thus for $\| \lambda \| = \rho$, there is $\epsilon_0(\rho, G)$ such that the equation (1) has the only solution $x = 0$ if $\| x \| \leq \epsilon_0$.

Definition: $(0,0)$ *is a point of "linearized local bifurcation" if and only if for all $G(\lambda, x) = o(\| x \|)$, the bifurcation equation has a solution with $\| x \| = \epsilon, \| \lambda \| \leq \rho$, for all $\epsilon \leq \epsilon_0(\rho, G)$.*

Remark 1.1. If $B(\lambda)$ is linear in λ, that is $B(\lambda) = \lambda_1 B_1 + \ldots + \lambda_k B_k$, such that $\| B(\lambda)x \| \geq C \| \lambda \| \| x \|$ (i.e. a nondegenerate "quadratic form"), then let $C_k = k$ for $k = 1, 2, 4, 8$, $C_k = 4$ for $k = 3$, $C_k = 8$ for $k = 5, 7$, $C_{k+8} = 16C_k$ for $k \geq 8$; one now has the following restriction (Alexander-Yorke 1978): $d = rC_k$ with r as a positive integer.

In fact let $d = (2a+1)2^b$, with $b = c+4e, 0 \leq c \leq 3$. Define the Hurwitz-Radon number $\rho(d)$ as $2^c + 8e$. Then it is known that there are at most $\rho(d) - 1$ linearly independent tangent vector fields on S^{d-1}. Note that here $x, B_1^{-1}B_2x, \ldots, B_1^{-1}B_kx$ are linearly independent vector fields and that, by the Gram-Schmidt orthogonalization, one has $k - 1$ tangent vector fields, hence $k \leq \rho(d)$. Since $C_{\rho(d)}$ is easily computed, with $d = (2a+1)C_{\rho(d)}$, and $C_{\rho(d)} = 2^\alpha C_k$, with $\alpha \geq 1$ if $k \equiv 1, 2, 4, 8$, [8], the result follows easily. In particular, if $d = k$, then, since $C_k > k$ for $k \neq 1, 2, 4, 8$, the only possibility is $k = 1, 2, 4, 8$ where $B(\lambda)x$ is a multiplication on $\mathbb{R}^k$, corresponding to $\mathbb{R}$, $\mathbb{C}$, $\mathbb{H}$ and the Cayley numbers respectively.

From the topologoical point of view, one may look for solution of the bifurcation equation, in the following ways:

1) Look for solutions of $B(\lambda)x_0 + G(\lambda, x_0) = 0$, with x_0 fixed, $\| x_0 \| = \epsilon$, in $B_1 = \{\lambda : \| \lambda \| \leq \rho\}$.
2) Look for solutions of $(B(\lambda)x + G(\lambda, x), x - x_0) = (0, 0)$ in $B_2 = \{(\lambda, x) : \| \lambda \| \leq \rho, \| x \| \leq 2\epsilon)$.
3) Study $(B(\lambda)x + G(\lambda, x), \| x \| - \epsilon) = (0, 0)$ in $B_3 = B_2$.
4) Find zeros of $B(\lambda)x + G(\lambda, x)$ in $B_4 = B_1 \times S^{d-1}$, $S^{d-1} = \{x : \| x \| = \epsilon\}$.

The topological facts used here will be the following:

a) *Let $F : S^m = \partial B^{m+1} \to \mathbb{R}^n \backslash \{0\}$. Then any extension of F to B^{m+1} has a zero if and only if F is not deformable to a constant map.*

b) *(Borsuk extension theorem). Let $A \subset X$ be closed subsets of $\mathbb{R}^{m+1}$ and assume $F_0, F_1 : A \to \mathbb{R}^n \backslash \{0\}$ are homotopic maps; then F_0 has nonzero extension to X if and only if F_1 does have a nonzero extension to X, and the extensions are homotopic.*

Proof. a): If F extends to $\tilde{F}$, define the homotopy $F(t, x) : I \times S^m \to \mathbb{R}^n \backslash \{0\}$, by $F(t, x) = \tilde{F}((1 - t)x)$, deforming radially F to $\tilde{F}(0)$. On the other hand, if $F(t, x)$ deforms $F(x)$, for $t = 1$, to the constant $F(0, x)$, then $\tilde{F}(x) = F(\| x \|, x/ \| x \|)$ is an appropriate deformation.

b): Let $\tilde{F}_0 : X \to \mathbb{R}^n \backslash \{0\}$ be the extension of F_0 and $F(t, x) : A \to \mathbb{R}^n \backslash \{0\}$ be the homotopy. Let, by Tietze's extension theorem, $G(t, x) : I \times X \to \mathbb{R}^n$, be any extension of the map defined as $F(t, x)$ on $I \times A$ and $\tilde{F}_0$ on $\{0\} \times X$ and let $B = \{x \in X : \exists t \text{ such that } G(t, x) = 0\}$. Then $A \cap B = \phi$, A and B are closed subsets of X. Let $\varphi(x) : X \to [0, 1]$, be a partition function such that $\varphi(B) = 0, \varphi(A) = 1$. Define $\tilde{G}(t, x) = G(\varphi(x)t, x)$. Then, if $\tilde{G}(t, x) = 0$ for some t, x belongs to $B, \varphi(x) = 0$, but $G(0, x) = \tilde{F}_0(x) \neq 0$. Furthermore $\tilde{G}(0, x) = \tilde{F}_0(x), \tilde{G}(1, x) = F_1(x)$ if $x \in A$ and $\tilde{G}$ is a homotopy from $\tilde{F}_0(x)$ to $\tilde{F}_1(x) = \tilde{G}(1, x)$.   Q.E.D.

In the four formulations of the bifurcation problem, the maps are nonzero on $A = \partial B_i$, $i = 1, \ldots, 4$, and one will have a solution if any extension to B_i has a zero. From (b), this should be true for any valid deformation on ∂B_i; in particular one may deform $G(\lambda, x)$ to 0 and replace $B(\lambda)$ by $\| \lambda \| \rho^{-1} B(\lambda \rho / \| \lambda \|)$, via a linear homotopy on ∂B_i. From (a), in the cases (1)–(3), one has a ball and one will need that the map is not homotopic to a constant.

In case (1), $B(\lambda)x_0$ maps S^{k-1} into $\mathbb{R}^d \backslash \{0\}$ and defines an element of the group $\Pi_{k-1}(S^{d-1})$ of all homotopy classes of such maps. In case (2), $(B(\lambda)x, x - x_0)$ is easily seen to be linearly deformable to $(B(\lambda)x_0, x - x_0)$ and to $(B(\lambda)x_0, x) = \Sigma^d(B(\lambda)x_0)$, the d-fold suspension.

In general if $F(\lambda) : S^{k-1} \to \mathbb{R}^d\backslash\{0\}$, one defines the suspension of $F, \Sigma F : S^k \to \mathbb{R}^{d-1}\backslash\{0\}$, by $(\Sigma F)(\lambda, \mu) = (\|\lambda\| F(\lambda/\|\lambda\|), \mu)$.

Remark 1.2. Facts about $\Pi_{k-1}(S^{d-1})$.

1) Let $F : S^{k-1} \to \mathbb{R}^d\backslash\{0\}$; then F is homotopic to $F(\lambda)/\|F(\lambda)\|$ and one may deform F (via a rotation and a scaling which are deformable to the identity) such that $F(0, \ldots, 0, -1) = (1, 0, \ldots, 0)$.

2) Let $\lambda = (\tilde{\lambda}, \lambda_k)$; then one may pull the sphere over itself in such a way that the south hemisphere goes to the south pole by the deformation $(\varphi(t, \lambda_k)\tilde{\lambda}, \lambda_k + t(|\lambda_k| - 1))$, where $\varphi(t, \lambda_k) = [1 - t(2\lambda_k + t(1 - |\lambda_k|))/(1 + |\lambda_k|)]^{1/2}$ is a factor such that the map has norm 1. From this deformation one may show easily that $\Pi_{k-1}(S^{d-1}) \cong [\tilde{F} : (B^{d-1}, S^{d-2}) \to (\mathbb{R}^d\backslash\{0\}, (1, 0, \ldots, 0))]$, the class of homotopy maps from $B^{d-1} = \{\tilde{\lambda} : \|\tilde{\lambda}\| \le 1\}$ into $\mathbb{R}^d\backslash\{0\}$, sending the boundary of B^{d-1} into the fixed point $(1, 0, \ldots, 0)$.

3) $\Pi_{k-1}(S^{d-1})$ is an abelian group.

4) $\Pi_{k-1}(S^{d-1}) = 0$ if $k < d$, $\mathbb{Z}$ if $k = d$ where the class $[F]$ is characterized by its degree. Furthermore $\Pi_{k-1}(S^1) = 0$ if $k > 2$; $\Pi_3(S^2) \cong \mathbb{Z}$, generated by the Hopf map $\eta(\lambda_1, \lambda_2) = (2\bar{\lambda}_1\lambda_2, |\lambda_1|^2 - |\lambda_2|^2)$, where $\lambda_1, \lambda_2 \epsilon \mathbb{C}$; $\Pi_{n+1}(S^n) \cong \mathbb{Z}_2$ for $n > 2$, generated by the suspension of the Hopf map.

5) **Freudenthal suspension theorem.** $\Sigma : \Pi_{k-1}(S^{d-1}) \to \Pi_k(S^d)$ is onto for $k = 2d - 2$ and an isomorphism for $k < 2d - 2$. Hence the second formulation will give the same result as the first, if $k < 2d - 2$.

Note that in the four formulations one may replace $B(\lambda)$ by $\|\lambda\|$ $\rho^{-1}B(\lambda\rho/\|\lambda\|)$ and, if $B(t, \lambda) : I \times S^{k-1} \to GL(\mathbb{R}^d)$ is a family of invertible (for $\|\lambda\| = \rho$) matrices, then one may deform $\|\lambda\| \rho^{-1}B(0, \lambda\rho/\|\lambda\|)$ to $\|\lambda\| \rho^{-1}B(1, \lambda\rho/\|\lambda\|)$, i.e. in each of the four formulations, the topological invariants will depend only on the homotopy class of $B(\lambda) : S^{k-1} \to GL(\mathbb{R}^d)$, that is, of $[B(\lambda)]$ in $\Pi_{k-1}(GL(\mathbb{R}^d))$.

The first approach corresponds to the evaluation map:

$$P_* : \Pi_{k-1}(GL(\mathbb{R}^d)) \to \Pi_{k-1}(S^{d-1}),$$

with $P_*[B(\lambda)] = [B(\lambda)x_0]$.

The third approach corresponds to the Whitehead's $J-$homomorphism: $J : \Pi_{k-1}(GL(\mathbb{R}^d)) \to \Pi_{k+d-1}(S^d)$, with $J[B(\lambda)] = [B(\lambda)x, \|x\| - \epsilon] = [B(\lambda)x, \|x\| - \|\lambda\|] = [B(\lambda)x, \rho - 2\|\lambda\|]$, via linear deformations.

Theorem 1.1. a) *If $P_*[B(\lambda)] \ne 0$ then one has bifurcation in all directions in S^{d-1}.*

b) *If $J[B(\lambda)] \ne 0$, then one has local bifurcation.*

Proof. It is enough to note that in (a), one may rotate x_0 to any other direction x_1 in S^{d-1} without changing $P_*[B(\lambda)]$. The rest of the proof was already given.   Q.E.D.

Remark 1.3. Facts about $\Pi_{k-1}(GL(\mathbb{R}^d))$.

1) $\Pi_0(GL(\mathbb{R}^d)) \cong \mathbb{Z}_2$, where $B(\lambda)$ is nontrivial if and only if $\det B(\lambda)$ changes sign as λ crosses 0. In this case $J[B(\lambda)] = i_- - i_+$, as seen in Section I.

2) If $\det B(\lambda) > 0$, then $\Pi_1(GL_+(\mathbb{R}^d)) \cong \mathbb{Z}$ if $d = 2$, generated by $\begin{pmatrix} \lambda_1 & -\lambda_2 \\ \lambda_2 & \lambda_1 \end{pmatrix}$, and $\mathbb{Z}_2$ if $d > 2$, generated by $\begin{pmatrix} \lambda_1 & -\lambda_2 & 0 \\ \lambda_2 & \lambda_1 & 0 \\ 0 & 0 & I \end{pmatrix}$.

3) $\Pi_{k-1}(GL_+(\mathbb{R}^d))$ is an abelian group with $[B(\lambda)] + [D(\lambda)] = [B(\lambda)D(\lambda)]$.

4) **The Gram-Schmidt process.** If B is in $GL(\mathbb{R}^d)$, then the columns $B_1, \ldots, B_d$ of B can be orthonormalized by taking $D_1 = B_1/\|B_1\|$, $D_2' = B_2 - (B_1 \cdot B_2)D_1$, $D_2 = D_2'/\|D_2\|$, etc... and one may write $BP(B) = D$, where D is in $O(d)$ with columns $D_1, D_2, \ldots, D_d$, and $P(B)$ is an upper triangular matrix with $\| D_i' \|^{-1}$ in the diagonal. Hence, $GL(\mathbb{R}^d) \cong O(d) \times \mathbb{R}^{(d^2+d)/2}$ (the number of nonzero terms in $P(B)$), and $\Pi_{k-1}(GL_+(\mathbb{R}^d)) \cong \Pi_{k-1}(SO(d))$, by deforming the off diagonal terms in $P(B(\lambda))$ to 0 and the diagonal terms to 1. holahola

5) **Covering homotopy lemma.** *Let $B(\lambda) : S^{k-1} \to GL(\mathbb{R}^d)$ be such that its first column $B_1(\lambda)$ is homotopic, in $\Pi_{k-1}(S^{d-1})$, to $\tilde{B}_1(\lambda)$ via $B_1(t, \lambda)$. Then there is $B(t, \lambda)$ in $GL(\mathbb{R}^d)$ such that $B(0, \lambda) = B(\lambda)$ and the first column of $B(t, \lambda)$ is $B_1(t, \lambda)$.*

The proof can be found in any book on homotopy theory. It is also valid if one replaces $GL(\mathbb{R}^d)$ by $SO(d)$ or by $GL(\mathbb{C}^d)$ or $U(d)$.

6) Let $\Pi_k(SO(d), SO(d-1)) = \{B(\lambda) \in SO(d)$, for $\| \lambda \| \leq \rho$, and $B(\lambda)|_{S^{k-1}} = \begin{pmatrix} C(\lambda) & 0 \\ 0 & 1 \end{pmatrix}\}$ and let $P_*' : \Pi_k(SO(d), SO(d-1)) \to \Pi_k(S^{d-1})$ be defined by $P_*'[B(\lambda)] = [B(\lambda)(0, \ldots, 0, 1)^T]$, (note that for $\| \lambda \| = \rho$, one has $(0, \ldots, 0, 1)^T$, that is, via a rotation, one has an element of $\Pi_k(S^{d-1})$ as in fact (2) of Remark 1.2).

Then P_*' is an isomorphism.

Proof. a) P_*' is clearly a morphism.

b) P_*' is one-to-one: in fact if $P_*'[B_0(\lambda)] = P_*'[B_1(\lambda)]$, i.e. if $B_0(\lambda)x_d$ is homotopic to $B_1(\lambda)x_d$ via $h(t, \lambda)$ where $x_d = (0, \ldots, 0, 1)^T$, then $B_1^{-1}(\lambda) B_0(\lambda)x_d$ is homotopic to x_d, and from the covering homotopy lemma, there is $B(t, \lambda)$ in $SO(d)$ such that $B(t, \lambda)x_d = B_1^{-1}(\lambda)h(t, \lambda)$ and $B(0, \lambda) = B_1^{-1}(\lambda)B_0(\lambda)$. For $\| \lambda \| = \rho$, $h(t, \lambda) = x_d$ and $B(t, \lambda)x_d = x_d$. For $\| \lambda \| \leq \rho$, $B(1, \lambda)x_d = x_d$, i.e. $B(1, \lambda)$ belongs to $SO(d-1)$. Thus, $B(0, \lambda)$ is deformable to $B(1, \lambda) = \begin{pmatrix} C(\lambda) & 0 \\ 0 & 1 \end{pmatrix}$. Next $C(\lambda)$ is deformable radially to $C(0)$ and then to I; thus $[B_1^{-1}(\lambda)B_0(\lambda)] = [I] = 0$.

c) P_*' is onto: Let $F(\lambda) : B^k \to S^{d-1}$ with $F|_{S^{k-1}} = x_d$. Let $A(t) \in SO(d)$, be such that $A(0) = I$ and $A(1)F(0) = x_d$. Then the deformation $A(t(\rho - \|\lambda\|)/\rho)F(\lambda)$ enables us to assume that $F(\lambda)$ has the property that $F(0) = x_d$. Define then $F(t, \lambda) = F(t\lambda)$ which deforms $F(\lambda) = x_d$ on S^{k-1} to $F(0) = x_d$. In the covering homotopy lemma, take $B(\lambda) = I$; then there is $B(t, \lambda) : I \times S^{k-1} \to SO(d)$, such that $B(0, \lambda) = I$, $B(t, \lambda)x_d = F(t\lambda)$. Define $\tilde{B}(\lambda) = B(\|\lambda\|/\rho, \lambda\rho/\|\lambda\|)$. Then $\tilde{B}(0) = I$ and $\tilde{B}(\lambda)$ is continuous for $\|\lambda\| \le \rho$. Furthermore $\tilde{B}(\lambda)x_d = F(\lambda)$ and for $\|\lambda\| = \rho$, $\tilde{B}(\lambda)x_d = x_d$. Q.E.D.

7) For $C(\lambda)$ in $GL_+(\mathbb{R}^{d-1})$, let $i_*C(\lambda) = \begin{pmatrix} C(\lambda) & 0 \\ 0 & 1 \end{pmatrix}$, $P_* : GL(\mathbb{R}^d) \to S^{d-1}$ be the evaluation map $B(\lambda)x_d$ and if $[F(\lambda)] \in \Pi_k(S^{d-1})$, let $[B(\lambda)] = P_*'^{-1}[F(\lambda)]$, with $B(\lambda)|_{S^{k-1}} = \begin{pmatrix} C(\lambda) & 0 \\ 0 & 1 \end{pmatrix}$. Define $\delta[F(\lambda)] = C(\lambda)|_{S^{k-1}}$. Then the following sequence

$$\to \Pi_k(S^{d-1}) \xrightarrow{\delta} \Pi_{k-1}(GL(\mathbb{R}^{d-1})) \xrightarrow{i_*} \Pi_{k-1}(GL(\mathbb{R}^d)) \xrightarrow{P_*} \Pi_{k-1}(S^{d-1}) \to$$

is exact (i.e. at each point the image of the left morphism is equal to the kernel of the right morphism). The proof is classical and relies on the previous facts.

8) $\Pi_{k-1}(GL(\mathbb{R}^d)) \xrightarrow{i_*} \Pi_{k-1}(GL(\mathbb{R}^{d+1}))$ is an isomorphism for $k < d$ and onto if $k = d$. (This follows easily from the exact sequence). In particular if $k \le d$, $B(\lambda)$ is deformable to $\begin{pmatrix} C(\lambda) & 0 \\ 0 & I \end{pmatrix}$, where $C(\lambda)$ is in $GL(\mathbb{R}^k)$. Note that $C(\lambda)$ is not unique and can be replaced by $C_1(\lambda)C(\lambda)$ for any $C_1(\lambda) = \delta[F]$. Remark also that if $k = 1$, then $B(\lambda)$ is deformable to $\begin{pmatrix} \det B(\lambda) & 0 \\ 0 & I \end{pmatrix}$.

9) There is also an exact sequence for $GL(\mathbb{C}^d)$

$$\to \Pi_k(S^{2d-1}) \xrightarrow{\delta} \Pi_{k-1}(GL(\mathbb{C}^{d-1})) \xrightarrow{i_*} \Pi_{k-1}(GL(\mathbb{C}^d)) \xrightarrow{P_*} \Pi_{k-1}(S^{2d-1}) \to .$$

Then, as before, $\Pi_{k-1}(GL(\mathbb{C}^d)) \to \Pi_{k-1}(GL(\mathbb{C}^{d+1}))$ is an isomorphism for $k \le 2d$ and onto for $k = 2d + 1$. Furthermore, for $k \le 2d$, then (Bott periodicity theorem), $\Pi_{k-1}(GL(\mathbb{C}^d)) \cong \mathbb{Z}$ if k is even, 0 if k is odd. In particular if $k = 2$, any $B(\lambda)$ is deformable to $\begin{pmatrix} a(\lambda) & 0 \\ 0 & I \end{pmatrix}$ with $a(\lambda) : S^1 \to \mathbb{C}\backslash\{0\}$. Hence $\det B(\lambda)$ is deformable to $a(\lambda)$, and $B(\lambda)$ to $\begin{pmatrix} \det B(\lambda) & 0 \\ 0 & I \end{pmatrix}$.

Also $B(\lambda)$ and $D(\lambda)$ are homotopic if and only if $\det B(\lambda)$ is homotopic to $\det D(\lambda)$ as complex maps from S^1 into $\mathbb{C}\backslash\{0\}$, i.e. if and only they have the same winding number.

10) **Theorems of Bott and Adams.** *If $k < d$, then $\Pi_{k-1}(GL(\mathbb{R}^d))$* is

  a) $\mathbb{Z}_2$ *if $k \equiv 1$ or 2, [8] and J is one-to-one.*
  b) $\mathbb{Z}$ *if $k \equiv 0$ or 4, [8], and $\mathrm{Im}\, J$ is a cyclic group of order $m(k)$, the denominator of B_n/k, where $k = 4n$ and B_n is the Bernouilli number in the series $x/(e^x - 1) = 1 - x/2 + \Sigma_1^\infty (-1)^{n-1} B_n x^n/(2n)!$*
  c) 0 *if $k \equiv 3, 5, 6$ or 7, [8].*

In particular if $B(\lambda) = \lambda_1 B_1 + ... + \lambda_k B_k$ with $\| B(\lambda)x \| \geq C\|\lambda\|\|x\|$, then (Alexander-Yorke, 1978) $[B(\lambda)] \equiv r$, [2], where $d = rC_k$ for $k \equiv 1, 2, 4$, or 8, [8]. Furthermore $J[B(\lambda)] \neq 0$ if r is odd, i.e. if $k = \rho(d)$, since $m(k)$ is divisible by 8.

2. Necessary Conditions for Linearized Local Bifurcation

The fourth approach will give the answer to the question of linearized bifurcation, which will be converted into an equivalent extension problem. As before $B(\lambda)$ will be assumed to be invertible for $0 < \| \lambda \| \leq \rho$, $B(0) = 0$ and $B(\lambda)$ is supposed to be C^1; hence $\| B(\lambda) \| \leq A\|\lambda\|$ for some A. Let $S^{d-1} = \{\eta : \|\eta\| = 1\}$, $B^k = \{\lambda : \|\lambda\| \leq \rho\}$ and $S^{k-1} = \partial B^k$.

Theorem 2.1. $B(\lambda)\eta : S^{k-1} \times S^{d-1} \to \mathbb{R}^d\backslash\{0\}$ *extends to $B(\lambda, \eta) : B^k \times S^{d-1} \to \mathbb{R}^d\backslash\{0\}$, if and only if there is a $G(\lambda, x)$ continuous on $B^k \times \mathbb{R}^d$ such that*

  a) $G(\lambda, x) = 0(\| x \|^2)$ *for fixed λ and $G(\lambda, x) = o(\| x \|)$ uniformly in λ,*
  b) *There is $\rho_1, 0 < \rho_1 \leq \rho$, such that if $F(\lambda, x) = B(\lambda)x + G(\lambda, x) = 0$, for $\| \lambda \| \leq \rho_1$, then $x = 0$, and there is $\epsilon_0 > 0$, such that if $F(\lambda, x) = 0$ for $\| x \| \leq \epsilon_0, \|\lambda\| \leq \rho$ then $x = 0$.*

Proof. If: on $S^{k-1} \times S^{d-1}$, $F(\lambda, \epsilon\eta) \equiv B(\lambda)\epsilon\eta + G(\lambda, \epsilon\eta)$ is deformable to $B(\lambda)\eta$ and extends to $B^k \times S^{d-1}$ if $\epsilon \leq \epsilon_0$. From Borsuk's extension theorem, $B(\lambda)\eta$ has also a nonzero extension.

Only if. Define $D(\lambda) = \|\lambda\|^{-1/2} B(\lambda)$. Then $\| D(\lambda) \| \leq A\|\lambda\|^{1/2}$. Further, there are constants such that, on $B^k \times S^{d-1}$, $B \leq \|B(\lambda, \eta)\| \leq C$. Let D be such that $\| B(\lambda)^{-1} \| \leq D/|\det B(\lambda)|$ and define $\tilde{g}(\| \lambda \|) = D^{-1} \min_{\|\lambda\| \leq \|\mu\| \leq \rho} |\det B(\mu)|$ and $g(\| \lambda \|) = \rho^{-1} \int_0^{\|\lambda\|} \tilde{g}(r)dr$. It is easy to prove (see Ize (1988)) that $\tilde{g}(0) = 0, \tilde{g}$ is continuous and nondecreasing, g is strictly increasing, and $\| B(\lambda)^{-1} \| \leq \tilde{g}(\| \lambda \|)^{-1} \leq g(\| \lambda \|)^{-1}$.

Define $\varphi(\|\lambda\|, \|x\|)$ in $\{(\lambda, x) : \|\lambda\| \leq \rho, (\lambda, x) \neq (0,0)\}$ as 1 if $\|x\| \leq (\|\lambda\|/\rho)^{1/2} g(\|\lambda\|)/(2C)$, (region III), 0 if $\|x\| \geq (\|\lambda\|/\rho)^{1/2} g(\|\lambda\|)/C$, (region I), and a linear, in $\|x\|$, interpolation between these values in the intermediate region II. Let

$$F(\lambda, x) = (\|\lambda\|/(\varphi\|\lambda\| + (1 - \varphi)\rho))^{1/2} B(\lambda(\varphi\|\lambda\| + (1 - \varphi)\rho)/\|\lambda\|)x$$
$$+ \|x\|^2 B(\lambda\rho/(\|\lambda\| + \|x\|^4), x/\|x\|),$$

where φ stands for $\varphi(\|\lambda\|, \|x\|)$. Note that φ is not continuous at the origin, but all limits are between 0 and 1.

The following properties are easy to derive:

1) $\|F(\lambda, x)\| \leq A(\rho\|\lambda\|)^{1/2}\|x\| + C\|x\|^2$
2)

$$\|F(\lambda, x) - B(\lambda)x\|$$
$$\leq \left(\begin{array}{c} C\|x\|^2 \text{in Region } III \\ 2A(\rho\|\lambda\|)^{1/2}\|x\| + C\|x\|^2 \text{ in Regions } I \text{ and } II \end{array} \right)$$
$$\leq \|x\|(2A\rho)^{1/2} h(\|x\|)^{1/2} + C\|x\|)$$

where $h(\|x\|)$ is the inverse of the increasing function

$$\|x\| = (\|\lambda\|/\rho)^{1/2} g(\|\lambda\|)/(2C).$$

3) $\|F(\lambda, x)\| \geq (\|\lambda\|/\rho)^{1/2} g(\|\lambda\|)\|x\| - C\|x\|^2$, hence $F(\lambda, x) \neq 0$ is regions II and III. Thus, any zero of F with $x \neq 0$ has $\varphi = 0$. There $\|F(\lambda, x)\| \geq B\|x\|^2 - A(\rho\|\lambda\|)^{1/2}\|x\|$ and F is not zero if $\|x\| > A(\rho\|\lambda\|)^{1/2}/B$.

4) Let $0 < \delta < \rho$ be such that $\|B(\lambda, \eta) - B(\tilde{\lambda})\eta\| \leq g(\rho)/2$ if $\|\lambda - \tilde{\lambda}\| \leq \delta$ for all $(\tilde{\lambda}, \eta)$ with $\|\tilde{\lambda}\| = \rho$, $\|\eta\| = 1$. Then, in the region I, one has

$$F(\lambda, x) = ((\|\lambda\|/\rho)^{1/2} + \|x\|) B(\lambda\rho/\|\lambda\|)x$$
$$+ \|x\|^2 (B(\lambda\rho/(\|\lambda\| + \|x\|^4),$$

$x/\|x\|) - B(\lambda\rho/\|\lambda\|)x/\|x\|)$. Hence $\|F(\lambda, x)\| \geq ((\|\lambda\|/\rho)^{1/2} + \|x\|)g(\rho)\|x\| - \|x\|^2 g(\rho)/2$ if $\|x\|^4 \leq (\delta/(\rho - \delta))\|\lambda\|$ and $F \neq 0$ in this subregion of I.

If (ρ_1, ϵ_0) is the intersection of the two curves $\|x\| = A(\rho\|\lambda\|)^{1/2}/B$ and $\|x\|^4 = (\delta/(\rho - \delta))\|\lambda\|$, i.e. $\rho_1 = (B/A)^4 \rho^2 \delta/(\rho - \delta)$, then the conclusions of the theorem hold for these numbers.   Q.E.D.

Remark 2.1. 1) If $B(\lambda) = \|\lambda\|^{1/2} D(\lambda\rho/\|\lambda\|)$, then there is no need to construct the function φ and $G(\lambda, x) = \|x\|^2 B(\lambda/(\|\lambda\| + \|x\|^4), x/\|x\|) = 0(\|x\|^2)$ uniformly in λ.

2) If $B(\lambda)$ is deformable, for $\|\lambda\| = \rho$ to I via $B(t, \lambda)$, one may take $B(\lambda, \eta) = B(\|\lambda\|/\rho, \lambda\rho/\|\lambda\|)\eta$.

3) If $B(\lambda)$ is real or complex analytic in λ (λ in $\mathbb{R}$ or $\mathbb{C}$ respectively), with $\det B(\lambda) = a_0\lambda^m + \dots$, with m even, $G(\lambda, x)$ is real analytic in x and λ, except in the complex case at $\lambda = 0$, where G is Hölder continuous in λ.

The proof of this last result was given in the real case in Section I, Remark 1.3, by deforming explicitly $B(\lambda)$ to $\wedge(\lambda) = \mathrm{diag}(\lambda^{p_1}, \dots, \lambda^{p_d})$, with Σp_i even. Clearly this deformation is also valid for the complex case. Then, if d is even, one may deform a block of the form $\begin{pmatrix} \lambda^{p+2n} & 0 \\ 0 & \lambda^p \end{pmatrix}$

$\begin{pmatrix} z_1 \\ z_2 \end{pmatrix}$ via $\begin{pmatrix} \lambda^n & nt \\ -nt & \bar{\lambda}^n/|\lambda|^{n-\epsilon} \end{pmatrix} \begin{pmatrix} \lambda^{p+n}z_1 + t\bar{z}_2 \\ \lambda^{p+n}/|\lambda|^{n+\epsilon}z_2 - t\bar{z}_1 \end{pmatrix}$, which is not 0 if $|z_1| + |z_2| \neq 0$ and $t > 0$, and where $\epsilon = 0$ if $n = 0$ and $\epsilon = p/2$ if $n > 0$ (and analytic in λ if $n = 0$ but C^ϵ if $n > 0$). Regarded as a 4×4 real matrix this gives a deformation $\wedge(t, \lambda)$ of $\wedge(\lambda)$, with $\wedge(t, \lambda)$ invertible for $t > 0$.

If d is odd, one will have a 3×3 block of the form $\mathrm{diag}(\lambda^p, \lambda^q, \lambda^r)$ with $p \leq q \leq r$, $p + q + r = 2m$, i.e. $r = p + q + 2n$, with $2n \geq -p$. Define

$$\wedge (t, \lambda)(z_1, z_2, z_3)^T$$
$$= \begin{pmatrix} 1 & 0 & n^2 t \\ n^2 t & \bar{\lambda}^n/|\lambda|^{n-\epsilon} & 0 \\ 0 & n^2 t & \lambda^n/|\lambda|^{n-\epsilon} \end{pmatrix} \begin{pmatrix} \lambda^p z_1 + t\bar{z}_3 \\ \lambda^{q+n}/|\lambda|^{n+\epsilon}z_2 + tz_1 \\ \lambda^{p+q+n}/|\lambda|^{-n+\epsilon}z_3 + t\bar{z}_2 \end{pmatrix}$$

considered as a real 6×6 matrix. Here $\epsilon = 0$ if $n = 0$ and $\epsilon = q/2$ if $n \neq 0$. It is clear that $\wedge(t, \lambda)$ is analytic in λ if $n = 0$ and C^ϵ if $n \neq 0$. The matrix is invertible if $t > 0$ and taking the conjugate of the third component, the vector gives $\begin{pmatrix} \lambda^p & 0 & t \\ t & \lambda^{q+n}/|\lambda|^{n+\epsilon} & 0 \\ 0 & t & \bar{\lambda}^{p+q+n}/|\lambda|^{-n+\epsilon} \end{pmatrix} \begin{pmatrix} z_1 \\ z_2 \\ \bar{z}_3 \end{pmatrix}$, which is zero, for $t > 0$, only if $z_j = 0$. Defining $\wedge(\lambda, x)x = \wedge(\|x\|^2, \lambda)x$, one has that $\wedge(\lambda, x)x = 0$ only if $x = 0$ and $\wedge(\lambda, x)x - \wedge(\lambda)x = 0(\|x\|^3)$. The rest of the construction is as in Section I.

Note that if $B(\lambda)\eta$ extends to $B(\lambda, \eta)$, this will be true for a fixed η, for example the south pole x_d, and $B(\lambda)x_d$ is deformable to the constant $B(0, x_d)$, i.e. $P_*[B(\lambda)] = 0$. From Remark 1.3 (7), it follows from the exact sequence, that $B(\lambda)$ is deformable to $\begin{pmatrix} C(\lambda) & 0 \\ 0 & 1 \end{pmatrix}$. Writing η as $(\tilde{\eta}, \eta_d)$, with $\tilde{\eta}$ in $\mathbb{R}^{d-1}$, one may extend $(C(\lambda)\tilde{\eta}, \eta_d)$ from $S^{k-1} \times S^{d-1}$ to $B^k \times B_-^d$, (B_-^d the south hemisphere), via $(\|\lambda\|C(\lambda/\|\lambda\|)\tilde{\eta}, \eta_d - \epsilon(1 - \|\lambda\|))$, for some $0 < \epsilon < 1$.

Note that, for simplicity, ρ is taken to be 1 from here on. This map will be extendable to the north hemisphere if and only if $(\|\lambda\|C(\lambda/\|\lambda\|)\tilde{\eta},$

$(1 - \|\tilde{\eta}\|^2)^{1/2} - \epsilon(1 - \|\lambda\|)$ is homotopic to a constant map on $\partial((\lambda, \tilde{\eta})$:
$\|\lambda\| \leq 1, \|\tilde{\eta}\| \leq 1)$, i.e. if it is 0 in $\Pi_{k+d-2}(S^{d-1})$. It is easy to see that the
linear deformation $(1 - t)((1 - \|\tilde{\eta}\|^2)^{1/2} - \epsilon(1 - \|\lambda\|)) + t(\epsilon - \|\tilde{\eta}\|^2)$ is valid
on this sphere. Thus this map will extend to the north hemisphere if and
only if $-J[C(\lambda)] = 0$. If this is the case, from Borsuk's extension theorem,
$B(\lambda)\eta$ will also extend. Note however that $C(\lambda)$ is not the only possibility
(see again Remark 1.3) and that one may have other ways to extend to the
south hemisphere. However one has the following result:

Theorem 2.2. $B(\lambda)\eta$ *extends to* $B^k \times S^{d-1}$ *if and only if there is a* $\tilde{C}(\lambda)$
such that $B(\lambda)$ *is deformable to* $\begin{pmatrix} \tilde{C}(\lambda) & 0 \\ 0 & 1 \end{pmatrix}$, *for* $\|\lambda\| = 1$, *and* $J[\tilde{C}(\lambda)] = 0$.

Proof. The "if" part has already been proved above. The "only if" will
follow from two lemmas, stated in a slightly more general form than needed
for the extension of a map $B(\lambda, \eta)$; here $B(\lambda)\eta$, from $S^{k-1} \times S^{d-1}$ to $S^{k-1} \times$
$S^{d-1} \cup B^k \times B^d_-$, B^d_- the south hemisphere of S^{d-1} with south pole x_d.

Lemma 2.1. $B(\lambda, \eta)$ *extends if and only if* $B(\lambda, x_d)$ *is deformable to a*
constant on S^{k-1}.

Proof. If $B(\lambda, \eta)$ extends via $\tilde{B}(\lambda, \eta)$, then $B(\lambda, x_d)$ deforms to $B(0, x_d)$.
On the other hand, let $B_t(\lambda, \eta) = B(\lambda, \varphi\tilde{\eta}, \eta_d + t(|\eta_d| - 1))$, where $\varphi(t, \eta_d)$
is such that $(\varphi\tilde{\eta}, \eta_d + t(|\eta_d| - 1))$ has norm 1 for η in S^{d-1} and corresponds
to the pulling of the south hemisphere to the south pole already used in
Remark 1.2 (2). In particular $\varphi(0, \eta_d) = 1, \varphi(1, \eta_d) = 0$ if $\eta_d \leq 0$ and
$\varphi(1, d) = 2(\eta_d/1 + \eta_d)^{1/2}$ if $\eta_d > 0$. Now $B_1(\lambda, \eta)$, which is $B(\lambda, 0, -1)$ if
$\eta_d \leq 0$, has an extension to $S^{k-1} \times S^{d-1} \cup B^k \times B^d_-$, by appending, on
the second term, the deformation of $B(0, \eta_d)$ to a constant. By Borsuk's
extension theorem, $B(\lambda, \eta)$ will also have an extension.   Q.E.D.

Lemma 2.2. *If* $B_0(\lambda, \eta)$ *and* $B_1(\lambda, \eta)$ *are two extensions of* $B(\lambda, \eta)$ *from*
$S^{k-1} \times S^{d-1}$ *to* $S^{k-1} \times S^{d-1} \cup B^k \times B^d_-$, *such that there is* $\tilde{B}_t(\lambda)$ *defined on*
B^k, *with* $\tilde{B}_i(\lambda) = B_i(\lambda, x_d)$ *for* $\|\lambda\| \leq 1$, $i = 0$ *or* 1, *and* $\tilde{B}_t(\lambda) = B(\lambda, x_d)$
for $\|\lambda\| = 1$, *there is a homotopy* $\tilde{B}_t(\lambda, \eta)$ *on* $S^{k-1} \times S^{d-1} \cup B^k \times B^d_-$, *such*
that $\tilde{B}_i(\lambda, \eta) = B_i(\lambda, \eta)$ *for* $\|\lambda\| \leq 1$ *and* $i = 0$ *or* 1, $\tilde{B}_t(\lambda, x_d) = \tilde{B}_t(\lambda)$
and $\tilde{B}_t(\lambda, \eta) = B(\lambda, \eta)$ *for* $\|\lambda\| = 1$. *In other words, any two extensions are*
characterized by the relative homotopy class of $B_i(\lambda, x_d)$ *for* $\|\lambda\| \leq 1$, *with*
fixed value $B(\lambda, x_d)$ *for* $\|\lambda\| = 1$.

Proof. On $I \times S^{k-1} \times S^{d-1} \cup \{0\} \times B^k \times B^d_- \cup \{1\} \times B^k \times B^d_- \cup I \times B^k \times \{x_d\}$,
consider the homotopy given by $F_\tau(t, \lambda, \eta) = (B(\lambda, \varphi\tilde{\eta}, \eta_d + \tau(|\eta_d| - 1))$,
$B_i(\lambda, \varphi\tilde{\eta}, \eta_d + \tau(|\eta_d| - 1)))$, $\tilde{B}_t(\lambda))$, where $\varphi(\tau, \eta_d)$ is as above. For $\tau =$

0, one has $(B(\lambda, \eta),\ B_0(\lambda, \eta),\ B_1(\lambda, \eta),\ \tilde{B}_t(\lambda))$ and for $\tau = 1$, the map $(B(\lambda, \varphi\tilde{\eta},\ \eta_d + (|\eta_d| - 1)),\ B_0(\lambda, x_d),\ B_1(\lambda, x_d),\ \tilde{B}_t(\lambda))$ which extends to $I \times S^{k-1} \times S^{d-1} \cup I \times B^k \times B_-^d$, via $B(t, \lambda, \eta) = B(\lambda, \eta)$ on S^{k-1} and $\tilde{B}_t(\lambda)$ on the second part. From Borsuk's extension theorem, the map for $\tau = 0$, will also have an extension. Q.E.D.

Proof of the Theorem. If $B(\lambda)\eta$ has an extension, then $P_*[B(\lambda)] = 0$ and $B(\lambda)$ is homotopic on S^{k-1} to $\begin{pmatrix} C(\lambda) & 0 \\ 0 & 1 \end{pmatrix}$ and $(\|\lambda\|C(\lambda/\|\lambda\|\tilde{\eta}, \eta_d)$ also has an extension. Assume then that $B(\lambda) = \begin{pmatrix} C(\lambda) & 0 \\ 0 & 1 \end{pmatrix}$ has the extension $B(\lambda, \eta)$, such that $F(\lambda) = B(\lambda, x_d)$ has norm 1. From the homotopy covering (Remark 1.3 (6)), there is $D(\lambda)$ in $SO(d)$ for each $\|\lambda\| \leq 1$, such that $D(\lambda) = \begin{pmatrix} C_1(\lambda) & 0 \\ 0 & 1 \end{pmatrix}$ if $\|\lambda\| = 1$ and $D(\lambda)x_d = F(\lambda)$. Define

$$\tilde{B}(\lambda, \eta) = D(\lambda) \begin{pmatrix} \|\lambda\|C_1^{-1}(\lambda/\|\lambda\|)C(\lambda/\|\lambda\|) & 0 \\ 0 & 1 \end{pmatrix}$$
$$\cdot \begin{pmatrix} \tilde{\eta} \\ (\eta_d - \epsilon(1 - \|\lambda\|))/(1 + \epsilon(1 - \|\lambda\|)) \end{pmatrix}.$$

Then $\tilde{B}(\lambda, \eta) = 0$ only if $\lambda = 0, \eta_1 = \epsilon$ since if $\lambda \neq 0$ on a zero in S^{d-1}. Now $\tilde{\eta} = 0$ and $\eta_d = \epsilon(1 - \|\lambda\|) < 1$ which is not possible in S^{d-1}. Furthermore for $\|\lambda\| = 1$, $\tilde{B}(\lambda, \eta) = B(\lambda)\eta$. For $\eta = x_d$, $\tilde{B}(\lambda, x_d) = F(\lambda)$. From these facts and Lemma 2.2, $\tilde{B}(\lambda, \eta)$ and $B(\lambda, \eta)$ are homotopic on $S^{k-1} \times S^{d-1} \cup B^k \times B_-^d$ relative to $S^{k-1} \times S^{d-1}$. But since $B(\lambda, \eta)$ extends to $B^k \times S^{d-1}$, i.e. to the north hemisphere, this is also the case for $\tilde{B}(\lambda, \eta)$. Thus, the class of $\tilde{B}(\lambda, \tilde{\eta}, (1 - \|\tilde{\eta}\|^2)^{1/2})$ on $\partial((\lambda, \tilde{\eta}) : \|\lambda\| \leq 1, \|\tilde{\eta}\| \leq 1)$ is 0. Deforming $D(\lambda)$ radially to $D(0)$ and then to I, the obstruction is the class of $(\|\lambda\|\tilde{C}(\lambda/\|\lambda\|))\tilde{\eta}, \eta_d - \epsilon(1 - \|\lambda\|))$, with $\tilde{C}(\lambda) = C_1^{-1}(\lambda)C(\lambda)$, i.e. $-J[\tilde{C}(\lambda)]$. Finally, since for $\|\lambda\| = 1$, $\tilde{B}(\lambda, \eta) = B(\lambda)\eta$ is also deformable to $\begin{pmatrix} C(\lambda) & 0 \\ 0 & 1 \end{pmatrix}\eta$, one has that $B(\lambda)$ has the properties required in the theorem, with $J[\tilde{C}(\lambda)] = 0$. Q.E.D.

Remark 2.2. If $B(\lambda)$ is deformable to $\begin{pmatrix} C(\lambda) & 0 \\ 0 & 1 \end{pmatrix}$, it is easy to see by a direct computation that $J[B(\lambda)] = -\Sigma \tilde{J}[C(\lambda)]$. Thus if $J[B(\lambda)] \neq 0$ one cannot have an extension. Now it is clear that $J[B(\lambda)]$ is a more "natural" invariant than the set of possible $J[C(\lambda)]$. In this sense one has the following result:

Theorem 2.3. a) *If $k \leq 2d - 2$ or $d = 2, 4, 8$ and one has the conditions $P_*[B(\lambda)] = 0$, $J[B(\lambda)] = 0$, there is an extension $B(\lambda, \eta)$ from $S^{k-1} \times S^{d-1}$ to $B^k \times S^{d-1}$.*

b) *If $k \leq 2d - 3$, $J[B(\lambda)] = 0$ is enough to guarantee the extension. This is also the case if $d = 2$ and any k, or $d = 4$ or 8 and $k \leq 2d - 2$.*

Idea of the Proof. The complete proof is given in Ize (1988) and depends on deeper topological results. Consider the commutative diagram

$$\to \Pi_k(S^{d-1}) \xrightarrow{\delta} \Pi_{k-1}(GL(\mathbb{R}^{d-1})) \xrightarrow{i_*} \Pi_{k-1}(GL(\mathbb{R}^d)) \xrightarrow{P_*} \Pi_{k-1}(S^{d-1})$$
$$\downarrow J \qquad\qquad\qquad \downarrow J \qquad\qquad\qquad \downarrow \Sigma^d$$
$$\Pi_{k+d-2}(S^{d-2}) \xrightarrow{-\Sigma} \Pi_{k+d-1}(S^d) \xrightarrow{H} \Pi_{k+d-1}(S^{2d-1})$$

where H is the generalized Hopf map.

It can be shown that Σ is one-to-one if $d = 2, 4, 8$, and from the Freudenthal suspension theorem, Σ and i_* are isomorphisms if $k \leq d - 2$, and onto if $k < d$. If $k \leq 2d - 2$ one may show that $Jo\delta$ is onto $ker\,\Sigma$. Hence if $[B(\lambda)] = i_*[C(\lambda)]$ and $J[B(\lambda)] = 0$, there is an α in $\Pi_k(S^{d-1})$ such that $J(\delta\alpha) = -J[C(\lambda)]$, i.e. $J[\delta\alpha C(\lambda)] = 0$ and $i_*[\delta\alpha C(\lambda)] = [B(\lambda)]$. This proves (a). If $k \leq 2d - 3$, then Σ^d is an isomorphism and if $J[B(\lambda)] = 0$ one has that $P_*[B(\lambda)] = 0$.   Q.E.D.

Remark 2.3. a) One has the following conjecture: $J \circ \delta$ is onto $ker\,\Sigma$.

b) If $k \geq 2d - 2$, one may have $P_*[B(\lambda)] \neq 0$ and $J[B(\lambda)] = 0$: see Ize (1988) for an example.

Remark 2.4. The positive index case. Assume $B(\lambda)$ is a $d \times d^*$ matrix, with $d > d^*$, such that $B(0) = 0$ and $B(\lambda)$ has maximal rank (i.e. d^*) for $0 < \|\lambda\| \leq \rho$. Suppose in fact that $B(\lambda)$ has a fixed $d^* \times d^*$ nonzero minor for $\lambda \neq 0$. Then one may write $B(\lambda)$ as $(B_1(\lambda) \;\; B_2(\lambda))$ where $B_1(\lambda)$ is an invertible $d^* \times d^*$ matrix for $\lambda \neq 0$.

Complement the bifurcation equation $B(\lambda)x + G(\lambda, x)$ by $C(\lambda)\tilde{x} - \epsilon_1$, where $C(\lambda)$ is an arbitrary k-family of $(d - d^*) \times (d - d^*)$ matrices with $C(0) = 0$ and $C(\lambda)$ invertible for $0 < \|\lambda\| \leq \rho$, and $\tilde{x}$ corresponds to the last $d - d^*$ coordinates of x. Hence one has a new linear part $\begin{pmatrix} B_1(\lambda) & B_2(\lambda) \\ 0 & C(\lambda) \end{pmatrix}$.

The J-invariant of this last matrix is $J\left[\begin{pmatrix} B_1(\lambda) & 0 \\ 0 & C(\lambda) \end{pmatrix} \right]$. Furthermore if $k \leq \min(d^*, d - d^*)$, one may deform $B_1(\lambda)$ and $C(\lambda)$ to the suspensions of two $k \times k$ matrices $\tilde{B}_1(\lambda)$ and $\tilde{C}_1(\lambda)$. It is easy to see that the deformation

$$\begin{pmatrix} t\tilde{B}_1 & (1-t)\tilde{B}_1\tilde{C}_1 \\ -(1-t)I & t\tilde{C}_1 \end{pmatrix} \text{ is valid in } \Pi_{k-1}(GL(\mathbb{R}^{2k})) \text{ and that}$$

$$J\left[\begin{pmatrix} B_1(\lambda) & B_2(\lambda) \\ 0 & C(\lambda) \end{pmatrix}\right] = \Sigma^{d-k} J[\tilde{B}_1\tilde{C}_1].$$

Then if $k \equiv 1, 2, 4$ or 8, [8], one may choose $\tilde{C}_1(\lambda)$ such that this last invariant is non-zero. Then one always has bifurcation (in fact, as it will be seen in Section III, a $(d - d^* + 1)$- "surface" bifurcating from $(0,0)$). This argument was already used in Section I, for the case $k = 1$.

3. Multiparameter Global Bifurcation

In this subsection, "necessary" and sufficient conditions for global bifurcation will be derived for the equation from $\wedge \times B$ into E

$$(A_0 - A(\lambda))x - g(\lambda, x) = 0.$$

In order to simplify one will assume that $B = E$, A_0 is of the form $\text{Id} - \text{Compact}$, $A(\lambda)$ and $g(\lambda, x)$ are compact, $A(0) = 0$ and $A_0 - A(\lambda)$ is invertible for $0 < \|\lambda\| \le \rho$, $g(\lambda, x) = o(\|x\|)$ and $\wedge = \mathbb{R}^k$. One may relax some of these hypotheses: for example by replacing compactness by $k-$set contractions ($k < 1$) and use the results of Alexander-Fitzpatrick (1975) for the sufficient conditions.

Let S be the closure in $\wedge \times E$ of the set of nontrivial solutions and $\mathcal{C}(g)$ the continuum in S bifurcating from $(0,0)$.

Definition. *$(0,0)$ is said to be a point of "linearized global bifurcation" if, for all g's, $\mathcal{C}(g)$ is either unbounded or $\mathcal{C}(g) \cap \{x = 0\}$ contains a point different from $(0,0)$.*

As in the local bifurcation section, the following topological facts will be useful:

    a) *If $A \subset X$ are closed subsets of $\wedge \times E$ and there is a homotopy $F_t(\lambda, x) = x - k_t(\lambda, x)$, with k_t compact, from A into $E\backslash\{0\}$, then $F_0(\lambda, x)$ extends to $\tilde{F}_0(\lambda, x) = x - \tilde{k}_0(\lambda, x)$, with $\tilde{k}_0$ compact, from X into $E\backslash\{0\}$, if and only if F_1 has a similar extension.*

    b) *If X is a ball and $A = \partial X$, then $F(\lambda, x) = x - k(\lambda, x)$ has a nonzero extension if and only if $F(\lambda, x)|_A$ is deformable to $(\tilde{x}, 1)$, where $x = (\tilde{x}, r)$ and all maps have the form $x - K(\lambda, x)$ with K compact.*

The proof of (a) is similar to the proof of the Borsuk's extension theorem, replacing Tietze's extension theorem by Dugundji's. This proves the "if" part of (b). For the "only if" part, if $x - k(\lambda, x)$ is nonzero on X, then by the properness of this map, $\|x - k(\lambda, x)\| \geq \epsilon$ for some $\epsilon > 0$. Now, $k(\lambda, x)$ is compact on X if and only if there are finite dimensional maps $k_N(\lambda, x) : X \to \mathbb{R}^N$ such that $k(\lambda, x) = \Sigma_0^\infty k_N(\lambda, x)$ and $\|k_N(\lambda, x)\| \leq 2^{-N}$ (see Geba-Granas, 1973, for example). For large $N, x - k_N(\lambda, x) \neq 0$ on X. By writing $x = x^N \oplus x_N$, with x_N in $\mathbb{R}^N$, then $(x^N, x_N - k_N(\lambda, x))$, as well as $(x^N, x_N - k_N(\lambda, x_N))$, are not zero on X. It is clear that this last map may be deformed, on A, to $(\tilde{x}, 1)$. As in the local case, the first result is an equivalence with an extension problem: let $B^k = \{\lambda : \|\lambda\| \leq \rho\}, S^{k-1} = \partial B^k, S = \{\eta \epsilon X, \|\eta\| = 1\}$ and $I = \{r, 0 \leq r \leq \epsilon_0\}$. Below ϵ will stand for any number, $0 < \epsilon < \epsilon_0$.

Theorem 3.1. *There is no linearized global bifurcation if and only if $((A_0 - A(\lambda))\eta, r - \epsilon)$, as a map from $\partial(B^k \times I) \times S$ into $E \times \mathbb{R}\backslash\{0\}$, has a nonzero extension (of the form $\eta - K(\lambda, r, \eta)$, K compact) to $B^k \times I \times S$.*

Proof. "Only if". If there is $g(\lambda, x)$ compact such that $\mathcal{C}(g)$ is bounded and does not return to the trivial solution, one may construct an open bounded Ω, with $\mathcal{C}(g) \subset \Omega \subset B = \{(\lambda, x) : \|x\|, \|\lambda\| \leq R\}, \Omega \cap \{x : \|x\| \leq \epsilon_0\} = B_\rho \times B_{\epsilon_0}, B_\rho = \{\lambda : \|\lambda\| \leq \rho\}, B_{\epsilon_0} = \{x : \|x\| \leq \epsilon_0\}$ and the only solution of the bifurcation equation on $\partial\Omega$ is for $x = 0$ ("Whyburn's lemma"), ϵ_0 is chosen such that for $\|\lambda\| = \rho$, the only solution with $\|x\| \leq \epsilon_0$ is $x = 0$. Thus, on $\partial\Omega \cup \{x = 0\}, ((A_0 - A(\lambda))x - g(\lambda, x), \|x\| - \epsilon) \neq 0$ for any $\epsilon > 0$. These maps are, on $\partial\Omega \cup \{x = 0\}$, homotopic one to the other; by varying ϵ and, for $\epsilon > R$, it has a nonzero extension to B. Hence, for any $\epsilon > 0$, one has a nonzero extension from $\partial\Omega \cup \{x = 0\}$ to $B\backslash\Omega$.

Choose $\epsilon < \epsilon_0$ and let $(F(\lambda, x), h(\lambda, x))$ be the map defined as $((A_0 - A(\lambda))x - g(\lambda, x), \|x\| - \epsilon)$ in $\Omega \cup \{x = 0\}$ and this nonzero extension on $B\backslash\Omega$. In particular $(F, h) \neq 0$ if $(\lambda, x) \notin B_\rho \times B_{\epsilon_0}$. Now on $\partial(B_\rho \times B_{\epsilon_0}) \cup \{x = 0\}$, the map $\alpha(t)(F, h)((tR + (1 - t)\rho)\lambda/\rho, (tR + (1 - t)\epsilon_0)x/\epsilon_0)$, where $\alpha^{-1}(t) = (tR + (1 - t)\epsilon_0)/\epsilon_0$ is chosen such that one has a map of the form Id-compact, is a valid homotopy from $(F(\lambda, x), \|x\| - \epsilon)$ to the map $(\epsilon_0/R)(F, h)(R\lambda/\rho, Rx/\epsilon_0)$ which has a nonzero extension to $B_\rho \times B_{\epsilon_0}$ by scaling the extension from ∂B to B. Hence, $(F(\lambda, x), \|x\| - \epsilon)$ and $((A_0 - A(\lambda))x, \|x\| - \epsilon)$ have nonzero extensions from $\partial(B_\rho \times B_{\epsilon_0}) \cup \{x = 0\}$ to $B_\rho \times B_{\epsilon_0}$.

Note that up to here the set $\{x = 0\}$ has played no role.

Let $(F(\lambda, r\eta), h(\lambda, r\eta))$ be the last extension, with $F(\lambda, 0) = 0, h(\lambda, 0) = -\epsilon$. It is clear that there is ϵ_1 such that if $r \leq \epsilon_1$, then $h(\lambda, r\eta) \leq -\epsilon/2$. Let $\varphi(r)$ be r for $\epsilon_1 \leq r \leq \epsilon_0$ and ϵ_1 for $r \leq \epsilon_1$. Then $F(\lambda, r\eta) = r\eta -$

$K(\lambda, r\eta) = (A_0 - A(\lambda))r\eta - \tilde{K}(\lambda, r\eta)$, where $\tilde{K}$ is compact, since $A_0 = I - T$, with T compact. Hence $((A_0 - A(\lambda))\eta - \tilde{K}(\lambda, r\eta)(\varphi(r), h(\lambda, r\eta))$ is nonzero on $B^k \times I \times S$, it extends $((A_0 - A(\lambda))\eta, r - \epsilon)$ on $\partial(B^k \times I) \times S$ and differs from η by a compact map. (Note that in finite dimension this last step is not necessary).

"**If**". Let $(A(\lambda, r, \eta), h(\lambda, r, \eta))$ be an extension of $((A_0 - A(\lambda))\eta, r - \epsilon)$ from $\partial(B^k \times I) \times S$ to $B^k \times I \times S$, with $A(\lambda, r, \eta) - \eta$ compact. Let $\varphi = \varphi(\|x\|)$ be a map which is 0 if $\|x\| \leq \tilde{\epsilon}$ and 1 if $\|x\| \geq 2\tilde{\epsilon}$, for $\tilde{\epsilon}$ small, $2\tilde{\epsilon} < \epsilon_0$. In $B_\rho \times B_{\epsilon_0}$ let $g(\lambda, x) = \|x\|((A_0 - A(\lambda))x/\|x\| - A(\lambda, \|x\|\varphi(\|x\|), x/\|x\|))$ and extend g by 0 outside this box. It is easy to check that $g(\lambda, x)$ is compact and that $g(\lambda, x) = o(\|x\|)$. (In the case of a finite dimensional space one may replace φ by 1).

Let $\Omega = \{\lambda, x) \in B_\rho \times B_{\epsilon_0} : h(\lambda, \varphi\|x\|, x/\|x\|) < 0\}$. It is clear that Ω is an open neighborhood of $\{x = 0\}$ and does not intersect the level $\|x\| = \epsilon_0$ (there $h = \epsilon_0 - \epsilon > 0$). Furthermore $\partial\Omega \subset h^{-1}(0) \cup (\partial B_\rho) \times B_{\epsilon_0}$. On $h^{-1}(0)$, one has that $(A_0 - A(\lambda))x - g(\lambda, x) = A(\lambda, \varphi\|x\|, x/\|x\|)$ is not zero and, on $\partial B_\rho, g = 0$ and the only solution is for $x = 0$. This implies that $\mathcal{C}(g) \subset \Omega$.   Q.E.D.

The proof of the preceding theorem was given *in extenso* so that the reader can see where the compactness of the maps was used. It is easy to see that one needs this compactness only in the "Whyburn lemma", since the extension argument can be done by only using continuous functions. However in this case and infinite dimensions, the topological invariants which will characterize the linearized global bifurcation, will be 0.

In the present case, for $\|\lambda\| = \rho, A_0 - A(\lambda)$ is deformable to $A_0 \oplus B(\lambda)$, as in Remark 1.1 of Section I, where A_0, on a complement of ker A is of the form Id-compact. Since the suspension is an isomorphism in higher dimensions, one may define $P_*[A_0 - A(\lambda)]$ and $J[A_0 - A(\lambda)]$ as, in the infinite dimension, the stable limit of these invariants on finite dimensional subspaces.

Theorem 3.2. a) *There is no linearized global bifurcation if and only if* $\Sigma P_*[A_0 - A(\lambda)] = 0$ *and* $J[A_0 - A(\lambda)] = 0$.

b) *If* dim $E > d$ *or* $k \leq 2$ dim $E - 2$, *the last condition reduces to* $J[A_0 - A(\lambda)] = 0$.

Proof. "Only if" If $\mathcal{C}(g)$, or the branch, $\mathcal{C}(g, \eta_0)$, in the direction of a fixed η_0, is bounded and does not return to $x = 0$, from the first part of the proof of the preceding theorem, one has that $((A_0 - A(\lambda))x, \|x\| - \epsilon)$ extends from $\partial(B_\rho \times B_{\epsilon_0})$ to $B_\rho \times B_{\epsilon_0}$; hence $J[A_0 - A(\lambda)] = 0$. Similarly $((A_0 - A(\lambda))t\eta_0, t - \epsilon)$ extends from $\partial(B_\rho \times I)$ to $B_\rho \times I$, where $I =$

$\{0 \le t \le \epsilon_0\}$. Thus, if $N = \dim E$, this last map is 0 in $\Pi_k(S^N)$ and is clearly deformable to $[\Sigma(A_0 - A(\lambda))\eta_0]$.

If the "**if**" part has been proved for $N = d$, with $P_*[B(\lambda)] = 0$, $J[B(\lambda)] = 0$, there is a $G(\lambda, x_1) = o(\|x_1\|)$ such that the component $\mathcal{C}(G)$ remains in $B_\rho \times B'_{\epsilon_0}$, with $B'_{\epsilon_0} = \{x_1 : \|x_1\| \le \epsilon_0\}$. Define $g(\lambda, x_1 + x_2)$ by $Qg(\lambda, x) = 0, (I - Q)g(\lambda, x) = -G(\lambda, x_1)$. Then, as in Section I,

$$F(\lambda, x) = (A_0 - A(\lambda))x - g(\lambda, x) = (A - QA(\lambda))H(\lambda, x_1, x_2)$$
$$\oplus B(\lambda)x_1 + G(\lambda, x) - (I - Q)A(\lambda)H(\lambda, x_1, x_2),$$

where $H(\lambda, x_1, x_2) = x_2 - (I - KQA(\lambda))^{-1}KQA(\lambda)x_1$. It is clear that, for ρ and ϵ_0 so small that this reduction is valid, the sets of zeros of $F(\lambda, x)$ and of $B(\lambda)x_1 + G(\lambda, x_1)$ are homeomorphic. Hence it will be enough to prove the "if" part for finite N.

Note also that if $d < N$, one may take η_0 such that $P\eta_0 = 0$, in which case $(A - A(\lambda))\eta_0 = (A_0 - QA(\lambda))\eta_0 \ominus (I - Q)A(\lambda)\eta_0$ which is deformable, replacing λ by $t\lambda$, to the constant $A_0\eta_0$, that is $P_*[B(\lambda)] = 0$. Furthermore if $k \le 2d - 2$, then Σ^d is an isomorphism and, from the diagram of Theorem 2.3, $HJ = \Sigma^d P_*$. Thus $P_*[B(\lambda)] = 0$ if $J[B(\lambda)] = 0$, giving (b).

It remains to prove the "if" part, when N is finite. If $P_*[B(\lambda)] = 0$, then $A_0 - A(\lambda)|_{\partial B_\rho}$ is deformable to $\begin{pmatrix} C(\lambda) & 0 \\ 0 & 1 \end{pmatrix}$. As in Theorem 2.2, $((A_0 - A(\lambda))\eta, r - \epsilon)$ is deformable on $\partial(B_\rho \times I) \times S^{N-1}$ to $(\|\lambda\|C(\lambda\rho/\|\lambda\|)\tilde{\eta}, \eta_d, r - \epsilon)$, where $\eta = (\tilde{\eta}, \eta_d)$. This map extends to the south hemisphere, $\eta_d \le 0$, by

$$(\|\lambda\|C(\lambda\rho/\|\lambda\|)\tilde{\eta}, \eta_d - \epsilon(\rho - \|\lambda\|), r - \epsilon)$$

and one has an obstruction to the extension to the north hemisphere given by the class of this map on $\partial(B_\rho \times I \times B^{N-1})$, $B^{N-1} = \{\tilde{\eta}/\|\tilde{\eta}\| \le 1\}$ and $\eta_d = (1 - \|\tilde{\eta}\|^2)^{1/2}$. Since this map is deformable to $(\|\lambda\|C(\lambda\rho/\|\lambda\|)\tilde{\eta}, 1 - 2\|\tilde{\eta}\|^2, r - \epsilon)$, its class is $\Sigma J[C(\lambda)] = -J[A_0 - A(\lambda)] = 0$, one has the desired extension.

It remains to look at the case $N = d$. By scaling λ, one may assume that $\rho = 1$ and that $\mu = (r - \epsilon_0/2)/(\epsilon_0/2)$ varies from -1 to 1. One needs to show that $(\|\lambda\|B(\lambda/\|\lambda\|)\eta, \mu)$ extends from $\partial(B_1 \times I) \times S^{d-1}$ to $B_1 \times I \times S^{d-1}$. Take $\eta_0 = (0, \ldots, 0, 1)^T$; then, since $\Sigma[B(\lambda)\eta_0] = 0$, $(\|\lambda\|B(\lambda/\|\lambda\|)\eta_0, \mu)$ extends from the annulus $B_1 \times I - B_{1/2} \times \{|\mu| \le 1/2\}$ to $B_1 \times I$ via $(B_d(\lambda, \mu), h(\lambda, \mu))$. Let $B(\lambda, \mu)$ be the matrix $\|\lambda\|B(\lambda/\|\lambda\|)$ where the last column has been replaced by $B_d(\lambda, \mu)$ and let S be the set $\{(\lambda, \mu) : h(\lambda, \mu) = 0, \det B(\lambda, \mu) = 0\}$. Clearly S is contained in $B_{1/2} \times \{|\mu| < 1/2\}$. Let $\varphi(\lambda, \mu)$ be a partition function with value 0 on S and 1 on the complement of $B_{1/2} \times \{|\mu| < 1/2\}$.

Consider the function $F(\lambda, \mu, \eta) = (B(\lambda, \mu) \begin{pmatrix} \varphi(\lambda, \mu)\bar{\eta} \\ \eta_d - (1 - \|\tilde{\lambda}\|)/2 \end{pmatrix},$
$h(\lambda, \mu))$ where $\|\tilde{\lambda}\| = \max(\|\lambda\|, |\mu|)$. Clearly if $F(\lambda, \mu, \eta) = 0$ then $h(\lambda, \mu) = 0$ and, if $\det B(\lambda, \mu) = 0$, then $\varphi(\lambda, \mu) = 0$ and $\eta_d = (1 - \|\tilde{\lambda}\|)/2 \geq 1/4$; or $B(\lambda, \mu)$ is invertible, $\|\tilde{\lambda}\| \leq 1/2$ and $\eta_d > 1/4$. In particular F has no zero on the south hemisphere and the obstruction for an extension to the north hemisphere is the class of $F(\lambda, \mu, \bar{\eta}, (1 - \|\bar{\eta}\|^2)^{1/2})$ on $\partial(B_1 \times I \times B^{d-1})$.

One may perform the linear deformation $t((1-\|\bar{\eta}\|^2)^{1/2}-(1-\|\tilde{\lambda}\|)/2)+ (1 - t)(2\|\lambda\| - 1)$, since if one has a zero on the boundary then, if $\|\lambda\| = 1$, $h = \mu = 0$, $\|\lambda\| = 1$, $\varphi = 1$, $B(\lambda, \mu) = B(\lambda)$ and $\bar{\eta} = 0$ and the linear term has value 1. On the other hand if $\|\bar{\eta}\| = 1$, then $h = 0$, $\varphi = 0$; hence $\|\tilde{\lambda}\| < 1/2$, $\|\lambda\| < 1/2$ and the linear term is negative.

Replace next $(2\|\lambda\|-1, h)$ by $((1-t)(2\|\lambda\|-1)+th, (1-t)h-t(2\|\lambda\|-1))$ and $\varphi(\lambda, \mu)$ by $\varphi(\lambda, \mu, t)$ where φ is 0 if $\det B(\lambda, \mu) = 0$ and $(1 - t)h - t(2\|\lambda\| - 1) = 0$, $\varphi(\lambda, \mu, 0) = \varphi(\lambda, \mu), \varphi(\lambda, \mu, 1) = 1$ and $\varphi(\lambda, \mu, t) = 1$ if $\|\lambda\| \geq 1/2$ (note that for $t = 1$ and $2\|\lambda\| - 1 = 0$ or if $\|\lambda\| \geq 1/2$, then $B(\lambda, \mu)$ is invertible, so that φ is well defined). The zeros of the deformation are for $\mu = 0, \|\lambda\| = 1/2$ and $\bar{\eta} = 0$; hence inside $B^k \times I \times B^{N-1}$. For $t = 1$, the map $(B(\lambda, \mu) \begin{pmatrix} \bar{\eta} \\ h \end{pmatrix}, 1 - 2\|\lambda\|)$ can be deformed via $((t\|\lambda\|B(\lambda/\|\lambda\|) + (1 - t)B(\lambda, \mu)) \begin{pmatrix} \bar{\eta} \\ (1-t)h + t\mu \end{pmatrix}, 1 - 2\|\lambda\|)$, since for $\|\lambda\| = 1/2, B(\lambda, \mu) = \|\lambda\|B(\lambda/\|\lambda\|)$ and $h = \mu$. Thus, the obstruction is just $J[B(\lambda)] = 0$ by hypothesis. Hence one has the extension.   Q.E.D.

Remark 3.1. a) If $\Sigma^l[(A_0 - A(\lambda))\eta_0] \neq 0$ for some l, $1 \leq l \leq N$, hence $N = d$, and there is global bifurcation in the direction η_0, by applying the argument of the proof of the "only if" part of the preceding theorem to $(A_0 - A(\lambda))r\eta_0 - g(\lambda, r\eta_0)$ in $\mathbb{R}^k \times \mathbb{R}$: the local invariant will be the class of $((A_0 - A(\lambda))\eta_0, r - \epsilon)$ on $\partial(B_k \times I)$, i.e. $\Sigma P_*[A_0 - A(\lambda)]$. Since η_0 can be moved in any direction, this will be true for all directions in S^{d-1}. Now, by looking at $(F(\lambda, r, r\mu_2, \ldots, r\mu_l, 0, \ldots, 0), \mu_2, \ldots, \mu_l, r - \epsilon)$ on $\partial(\|\lambda\| \leq \rho, |r| \leq 2\epsilon, |\mu_i| \leq \rho)$ one obtains $\Sigma^l P_*[B(\lambda)]$ and, by using the results of next section, a connected subset Σ of $\mathcal{C}(g)$, bifurcating globally from $(0,0)$, such that the local covering dimension at each point of Σ is at least l.

b) The global invariants are the suspension of the invariants for $B(\lambda)$. Thus, if $\Sigma P_*[B(\lambda)] \neq 0$ or $J[B(\lambda)] \neq 0$, one will have a "global continuum" in the sense that it may go to the boundary of the set where the Liapunov-Schmidt reduction is valid. For example take $A_0 - A(\lambda) = \begin{pmatrix} \lambda & -1 \\ 0 & \lambda \end{pmatrix}$, $\lambda \epsilon \mathbb{C}$, so that $k = 2$, $N = 4$, $d = 2$ and $B(\lambda) = \lambda^2$. Then $B(\lambda)\eta_0$ has degree 2, as well as $\Sigma^2 B(\lambda)\eta_0$. On the other hand, $J[B(\lambda)] = 2$, $\Sigma J[B(\lambda)] = J[A_0 - A(\lambda)] = 0$ and $[(A_0 - A(\lambda))\eta_0] = 0$. That is the local invariants are nontrivial, while the global invariants are 0. If $r^2 = |z_1|^2 + |z_2|^2$, consider the system $(\lambda z_1 - \varphi(r^2)z_2 + r^2\bar{z}_2, \lambda z_2 - r^2\bar{z}_1) = (0,0)$, where $\varphi(t)$ is a nonincreasing function, with values 1 if $t < \frac{1}{2}$, 0 if $t > 1$ and $\varphi(t)$ is concave for $t \leq t_0$, with $t_0 = \varphi(t_0)$. Multiplying the first term by z_2 and the second by $-z_1$, one has that $r^4 = \varphi(r^2)z_2^2$, that is z_2 is real and $|\lambda|^2\bar{z}_1 = r^2(\varphi(r^2) - r^2)\bar{z}_1$; also either $z_1 = z_2 = 0$ or $|\lambda|^2 = r^2(\varphi(r^2) - r^2)$ (hence $r^2 \leq t_0$). Since $t(\varphi(t) - t)$ has, in $[0, t_0]$, a unique maximum at $t = 1/2$, one has that $|\lambda| \leq 1/2$ and the set of solutions is bounded: it is given, as a function of r, as $|\lambda| = r(\varphi(r^2) - r^2)^{1/2}$, $|z_1| = |\lambda|/\varphi(r^2)$, $z_2 = \pm r^2/(\varphi(r^2))^{1/2}$ and if $\lambda = |\lambda|e^{i\varphi}$, then $z_1 = (sign\, z_2)|z_1|e^{-i\varphi}$ is the solution, i.e. this set is two-dimensional. Note that on the solution set, $\frac{\partial}{\partial z_2}(z_2(\varphi(r^2) - r^2) - \bar{\lambda}\bar{z}_1) = \varphi - r^2 + 2r^4(\varphi' - 1)/\varphi$ is 0 at $r^2 = 1/2$; there the Liapunov-Schmidt reduction fails.

4. A Summation Formula and a Generalized Degree

The last result in this section will relate the local invariants on a bounded continuum. Let $\mathcal{C}$ be the continuum branching from $(0,0)$ and let $\mathcal{C}(\eta_0)$ be the continuum for $F(\lambda, t\eta_0), t > 0$. Assume all bifurcation points, $(\lambda_i, 0)$, on $\mathcal{C}$ or $\mathcal{C}(\eta_0)$ have the property that $A - A(\lambda)$ is invertible for λ close to $\lambda_i, \lambda \neq \lambda_i$. Thus, if $\mathcal{C}$ is bounded there is only a finite number of such bifurcation points.

Theorem 4.1. *Assume $\mathcal{C}$ or $\mathcal{C}(\eta_0)$ is bounded, with the above hypothesis. Then*

$$\Sigma_I J[A_0 - A(\lambda)]_i = 0 \quad for \ (\lambda_i, 0) \in \mathcal{C} \ and$$

$$\Sigma_I \Sigma[A_0 - A(\lambda\eta_0)]_i = 0 \quad for \ (\lambda_i, 0) \in \mathcal{C}(\eta_0).$$

Proof. The proof is very similar to the preceding results: construct a bounded Ω such that $\mathcal{C}$ (or $\mathcal{C}(\eta_0)$) is contained in Ω and $F(\lambda, x)$ (or $F(\lambda, t\eta_0)$) is nonzero on $\partial\Omega$ if $x \neq 0$ (or $t \neq 0$). Complement $F(\lambda, x)$ by $\|x\| - \epsilon$ (or by $t - \epsilon$) and one gets a map which has a nonzero extension from $\cup_I \partial B_i$ to $B \backslash \cup_I B_i$, where $\Omega \subset B = \{(\lambda, x) : \|\lambda\|, \|x\| \leq R\}$ (respectively $\|\lambda\| \leq R, 0 \leq t \leq R$) and $B_i = \{(\lambda, x), \|\lambda - \lambda_i\| \leq \rho_i, \|x\| \leq 2\epsilon\}$ (respectively $0 \leq t \leq 2\epsilon_i$), as was done in Theorem 3.1.

In Ize (1982) a special degree was constructed for this situation, while in Ize (1988) obstruction theory was used to prove the finite dimensional equivalent of the theorem. However I shall take this opportunity to introduce a new degree, defined in the equivariant case in Ize–Massabó –Vignoli (1989). For this degree, one has that

$$\deg(F, \|x\| - \epsilon; B) = \Sigma_I \deg(F, \|x\| - \epsilon; B_i) = \Sigma_I J[A - A(\lambda)]_i = 0,$$

since on ∂B one has an extension to B. (The sums are in principle "up to two suspensions", but in this case one does not need them). Q.E.D.

In order to define this generalized degree, let us go to the general situation of maps $f(u)$ from B into E, two Banach spaces, such that $f(u) \neq 0$ for u on the boundary of an open bounded subset Ω of B. In the infinite dimensional case, the following simplifying assumptions will be made:

1) $B = \mathbb{R}^M \times G, E = \mathbb{R}^N \times G$
2) $f(u) = f(x, y) = (f_1(x, y), y - f_2(x, y))$, where f_2 is compact.

Consider then a closed ball B_R centered at the origin with radius R such that $\Omega \subset B_R$. Let $\tilde{f} : B_R \to E$ be a continuous extension of f, with the usual compactness properties. Let N be a bounded open neighborhood of $\partial\Omega$ such that $\tilde{f}(u) \neq 0$ on $\bar{N}$. Let $\varphi : B_R \to [0, 1]$ be a partition function such that $\varphi(u) = 0$ if u is in Ω and $\varphi(u) = 1$ if u is outside $\Omega \cup N$. Let $\tilde{F} : [0, 1] \times B_R \to \mathbb{R} \times E$ be the map defined by

$$\tilde{F}(t, u) = (2t + 2\varphi(u) - 1, \tilde{f}(u)).$$

Since $\tilde{F}(t, u) = 0$ only if $f(u) = 0$, u in Ω, and $t = 1/2$, $\tilde{F}$ can be regarded as a map from $S^B = \partial([0, 1] \times B_R)$ into $S^E \cong \mathbb{R} \times E \backslash \{0\}$. Hence $\tilde{F}$ defines an element of $\Pi_{S^B}(S^E)$, the homotopy group of maps between these two spheres, with the compactness assumptions in case of infinite dimensions; in this case, as for the Leray-Schauder degree, one may approximate $\tilde{F}$ by finite dimensional maps and, through the Freudenthal suspension theorem, $\Pi_{S^B}(S^E)$ is the stable limit of $\Pi_{S^{M+v}}(S^{N+v})$ when v goes to ∞. It is also easy to prove that $[\tilde{F}]$ is independent of $\varphi, N, \tilde{f}$ and B_R (see Ize et al. (1989), Propositions 2.1 and 3.1).

Definition 4.1. *Given* $f : \bar{\Omega} \to E$, *with* $f(u) \neq 0$ *on* $\partial\Omega$, *the generalized degree of* f *will be defined as* $\deg(f; \Omega) = [\tilde{F}]$.

Properties of the degree.

(P1): Existence. *If* $\deg(f; \Omega)$ *is nontrivial, then there exists* $u \in \Omega$ *such that* $f(u) = 0$.

(P2): Homotopy invariance. *If* $f(\tau, u) : [0, 1] \times \bar{\Omega} \to E$ *is nonzero on* $[0, 1] \times \partial\Omega$, *then* $\deg(f(\tau, \cdot); \Omega)$ *is constant.*

(P3): Excision. *If* $f : \bar{\Omega} \to E$ *is such that* $f(u) \neq 0$ *on* $\bar{\Omega}\backslash\Omega_0$, *for some open* $\Omega_0 \subset \Omega$, *then* $\deg(f; \Omega) = \deg(f; \Omega_0)$.

(P4): Suspension property. *If there is* $\tilde{f}$, *extension of* f *to* B_R, *such that* $\tilde{f}(u) \neq 0$ *on* $B_R\backslash\Omega$ *(in particular if* $\Omega = B_R$), *then* $\deg(f; \Omega) = \deg(\tilde{f}; \Omega) = \Sigma[\tilde{f}]$.

(P5): Additivity (Up to one suspension). *If* $\Omega = \Omega_1 \cup \Omega_2, \Omega_i$ *open such that* $\bar{\Omega}_1 \cap \bar{\Omega}_2 = \phi$, *then* $\Sigma \ \deg(f; \Omega) = \Sigma \ \deg(f; \Omega_1) + \Sigma \ \deg(f; \Omega_2)$, *where* Σ *is the suspension.*

(P6): The Leray-Schauder degree. *If* $B = E$, *then* $\deg(f; \Omega) = \deg_{LS}(f; \Omega)$.

(P7): Hopf property. *If* Ω *is a ball and* $\deg(f_1; \Omega) = \deg(f_2; \Omega)$ *and if* Σ *is one-to-one, then* $f_1|_{\partial\Omega}$ *is homotopic to* $f_2|_{\partial\Omega}$.

(P8): Universality property. *If* $\Delta(f; \Omega)$ *is any other degree with properties* (P1)–(P3) *and* Σ *is one to one, then, if* $\Delta(f; \Omega)$ *is nontrivial, this is also the case for* $\deg(f; \Omega)$.

Proof. The proof of the properties (P 1)–(P 4) are easy and left to the reader (see also Ize et al (1989).

To prove (P5), note that any map $F(t, u) = (h(t, u), f(t, u))$ defined on $\partial(I \times B_R)$ can be deformed on $(\partial I) \times B_R$ to $(h(t, 0), f(t, 0))$, with $t = 0$ or 1 and then to $(1, 0)$. From Borsuk's extension theorem, one may assume that $F(t, u)$ is (1,0) for $t = 0$ or 1. Then $[F] + [G]$ is defined as $F(2t, u)$ for $0 \leq t \leq 1/2$, $G(2t - 1, u)$ for $1/2 \leq t \leq 1$.

Take $N = N_1 \cup N_2$, with $\bar{N}_1 \cap \bar{N}_2 = \phi$ and let $\varphi, \varphi_1, \varphi_2$ denote the partition functions associated to N, N_1, N_2. Then $\tilde{F}(t, u) = (2t + 2\varphi(u) - 1, \tilde{f}(u))$ is deformable to $F(t, u)$ defined as $(2t + (1 - 2t)(2\varphi(u) - 1, \tilde{f}(u))$ for $0 \leq t \leq 1/2$ and $(1, \tilde{f}(u))$ for $1/2 \leq t \leq 1$, by replacing $2t + 2\varphi - 1$ by $2t + (1 - 2t\tau)(2\varphi - 1)$ for $0 \leq t \leq 1/2$ and by $(1 - \tau)(2t + 2\varphi)$ for $1/2 \leq t \leq 1$. Now, if one changes t by $1 - t$ in the above formula, one gets $-[F]$. Hence,

$$[F] - [F_1] = \begin{bmatrix} (2t + (1 - 2t)(2\varphi - 1), \tilde{f}(u)) & 0 \leq t \leq 1/2 \\ (2(1 - t) + (2t - 1)(2\varphi_1 - 1), \tilde{f}(u)) & 1/2 \leq t \leq 1 \end{bmatrix},$$

where $[F_1] = \deg \ (f; \Omega_1)$.

Consider the homotopy $H_\tau(t, u)$ defined as $(1, \tilde{f}(u))$ if $u \notin \bar{\Omega} \cup \bar{N}$, on $\bar{\Omega}_1 \cup \bar{N}_1$ as $(h_1(t, \tau, u), \tilde{f}(u))$ where

$$h_1(t, \tau, u) = \begin{pmatrix} 2t + (1 - 2t)(2\varphi_1 - 1) & if & 0 \leq 2t \leq \tau \\ \tau + (1 - \tau)(2\varphi_1 - 1) & if & \tau \leq 2t \leq 2 - \tau \\ 2(1 - t) + (2t - 1)(2\varphi_1 - 1) & if & 2 - \tau \leq 2t \leq 2, \end{pmatrix}$$

and, on $\bar{\Omega}_2 \cup \bar{N}_2$, as $(h_2(t, \tau, u), \tilde{f}(u))$ with

$$h_2(t, \tau, u) = \begin{pmatrix} 2t + (1 - 2t)(2\varphi_2 - 1) & if & 0 \leq t \leq 1/2 \\ 1 & if & 1/2 \leq t \leq 1 \end{pmatrix}.$$

Clearly $[H_1] = [F] - [F_1]$. On the other hand, $H_0(t, u) = (2\varphi_1 - 1, \tilde{f}(u))$ if $u \in \bar{\Omega}_1 \cup \bar{N}_1$ (and hence H_0 is nonzero there), while $H_0(t, u)|_{\bar{\Omega}_2 \cup \bar{N}_2} =$

$F_2(t, u)$. That is, H_0 is an extension of F_2 which is nonvanishing on $I \times (B_R \backslash (\bar{\Omega}_2 \cup \bar{N}_2))$.

From (P4), $\deg(H_0; I \times B_R) = \deg(F_2; I \times \bar{\Omega}_2) = \deg(F_2; I \times B_R) = \Sigma[F_2]$, by (P3). Finally $\deg(H_0; I \times B_R) = \Sigma(H_0) = \Sigma[H_1] = \Sigma([F] - [F_1])$.

For (P6) it is enough to note that $\deg(f; \Omega) = [F] = \deg_{LS}(F; I \times B_R) = \deg_{LS}(F; I \times \Omega) = \deg_{LS}((2t - 1, f); I \times \Omega) = \deg_{LS}(f; \Omega)$, where the properties of excision and product for the Leray-Schauder degree are used.

(P7) is essentially by definition and (P8) follows from the fact that $\Delta(F; I \times B) = \Delta((2t - 1, f); I \times \Omega) = \Sigma \Delta(f; \Omega)$, where the first equality comes from (P3) and the second is the suspension. Hence, if $\deg(f; \Omega) = 0$, then F has a nonzero extension from $\partial(I \times B)$ to $I \times B$. By (P1), $\Delta(F; I \times B)$ must be trivial and, since Σ is one-to-one, $\Delta(f; \Omega)$ is also trivial. Q.E.D.

Remark 4.1. In the proof of (P4), if Ω_2 is a ball, then, since H_0 is nonzero on $I \times (B_R \backslash \bar{\Omega}_2 \cup \bar{N}_2)$, the class of H_0 on $\partial(I \times B_R)$ is the same as the class of H_0 an $\partial(I \times \bar{\Omega}_2)$ by a radial retraction. That is, $[H_0] = [F_2]$. But $[H_0] = [F] - [F_1]$. Hence in this case the addition formula is true without suspension.

In general, consider the following example, taken from Ize et al. (1992), close to Romero Ruiz del Portal (1990). Let $f(x_1, x_2, x_3) = f_1 + if_2 = (x_1^2 + x_2^2 - 1 + ix_3)((x_1 - 1)^2 + x_3^2 - 1 + ix_2)$. The zeros of f are the two linked circles $S_1 = \{x_1^2 + x_2^2 = 1, x_3 = 0\}$ and $S_2 = \{(x_1 - 1)^2 + x_3^2 = 1, x_2 = 0\}$. Take $B = \{(x_1, x_1, x_3) : x_1^2 + x_2^2 + x_3^2 < 4\}$ and Ω_j two small disjoint tubular neighborhoods of S_j. Then one has $\deg(f; \Omega) = \deg(f; B) = \Sigma[f]$, by (P3). But $[f] \in \Pi_2(S^1) = 0$, hence $\deg(f; B) = 0$.

On Ω_1, one may perform the deformation $(x_1^2 + x_2^2 - 1 + ix_3)(\tau(x_1 - 1)^2 + x_3^2 - 1) - (1 - \tau)x_1 + ix_2)$. On $\partial \Omega_1$ the first factor is nonzero; thus a zero would have $x_2 = 0, x_1$ close to ± 1 and x_3 close to 0. If x_1 is close to 1, the deformed term is negative, while if x_1 is close to -1, the deformed term is positive. Hence, $\deg(f; \Omega_1) = \deg((x_1 - ix_2)(x_1^2 + x_2^2 - 1 + ix_3); \Omega_1) = \deg(\bar{z}(|z|^2 - 1 + ix_3); \Omega_1)$, where $z = x_1 + ix_2$. Now in $(2t + 2\varphi - 1, f)$ one may deform φ to $(|z|^2 - 1)^2$, and $(|z|^2 - 1 + ix_3)$ to $(|z|^2 - 1)(1 + \tau)(1 - \tau(2t - 1)) + ix_3$ and obtain $(2t - 1 + 2(|z|^2 - 1)^2, \bar{z}(4(|z|^2 - 1)(1 - t) + ix_3))$. By performing the rotation

$$\begin{pmatrix} \tau & -(1 - \tau) + 2\tau(|z|^2 - 1) \\ 1 - \tau - 2\tau(|z|^2 - 1) & 2\tau \end{pmatrix} \begin{pmatrix} 2t - 1 \\ |z|^2 - 1 \end{pmatrix}$$

one arrives at $\deg(f; \Omega_1)\eta = [1 - |z|^2, \bar{z}(2t - 1 + ix_3)] = \eta$, where η is the Hopf map. Thus, $\deg(f; \Omega_1) = 1$.

Similarly for Ω_2, make the deformation $(\tau(x_1^2 + x_2^2 - 1) + (1 - \tau)(x_1 - 1) + ix_3)((x_1 - 1)^2 + x_3^2 - 1 + ix_2)$. The resulting map can be written as

$(y_1 - iy_2)(y_1^2 + y_2^2 - 1 + iy_3)$, with $y_1 = x_1 - 1, y_2 = -x_3, y_3 = x_2$ with a positive Jacobian and Ω_2 is sent into Ω_1. Then one has $\deg(f; \Omega_2) = 1$.

Thus, $\deg(f; \Omega_1 \cup \Omega_2) = 0 \neq \deg(f; \Omega_1) + \deg(f; \Omega_2) = 2$. Clearly, when one suspends, the equality holds.

Remark 4.2. In the bifurcation problem, the preceding remark implies that $0 = \Sigma_I \deg(F, \|x\| - \epsilon; B_i) = \Sigma_I(\Sigma[J(A - A(\lambda))])_i$. This last suspension can be removed by using the results mentioned in the above references. Note that $J[A_0 - A(\lambda)]$ is related to $J[B(\lambda)]$ via suspensions and an orientation factor, as in Section I, Remark 1.1, or as in Ize (1988), Remark 4.7.

Remark 4.3. A generalized degree theory, topologically equivalent to the one presented here, is given, for the nonequivariant case in Geba et al. (1986): there the Alexandroff compactification is used and the additivity property is proved in the stable case only, by using cohomotopy operations. This approach was extended in F. Romero (1990). Note that the extension to the more general class of maps (k—set contraction, condensing, A proper) seems to be straightforward.

III. STRUCTURE AND DIMENSION
OF GLOBAL BRANCHES

In this section, I shall consider equations of the form $f(\lambda, x) = 0$, where $F : \wedge \times B \to E$ is a continuous map. B, E and $\wedge$, the parameter space, are Banach spaces and $\wedge$ may be infinite dimensional. As mentioned in the preceding sections, a first, rather natural, approach to this problem is to reduce it to a one-dimensional parameter problem (whenever possible) and to obtain, in the case of bifurcation problems, a family of global branches satisfying the global alternative. The problem is then to understand how these one-dimensional threads are woven together. In fact, if $F(\lambda, x) = (A_0 - A(\lambda))x - g(\lambda, x)$, such that $A_0 - A(\lambda_i)$ is invertible for λ_1 and λ_2 close to 0, and these operators have different indices (hence $A_0, A(\lambda)$ etc... have the right compactness properties for the Leray-Schauder degree) then, on any line connecting λ_1 to λ_2, one has a point of global bifurcation. Alexander and Antman (1981) have proved, using cohomology theories, that in fact one has a "surface" of local dimension at least $\dim \wedge$ bifurcating from (0,0).

For problems of the form $x - g(\lambda, x)$, λ in $\mathbb{R}^n$ and g compact, the Leray-Schauder continuation principle was studied by Massabó and Pejsachowicz (1984) and extensions of the same problem to infinite dimensional parameter spaces were given in Alexander-Antman (1983) and Alexander et al. (1982). For A-proper maps these results were extended by Fitzpatrick et al. (1983 and 1986) and Milojevic (1986). For the case where $J[A - A(\lambda)]$ is nontrivial on a subspace of $\wedge$, one may consult Bartsch (1986). All the above papers are based on Cech cohomology. The approach given here uses only very simple facts of point-set topology and enables one to treat easily more general problems. Most of the results given here are taken from Ize et al. (1985), and, for the equivariant case, from Ize et al (1986).

In order to follow the definitions and results which will be presented next, consider the following example: let S be the set of nontrivial solutions of the equation $F(\lambda, x) = 0$ and suppose that $A_0 - A(\lambda)$ is invertible for $\lambda = (\tilde{\lambda}, \tilde{\tilde{\lambda}}), \|\tilde{\lambda}\| = \rho, \tilde{\tilde{\lambda}} = 0$. Then if $J[A_0 - A(\tilde{\lambda}, 0)] \neq 0$ one expects a global branch bifurcating in the subspace $\tilde{\wedge} \times \{0\}$, as well as another branch in $\tilde{\wedge} \times \{\tilde{\tilde{\lambda}}\}$ for each fixed $\tilde{\tilde{\lambda}}$, small enough. In fact, if the map $(\tilde{\tilde{\lambda}}, \|x\| - \epsilon)$ is "0-epi" on S, one has a subset of S with the properties of global bifurcation and of local dimension, at each point, at least dim $\tilde{\wedge} + 1$.

1. 0-Epi Maps

In what follows $\tilde{E} = \wedge \times B$ and G are Banach spaces, U is an open subset of $\tilde{E}$ and S, not necessarily bounded and unless otherwise specified, will stand for an arbitrary subset of $\tilde{E}$. The notion of 0-epi maps (and later on sectionally 0-epi maps for the infinite dimensional parameter case) grows out of the concepts of essential maps, defined in Granas (1962) and epi-maps used by Furi et al. (1980).

Definition 1.1. a) *Let $g : U \to G$ be a continuous map. Then g is said to be **admissible** on S and U if there is an open and bounded subset V_0 such that $g^{-1}(0) \cap S \subset V_0 \subset \bar{V}_0 \subset U$.*

b) *If $g : \bar{U} \to G$ is continuous, then g is said to be **admissible** on S and $\bar{U}$ if $g^{-1}(0) \cap S$ is a bounded subset of U.*

Definition 1.2. a) *If g is admissible on S and U (resp. S and $\bar{U}$), g is said to be **0-epi** on $S\&U$ (resp. $S\&\bar{U}$) if the equation $g(x) = h(x)$ has a solution in $X \cap U$ for any compact map $h : \tilde{E} \to G$ with supp h bounded and contained in U (resp. supp $h \subset \bar{U}$ and $h|\partial U = 0$).*

b) *If g is admissible on S and U (resp. S and $\bar{U}$), then g is said to be **0-essential** on S and U (resp. S and $\bar{U}$) if for any open and bounded set V, with $g^{-1}(0) \cap S \subset V \subset \bar{V} \subset U$ (resp. $g^{-1}(0) \cap S \subset V \subset U$), any continuous extension $\tilde{g} : \bar{V} \to G$ of $g : \partial V \to G$ with $g - \tilde{g}$ compact on $\bar{V}$, has a zero on $S \cap V$.*

Proposition 1.1. a) *The map g is 0-epi on S and U (resp. S and $\bar{U}$) if and only if g is $0-$essential on S and U (resp. S and $\bar{U}$).*

b) *If U is bounded and $g : \bar{U} \to G$ is nonzero on $S \cap \partial U$, then g is 0-essential on S and $\bar{U}$ if and only if any extension $\tilde{g} : \bar{U} \to G$ of $g|_{\partial U}$, with $g - \tilde{g}$ compact on $\bar{U}$, has a zero on $S \cap U$.*

c) *If $g : \bar{U} \to G$ is admissible and such that, for any closed and bounded subset D of ∂U, $\operatorname{dist}(g(\bar{S} \cap D), 0) > 0$ whenever $\bar{S} \cap D \neq \phi$, then g is 0-epi on S and $\bar{U}$ if and only if g is 0-epi on S and U.*

d) *If U is bounded and $S \cap \bar{U}$ is closed, let $g : \bar{U} \to \mathbb{R}^n$ be bounded and $g^{-1}(0) \cap S \subset U$. Then g is 0-epi on $S\&\bar{U}$ if and only if $g/\|g\| : S \cap \partial U \to S^{n-1}$ has no extension to $S \cap \bar{U}$.*

Proof. (a): **"if"**. Let h be as in Definition 1.2 (a) and set $V = \{x \in \bar{E} : h(x) \neq 0\} \cup V_0$. Then V is open, bounded and $\bar{V} = \operatorname{supp} h \cup \bar{V}_0$. Since $h|_{\partial V} = 0$, then $\tilde{g} = (g - h)|_{\bar{V}}$ is an extension of $g|_{\partial V}$ and, as such, $\tilde{g}$ has a zero on $S \cap V$.

(a): **"Only if"**. Let V and $\tilde{g}$ be as in Definition 1.2 (b). Take h to be $g - \tilde{g}$ on $\bar{V}$ and 0 outside V. Then $\operatorname{supp} h \subset \bar{V}$ and the equation $g(x) = h(x)$ has a solution $x \in S \cap U$. Since x cannot be in $\bar{E} \backslash \bar{V}$, it has to be in $S \cap V$.

(b) **"Only if"**. Take $V = U$.

(b) **"if"**. If there is an open set V with $g^{-1}(0) \cap S \subset V \subset U$ and $\tilde{g} : \bar{V} \to G$ extends $g|_{\partial V}$, with $g - \tilde{g}$ compact and $\tilde{g} \neq 0$ on $S \cap V$, then the map g_1 defined as $\tilde{g}$ on $\bar{V}$ and g on $\bar{U} - V$ is non-zero on $S \cap \bar{U}$. Moreover, $g - g_1$ is compact on $\bar{U}$ and g_1 extends $g|_{\partial U}$.

(c) The proof is given in Ize et al (1985), Proposition 2.3.

(d) See Proposition 3.1 in the above reference.   Q.E.D.

Properties of 0-epi maps

(P1). Existence. *If g is 0-epi on S and U (resp. S and $\bar{U}$), then $g^{-1}(0) \cap S \neq 0$.*

(P2). Localization with respect to U. *If g is 0-epi on S and U, then g is 0-epi on S and V, for any open subset of U such that g is admissible for S and V. (The same is true for $\bar{U}$ and $\bar{V}$). If $\bar{V} \subset U$, then g is also 0-epi on S and $\bar{V}$.*

(P3). Localization with respect to the boundary. *If g is 0-epi on S and $\bar{U}$ and $g^{-1}(0) \cap S \subset V_0 \subset \bar{V}_0 \subset U$, then g is 0-epi on S and U.*

(P4). Extension of the set S. *If g is 0-epi on S_1 and U and $S_2 \supset S_1$ is such that g is admissible on S_2 and U, then g is 0-epi on S_2 and U and also on $(S_2 \backslash U^C)$ and U.*

(P5). Normalization. *If $i : U \to \tilde{E}$ is the inclusion and $0 \notin \partial U$, then i is 0-epi on S and U if and only if $0 \in S \cap U$ and the connected component of 0 in U is contained in S. In particular if U is connected, then $S \cap U = U$.*

(P6). Semi-excision property. *If A is a closed (in U) subset of U such that $\bar{S} \cap A = \phi$, then g is 0-epi on S and $(U \backslash A)$ if and only if g is 0-epi on S and U. (In particular if V is open and $\mathring{S} \cap \bar{U} \subset V \subset U$, take $A = V^C \cap U$, and then $U \backslash A = V$).*

The proofs of these properties can be found in the above reference.

Remark 1.1. a) If one has any degree theory for which $\deg(g - h; U) = \deg(g; U)$ for any compact map h with bounded support contained in U, taking $S = \tilde{E}$, if $\deg(g; U) \neq 0$ then g is 0-epi on U. If $g : \bar{U} \to \mathbb{R}^n$, with U open bounded and **connected** in $\mathbb{R}^n$, then g is 0-epi on U if and only if $\deg(g; U) \neq 0$ (see Remark 2.7 in the above reference). Here the connectedness of U is essential, as seen below.

b) The above properties (together with the homotopy property) are very close to those enjoyed by any degree theory, except the (full) excision property: if $g : \bar{U} \to \mathbb{R}^n$, with U open and bounded in $\mathbb{R}^n$, such that $g^{-1}(0) \subset V \subset U$, V open, then $\deg(g; V) = \deg(g; U)$. If this degree is nonzero, then g is 0-epi on V and on U. This is not true in general for 0-epi maps, even in this simple case. For example, let $V \subset \mathbb{R}^{n+m}$ be defined by $\{x \in \mathbb{R}, y \in \mathbb{R}^M : (\|y\| - a)^2 + \|x\|^2 < r^2\}$, with $a > r$, i.e. V has the type of $B^n \times S^{m-1}$ (hence not connected if $m = 1$). Let $g : \mathbb{R}^{n+m} \to \mathbb{R}^{n+1}$ be the map $g(x, y) = (x, \|y\| - a)$ and $B = \{(x, y) \in \mathbb{R}^{n+m} : \|x\|^2 + \|y\|^2 < R^2\}$, with $R > a + r$, so that $V \subset B$ and $g(x, y) \neq 0$ on $B \backslash V$. On $\partial B, g$ is deformable to $(0, R - a)$, hence, it is inessential and has a nonzero extension to B. If $g|_{\partial V}$ has a nonzero extension $\tilde{g}$ to $\bar{V}$, let $y = (y_1, \tilde{y}) \in \mathbb{R}^{n+1} : |x|^2 + (a \pm y)^2 < r^2\}$, the union of two disjoint balls. On their boundary $g(x, y_1) = (x, \pm y_1 - a)$ is a linear map with degree ± 1, hence essential. This contradiction implies that g is 0-epi on $\bar{V}$ (but not on $B \supset V$). Note that, if $m = 1$, then V is not connected.

The following property will be used for detecting 0-epi maps:

(P7): Products. *If $g : U \to G_1 \times G_2$, given by $g(x) = (g_1(x), g_2(x))$, is 0-epi on S and U, then g_2 is 0-epi on $g_1^{-1}(0) \cap S$ and U.*

For the proof take $h_2 : \tilde{E} \to G_2$ and $h = (0, h_2)$. Then the equations $g_2(x) = h_2(x)$ and $g_1(x) = 0$ are solvable in $S \cap U$.

Remark 1.2. If $S = \tilde{E}$, then the above property gives sufficient conditions for g_2 to be 0-epi on $g_1^{-1}(0)$ and U. Note that the converse is not true: let $\tilde{E} = \mathbb{R}^2$, $G_1 = G_2 = \mathbb{R}$, $g_1(x, y) = x^2 - 1, g_2(x, y) = y$. Then the index of (g_1, g_2) at $(\pm 1, 0)$ is ± 1, so that $\deg(g_1, g_2; U) = 0$, for any U containing the zeros. If U is connected then, from the preceding remark, (g_1, g_2) is not 0-epi on U. On the other hand, if $y - h(x, y) \neq 0$ on $g_1^{-1}(0) \cap U$, for some h

with bounded support contained in U, then $y - h(1, y) \neq 0$, but this is not possible since this map has degree 1 on $U \cap \{x = 1\}$. Hence, g_2 is 0-epi on $g_1^{-1}(0)$ and U.

The last of the basic properties is the homotopy property. Here one pays the price of a theory with few ingredients by restricting the possible deformations.

(P8): Homotopy-principle. *Assume that S is closed in U and that g is 0-epi on S and U. Let $h(t, x) : [0, 1] \times (S \cap U) \to G$ be continuous, with $h(0, x) = 0$ for x in $S \cap U$. Assume that:*

    (1) *there is an open and bounded set V_0 such that $A_0 \equiv \{x \in S \cap U : g(x) - h(t, x) = 0$ for some $t \in [0, 1]\} \subset V_0 \subset \bar{V}_0 \subset U$,*

    (2) *for any compact $k : E \to G$ with bounded support contained in U, let $A_k \equiv \{x \in S \cap U : g(x) - h(t, x) - k(x) = 0$ for some $t\}$ and let $\phi_k : \bar{E} \to [0, 1]$ be a continuous function such that $\phi_k(A_k) = 1$ and $\phi_k(V_k^C) = 0$, where $V_k = V_0 \cup \{x \in \bar{E} : k(x) \neq 0\}$. Assume that if M is a closed subset of $S \cap \bar{V}_k$ such that $M = g^{-1}(Co(h(\phi_k(\cdot), \cdot) + k(\cdot))(M) \cup \{0\})) \cap S \cap \bar{V}_k$. Then $h(\phi_k(\cdot), \cdot)(M)$ is precompact. (Here Co is the convex closure).*

Then, the map $g(\cdot) + \tilde{h}(\cdot)$ is 0-epi on S and U for any $\tilde{h}$ continuous extension to U of $h(1, \cdot)$.

The proof is given in Theorems 2.1 and 2.2 in Ize et al. (1985). The first condition is a natural control of the zeros of the deformation. The second condition is met in several well known cases.

Corollary 1.1. *If the homotopy h is compact on $I \times (S \cap U)$, condition (2) is satisfied. In particular, if $g : \bar{U} \to G$ is 0-epi on the bounded set S and $\bar{U}$, S closed, and $h : I \times (S \cap \partial U) \to G$ is compact with $h(0, x) = 0$, then, if $g(x) \neq h(t, x)$ on $I \times (S \cap \partial U)$, one has that $g - \tilde{h}$ is 0-epi on S and $\bar{U}$ for any extension $\tilde{h}$ to $\bar{U}$ of $h(1, \cdot)$ which is compact on $S \cap \bar{U}$.*

Note that this result is very close to Borsuk's extension theorem.

Definition 1.3. *Denote by $\alpha(A)$ the measure of noncompactness of a set A. If $g : S \cap U \to G$ and $h : I \times (S \cap U) \to G$ send bounded and closed (in $\bar{E}$) sets into bounded sets, then the map h is said to be **condensing with respect to g on $S \cap U$**, if for any bounded set $M \subset \bar{M} \subset S \cap U$, we have $\alpha(h(\cdot, \cdot)(I \times M)) < \alpha(g(M))$ when this last number is positive and $\alpha(h(\cdot, \cdot)(I \times M)) = 0$ if $\alpha(g(M)) = 0$.*

Corollary 1.2. *If h is condensing with respect to g on $S \cap U$, then condition (2) of property (P8) is satisfied.*

A more useful deformation is given in the following

Definition 1.4. *Let $g \equiv f - g_0 : S \cap U \to G$ and $h : I \times (S \cap U) \to G$ be such that f, g_0 and h send bounded and closed (in $\bar{E}$) sets into bounded sets. The homotopy $g_0(\cdot) + h(\cdot, \cdot)$ is said to be k-**set-contractive with respect to f on $S \cap U$**, if there is k, $0 \le k < 1$, such that for any bounded $M \subset \bar{M} \subset S \cap U$, then $\alpha((g_0(\cdot) + h(\cdot, \cdot))(I \times M)) \le k\alpha(f(M))$.*

Corollary 1.3. a) *Let $g \equiv f - g_0 : U \to G$ be 0-epi on S and U and let $g_0 + h$ be k-set-contractive with respect to f on $S \cap U$, with $h(0, x) = 0$ on $S \cap U$. Assume condition (1) holds. Then $f(\cdot) - g_0(\cdot) - \tilde{g}(\cdot)$ is 0-epi on S and U for any $\tilde{h}$ continuous extension to U of $h(1, \cdot)$.*

   b) *Let $g_0 + h$ be condensing with respect to f on $S \cap U$ (i.e. $k = 1$ and strict inequality). Assume that h is as in (a), that f is proper on bounded and closed subsets of $S \cap U$ and that for any open and bounded V, with $A_0 \subset V \subset \bar{V} \subset U$, there is τ_V for which $f - (1 - \tau)g_0$ is 0-epi on $S \cap \bar{V}$, for $0 \le \tau < \tau_V$. Then the conclusion of (a) holds.*

The following result gives the first topological consequences of the notion of 0-epi.

Proposition 1.2. *Let g be 0-epi on S and U (resp. S and $\bar{U}$). Then:*
(a) *either $S \cap \partial V \ne \phi$ or $g(S \cap \bar{V}) = G$ for any open and bounded V, with $g^{-1}(0) \cap S \subset V \subset \bar{V} \subset U$ (resp. $g^{-1}(0) \cap S \subset V \subset U$). In particular, if g sends bounded and closed (in $\bar{E}$) subsets of $S \cap U$ (resp. $S \cap \bar{U}$) into bounded sets of G, then $S \cap \partial V \ne \phi$.*

   (b) *Either $S \cap U$ is unbounded or there is a V as above such that $g(S \cap \bar{V}) = G$ or $(\overline{S \cap U}) \cap \partial U \ne \phi$. In particular, if $S \cap \bar{U}$ is closed, then $S \cap \partial U \ne \phi$.*

Proof. (a). If for some V, one has $S \cap \partial V = \phi$, then, from (P2), g is 0-epi on S and $\bar{V}$. If there is $p \notin g(S \cap \bar{V})$, let $h(t, x) = tp$. Then, from Corollary 1.1, $g(x) - p$ is 0-epi on S and $\bar{V}$, which contradicts (P1).

   (b) If $(\overline{S \cap U}) \cap \partial U = \phi$ and $S \cap U$ is bounded, then there is a bounded and open V such that $\overline{S \cap U} \subset V \subset \bar{V} \subset U$. Since $S \cap \partial V = \phi$, from (a), one gets $g(S \cap \bar{V}) = G$. Finally if $S \cap \bar{U}$ is closed and $(\overline{S \cap U}) \cap \partial U \ne \phi$, then $(\overline{S \cap U}) \cap \partial U \subset S \cap \partial U$ so that $S \cap \partial U \ne \phi$. Q.E.D.

Remark 1.3. Zero-epi maps on sections. When the parameter space is infinite dimensional and g is not proper, one may use a weaker notion of essentiallity. A **family of sections** of G is a collection $\mathcal{M}$ of finite dimensional subspaces of G, such that:

   (1) There is a fixed $M_0 \in \mathcal{M}$, such that if $M \in \mathcal{M}$ then $M_0 \subset M$.

(2) If $M_1, M_2 \in \mathcal{M}$, then there is $M_3 \in \mathcal{M}$ such that $M_1, M_2 \subset M_3$.
3) $\overline{\cup_{\mathcal{M}} M} = G$.

Definition 1.5. (a) *If g is admissible on S and U (resp. S and $\bar{U}$), g is said to be **sectionally 0-epi** on S and U (resp. S and $\bar{U}$) if and only if for any $M \in \mathcal{M}$, writing $G = M \oplus N, g = (g_M, g_N)$, g_M is 0-epi on $g_N^{-1}(0) \cap S$ and U (resp. $g_N^{-1}(0) \cap S$ and $\bar{U}$). One has a similar definition for sectionally 0-essential maps.*

Proposition 1.3. a) *The definitions of sectionally 0-epi and sectionally 0-essential are equivalent.*

b) *If $g : \bar{U} \to G$ is such that $g^{-1}(0) \cap S$ is bounded and for any closed and bounded subset D of ∂U one has, for any $M \in \mathcal{M}$, dist $(g_M(g_N^{-1}(0) \cap \bar{S} \cap D), 0) > 0$ whenever the first set is nonempty. Then g is sectionally 0-epi on S and $\bar{U}$ if and only if it is on S and U.*

c) *If g is proper on bounded and closed (in E) subsets of $S \cap U$ (resp. $S \cap \bar{U}$), then g is 0-epi on S and U (resp. S and $\bar{U}$) if and only if g is sectionally 0-epi on S and U (resp. S and $\bar{U}$).*

d) *The conclusions of Proposition 1.2. are valid for sectionally 0-epi maps, replacing G by $\cup_{\mathcal{M}} M$.*

e) *Properties (P1)–(P8) are true with the obvious modifications in the case of the homotopy principle.*

Below the following terminology will be used:

a) *If X is closed in E, $g : X \to G$ is said to be **sectionally bounded on X** if, for any $M \in \mathcal{M}$, $g_M(g_N^{-1}(0) \cap X)$ is bounded.*

b) *g is said to be **sectionally proper** on X if, for any compact $K \subset M \in \mathcal{M}$, $g^{-1}(K \times \{0\})$ is compact.*

Finally, assume that there are sequences of finite dimensional subspaces E_n and G_n of $\tilde{E}$ and G (not necessarily of the same dimension) and projections $Q_n : G \to G_n$. Let $g : U \to G$ be **not necessarily continuous**, $U_n = U \cap E_n, g_n = Q_n g|_{U_n}$ and assume that:

1) $g^{-1}(0) \cap S \subset V_0 \subset \bar{V}_0 \subset U, g_n^{-1}(0) \cap S \subset V_0 \cap E_n$ for all n's, for some fixed open and bounded V_0.

2) $\|Q_n\| \leq C$.

3) If for z in G one has a sequence $x_{n_k} \in S \cap U_{n_k} \cap V$, with V open and bounded such that $\bar{V} \subset U$, and $g_{n_k}(x_{n_k}) \to z$, then $\{x_{n_k}\}$ has a subsequence which converges to some $x \in S \cap U$, with $g(x) = z$.

Such a g will be said to be **A-proper 0-epi on S and U** if g_n is 0-epi on S and U_n, for infinitely many n's (hence g_n is continuous). It can be shown that g has the property of 0-epi maps in this case, without asking for its continuity.

2. Dimension

Here the notion of dimension which will be used is that of **covering dimension**: a normal topological space X has covering dimension equal to n, provided n is the smallest integer such that, if $\mathcal{U}$ is an open covering of X, there is a refinement $\mathcal{U}'$ of $\mathcal{U}$ such that no more than $n+1$ members of $\mathcal{U}'$ have a nonempty intersection.

Using standard properties of this dimension and Proposition 2.2 (d), one has

Proposition 2.1. *If g is sectionally 0-epi: on S and U (resp. S and $\bar{U}$) and sectionally bounded on bounded and closed (in $\bar{E}$) subsets of $S \cap U$ (resp. $S \cap \bar{U}$), then $\dim(S \cap \bar{V}) \geq \dim G$ and $\dim(S \cap \partial V) \geq \dim G - 1$, for any open bounded set V such that $g^{-1}(0) \cap S \subset V \subset \bar{V} \subset V$ (resp. $g^{-1}(0) \cap S \subset V \subset U$). In particular, $\dim(S \cap U) \geq \dim G$ and, if $S \cap \bar{U}$ is bounded, $\dim(S \cap \partial U) \geq \dim G - 1$.*

If two nonzero maps are different on a set of small dimension, there is a common nonzero extension. More precisely,

Lemma 2.1. *Let $A = A_1 \cup A_2$ be bounded, with A_i closed in A. Let $g, g_1, g_2 : A \to G$ be continuous maps which are sectionally bounded on A and such that:*

(1) $g_i - g$ is compact, $i = 1, 2$. (2) $g_i : A_i \to G \backslash \{0\}, i = 1, 2$. (3) there is a closed subset $B_i \subset A_i$ such that $g_i = g$ on $B_i, i = 1, 2$. (4) If $g_i - g$ are finite dimensional maps then $\dim\{x \in A_1 \cap A_2 : g_1(x) \neq \lambda g_2(x)$, for some $\lambda > 0\} < \dim G - 1$, or if $\dim G = \infty$, then $\dim(A_1 \cap A_2 \backslash (B_1 \cup B_2)) < \infty$ and g is proper. Then $g|_{B_1 \cup B_2}, g_i|_{A_i}$ extend to $\tilde{g}, \tilde{g}_i : A \to G \backslash \{0\}$ with $g - \tilde{g}, g_i - \tilde{g}_i$ compact on A.

From this technical result one may improve Proposition 1.2:

Proposition 2.2. *Let $g : \bar{U} \to G$ be sectionally 0-epi on S and $\bar{U}$ and sectionally bounded on bounded and closed subsets of $S \cap \bar{U}$. Assume that for any $p \in \cup_M M$. Then $g^{-1}([0, p]) \cap S$ is bounded (for example if g is sectionally proper on $S \cap \bar{U}$ or if $\|g(x)\| \to \infty$ when $\|x\| \to \infty$ on $S \cap \bar{U}$). Then $\dim(S \cap \partial U) < \dim G - 1$ implies $\cup_M M \subset g(S \cap \bar{U})$.*

Proof. Assume that $g(x) \neq p$ for some $p \in \cup_M M$ and all $x \in S \cap \bar{U}$. Take R so large that $B_R \supset g^{-1}([0, p]) \cap S$ and let ϕ be an increasing function such that $\phi(r) = 0$ if $0 \le r \le R$ and $\phi(r) = 1$ if $r \ge 2R$. On $S \cap \partial(\bar{B}_{2R} \cap \bar{U})$ the homotopy $g(t,x) = g(x) - t\phi(\|x\|)p$ is admissible, hence, from Corollary 1.1, $g - \phi p$ is sectionally 0-epi on $S \cap \bar{B}_{2R} \cap \bar{U} = A = A_2$. Take $A_1 = S \cap \partial(\bar{B}_{2R} \cap \bar{U})$, $B_1 = S \cap \bar{B}_R \cap \partial U$, $B_2 = \phi$, $g_1 = g - \phi p$, $g_2 = g - p$. From the lemma one has that g_1 has a nonvanishing extension to A. Q.E.D.

Theorem 2.1. *Let S be closed in U and let g be sectionally 0-epi on S and U. Assume that g is sectionally proper and sectionally bounded on bounded and closed (in $\tilde{E}$) subsets of $S \cap U$. Then there is a minimal closed (in U) subset Σ of $S \cap U$ such that:*

(a) *g is sectionally 0-epi on Σ. In particular $g^{-1}(0) \cap \Sigma \ne \phi$, Σ is either unbounded or $\bar{\Sigma} \cap \partial U \ne \phi$, $\dim \Sigma \ge \dim(\Sigma \cap \bar{V}) \ge \dim G$ and $\dim(\Sigma \cap \partial V) \ge \dim G - 1$ for any open and bounded set V such that $g^{-1}(0) \cap S \subset V \subset \bar{V} \subset U$.*

(b) *If $\Sigma = \Sigma_1 \cup \Sigma_2$, with Σ_i closed (in Σ) and proper subsets of Σ, then $\dim(\Sigma_1 \cap \Sigma_2) \ge \dim G - 1$. In particular Σ is connected and has dimension at each point at least $\dim G$. The same results hold, if $\bar{\Sigma} = \Sigma_1 \cup \Sigma_2$, for $\Sigma_1 \cap \Sigma_2 \cap V$. Then $\dim(\bar{W} \cap U) \ge \dim G$ for any open neighborhood W of p in $\bar{\Sigma}$.*

(c) *Σ is also minimal for any map g_1 homotopic to g via a finite dimensional homotopy, or for any g_1 homotopic to g if g and g_1 are proper on bounded and closed subsets of $S \cap U$.*

(d) *If g is defined on $\bar{U}$, $g(x) \ne 0$ on $\bar{\Sigma} \cap \partial U$, and g is sectionally proper and bounded on bounded and closed subsets of $\bar{\Sigma} \cap \bar{U}$, then g is sectionally 0-epi on $\bar{\Sigma}$ and $\bar{\Sigma}$ is minimal. Hence $\dim(\bar{\Sigma} \cap \partial V) \ge \dim G - 1$ for any open and bounded subset V with $g^{-1}(0) \cap \Sigma \subset V \subset U$. Proposition 2.1 and 2.2 apply for $\bar{\Sigma}$.*

Proof. (a). Since g is sectionally proper, $g^{-1}(0) \cap S$ is compact. Let $\mathcal{C} = \{C \subset S, C$ closed in $S \cap U$ and g be sectionally 0-epi on C and $U\}$. Then $S \cap U \in \mathcal{C}$. Let $\mathcal{C}'$ be a chain in $\mathcal{C}$ and $\Sigma = \cap_{\mathcal{C}'} C$. Since $g^{-1}(0) \cap C$ is a descending family of nonempty compact sets, $g^{-1}(0) \cap \Sigma$ is nonempty and compact. Let $h: \tilde{E} \to M \in \mathcal{M}$ be compact with bounded support contained in U. Let V_0 be such that $g^{-1}(0) \cap S \subset V_0 \subset \bar{V}_0 \subset U$ and set $V_1 = V_0 \cup \{x : h(x) \ne 0\}$, which is bounded, open and $\bar{V}_1 \subset U$. Let $g = (g_M, g_N)$, since g_M is proper on $g_N^{-1}(0) \cap S \cap \bar{V}_1$ and h is compact, $g_M - h$ is proper on this last set. Hence $(g-h)^{-1}(0) \cap C \cap \bar{V}_1$ is a descending family of nonempty compact sets; since $g(x) = h(x)$ has a solution in $C \cap U$; but $h(x) = 0$ and $g(x) \ne 0$ on $S \cap V_1^C$, that is the solution is in $C \cap \bar{V}_1$. Thus, $(g-h)^{-1}(0)$ intersects $\Sigma \cap \bar{V}_1$, i.e. g is sectionally 0-epi on $\Sigma \cap U$.

By Zorn's lemma, $\mathcal{C}$ has a minimal element, also denoted by Σ. Since g is sectionally 0-epi on $\Sigma \cup U$ then, by minimality, $\Sigma \subset S \cap U$. The rest of (a) comes from (P1), Propositions 1.2. and 2.1.

(b). If $\Sigma = \Sigma_1 \cup \Sigma_2, \Sigma_i$ proper closed subsets of Σ and $dim \, \Sigma_1 \cap \Sigma_2 <$ dim $G - 1$, then, by minimality of Σ, g is not sectionally 0-epi on $\Sigma_i and U$. That is there are open and bounded sets V_i, with $g^{-1}(0) \cap \Sigma_i \subset V_i \subset \bar{V}_i \subset U$ such that $g|_{\Sigma_i \cap \partial V_i}$ is nonzero and extends to $\tilde{g}_i : \Sigma_i \cap \bar{V}_i \to G \backslash \{0\}$, with $(\tilde{g}_i - g_i)(\bar{V}_i)$ bounded in some $M \in \mathcal{M}$. Let $V = V_1 \cup V_2, A = \Sigma \cap \bar{V}$, $A_i = (\Sigma_i \cap \bar{V}) \cup (\Sigma \cap \partial V), B_i = (\Sigma_i \cap (\bar{V} \backslash V_i)) \cup (\Sigma \cap \partial V)$. Take a compact finite dimensional extension k_j to A_j of the map defined as $\tilde{g}_i - g$ on $\Sigma_i \cap \bar{V}_i$ and as zero on $B_i, j \neq i$. Set g_i as g on $B_i, \tilde{g}_i$ on $\Sigma_i \cap \bar{V}_i, g + h_j$ on $A_j, j \neq i$. Since $g \neq 0$ on B_i, we have that $g_i \neq 0$ on A_i. Moreover g_1 and g_2 restricted to $A_1 \cap A_2$ differ on a subset of $\Sigma_1 \cap \Sigma_2 \cap \bar{V}$. Hence, from Lemma 2.1, $g|_{B_1 \cup B_2}$ has a nonvanishing extension $\tilde{g}$ to A with $(g - \tilde{g})(A)$ bounded in M. In particular, $g|_{\Sigma \cap \partial V}$ has an extension to $\Sigma \cap \bar{V}$, contradicting the fact that g is sectionally 0-epi on Σ.

In particular if Σ is not connected, $\Sigma = \Sigma_1 \cup \Sigma_2, \Sigma_1 \cap \Sigma_2 = \phi$ so that Σ_1, Σ_2 cannot be proper subsets of Σ. Finally if $p \in \Sigma$, for any open neighborhood W of p, let $\Sigma_1 = \Sigma \backslash W$ and $\Sigma_2 = \bar{W}$. Then $\dim(\Sigma_1 \cap \Sigma_2) = \dim(\partial W) \geq \dim G - 1$. Hence, $\dim \bar{W} \geq \dim G$ and the local dimension of Σ at p is at least $\dim G$. The rest of the proof is along the same lines and can be found in Theorem 3.1 of the above reference.   Q.E.D.

3. Application to Bifurcation Problems

If S is a subset of the zeros of a map $f(\lambda, x) : \tilde{E} = \wedge \times B \to E$ and one finds a map g which is sectionally 0-epi on S and U, then the preceding theorem will give the right kind of information. Now it is not easy to prove directly that g has this property. Thus the idea is to prove that (f, g) is 0-epi on some set U. Since one wishes to use only local information, the first step is to find a substitute to the excision property and play the scaling argument used in Section II. This approach will restrict the type of maps for which one has positive results. In this exposition I shall restrict the results to bifurcation. Similar statements concerning continuation problems can be found in Ize et al. (1985).

Proposition 3.1. Scaling property. *Let $\tilde{E} = E_1 \times E_2, G = G_1 \times G_2, g = (g_1, g_2)$ with $g_i(x_1, x_2) = f_i(x_1, x_2) - k_i(x_1, x_2)$ such that $f_i(t_1 x_1, t_2 x_2) = t_1^{x_i} t_2^{m_i} f_i(x_1, x_2)$ for $t_1, t_2 > 0$ and some nonnegative integers n_i and m_i. Assume that for any bounded set $M \subset E$, there exists $0 \leq k(M) < 1$ such that $\alpha(k_i(M)) \leq k(M) \alpha(f_i(M))$, (if $k(M)$ is not uniformly bounded by $k < 1$. Then $f = (f_1, f_2)$ is assumed to be proper on bounded and closed subsets of $\tilde{E}$). Let $B_i = \{x \in E_i : \|x_i\| < r_i\}, B_i^\epsilon = \{x \in E_i : \|x_i\| < r_i - \epsilon\}$ and $A \subset (E_1 \backslash B_1^\epsilon) \times \{0\}$ be a closed set such that if $(x_1, 0) \in A$, then $(t x_1, 0) \in A$ for any $t \geq 1$. Assume that $g^{-1}(0) \subset B_1^\epsilon \times B_2^\epsilon \cup A$. Then*

    (a) *If $g^{-1}(0) \cap (B_1 \times B_2 \setminus (B_1 \times \{0\}))$ is closed and $k < 1$, g is 0-epi on $E \setminus (A \cup (\bar{B}_1^\epsilon \times \{0\}))$ if and only if g is 0-epi on $B_1 \times B_2 \setminus (A \cup (\bar{B}_1^\epsilon \times \{0\}))$. If $k = 1$ and $f - (1 - \tau)(k_1, k_2)$ is 0-epi on the smaller set for $0 \le \tau < \tau_0$, g is 0-epi on the larger set if the zeros of $f - (1 - \tau)(k_1, k_2)$ remain in $B_1^\epsilon \times B_2^\epsilon \cup A$.*

    (b) *If U is an open set with $\bar{B}_1 \times \bar{B}_2 \subset U$ and $g : U \to G$ has the above properties relative to U and g_2 has an extension to E such that $g_2^{-1}(0) \subset V_1 \cup A$, where V_1 is an open bounded set with $\bar{B}_1 \times \bar{B}_2 \subset V_1 \subset \bar{V}_1 \subset U$, the above conclusions are valid with $\tilde{E}$ replaced by U.*

From this property and the previous results it is easy to derive the following:

Corollary 3.1. *Let U and g be as above with (g_1, g_2) 0-epi on $B_1 \times B_2 \setminus (A \cup \bar{B}_1^\epsilon \times \{0\})$, (if $k = 1$ then $f - (1 - \tau)(k_1, k_2)$ is 0-epi). Let $S = g_1^{-1}(0)$ and assume that g_2 is sectionally proper and bounded on closed (in $\tilde{E}$) and bounded subsets of $S \setminus (A \cup \bar{B}_1^\epsilon \times \{0\})$. Then there is a minimal Σ of the previous set such that g_2 is sectionally 0-epi on $\Sigma \cap (U \setminus (A \cup (\bar{B}_1^\epsilon \times \{0\}) \cap \bar{\Sigma}))$ and Σ has all the properties of Theorem 2.1.*

Remark 3.1. For A-proper 0-epi maps, if one has the same control on the zeros of g_n and g, one has the same equivalence as in (a): in this case there is no need for homogeneity or continuity of g. Moreover, if $S = g_1^{-1}(0)$ is closed in U and g_2 is continuous on S and sectionally proper and bounded on closed bounded subsets of S, then the equivalent to Corollary 3.1. holds.

    The application to bifurcation problems will require the following hypotheses: $\tilde{E} = \wedge \times B, \wedge = \tilde{\wedge} \times \tilde{\tilde{\wedge}}$, with dim $\tilde{\wedge} < \infty$ and $\tilde{\tilde{\wedge}}$ is equipped with the family of all finite dimensional subspaces $\tilde{\tilde{\wedge}}_M \supset \tilde{\tilde{\wedge}}_{M_0}$. Assume that $0 \in \bar{B}_1 \times \bar{B}_2 \subset U$ and $g_1 : U \to E$ has the form $g_1(\lambda, x) = f_1(x) - k_1(\lambda, x)$, with $k_1(\lambda, 0) = 0$ and $f_1(tx) = t^x f_1(x)$ for $t > 0$ and some $n > 0$. Furthermore, f_1 and k_1 send bounded sets into bounded sets. Assume that

    (1) f_1 is proper on $\bar{B}_2$.

    (2) If C is a bounded subset of $U_M = U \cap (\tilde{\wedge} \times \tilde{\tilde{\wedge}}_M \times B)$, then $\alpha(K_1(C)) \le k_M(C) \alpha(f_1(C))$, with $k_M(C) \le k \le 1$ for large M.

    (3) There is a continuous map $f : \wedge \to \tilde{\tilde{\wedge}}_{M_0}$, sending bounded sets into bounded sets, and $\epsilon > 0$, such that if $g_1(\tilde{\lambda}, f(\lambda), x) = 0$, for $\|x\| \le r_2$ and $r_1 - \epsilon \le \|\lambda\| \le r_1 + \epsilon$, then $x = 0$. Let $B_1^M = B_1 \cap \tilde{\wedge} \times \tilde{\tilde{\wedge}}_M$.

Let $\phi(r)$ be a nonincreasing map, with $\phi(r) = r_2/2$ for $r \le r_1$ and $\phi(r) = 0$ for $r \ge r_1 + \epsilon$. Define $g_M(\lambda, x) = (g_1(\tilde{\lambda}, \tilde{\tilde{\lambda}}_M, x), \tilde{\tilde{\lambda}}_M - f(\tilde{\lambda}, \tilde{\tilde{\lambda}}_M), \|x\| -$

$\phi(\|\tilde{\lambda}, \tilde{\lambda}_M\|))$ and set $S = \{(\lambda, x) \in U : g_1(\lambda, x) = 0, x \neq 0\}$. For the case $k = 1$, define $g_{m,\tau}(\lambda, x)$ the map similar to g_M but with k_1 replaced by $(1 - \tau)k_1$.

Theorem 3.1. *Assume that $g_M(\lambda, x)$ is 0-epi on $B_1^M \times B_2$ for M large enough (for the case of condensing maps, that $g_{M,\tau}$ is 0-epi for $0 \leq \tau \leq \tau_0$). Then S has a minimal closed (in S) subset Σ such that the map $(\tilde{\lambda} - f(\lambda), \|x\| - \phi(\|\lambda\|))$ is sectionally 0-epi on Σ $(U\backslash((\wedge \times \{0\}) \cap \bar{\Sigma}))$. Moreover Σ has the following properties:*

(a) *The set $\bar{\Sigma} \cap \{(\lambda, x) : \tilde{\lambda} = f(\lambda)\}$ intersects $B_1^\epsilon \times \{0\}$ and contains a closed connected subset, of local dimension at least 1, which is either unbounded or intersects either ∂U or $\wedge \times \{0\}$ outside $\bar{B}_1 \times \{0\}$.*

(b) *If $\Sigma = \Sigma_1 \cup \Sigma_2$, with Σ_i proper closed subsets of Σ, then $\dim(\Sigma_1 \cap \Sigma_2) \geq \dim \tilde{\wedge}$. In particular Σ is connected and $\bar{\Sigma}$ has dimension at each point at least $\dim \tilde{\wedge} + 1$.*

(c) *If the estimates of (3) hold for a family of mappings $f_t(\lambda)$, then Σ is also minimal for the family and has the same properties.*

(d) *Structure of the set of bifurcation points. Let $\Sigma_0 = \bar{\Sigma} \cap (\wedge \times \{0\})$ and $U_0 = U \cap (\wedge \times \{0\})$ and assume $\dim \tilde{\wedge} \geq 1$. Then $\tilde{\lambda} - f(\lambda)$ is 0-epi on Σ_0 and $(U_0 - A_0)$, where $A_0 = \{\lambda \in \wedge : \tilde{\lambda} = f(\lambda), \|\lambda\| \geq r_1\}$. In particular Σ_0 has a minimal closed (in $U_0 - A_0$) subset Σ_1, such that $\tilde{\lambda} - f(\lambda)$ is 0-epi on $\Sigma_1 \cap (U_0 - \bar{\Sigma}_1 \cap A_0)$. Σ_1 meets the zeros of $\tilde{\lambda} - f(\lambda)$ inside $B_1, \bar{\Sigma}_1$ is either unbounded or meets $\partial U_0 \cup A_0$. Σ_1 is connected and $\bar{\Sigma}_1$ has dimension at each point at least $\dim \tilde{\wedge}$.*

Proof. Let $A = (\wedge \backslash B_1^\epsilon) \times \{0\}$. Then it is easy to see that f_1 and g_1 are proper on closed and bounded subsets of U_M, hence $\bar{S} \cap C$ is compact for any bounded and closed subset of U_M. Furthermore the zeros of $(g_1, \tilde{\lambda} - f(\lambda), \|x\| - \phi(\|\lambda\|))$, which are those of g_M, for $M \geq M_0$, are the zeros of g_1 in $B_1^\epsilon \times B_2^\epsilon$ with $\tilde{\lambda} = f(\lambda), \|x\| = r_2/2$, together with the set of λ's with $\tilde{\lambda} = f(\lambda), \|\lambda\| > r_1 + \epsilon, x = 0$. Hence, from Corollary 3.1, g_M is 0-epi on $U_M\backslash(\wedge_M \times \{0\})$. From property (P7), it follows that $(\tilde{\lambda} - f(\lambda), \|x\| - \phi(\|\lambda\|))$ is sectionally 0-epi on S and $(U\backslash(\wedge \times \{0\}))$. Then, from Corollary 3.1 and Theorem 2.1, one has Σ with properties (b) and (c).

Finally $(\|x\| - \phi(\|\lambda\|))$ is 0-epi on $\Sigma \cap \{\lambda : \tilde{\lambda} = f(\lambda)\}$. Hence, by Theorem 2.1, there is a minimal Σ_1 of the above set, such that $(\|x\| - \phi(\|\lambda\|))$ is 0-epi on Σ_1 and $(U\backslash((\wedge \times \{0\}) \cap \bar{\Sigma}))$. Moreover Σ_1 meets the level $\|x\| = r_2/2$ inside $B_1^\epsilon \times B_2^\epsilon$ and Σ_1 is either unbounded or $\bar{\Sigma}_1$ intersects ∂U or $(\wedge \times \{0\}) \cap \bar{\Sigma}$. Finally $\bar{\Sigma}_1$ has to intersect $B_1^\epsilon \times \{0\}$ (if not it would be at a positive distance and one could lower the level $r_2/2$ so that the map $\|x\| - \tau\phi(\|\lambda\|)$ would be nonzero in $B_1^\epsilon \times B_2^\epsilon$) and, if $\bar{\Sigma}_1$ is bounded and does not intersect ∂U, then $\bar{\Sigma}_1$ has to intersect $\wedge \times \{0\}$ outside $\bar{B}_1$ (if not, it would be at a positive distance, d, of $\wedge \times \{0\} \backslash (B_1 \times \{0\})$. Let V be a bounded neighborhood of $\bar{\Sigma}_1$ in U intersected with the set $\{(\lambda, x) : \|x\| \geq \min(d, r_2/2)/2\}$. But then on $\Sigma_1 \cap \partial V$, which is contained in $B_1 \times B_2$, the map $\|x\| - \phi(\|\lambda\|)$ is negative, contradicting Proposition 1.1. (d), since this map is 0-epi on Σ_1 and V).

The proof of (d) is given in Ize et al. (1985), Proposition 4.7. Q.E.D.

Remark 3.2. For $A-$proper maps, one may replace all the assumptions by asking only for (3) and that $g_1^{-1}(0) \cap (\tilde{\wedge} \times \tilde{\wedge}_M) \cap C$ is compact, for any large M, for any C bounded and closed subset of U. Then if g_M is $A-$proper 0-epi on $B_1 \times B_2$, for $M \geq M_0$, one has the same conclusions.

Theorem 3.1. reduces the study to a "local" situation, i.e. on $B_1 \times B_2$. One may reduce the dependence to a "leading" term if one has the following hypotheses: Assume that

1') g_1 and U satisfy properties (1) and (2) (in case $k = 1$, then $k_M(tC) \leq k_M(C)$ for any C in B_2 and t in $[0,1]$).

2') $g_1(\lambda, x) = f(x) - T(\lambda)(x) - h_1(\lambda, x)$, where $T(\lambda)(x)$ and $f_1(x)$ are homogeneous of degree n and $h_1(\lambda, x) = o(\|x\|^n)$, uniformly in λ in $\tilde{\wedge} \times \tilde{\wedge}_M$.

3') $\tilde{\bar{\lambda}} = f(\lambda)$ is solvable in the form $\tilde{\bar{\lambda}} = \tilde{f}(\tilde{\lambda})$, with $\tilde{f}$ continuous and $f_1 - T(\tilde{\lambda}, \tilde{f}(\tilde{\lambda}))$ is one-to-one for $r_1 - \epsilon \leq \|\lambda\| \leq r_1 + \epsilon$.

Then, under these hypotheses, and for small r_2, g_M is 0-epi on $B_1^M \times B_2$ if and only if $(f_1(x) - T(\tilde{\lambda}, \tilde{f}(\lambda))(x), \lambda_M - f(\lambda), \|x\| - r_2/2)$ is 0-epi on $B_1^M \times B_2$. If $k = 1$, replace k_1 by $(1 - \tau)k_1$ with $0 < \tau < \tau_0$.

The proof of this homotopy can be found in Lemma 4.1. of Ize et al. (1985). A particular case is when $f_1(x) = Ax$ is a Fredholm operator of index 0, with $n = 1$, and $T(\lambda)$ is linear with $\|T(\lambda)\| \to 0$ as $\lambda \to 0$, uniformly on $\wedge_M$. Hypothesis (1) is satisfied since A is Fredholm and hypothesis (2) is true for $A - T(\lambda)$, since $T(\lambda)$ is a contraction for λ small, λ in $\wedge_M$. For $h_1(\lambda, x)$, this will also be the case if $\|h_1(\lambda, x_1) - h_1(\lambda, x_2)\| \leq C_M(\max(\|x_1\|, \|x_2\|))^\beta \|x_1 - x_2\|$ with $\beta > 0$ and $\|x_1\|, \|x_2\| \leq r_2$ small enough, and λ in $\wedge_M$. One may also use the weak equivalence between maps from S^{M-1} into the set of maps of compact type and the set of maps of condensing type given in Alexander-Fitzpatrick (1979). Then, from Section II, one has

Proposition 3.2. *Assume there is a continuous family $f_t(\lambda)$ such that, for $\|\lambda\| \leq r_1$, the zeros of $\tilde{\lambda} = f_t(\lambda)$ have the form $\tilde{\lambda} = \tilde{f}_t(\tilde{\lambda})$, with $\tilde{f}_0(\tilde{\lambda}) = 0$ and suppose that, for $\|\lambda\| = r_1$, $A - T(\tilde{\lambda}, \tilde{f}_t(\tilde{\lambda}))$ is invertible. Then g_M is 0-epi on $B_1^M \times B_2$ if and only if $J[A - T(\tilde{\lambda}, 0)] \neq 0$.*

Note that if A has positive Index, one may use Remark 2.4 of Section II.

Remark 3.3. If $B = \mathbb{R}^{N_1} \times G$, $E = \mathbb{R}^{N_2} \times G$ and $g_1(\lambda, x, y) = (f_1(\lambda, x, y), y - f_2(\lambda, x, y))$ where f_2 is compact on finite dimensional subspaces of $\wedge$, then, as in Section II, Subsection 4, one may define the generalized degree of g_M with respect to any bounded open subset of $\wedge_M \times B$. Now g_M will be 0-epi on $B_1^M \times B_2$ if and only if $\deg(g_M; B_1^M \times B_2)$ is nontrivial. For example, if $\dim \tilde{\wedge} = 1$, one may have the following situation:

Proposition 3.3. *Assume there are two points λ_1 and λ_2 (taken to be of the form $(-r_1, 0)$ and $(r_1, 0)$) and $\epsilon > 0$, such that*

(1) *If $g_1(\lambda, x) = 0$ for $\|x\| \leq 2\epsilon$, $\|\lambda - \lambda_i\| \leq \epsilon$ and λ on the curve $\tilde{\lambda} = \tilde{f}(\tilde{\lambda})$, going through λ_i, then $x = 0$.*

(2) *Σ Ind $(g_1(\lambda_1, x); 0) \neq \Sigma$ Ind $(g_1(\lambda_2, x); 0)$, for the stable generalized index.*

(3) *The above curve is part of a curve $\tilde{\lambda} = f(\lambda) \in \wedge_{M_0}$ such that, between λ_1 and λ_2, it lies in the box $\{|\tilde{\lambda}| \leq r_1, |\tilde{\tilde{\lambda}}| \leq R\}$ and, outside the portion between $\lambda_1 - \epsilon$ and $\lambda_2 + \epsilon$ it does not enter the box $\{|\tilde{\lambda}| \leq r_1 + \epsilon, \|\tilde{\tilde{\lambda}}\| \leq R + \epsilon\}$.*

   (a) *Then the set of nontrivial solutions has a minimal closed subset Σ with the properties of Theorem 3.1. In particular Σ is connected, has dimension at each point of at least $\dim \wedge$. Moreover $\bar{\Sigma} \cap \{\tilde{\lambda} = f(\lambda)\}$ intersects the curve $\{\tilde{\lambda} = f(\lambda), x = 0\}$ strictly between λ_1 and λ_2 and, if bounded, again meets that curve outside this portion.*

   (b) *If $\Sigma_0 = \Sigma \cap (\wedge \times \{0\})$, $U_0 = U \cap (\wedge \times \{0\})$. Suppose U_0 is convex and $\dim \tilde{\wedge} \geq 1$. Then Σ_0 has a closed connected minimal subset Σ_1 of local dimension at least $\dim \tilde{\wedge}$, which disconnects U_0 between λ_1 and λ_2. Also if $\Sigma_1 = A_1 \cup A_2$, with A_1 and A_2 two proper closed subsets of Σ_1, then $\dim (\Sigma_1 \cap \Sigma_2) \geq \dim \wedge -2$ (if $\dim \wedge = 2$, and $A_1 \cap A_2 \neq \phi$).*

   (c) *For any pair of points μ_1, μ_2, with μ_i in the connected component of λ_i in $U_0 - \Sigma_0$, either $\bar{\Sigma}$ meets ∂U above the segment $[\mu_1, \mu_2]$ (this includes the case when Σ is unbounded above this segment) or Σ covers μ_1 or μ_2. If the first part of this alternative fails, for* *example with a priori bounds, then Σ covers at least one of the components, i.e. for any μ in that component there is an x such that (μ, x) is in Σ.*

Proof. Take $B_1 = \{|\tilde{\lambda}| < r_1 + \epsilon/2, \|\tilde{\tilde{\lambda}}\| < R\}$, $B_2 = \{\|x\| < 2\epsilon\}$, $\phi(\lambda) = \epsilon$ on $\{|\tilde{\lambda}| < r_1, |\tilde{\tilde{\lambda}}| < R\}$ and $\phi(\lambda) = 0$ outside the second box of (3). Deform g_M on $\partial(B_1^M \times B_2)$ to $(g_1(\tilde{\lambda}, f(\lambda), x), \tilde{\tilde{\lambda}}_M - f(\lambda), \|x\| - \phi(\lambda))$. Then the homotopy $(1-t)(\|x\| - \phi(\lambda)) + t(r_1 - |\tilde{\lambda}|)$ is valid since this term is positive if $\|x\| > \epsilon$ and $|\tilde{\lambda}| < r_1$, while, if $\|x\| \leq 2\epsilon$ and $r_1 \leq |\tilde{\lambda}| \leq r_1 + \epsilon/2$, $x = 0$ but $\phi(\lambda) > 0$ there. Then $\deg(g_M; B_1^M \times B_2) = \deg(g_M; B_{+1}^M \times B_2) + \deg(g_M; B_{-1}^M \times B_2)$, up to one suspension by Remark 4.1. of Section II, $B_{\pm 1}$ are the balls $\{|\tilde{\lambda} \mp r_1| < \epsilon < 2, \|\tilde{\tilde{\lambda}}\| < R\}$. On these balls it is easy to see that one may deform g_M to $(g_1(\pm r_1, 0, x), \tilde{\tilde{\lambda}}_M, r_1 - |\tilde{\lambda}|)$; hence $\deg(g_M; B_1^M \times B_2) = \eta \Sigma^{m+1} (\text{Ind}(g_1(\lambda_1, x); 0) - \text{Ind}(g_1(\lambda_2, x); 0))$, where η is an orientation factor and Σ is the suspension, as in Remark 1.5 of Section I.

For the proof of (b) and (c), see Proposition 4.8 in Ize et al. (1985). Q.E.D.

Remark 3.4. It is clear that the notion of 0-epi maps on S and U is the right one for results on dimension and global alternative. However, when trying to prove that this is the case for the zeros of $g_1(\lambda, x)$ and using only a local information, one is reduced to considering maps which are essentially "condensing" or "A-proper" with respect to a leading proper term. For this class of maps it should be possible to define, along the lines of Section II, Subsection 4, a generalized degree with the right properties. Now, the generalized degree of Section II can in principle, be defined for a larger class of functions, for example just continuous. However, in this last case, all the local degrees would vanish in infinite dimensions since $GL(B)$ is contractible.

Similarly if one asks, for 0-epi maps, that the extension $\tilde{g}$ is such that $g - \tilde{g}$ is continuous (instead of compact), then there will be very few 0-epi maps: in fact, the identity, if $\dim \tilde{E} = \infty$, would not be 0-epi (in this sense) on $\tilde{E}$ and B_1, (B_1 the unit ball), since the identity on ∂B_1 has a nonzero continuous extension given by any continuous retraction of B_1 onto ∂B_1.

On the other hand, the scaling property is a "natural" homotopy to consider. But if one wishes to have this homotopy admissible, one is lead to these "condensing" maps. In fact, if one does not have these restrictions on the measure of noncompactness, there are examples where the global alternative, or the global continuation principle, fails. Thus, one may start from a local situation where Proposition 3.2. may be applied (under very mild smoothness conditions) and obtain a local branch of the right local dimension dim $\tilde{\tilde{\wedge}}+1$, in the neighborhood where on still has the condensing properties (or where the Liapunov-Schmidt reduction works), but which disappears in the middle of nowhere.

C.A. Stuart (1992) gives the following example: consider the nonlinear Sturm-Liouville problem in $\mathbb{R}$,

$$-u'' + q(x)u + g(x, u^2)u = \lambda u,$$

where $g(x, 0) = 0, g(x, s) \geq 0, g$ is continuous except at a finite number of points $x_1, \ldots, x_M$ where it has bounded jumps and $0 < A \leq q + g(x, s) \leq B(s)$ for some continuous map, and $g(x, s)$ is nonincreasing in s. Then, under some conditions on q, one has the bifurcation of positive solutions from the infimum $\wedge < 0$ of the spectrum of $-u'' + qu$, which is a simple eigenvalue (see Section I). The branch is a continuous curve $v(\lambda)(x)$ in $H^2(\mathbb{R})$, with $v(\lambda)(x) \leq v(\mu)(x)$ if $\wedge \leq \lambda \leq \mu < 0$. Furthermore (Theorem 5.8 of the paper), there is $\bar{v} \in H^2(\mathbb{R}), \bar{v}(x) > 0$, such that $\lim_{\lambda \to 0-} \|v(\lambda) - \bar{v}\|_{H^2} = 0$ and $\bar{v}$ is solution of the problem with $\lambda = 0$, which corresponds to the infimum of the essential spectrum of the operator $-d^2/dt^2 + q + g(x, \bar{v}^2)$. There the map ceases to be a Fredholm map and the associated integral operator is not proper.

In $l^2 = \{x = (x_1, x_2, \ldots) : \|x\|^2 = \Sigma x_i^2 < \infty\}$, consider the following continuation problem

$$F(\lambda, x) = \begin{cases} x - \lambda((1 - \|x\|^2)^{1/2}, x_1, x_2, \ldots) & \text{if} \quad \|x\| \leq 1 \\ x - \lambda(0, x_1, x_2, \ldots) & \text{if} \quad \|x\| \geq 1. \end{cases}$$

For $\lambda = 0, F(0, x)$ has Leray-Schauder index 1 at the origin and, from the corresponding results in Ize et al. (1985) for continuation problems (Theorem 4.1), $(F(\lambda, x), \lambda)$ in 0-epi on $B_1 \times B_2$, where $B_1 = \{x, \|x\| < 1\}$, $B_2 = \{\lambda, |\lambda| < 1\}$, for which $F(\lambda, x) - x$ is a strict k-set contraction. Now the zeros of F are easily seen to be of the form $x(\lambda) = (\lambda k, \lambda^2 k, \ldots)$ where $k = (1 - \|x\|^2)^{1/2}$, if $\|x\| \leq 1$, and $k = 0$ if $\|x\| \geq 1$, in which case there are no zeros (hence the zeros are contained in B_1). Furthermore $\|x(\lambda)\|$ is finite only if $|\lambda| < 1$, in which case $\|x(\lambda)\| = |\lambda|$. Thus the branch going through (0,0) exists only for $-1 < \lambda < 1$. At $|\lambda| = 1$, the map loses its properness.

A final example is due to E.N. Dancer (personal communication) and settles also the case where the map is proper. Let $T(x) : l^2 \to l^2$ be defined as $Tx_n = |x_n| + 2^{-n}, n \geq 1$. It is clear that $T(x)$ is continuous, proper and has no zeros. Let $H(x, \lambda)$ be defined as $(1 - \lambda^2)x + \lambda^2 T(x)$. For $\lambda = 0, H(x, 0)$ has a Leray-Schauder index equal to 1. $H(x, \lambda) = 0$ implies $x_n < 0, x_n = 2^{-n}\lambda^2/(2\lambda^2 - 1)$, which gives a curve in $l^2 \times \mathbb{R}$ which blows up at $\lambda^2 = 1/2$. Since $\lambda^2\|x - T(x)\| \leq 2\lambda^2\|x\| + \lambda^2/3$, it is easy to see that one loses the condensing character at $\lambda^2 = 1/2$. However one may use the following result of R.D. Anderson: Proper homotopies in l^2- manifolds (Comp. Math. **23**, 1971, 153-157). If $H(x, t)$ is a homotopy in l^2, with $H(x, 0)$ and $H(x, 1)$ proper maps, then there is a proper homotopy $G(x, t)$ such that $G(x, 0) = H(x, 0), G(x, 1) = H(x, 1)$. One then has $G(x, \lambda^2)$ connecting x and $T(x)$ such that $G^{-1}(0)$ is compact. Hence $G^{-1}(0)$, which is the branch going through 0, is bounded in $\|x\|$ and never touches the slices at $|\lambda| = 1$.

IV. EQUIVARIANT BIFURCATION

Many problems in applied mathematics have symmetries. For example, the constitutive equations of continuum mechanics must be independent of the reference systems, i.e., rotations and translations cannot change the phenomenon. In evolution problems, a change in the origin in time should not change the nature of the equations. Furthermore if the domain, the boundary or initial conditions and the forcing terms have part of this symmetry, then the problem, stated in the form

$$F(\lambda, x) = 0,$$

from $\wedge \times B$ into E, should have the property that F is **equivariant**, i.e.

$$F(\gamma\lambda, \gamma x) = \tilde{\gamma} F(\lambda, x)$$

for $\gamma \in \Gamma$ the group of symmetries. Here I am assuming that $\wedge, B$ and E are Banach spaces and that the group Γ is a compact Lie group (for technical reasons) which acts linearly through isomorphisms on $\wedge, B$, denoted by $\gamma\lambda, \gamma x$ and on E, denoted by $\tilde{\gamma} F$, since the action on E may be different from the action on B (on the parameter space one may have $\gamma\lambda = \lambda$).

Examples. 1) $\Gamma = \mathbb{Z}_2$. (a). $F(\lambda, -x) = -F(\lambda, x)$, i.e. odd functions and Γ acts freely on B and E: i.e. there are no points fixed by Γ except the origin. (b). $F(\lambda, -x) = F(\lambda, x)$: Γ acts trivially on E. (c) $B = B_1 \times B_2, E = E_1 \times E_2, F(\lambda, x_1, x_2) = (F_1, F_2)(\lambda, x_1, x_2)$, with $F_1(\lambda, x_1, -x_2) = F_1(\lambda, x_1, x_2), F_2(\lambda, x_1, -x_2) = -F_2(\lambda, x_1, x_2)$, i.e. B_1 and E_1 are fixed under Γ. In particular $F_2(\lambda, x_1, 0) = 0$, that is F sends B_1 into E_1.

2) $\Gamma = S^1$: $2\pi-$periodic solutions to autonomous differential equations of the form $\frac{dX}{dt} = f(X, \lambda), X \in \mathbb{R}^n, \lambda \in \mathbb{R}^k$. For example, if one wishes to study the classical Hopf bifurcation problem: find periodic solutions to $\frac{dX}{d\tau} = g(X, \lambda_1), \lambda_1 \in \mathbb{R}$. Then, by letting $t = \nu\tau$, one has $\frac{dX}{dt} = g(X, \lambda_1)/\nu$ and periodic solutions, of period $2\pi/\nu$, of the first equation correspond to

2π−periodic solutions of the second. Here ν is the frequency. Now, $S^1 = R/2\pi$ acts on the space of 2π−periodic functions through $X(t) \to X(t+\phi)$, that is the time shift (the same result will hold for spatial periodicity). Since the equation is autonomous, it commutes with the time shift. One may also set the problem in terms of Fourier series. Write $X(t) = \Sigma_{-\infty}^{\infty} X_n e^{int}$, with $X_n \in \mathbb{C}^n$, $X_0 \in \mathbb{R}^n$, $X_n = \bar{X}_{-n}$ (since $X(t) \in \mathbb{R}^n$). For the series of the Fourier coefficients $(X_0, X_1, X_2, \ldots)$ one has the equivalent equation

$$i\nu n X_n - f_n(X_0, X_1, X_2, \ldots, \nu, \lambda) = 0, n = 0, 1, 2, \ldots$$

with $f_n(X_0, X_1, \ldots) = \frac{1}{2\pi} \int_0^{2\pi} f(X(t), \nu, \lambda) e^{-int} dt$. Here the action of S^1 is of the form: $e^{i\phi}(X_0, X_1, X_2, \ldots) = (X_0, e^{i\phi} X_1, e^{2i\phi} X_2, \ldots)$ and $f_n(X_0, e^{i\phi} X_1, e^{2i\phi} X_2, \ldots) = e^{in\phi} f_n(X_0, X_1, X_2, \ldots)$ i.e. the equation is equivariant.

3) $\Gamma = \Gamma_0 \times S^1$: Find 2π−periodic solutions of $\frac{dX}{dt} = f(X, \lambda)$ where $f(\gamma_0 X, \lambda) = \gamma_0 f(X, \lambda)$, for γ_0 in Γ_0. This is the case of many problems in mechanics.

There are many ways to handle such problems. Although many of these problems have been treated by looking at special solutions (i.e. with a given symmetry), one may say that a systematic approach using group theory is relatively recent. For bifurcation problems, Sattinger (1979) may have been the first to use equivariance in order to study the leading terms of the bifurcation equation. Golubitsky and his collaborators have used equivariant singularity theory in order to classify all the possible generic situations. Iooss and Chossat have applied these ideas to quite complicated problems in fluid dynamics. Vanderbauwhede (1982) has clarified many aspects of the analogous of the Crandall-Rabinowitz result, the "equivariant branching lemma" due to Cicogna (1984) and Cerami (1986).

From the topological point of view, the situation is simpler. In the nongeneric situation, there are few results and research is still going on. These results will be mentioned in this chapter, but let me comment that the Hopf bifurcation problem was treated by Alexander and Yorke (1978) as a two parameter problem without taking into account the symmetry (see also Ize 1976). Chow et al. (1978) used the Fuller degree, a rational introduced by Fuller in 1967 for the study of periodic solutions, but the fact that these results depended more on the symmetry than on the differential equation was recognized only later in Ize et al. (1986) and subsequent papers.

Example (3) was studied first by Fiedler (1988).

1. Preliminaries

In this first subsection I will collect some of the most useful definitions and properties of actions of compact Lie groups on Banach spaces. Let E be a Banach space and Γ be a compact Lie group, E is a Γ-space if there is a continuous homomorphism, or representation, $\Gamma \to GL(E)$.

Definition 1.1. *If $x \in E$, a Γ-space, the* **isotropy subgroup** *of Γ at x is the set $\Gamma_x = \{\gamma \in \Gamma : \gamma x = x\}$, which is a closed subgroup of Γ.*

Definition 1.2. *The action of Γ is* **free** *if $\Gamma_x = \{e\}, \forall x \in E\backslash\{0\}$. The action is* **semi-free** *if $\Gamma_x = \{e\}$ or Γ, $\forall x \in E$.*

Definition 1.3. *$x \in E$ is a* **fixed point** *if $\Gamma_x = \Gamma$. The* **subspace of fixed points** *of Γ on E is denoted by E^Γ. If H is a subgroup of Γ, then $E^H = \{x \in E : \gamma x = x, \forall \gamma \in H\}$. E^H is a closed linear subspace.*

Definition 1.4. *The* **orbit** *of x under Γ is the set $\Gamma(x) = \{\gamma x \in E : \gamma \in \Gamma\}$.*

It is easy see that $\Gamma(x)$ is homeomorphic to Γ/Γ_x, that the orbits form a partition of E. The set E/Γ is the **orbit space** of E with respect to Γ.

Definition 1.5. *Two points x and y have the same* **orbit type** *H if there are γ_0 and γ_1 such that $H = \gamma_0^{-1}\Gamma_x\gamma_0 = \gamma_1^{-1}\Gamma_y\gamma_1$.*

If E is finite dimensional then it is clear that there are only a finite number of orbit types.

Definition 1.6. *If H is a subgroup of $\Gamma, H < \Gamma$, the* **normalizer of H** *in Γ is $N(H) = \{\gamma \in \Gamma = \gamma^{-1}H\gamma \subset H\}$ and the* **Weyl group** *of H is $W(H) = N(H)/H$.*

If $x \in E^H$, then $\gamma x \in E^H$ for any $\gamma \in N(H)$, since $\gamma_1\gamma x = \gamma\gamma_2 x = \gamma x$ for some γ_1 and γ_2 in H; hence γx is fixed by the action of H. Furthermore if $H = \Gamma_x$ for some x and $\gamma x \in E^H$ for some γ, then it is easy to see that γ belongs to $N(H)$, i.e. $N(H)$ is the largest group which leaves E^H invariant.

Definition 1.7. *An isotropy subgroup H is* **maximal** *if H is not contained in a proper isotropy subgroup of H.*

Note that if $H < K$, then $E^K \subset E^H$. It has been remarked by Golubitsky that *if H is a maximal isotropy subgroup and $E^\Gamma = \{0\}$ then $W(H)$*

acts freely on $E^H \backslash \{0\}$. In fact, if $\gamma x = x$ for some $x \neq 0$, in E^H, and some γ in $N(H) \backslash H$, then $\Gamma_x \supset H \cup \{\gamma\}$. Hence, from the maximality of H, one has $\Gamma_x = \Gamma$, but then $x \in E^\Gamma = \{0\}$.

Now the groups which act freely on Euclideans spaces have been completely classified: a reduced number of finite groups, S^1 and $N(S^1)$ in S^3 and S^3 (see Bredon, p. 153).

Definition 1.8. $\Omega \subset E$ *is* Γ*−invariant if for any* $x \in \Omega, \Gamma(x) \subset \Omega$. *if* B *and* E *are* Γ*−spaces, then* $f : B \to E$ *is* Γ*−equivariant if* $f(\gamma x) = \tilde{\gamma} f(x)$. *On the other hand,* f *is* Γ*- invariant if* $f(\gamma x) = f(x)$.

One of the main reasons for working with compact Lie groups and linear actions is the existence of an integration on Γ, the Haar integral, such that $\int_\Gamma d\gamma = 1$, which is Γ−invariant on the class of continuous real valued functions g on Γ, under both left and right actions, i.e.

$$\int_\Gamma g(\gamma'^{-1}\gamma)d\gamma = \int_\Gamma g(\gamma)d\gamma = \int_\Gamma g(\gamma\gamma')d\gamma.$$

As a first consequence, if E is a Banach Γ−space, one may define a new norm $|||x||| = \int_\Gamma \|\gamma x\| d\gamma$. Then $|||\gamma' x||| = |||x|||$, i.e. the action of Γ is an isometry. In the rest of the section, the action will be assumed to be an isometry. In particular the ball $\{x : \|x\| < R\}$ is Γ−invariant. Using Pettis integrals and standard averaging, one has the following result:

Gleason's Lemma. *If* B *and* E *are* Γ*−spaces and* $f(x)$ *is a continuous map, then* $\bar{f}(x) \equiv \int_\Gamma f(\gamma x)d\tilde{\gamma}$ *is* Γ*−invariant and* $\tilde{f}(x) \equiv \int_\Gamma \tilde{\gamma}^{-1} f(\gamma x)d\tilde{\gamma}$ *is* Γ*−equivariant. Furthermore, if* f *is compact, then* $\bar{f}$ *and* $\tilde{f}$ *are compact.*

Proof. From the change of variables $\gamma\gamma'$, one has

$$\int_\Gamma f(\gamma\gamma' x)d\tilde{\gamma} = \int_\Gamma f(\gamma'' x)d\tilde{\gamma}''$$

and $\int_\Gamma \tilde{\gamma}^{-1} f(\gamma\gamma' x)d\tilde{\gamma} = \int_\Gamma \tilde{\gamma}^{-1} f(\gamma'' x)d\tilde{\gamma}''$ with $\tilde{\gamma}^{-1} = \tilde{\gamma}'\tilde{\gamma}''^{-1}$. See Bredon p. 36. The continuity of $\bar{f}$ and $\tilde{f}$ follows from the compactness of Γ. Finally if $f(x) = \Sigma f_N(x)$, with $f_N(x) \in E_N$, a finite dimensional space, and $\|f_N(x)\| \leq 2^{-N}$, then the averaged maps $\bar{f}_N$ and $\tilde{f}_N$ belong to ΓE_N and satisfy the same bounds since $\tilde{\gamma}$ is an isometry. Furthermore ΓE_N is a finite dimensional Γ−space since Γ is compact. Q.E.D.

Among the consequences of this result one has the following properties.

(P1). Invariant Urysohn functions. *If A and B are closed* $\Gamma-$*invariant subsets of E, with* $A \cap B = \phi$*, then there is a continuous* $\Gamma-$*invariant function* $\bar{\phi} : E \rightarrow [0,1]$*, such that* $\phi(x) = 0$ *if* $x \in A$ *and* $\phi(x) = 1$ *if* $x \in B$.

Indeed let ϕ be any Urysohn function relative to A and B, then $\bar{\phi}(x) = \int_\Gamma \phi(\gamma x)d\gamma$ has the required properties.

(P2). Invariant neighborhood. *If* $A \subset E$ *is a* $\Gamma-$*invariant closed set and* $U \supset A$ *is open and* $\Gamma-$*invariant, then there is a* $\Gamma-$*invariant open subset V such that* $A \subset V \subset \bar{V} \subset U$.

In fact, let $\bar{\phi} : E \rightarrow [0,1]$, be a $\Gamma-$invariant Urysohn function with $\bar{\phi}|_A = 0, \bar{\phi}|_{U^c} = 1$. Then $V = \phi^{-1}([0,1/2])$ has the required properties.

(P3). Dugundji-Gleason extensions. *Let* $A_1 \subset A_2$ *be* $\Gamma-$*invariant closed subsets of B. If* $f : A_1 \rightarrow F$ *is a* $\Gamma-$*equivariant map, then there is a* $\Gamma-$*equivariant extension* $\tilde{f} : A_2 \rightarrow F$ *of f. Furthermore* $\tilde{f}$ *is compact if f is compact.*

This follows from Dugundji's extension theorem.

(P4). Equivariant Borsuk homotopy extension theorem. *If* $A_1 \subset A_2$ *are* $\Gamma-$*invariant closed subset of* $\mathbb{R}^M \times G$*, where* $\mathbb{R}^M$ *and G are* $\Gamma-$*spaces, let* $F_0, F_1 : A_1 \rightarrow \mathbb{R}^N \times G\backslash\{0\}$ *be* $\Gamma-$*equivariant maps which are* $\Gamma-$*homotopic and of the form* $(f_1(x,y), y - f_2(x,y))$ *with* f_2 *compact. Then* F_0 *extends* $\Gamma-$*equivariantly to* A_2 *without zeros if and only if* F_1 *does. If this is the case the extensions are* $\Gamma-$*homotopic. All maps have the same compactness property.*

For the proof it is enough to check that, in the Borsuk extension theorem, given in Section II, one may use invariant Urysohn's functions and obtain equivariant maps.

(P5). Equivariant projections. *If* $E_0 \subset E$ *is a closed invariant subspace and if P is a continuous projection from E into* E_0*, then* $\tilde{P}$ *defined by* $\tilde{P}x = \int_\Gamma \gamma^{-1}P\gamma x d\gamma$ *is a* $\Gamma-$*equivariant projection on* E_0*. If* $E_0 = E^\Gamma$*, then* $\bar{P}x = \int_\Gamma \gamma x d\gamma$ *is a* $\Gamma-$*invariant projection onto* E^Γ.

The first part is clear since $\int_\Gamma d\gamma = 1$, for the second, notice that $\bar{P}x \in E^\Gamma$ and $\bar{P}x = x$ for x in E^Γ.

The last set of preliminary results will be on finite dimensional representations, i.e. $\mathbb{R}^n$ with the action of Γ given by $\gamma \in GL(\mathbb{R}^n)$.

Definition 1.9. *Two* $n-$*dimensional representations are equivalent if there is a T in* $GL(\mathbb{R}^n)$ *such that* $\tilde{\gamma}T = T\gamma$*, i.e. T is equivariant. Then every finite dimensional representation is equivalent to an orthogonal representation, i.e. with* $\tilde{\gamma}$ *in* $O(n)$.

In fact the bilinear form $B(x, y) = \int_\Gamma (\gamma x, \gamma y) d\gamma$ is positive definite, symmetric and invariant. Hence there is a positive definite matrix A such that $B(x, y) = (Ax, y)$. One may define a positive symmetric T such that $T^2 = A$ (diagonalize A). Hence $B(x, y) = (Tx, Ty)$ and $B(\gamma x, \gamma y) = B(x, y)$ implies $(T\gamma T^{-1} x, T\gamma T^{-1} y) = (x, y)$ that is $T\gamma T^{-1} \in O(n)$ for all $\gamma \in \Gamma$.

Definition 1.10. *A representation E of Γ is said to be* **irreducible** *if E has no proper invariant subspace.*

This implies that $E^\Gamma = \{0\}$ or Γ acts trivially on E and dim $E = 1$.

If E is a finite dimensional representation of Γ, then there are invariant subspaces $E_1, \ldots, E_k$ which are irreducible and such that $E = E_1 \oplus \ldots \oplus E_k$.

From the previous result it is enough to consider the case where the representation is orthogonal. Then, if E_1 is Γ−invariant, the orthogonal complement $E_1^\perp$ is also Γ−invariant, since $(\gamma x, \gamma y) = (x, \gamma^T y) = (x, \gamma^{-1} y)$, that is, if $x \in E_1^\perp$ and $y \in E_1$ (hence also $\gamma^{-1} y \in E_1$), this scalar product is 0 and $\gamma x \in E_1^\perp$. Applying this argument a finite number of times one obtains a complete reduction of E. Q.E.D.

Schur's Lemma. *If $\mathbb{R}^n$ and $\mathbb{R}^m$ are irreducible representations of Γ and there is a linear equivariant map: $A : \mathbb{R}^n \to \mathbb{R}^m$ such that $A\gamma = \tilde{\gamma} A, \forall \gamma \in \Gamma$, then either $A = 0$ or $n = m$ and A is nonsingular.*

Proof. Since ker A and Range A are invariant Γ−subspaces, then either ker $A = \mathbb{R}^n$ and $A = 0$ or ker $A = \{0\}$ and A is one-to-one. In this last case one cannot have Range $A = \{0\}$, hence A is onto. Q.E.D.

Note that if $\mathbb{R}^m$ is not irreducible, then one would have that either $A = 0$ or A is one to one onto Range A and this last subspace is irreducible, since A^{-1} is clearly equivariant.

Corollary. *If $\mathbb{R}^n$ is an irreducible represention of Γ and $A : \mathbb{R}^n \to \mathbb{R}^n$ is an equivariant map, i.e. $A\gamma = \gamma A$, such that A has a real eigenvalue λ, then $A = \lambda I$. In general $T^{-1} AT = \mu I + \nu B$, with $T = I$ if the representation is orthogonal, and $B^2 = -I, B + B^T = 0$.*

Proof. In fact $A - \lambda I$ is equivariant with a nontrivial kernel, hence it must be 0. If the representation is not orthogonal, then there is T such that $\gamma_1 = T^{-1} \gamma T$ is orthogonal. Clearly $T^{-1} AT\gamma_1 = T^{-1} A\gamma T = T^{-1} \gamma TT^{-1} AT = \gamma_1 T^{-1} AT$, i.e. $T^{-1} AT$ is equivariant. Assuming that the representation is orthogonal, then $A + A^T$ and $A^T A$ are self-adjoint and equivariant. $A +$

$A^T = 2\mu I$ from the first part, that is, $(A-\mu I)^T + (A-\mu I) = 0$. Furthermore $(A - \mu I)^T (A - \mu I) = \nu^2 I$ (this matrix is positive or identically 0, from the first part, if it has a kernel). Hence, if $B = (A - \mu I)/\nu$, for $\nu \neq 0$, then $B^T + B = 0, B^T B = I$, i.e. $B^2 = -I$. If $\nu = 0$, then $A - \mu I$ is nihilpotent, so it must have a nontrivial kernel and $A = \mu I$.   Q.E.D.

The above result will give a complete classification of linear equivariant operators between finite dimensional Γ–spaces. Let $\mathbb{R}^n = V_1 \oplus \ldots \oplus V_k$ and $\mathbb{R}^m = W_1 \oplus \ldots \oplus W_l$ be the decomposition of $\mathbb{R}^n$ and $\mathbb{R}^m$ in irreducible subspaces. Let $P_i : \mathbb{R}^n \to V_i$ and $Q_j : \mathbb{R}^m \to W_j$ be the equivariant projections, i.e. $\gamma P_i = P_i \gamma$ and $\tilde{\gamma} Q_j = Q_j \tilde{\gamma}$. Assume there is a linear map $A : \mathbb{R}^n \to \mathbb{R}^m$, such that $A\gamma = \tilde{\gamma} A$. Let $A_{ij} = Q_j A P_i : V_i \to W_j$. Then $A_{ij}\gamma = \tilde{\gamma} A_{ij}$ and from Schur's lemma, either $A_{ij} = 0$ or A_{ij} is an isomorphism, dim V_i = dim W_j, with V_i and W_j are equivalent representations. If one considers all possible A's, it follows that one has to consider only the representations in $\mathbb{R}^m$ which are equivalent to those of $\mathbb{R}^n$. Furthermore, since an equivalent representation amounts to a choice of bases (in $\mathbb{R}^n$ and $\mathbb{R}^m$) and since ker A and Range A are also representations of Γ (and also a complement of ker A), the problem can be reduced to the study of A on $\mathbb{R}^n = V_1 \oplus \ldots \oplus V_k$, with $A\gamma = \gamma A$, with $A_{ij} = 0$ if V_i and V_j are not equivalent.

As before, one may assume that $\gamma \in O(n)$, again a choice of basis. Then $A_{ij} = \mu_{ij} I + \nu_{ij} B_{ij}$, with $B_{ij} + B_{ij}^T = 0, B_{ij}^2 = -I$.

The following result follows the proof of Frobenius theorem, as given in Pontryagin's book (Topological groups (1939), p. 160).

Theorem 1.1. *Let V be an irreducible orthogonal representation. Then either,*

(a) *for any A, with $\gamma A = A\gamma$, and $A = \mu I$, i.e. V is an absolutely irreducible representation, or,*

(b) *there is only one B, equivariant, such that $B + B^T = 0, B^2 = -I$ and $A = \mu I + \nu B$, for any equivariant A. In this case V has a complex structure, for which $A = (\mu + i\nu)I$, or*

(c) *there are precisely B_1, B_2, B_3, with the same properties and $B_i B_j + B_j B_i = 0, i \neq j, B_3 = B_1 B_2, V$ has a quaternionic structure and any A, with $\gamma A = A\gamma$, can be written as $A = \mu I + \nu_1 B_1 + \nu_2 B_2 + \nu_3 B_3 = qI$ where $q = \mu + \nu_1 i_1 + \nu_2 i_2 + \nu_3 i_3 \in \mathbb{H}$.*

Proof. Let $\mathcal{C} = \{A : A\gamma = \gamma A, \forall \gamma \in \Gamma\}, D = \{A \in \mathcal{C} : A = kI\}$, $I = \{A \in \mathcal{C} : A^2 = -k^2 I, A + A^T = 0\}$. If $A \in I$, then $\alpha A \in I$, for $\alpha \in \mathbb{R}$. Also if A_1 and $A_2 \in I$, then $A_1 + A_2 = \mu I + \nu B$, for some B in I. However $A_1 A_2^T + A_2^T A_1$ is equivariant and self-adjoint, hence from Schur's

Lemma, it belongs to D. That is $A_1A_2 + A_2A_1 = kI$. Similarly $BA_1 + A_1B$ is self-adjoint, hence equal to αI. But $A_1A_2 + A_2A_1 = kI = A_1(\mu I + \nu B - A_1) + (\mu I + \nu B - A_1)A_1 = 2\mu A_1 + \nu \alpha I + 2k_1^2 I$, where $A_1^2 = -k_1^2 I$. Hence $2\mu A_1 = (k - \nu\alpha - 2k_1^2)I$, that is, if $\mu \neq 0$, $A_1 \in D \cap I = \{0\}$. Thus, $A_1 + A_2$ belongs to I.

Let $B_1, B_2 \in I$, be such that $B_1^2 = B_2^2 = -I$. Then $B_1B_2 \in C$ and, as such, $B_1B_2 = \mu I + \nu B$, for some B in I. Multiplying by B_1, one has $\mu B_1 + B_2 = -\nu B_1 B$ and, from the above, $\nu B_2 B \in I$. Now if $\nu = 0$, $B_2 = -\mu B_1, B_2^2 = -I = -\mu^2 I$, that is $\mu = \pm 1$ and $B_1 = \pm B_2$.

If $\nu \neq 0$, then $B_1 B \in I$. Let $B_1' = B_1, B_2' = B, B_3' = B_1 B = B_1' B_2'$. Then $B_3' + B_3'^T = 0$, because B_3' belongs to I, and $B_3' B_3'^T = B_1 B B^T B_1^T = I$, hence $B_3'^2 = -I$. Furthermore $B_1' B_2' = B_3' = -B_3'^T = -B_2' B_1', B_1' B_3' = -B_2' = -B_3' B_1'$ and $B_3' B_2' = -B_1' = -B_2' B_3'$.

Dropping the primes, one has B_i with the commutation properties of the theorem. Now these B_1, B_2, B_3 are linearly independent in the linear space I, since, if $\lambda_1 B_1 + \lambda_2 B_2 + \lambda_3 B_3 = 0$. Then multiplying by B_1, one has $\lambda_2 B_3 - \lambda_3 B_1 = \lambda_1 I$, but, from the fact that if A_1 and A_2 are in I, then $A_1 + A_2 = \nu B$ (as seen above), one has $\lambda_1 = 0$ and $B_i, i = 1, 2, 3$ are linearly independent.

Finally, suppose that there is a B in I, with $B^2 = -I$, which is not a linear combination of B_1, B_2, B_3. Then $B_j B = \mu_j I + \nu_j \tilde{B}_j$ and $BB_j = \mu_j I - \nu_j \tilde{B}_j$. Let $\hat{B} = \alpha(B + \mu_1 B_1 + \mu_2 B_2 + \mu_3 B_3)$; then $\hat{B}$ is in I and it is easy to see that $\hat{B}^2 = -\alpha^2(1 - \mu_1^2 - \mu_2^2 - \mu_3^2)I = -k^2 I$, hence, either $\hat{B}^2 = 0$ and $\hat{B}$ has a nontrivial kernel (in which case, from Schur's lemma $\hat{B} = 0$) or one may choose α such that $k^2 = 1$ and $\hat{B}^2 = -I$. Now $B_j \hat{B} = \alpha(\nu_j \tilde{B}_j \pm \mu_k B_l \pm \mu_l B_k) = -\hat{B}B_j \in I$. Thus, $(B_1\hat{B})B_3 = -(\hat{B}B_1)B_3 = \hat{B}B_2$ and $B_1(\hat{B}B_3) = -B_1 B_3 \hat{B} = B_2 \hat{B} = -\hat{B}B_2$. That is $\hat{B}B_2 = 0$ which is not possible unless $\hat{B} = 0$ since both are isomorphisms. Hence $\hat{B} = 0$ and $B = -\mu_1 B_1 - \mu_2 B_2 - \mu_3 B_3$.

Note that, because of the associativity of the product of matrices, there is no equivalent to the Cayley numbers. It remains to make explicit the structure of V. Let $B \in I$ be such that $B^2 = -I$, then there is a basis of V such that $B = \begin{pmatrix} 0 & -I \\ I & 0 \end{pmatrix}$ (n is even, $n = 2m$, since $(\det B)^2 = (-1)^n$) : in fact take a unit vector e_1, then Be_1 is orthogonal to e_1 (since $B + B^T = 0$). Choose e_2 orthogonal to $\{e_1, Be_1\}$; then Be_2 is orthogonal to $\{e_1, Be_1, e_2\}$ and so on... On the basis $\{e_1, e_2, \ldots, e_m, Be_1, \ldots, Be_m\}$ then B has the above form, $V \cong \mathbb{C}^m$ by putting $z_j = x_j + ix_{m+j}, Z = X + iY$. If $\gamma = \begin{pmatrix} \gamma_1 & \gamma_2 \\ \gamma_3 & \gamma_4 \end{pmatrix}$, then $\gamma B = B\gamma$ implies $\gamma = \begin{pmatrix} \gamma_1 & -\gamma_2 \\ \gamma_2 & \gamma_1 \end{pmatrix}, \gamma \begin{pmatrix} X \\ Y \end{pmatrix} = \hat{\gamma}Z = (\gamma_1 + i\gamma_2)(X + iY), B \begin{pmatrix} X \\ Y \end{pmatrix} = \tilde{B}(X + iY) = iI(X + iY)$, where $\tilde{B} = iI$.

Thus if $A \in C$, $A = \mu I + \nu B = (\mu + i\nu)I = \lambda I$, with $\lambda \in \mathbb{C}$.

For the third possibility, let $B_1 = \begin{pmatrix} 0 & -I \\ I & 0 \end{pmatrix}$ be as before, defining a complex structure on V. Take a unit vector e_1; then $(e_1, B_1e_1, B_2e_1, B_3e_1)$ are orthonormal. Next take e_2 orthogonal to that set. It is easy to see that the vectors $(e_2, B_1e_2, B_2e_2, B_3e_2)$ are all orthogonal to the first set and among themselves by using the relations of commutation of the B_j's. This implies that dim $V = 4m$, and, on the basis

$$\{e_1, ...e_m, B_1e_1, ...B_1e_m, B_2e_1, \ldots, B_2e_m, B_3e_1, \ldots B_3e_m\},$$

B_j have the form of the Pauli matrices:

$$B_1 = \begin{pmatrix} 0 & -I & 0 & 0 \\ I & 0 & 0 & 0 \\ 0 & 0 & 0 & -I \\ 0 & 0 & I & 0 \end{pmatrix}, \quad B_2 = \begin{pmatrix} 0 & 0 & -I & 0 \\ 0 & 0 & 0 & I \\ I & 0 & 0 & 0 \\ 0 & -I & 0 & 0 \end{pmatrix},$$

$$B_3 = \begin{pmatrix} 0 & 0 & 0 & -I \\ 0 & 0 & -I & 0 \\ 0 & I & 0 & 0 \\ I & 0 & 0 & 0 \end{pmatrix}.$$

Then if $X = (X_0, X_1, X_2, X_3)^T$ is written as $\hat{X} = X_0 + i_1X_1 + i_2X_2 + i_3X_3$, an element of $\mathbb{H}^m$, with $i_j^2 = -1$, $i_ji_k + i_ki_j = 0$, $i_1i_2 = i_3$, B_jX can be written as $i_j\hat{X}$. Using the relation $\gamma B_j = B_j\gamma$, where γ is written as a $(4m \times 4m)$ matrix (γ_{kl}), $k, l = 1, \ldots, 3$, then $\gamma = \begin{pmatrix} \gamma_0 & -\gamma_1 & -\gamma_2 & -\gamma_3 \\ \gamma_1 & \gamma_0 & \gamma_3 & -\gamma_2 \\ \gamma_2 & -\gamma_3 & \gamma_0 & \gamma_1 \\ \gamma_3 & \gamma_2 & -\gamma_1 & \gamma_0 \end{pmatrix}$ can be written as $\hat{\gamma} = \gamma_0 + i_1\gamma_1 + i_e\gamma_2 + i_3\gamma_3$ acting on $\hat{X}$ on the **right**: $\gamma X = \hat{X}\hat{\gamma} = (X_0 + i_1X_1 + i_2X_2 + i_3X_3)(\gamma_0 + i_1\gamma_1 + i_2\gamma_2 + i_3\gamma_3)$ and any A in C can be written as $A = qI, q = \mu + \nu_1i_1 + \nu_2i_2 + \nu_3i_3 \in \mathbb{H}$, I is the identity in $\mathbb{H}^m$, that is $AX = q\hat{X}$. Then $A(\gamma X) = q\gamma X = q\hat{X}\hat{\gamma}$ and $\gamma AX = (AX)\hat{\gamma} = q\hat{X}\hat{\gamma}$.   Q.E.D.

Take now $V = V_1 \oplus ... \oplus V_k$, where V_j are irreducible but with equivalent actions. Then, if $\gamma A = \tilde{\gamma}A$, A maps V into V and is similar to a matrix $\tilde{A} : (V_1)^k \to (V_1)^k$ such that $\gamma\tilde{A} = \tilde{A}\gamma$ and γ acts orthogonally on V_1. Note that the similarity depends only on the actions, not on A, hence assume that there are bases in V and a norm so that $\gamma \in O(V)$ has the form $\begin{pmatrix} \gamma & 0 \\ 0 & \gamma \end{pmatrix}_{k \times k}$ since V_i are Γ-invariant. Let $m = dim_K V_i$, where $K = \mathbb{R}, \mathbb{C}$

or $\mathbb{H}$. Then $\gamma|_{V_i}$ can be written as above, when considering the real matrix, or as $\hat{\gamma}$ for the K–structure. $A_{ij} : V_i \to V_j$ is $A_{ij} = \lambda_{ij} I$, with $\lambda_{ij} \in K$ and I is the identity in K^m, i.e. $A = (A_{ij})_{1 \leq i,j \leq k}$, with $\lambda_{ij} \in K$ and I is the identity in K^m, i.e. $A = (A_{ij})_{1 \leq i,j \leq k}$, with $\gamma A_{ij} = A_{ij} \gamma$, on the basis of Theorem 1.1.

Take a new basis for V by ordering the bases of $V_1, \{e_{11}, e_{12}, ...e_{1m}\}$, of $V_2, \{e_{21}, e_{22}, ..., e_{2m}\}, ...,$ of $V_k, \{e_{k1}, e_{k2}, ...e_{km}\}$ in the following way: $\{e_{11}, e_{21}, ..., e_{k1}, e_{12}, e_{22}, ..., e_{k2}, ..., e_{1m}, e_{2m}, ...e_{km}\}$. It is easy to see that, on the new basis, A has the form $A = \begin{pmatrix} \wedge & & 0 \\ & \wedge & \\ 0 & & \wedge \end{pmatrix}$, where $\wedge = (\lambda_{ij})_{1 \leq i,j \leq k}$

is repeated m–times on the diagonal. On the other hand, if $\gamma : V_l \to V_l$ had the form $(\gamma_{ij})_{1 \leq i,j \leq k}$, on the new basis, $\gamma|_V = (\gamma_{ij} I)_{1 \leq i,j \leq m}$, where I is the identity in K^k. The relation $\gamma A = A\gamma$ is maintained in the new basis, since $\gamma_{ij}\wedge = \wedge\gamma_{ij}$, if $K = \mathbb{R}$ or $\mathbb{C}$ and $\gamma_{ij} \in K$, and, for the quaternionic case, because the action is on the right and γq has to be interpreted as $\hat{q}\hat{\gamma}$ (one may also go back to 4×4 real matrices, where γ_{ij} is as above and commutes with q). This last equivariance will be important when considering Γ–equivariant deformations of A : any deformations of $\wedge$ as a $k \times k$ matrix over K, will give rise, by repeating the deformation on the m replicae, to a Γ–deformation of A. Note that Werner (1991) gives a similar result. The above considerations give a proof of the following result

Theorem 1.2. *Let* $V = \mathbb{R}^d$ *be decomposed as* $V_1^{\mathbb{R}} \oplus .. \oplus V_1^{\mathbb{R}} \oplus V_2^{\mathbb{R}} \oplus ... \oplus V_2^{\mathbb{R}} \oplus ... \oplus V_1^{\mathbb{C}} \oplus ... \oplus V_1^{\mathbb{C}} \oplus ... \oplus V_1^{\mathbb{H}} \oplus ... \oplus V_1^{\mathbb{H}} \oplus ...,$ *where* $V_i^{\mathbb{R}}$ *are the absolutely irreducible representations of real dimension* m_i *and repeated* n_i *times,* $V_j^{\mathbb{C}}$ *are complex representations of complex dimension* m_j *and repeated* n_j *times and* $V_l^{\mathbb{H}}$ *are quaternionic representations of dimension (over* $\mathbb{H}$*)* m_l *and repeated* n_l *times. Then there are bases of* V *such that any matrix* A *with* $\gamma A = A\gamma$, *has the block diagonal form:* $A = \text{diag}(A_1^{\mathbb{R}}, ..., A_1^{\mathbb{R}}, A_2^{\mathbb{R}}, ..., A_2^{\mathbb{R}}, A_1^{\mathbb{C}} ..., A_1^{\mathbb{C}}, ..., A_1^{\mathbb{H}}, ..., A_1^{\mathbb{H}}, ...),$ *where* $A_i^{\mathbb{R}}$ *are real matrices of dimension* n_i, *repeated* m_i *times,* $A_j^{\mathbb{C}}$ *are complex matrices of complex dimension* n_j, *repeated* m_j *times, and* $A_l^{\mathbb{H}}$ *are quaternionic matrices of dimension (over* $\mathbb{H}$*)* n_l *and repeated* m_l *times.*

Remark 1.1. If Γ is abelian, then the irreducible representations of Γ are either one-dimensional and Γ acts trivially or as $\mathbb{Z}_2$, or two-dimensional and Γ acts as $\mathbb{Z}_n, n \geq 3$, or S^1 (for a direct proof, see for example, Ize-Vignoli 1993). In this case there are no quaternionic components.

As a first consequence of Theorem 1.2, consider $GL_\Gamma(\mathbb{R}^d)$, the set of all invertible matrices on $\mathbb{R}^d$, which commute with Γ (one may suppose that the action is orthogonal). This set is clearly a group under composition and is closed under transposition. Consider the group $\Pi_{k-1}(GL_\Gamma(\mathbb{R}^d))$ of all equivariant homotopy classes of maps from S^{k-1} into $GL_\Gamma(\mathbb{R}^d)$, with positive determinant if $k > 1$.

Theorem 1.3. *The group $\Pi_{k-1}(GL_\Gamma(\mathbb{R}^d))$ is isomorphic to the product $\Pi_i[\Pi_{k-1}(GL(V_i^{\mathbb{R}}))]\Pi_j[\Pi_k(GL(V_j^{\mathbb{C}}))]\Pi_l[\Pi_{k-1}(GL(V_l^{\mathbb{H}}))]$. In particular:*

  (a) *If $k < \dim V_i^{\mathbb{R}}$, then $\Pi_{k-1}(GL(V_i^{\mathbb{R}}))$ is $\mathbb{Z}$ if $k \equiv 0$ or 4 [Mod 8], $\mathbb{Z}_2$ if $k = 1$ or 2 [mod 8], 0 if $k \equiv 3, 5, 6, 7$ [mod 8],*
  (b) *if $k \leq 2\dim_{\mathbb{C}} V_j^{\mathbb{C}}$, then $\Pi_{k-1}(GL(V_j^{\mathbb{C}}))$ is $\mathbb{Z}$ if k is even and 0 if k is odd.*
  (c) *if $k \leq 4\dim_{\mathbb{H}} V_l^{\mathbb{H}} + 2$, then $\Pi_{k-1}(GL(V_l^{\mathbb{H}}))$ is $\mathbb{Z}$ if $k \equiv 0$ or 4 [mod 8], $\mathbb{Z}_2$ if $k \equiv 5$ or 6 [mod 8], 0 if $k \equiv 1, 2, 3, 7$ [mod 8].*

Proof. From the preceding theorem it is enough to see that the mappings

$$A(\lambda)\xrightarrow{I} (A_1^{\mathbb{R}}(\lambda), A_2^{\mathbb{R}}(\lambda), \ldots, A_1^{\mathbb{C}}(\lambda), \ldots, A_1^{\mathbb{H}}(\lambda), \ldots)\xrightarrow{J}$$
$$\mathrm{diag}(A_1^{\mathbb{R}}(\lambda), \ldots, A_1^{\mathbb{R}}(\lambda), A_2^{\mathbb{R}}(\lambda), \ldots,$$
$$A_2^{\mathbb{R}}(\lambda), \ldots, A_1^{\mathbb{C}}(\lambda), \ldots, A_1^{\mathbb{C}}(\lambda), \ldots, A_1^{\mathbb{H}}(\lambda), \ldots, A_1^{\mathbb{H}}(\lambda), \ldots)$$

where, in J, one repeats according to the dimensions of $V_i^{\mathbb{R}}, V_j^{\mathbb{C}}, V_l^{\mathbb{H}}$, are morphisms and $JI = Id$, $IJ = Id$. Hence $GL_\Gamma(\mathbb{R}^d)$ is isomorphic to the product and this induces isomorphisms at the homotopy level, since deformation in the product can be lifted to equivariant deformations in $GL_\Gamma(\mathbb{R}^d)$. Q.E.D.

Remark 1.2. If $k = 1$, then, since $GL(V^{\mathbb{C}})$ and $GL(V^{\mathbb{H}})$ are connected, one is left only with $V^{\mathbb{R}}$. If $k = 2, GL(V^{\mathbb{H}})$ is simply connected and one has to consider only the real and complex representations. In the first case $A_i^{\mathbb{R}}(\lambda)$ is deformable to $\begin{pmatrix} \det A_i^{\mathbb{R}}(\lambda) & 0 \\ 0 & I \end{pmatrix}$ and in the second, $A_j^{\mathbb{C}}(\lambda)$ is deformable to $\begin{pmatrix} \det A_j^{\mathbb{C}}(\lambda) & 0 \\ 0 & I \end{pmatrix}$ and is characterized by the winding number of $\det_{\mathbb{C}} A_j^{\mathbb{C}}(\lambda)$.

2. Consequences of Symmetry

In this subsection, I will give some of the simplest consequences, good and bad, of the presence of symmetries. As before $F : \wedge \times B \to E$ is an equivariant map, i.e. $F(\gamma\lambda, \gamma x) = \tilde\gamma F(\lambda, x)$, where in infinite dimensions, $\wedge \times B = \mathbb{R}^M \times G, E = \mathbb{R}^N \times G$, all Γ-spaces and $F = (F_1, F_2)$ where F_2 has the form of Id-compact.

Since the zeros of F come in orbits, i.e. if $F(\lambda, x) = 0$ then $F(\gamma\lambda, \gamma x) = 0$, the local study of F near $(\gamma\lambda, \gamma x)$ can be reduced to the study of F near (λ, x). In particular, consider $F(\lambda, x + y) - F(\lambda, x)$ and $F(\gamma\lambda, \gamma x + \gamma y) - F(\gamma\lambda, \gamma x)$ and linearize; then, since $\|\gamma x\| = \|x\|$, one obtains

$$D_x F(\gamma\lambda, \gamma x)\gamma = \tilde{\gamma} D_x F(\lambda, x) \quad , \quad \forall \gamma \in \Gamma. \tag{1}$$

This implies that, if $B = E = \mathbb{R}^d$ and $D_x F(\lambda, x)$ is invertible at the zero (λ, x). Then $D_x F(\gamma\lambda, \gamma x)$ is conjugate to $D_x F(\lambda, x)$ with the same determinant: the local index will be the same at each point of the orbit. The contribution to a Brouwer degree (or a Leray-Schauder degree) will be $|\Gamma(\lambda, x)|$ Sign det $D_x F(\lambda, x)$, where $|\Gamma(\lambda, x)|$ is the number of points in the orbit (it has to be finite).

Also, assuming that Γ acts trivially on Λ, if the dimension of the orbit is positive, one may choose a differentiable $\gamma(t)$ with $\gamma(0) = I, \gamma'(0) \neq 0$ such that $F(\lambda, \gamma(t)x) = 0$. Differentiating with respect to t and evaluating at $t = 0$, one has

$$D_x F(\lambda, x)\gamma'(0)x = 0.$$

Hence $\gamma'(0)x$ will be in the kernel of $D_x F(\lambda, x)$ for each direction $\gamma'(0)$ such that $\gamma'(0)x \neq 0$. Since the orbit is a differentiable manifold, this will be true for each direction tangent to the orbit. Hence, one has in this case a high dimensional kernel. Taking a normal hyperplane, the normal bundle to the orbit, one may construct a Poincaré section and a tubular neighborhood of the orbit. On that section one may define a generalized index if the situation is stable under perturbations.

Another consequence of (1), is that for $\gamma \in \Gamma_{(\lambda, x)} = H$, one has $D_x F(\lambda, x)\gamma = \tilde{\gamma} D_x F(\lambda, x)$, i.e. $D_x F(\lambda, x)$ is equivariant with respect to H and as such it has the structure given in Theorem 1.2.

Remark 2.1. An easy byproduct of Theorem 1.2 is the "mod-p" formula:

If the finite group Γ acts freely on $E \setminus \{0\}, E = \mathbb{R}^d$, then the degree of any Γ−equivariant map with respect to the ball $B(0, r)$ (or any invariant neighborhood of 0) is $1 + k|\Gamma|$.

Proof. One may approximate f by an equivariant $\tilde{f}$ with $D\tilde{f}(x)$ invertible for each zero x of $\tilde{f}$: In fact B/Γ is a smooth manifold since the action is free and one may use Sard's lemma there. If $x \neq 0$, the contribution to the degree is $|\Gamma|$ Sign det $D\tilde{f}(x)$, i.e. a multiple of $|\Gamma|$. Now $\gamma\tilde{f}(0) = \tilde{f}(0)$, that is $\tilde{f}(0) \in E^\Gamma = \{0\}$, hence $\tilde{f}(0) = 0$. The index of 0 is Sign det $D\tilde{f}(0)$, with $\gamma D\tilde{f}(0) = D\tilde{f}(0)\gamma$, i.e. $D\tilde{f}(0)$ has the structure of Theorem 1.2. The complex and quaternionic matrices have positive (real) determinant and Sign det $D\tilde{f}(0)$ is Sign$(\det A_1^{\mathbb{R}})^{n_1} .. (\det A_i^{\mathbb{R}})^{n_i} ...$, where only the matrices $A_i^{\mathbb{R}}$ with negative determinant and $V_i^{\mathbb{R}}$ of odd dimension n_i, will count.

Now, as shown below, if n_i is odd, then $\Gamma \cong \mathbb{Z}_2$, hence $|\Gamma| = 2$ and $1 + k|\Gamma| = -1 + (k+1)|\Gamma|$; hence the formula holds. On the other hand if $\Gamma \neq \mathbb{Z}_2$, then n_i is even and the index at 0 is 1.

It remains to prove that if n_i is odd, then $\Gamma \cong \mathbb{Z}_2$. Let $\gamma \neq I$ and consider the deformation on $\partial B(0, r)$, $t\gamma x - (1-t)x = g(t, x)$. If $g(t, x) = 0$ for $x \neq 0$, then x is an eigenvector of γ with eigenvalue $(1-t)/t > 0$. Since $\gamma^T \gamma = I$, this implies that $(1-t)/t = 1$ and x is a fixed point of γ. This is not possible since the action is free. Hence $\deg(\gamma x; B(0, r)) = \det(-x; B(0, r)) = (-1)^{n_i} = \text{Sign det } \gamma$. Using this argument for γ^2 (if $\gamma^2 \neq I$), then $\text{Sign}(\det \gamma)^2 = 1 = (-1)^{n_i}$ and, if n_i is odd, this implies $\gamma^2 = I, \forall \gamma \in \Gamma$. Then $(\gamma - I)(\gamma + I) = 0$, but $\gamma \neq I, \gamma - I$ is invertible (free action), and $\gamma = -I$. Thus $\gamma|V_i^{\mathbb{R}}$ is I or $-I$. In general $\mathbb{R}^d = V_1 \oplus \ldots \oplus V_k$. If $\dim V_i$ is odd, then $\Gamma|_{V_i}$ acts as $\pm I$, as well as on $V_i \oplus V_j$ with $\dim V_j$ even. It is then easy to conclude that Γ acts as $\pm I$ on $\mathbb{R}^d$ if d is odd.   Q.E.D.

In case Γ has positive dimension, that is $\Gamma \cong S^1$ or $N(S^1)$ in S^3 or S^3 (see Subsection 1), then Γ has subgroups $\mathbb{Z}_p$, for any prime p. From the above result $\deg(f; B(0, r)) = 1 + kp \equiv 1 \ [\text{mod } p]$, all p's. Then $\deg(f; B(0, r)) = 1$. This result was proved, by different methods, in Marzantowicz (1981), Nirenberg (1981) and Ize (1985) for the case of S^1. For related results, of Borsuk-Ulam type, there is a vast literature, see for example Steinlein (1985), Nussbaum (1977), Rubinstein (1976), Komiya (1988), Wang (1989), Bartsch (1991), Rabier (1991) and so on...

Remark 2.2. Variational problems. There are many papers on equivariant bifurcation for variational problems: see the other contributions in this volume. Here I would like to point out a very simple property of such problems. Assume $J : E \to \mathbb{R}$ is an invariant functional, i.e. $J(\gamma x) = J(x)$, for all $\gamma \in \Gamma$. If one looks for critical points of J, i.e. zeros of $f(x) = \nabla J(x)$, then it is easy to see that

$$f(\gamma x) = \gamma f(x).$$

Now, if Γ has positive dimension and one takes a path $\gamma(t)$ with $\gamma(0) = I$, then differentiating the identity $J(\gamma(t)x) = J(x)$, one obtains

$$\nabla J(x) \cdot \dot{\gamma}(0)x = 0,$$

that is $\dot{\gamma}(0)x$ is orthogonal to the field $\nabla J(x)$. The same thing happens for integrals of the differential equation $\frac{dX}{dt} = g(X)$, i.e. if there is a real function $V(X)$ such that $\nabla V(X)$ is orthogonal to $g(X)$, then $V(X)$ is conserved along solutions of the equation, invariant with respect to translations in the case of period orbits and the action of S^1.

Geometrically, the orthogonality of $f(x)$ and $\dot{\gamma}(0)x$ is a restriction on the image of the field and can be seen as a reduction in the number of "free" equations. From the analytical point of view, one may use some analogue of the implicit function theorem and reduce the number of variables, or one may use, as in conditioned variational problems, a "Lagrange multiplier". More precisely, one may add a new variable μ and look for zeros of the equation

$$f(x) + \mu\dot{\gamma}(0)x = 0.$$

In fact, if $f(x) = 0$, then $\mu = 0$ gives a solution of the above equation. Conversely if (μ, x) is a solution, then multiplying by $\dot{\gamma}(0)x$, one has $\mu\|\dot{\gamma}(0)x\|^2 = 0$, hence $f(x) = 0$ and $\mu\dot{\gamma}(0)x = 0$. Hence, $\mu = 0$ if $\dot{\gamma}(0)x \neq 0$. The same argument implies that if $f(x) = 0$ and $\dot{\gamma}(0)x \neq 0$, then the solution x is unstable under small perturbations of the form $\epsilon\dot{\gamma}(0)x$ with $\epsilon \neq 0$. In this case, methods based on the implicit function theorem, or on a degree theory for $f(x)$ only, will fail unless one extends the number of unknowns by adding μ. This approach was used in Alexander-Yorke (1978).

Now, in general, $\dot{\gamma}(0)$ is not equivariant with respect to Γ, unless $\dot{\gamma}(0)$ belongs to the center of the Lie algebra of Γ (this is the case if Γ is abelian). Hence the augmented equation does not have the full symmetry of Γ, but the reduced symmetry of the centralizer of $\dot{\gamma}(0)$. In any case, by differentiating $\gamma^T(t)\gamma(t) = I$, one has that $\dot{\gamma}^T(0) + \dot{\gamma}(0) = 0$ and, since $\gamma(t)$ maintains any irreducible representation V_i of Γ, $\dot{\gamma}(0)$ sends V_i into itself. Thus, if $\dot{\gamma}(0)$ commutes with Γ, then $\dot{\gamma}(0)$ has the form given in Theorem 1.2., i.e. $\dot{\gamma}(0)|_{V_i}$ is 0 if $V_i = V_i^{\mathbb{R}}$, or $i\beta_j I$ if V_i is a complex representation, or $\beta_1 i_1 + \beta_2 i_2 + \beta_3 i_3$ for a quaternionic representation, with the same coefficients β_k for equivalent representations.

On the other hand, this argument can be repeated for each subgroup $\gamma(t)$ and one obtains $\dot{\gamma}_j(0)$ for each $j = 1, \dots$ dim Γ, or, if one wishes to maintain the symmetry, up to the dimension of the center of Γ. Considering the equation $f(x) + \Sigma\nu_j\dot{\gamma}_j(0)x = 0$, one obtains a problem with several parameters. If one wishes to conclude that $\nu_j = 0$ on a zero of this equation, that is, if one asks for the full symmetry, that $\dot{\gamma}_j(0)x$ are linearly independent. Hence the number of parameters is limited to one, for each class of nonequivalent complex representations for which the component of x is nonzero, and three for each class of nonequivalent quaternionic representations with nonzero component for x. In the case of first integrals, one has the same problem (Frobenius conditions) of independence. The optimal strategy in such problem is an open question.

Another important consequence of the symmetry is the **stratification of the space**, i.e. if $f(\gamma, x) = \tilde{\gamma}f(x)$, for $f : B \to E$, then, if $H < \Gamma$, and x is in B^H, one has that $f(\gamma x) = f(x) = \tilde{\gamma}f(x)$ for any γ in H. Thus, $f(x)$ belongs to E^H. In other words, f sends B^H into E^H. Hence, one may look for solutions with a given symmetry, i.e. in B^H (for example, radial solutions, and so on...). Now, since $N(H)$ is the largest subgroup of Γ which keeps B^H invariant, one has that $f^H = f|_{B^H}$ is an $N(H)$-equivariant map from B^H into E^H.

Note that if f is C^1, at some $\bar{x} \in E^H$ then Df has a diagonal structure. In fact if B^H and E^H have topological complements, then, from the equivariant projections, one may take these complements as $N(H)$−spaces (hence also H−spaces). If $B = B^H \oplus B_1, x = x_0 \oplus x_1, E = E^H \oplus E_1, f = f^0 \oplus f^1$, then $Df(\bar{x}) = \begin{pmatrix} f^0_{x_0} & f^1_{x_0} \\ f^0_{x_1} & f^1_{x_0} \end{pmatrix}$. From the fact that $f^1(x_0, 0) = 0$, one has that $f^1_{x_0} = 0$ and since, for γ in

$$H, Df(\bar{x}) \begin{pmatrix} x_0 \\ \gamma x_1 \end{pmatrix} = \begin{pmatrix} f^0_{x_0} x_0 \\ f^0_{x_1} x_0 + f^1_{x_1} \gamma x_1 \end{pmatrix} = \begin{pmatrix} I & 0 \\ 0 & \tilde{\gamma} \end{pmatrix} Df(\bar{x}) \begin{pmatrix} x_0 \\ x_1 \end{pmatrix},$$

one has that $\tilde{\gamma} f^1_{x_1} = f^1_{x_1} \gamma$ and $\tilde{\gamma} f^0_{x_1} x_0 = f^0_{x_1} x_0$, i.e. $f^0_{x_1} x_0 \in E^H \cap E_1 = \{0\}$. Thus

$$Df(\bar{x}) = \begin{pmatrix} Df^H & 0 \\ 0 & f^1_{x_1} \end{pmatrix}$$

where $f^1_{x_1}$ is H−equivariant.

It is now convenient to reduce the study to the smallest possible B^H, i.e. to the largest H, in particular to maximal isotropy subgroups, where one knows that $W(H)$ acts freely on B^H and are completely classified. Furthermore if one decomposes E^H into irreducible representations of $W(H)$, one may determine not only the linear terms, but also higher order terms if the dimension is low. These ideas have been used by Sattinger, Iooss, Golubitsky and coworkers, in order to give normal forms or equivariant singularities. The information obtained this way is very precise, but from the requirements of genericity and low dimension, it does not allow for a complete study of stability, symmetry breaking, or period doubling, when one has to consider perturbations with a symmetry different from the given solutions. Hence, in these cases, it is convenient not to fix a priori the symmetry of the solution and to treat the complete equivariant problem. Then one will have a more general vision, but certainly less precise. This point of view will be adopted in the next sections.

A final comment will be made on the particular bifurcation problem $F(\lambda, x) = 0$, with $F : \wedge \times B \to E$, dim $\wedge = k$, and $F(\lambda, \gamma x) = \tilde{\gamma} F(\lambda, x)$. As seen above, if, at (λ_0, x_0), $D_x F(\lambda, x)$ is invertible, then the orbit of x, homeomorphic to Γ/Γ_x, is finite. Let $H = \Gamma_x$. Then it is clear that $D_x F^H$, which is H−equivariant, is also invertible at (λ_0, x_0). From the implicit function theorem, one has that, near (λ_0, x_0), the solutions are given by $(\lambda, x(\lambda))$ and, by the uniqueness, $x(\lambda) \in E^H$ and $H < \Gamma_{x(\lambda)}$. By reversing the argument, one has that $H = \Gamma_{x(\lambda)}$ and by looking at H−equivariant maps, we may suppose that $x(\lambda)$ is a branch of stationary points. The change of coordinates which takes $x(\lambda)$ into 0 is admissible and one may consider the bifurcation problem as

$$F(\lambda, x) = Ax - A(\lambda)x - g(\lambda, x)$$

where $A - A(\lambda)$ is Γ-equivariant, as well as $g(\lambda, x)$, from formula (1).

Now, if A is a Fredholm operator of index 0, $A(\lambda) = 0(\|\lambda\|), g(\lambda, x) = o(\|x\|)$, then one may choose Γ-equivariant projections P and Q, on ker A and Range A, i.e. $\gamma P = P\gamma$ and $\tilde{\gamma}Q = Q\tilde{\gamma}$. Then, the pseudoinverse is clearly equivariant, i.e. $\gamma K = K\tilde{\gamma}$, and the bifurcation equation, obtained in Section I,

$$B(\lambda)x_1 + G(\lambda, x_1) = 0$$

is Γ-equivariant, i.e. $B(\lambda)\gamma = \tilde{\gamma}B(\lambda)$ and $G(\lambda, \gamma x_1) = \tilde{\gamma}G(\lambda, x_1)$, since the equation $x_2 = (I - KQA(\lambda))^{-1}KQ(A(\lambda)x_1 + g(\lambda, x))$ is Γ-equivariant and uniquely solvable for λ and x_1 small, hence $x_2(\lambda, \gamma x_1) = \gamma x_2(\lambda, x_1)$.

Furthermore $B(\lambda)$ has the block diagonal structure of Theorem 1.2. Since ker A and coker A are finite dimensional, one may renorm them so that γ and $\tilde{\gamma}$ are orthogonal and equivalent. Also, if $H < \Gamma$, then any irreducible representation V_i of Γ in $\mathbb{R}^d$ is a representation of H. On equivalent representations, say k of them, one may choose vectors in the basis by grouping those which are in V_i^H and repeating them on V_j^H. Then, if $B(\lambda)$ on $V_1 \oplus ... \oplus V_k$ is $\text{diag}(A_i(\lambda), ...A_i(\lambda))$, where the $k \times k$ matrix is repeated $\dim_K V_i$ times, one has that $B(\lambda)^H$, on $V_1^H \oplus ... \oplus V_k^H$, corresponds to $\dim_K V_i^H$ blocks $A_i(\lambda)$.

Note that V_i^H is $N(H)$-invariant. Hence if Γ is abelian, $V_i^H = V_i$ and there is only one isotropy type for each irreducible representation (which is of dimension 1 or 2). In general, if $x \in V_i$, then the space generated by $\Gamma(x)$ is Γ-invariant and, as such, $\langle \Gamma(x) \rangle = V_i$. Note also that V_i and V_j may be non-equivalent representations of Γ, but V_i^H and V_j^H may be equivalent $N(H)$-representations. Also V_i^H may be not irreducible for $N(H)$ and may have a complex or quaternionic structure, even if V_i has not that structure for Γ.

Note that if one knows that $g(0, 0) = 0$, but not that $g(\lambda, 0) = 0$, for $\lambda \neq 0$, then the bifurcation will hold, if g is C^1 for example, with $G(\lambda, x_1)$ such that $G(0, 0) = 0$, but not necessarily $G(\lambda, 0) = 0$ and $x_2(0x_1) = 0(x1)$. If one has that (Coker $A^\Gamma = \{0\}$, then $G(\lambda, \gamma x_1) = \tilde{\gamma}G(\lambda, x_1)$ forces $G(\lambda, 0) = 0$ and $x_2(\lambda, 0) \in E_1^\Gamma$, that is $x(\lambda) = x_2(\lambda, 0) \in E^\Gamma$.

A first result for bifurcation will be the following:

Theorem 2.1. *It for some* $H < \Gamma, B(\lambda)^H$ *is invertible for* $|\lambda| \neq 0$ *and* $J[B(\lambda)^H] \neq 0$, *then one has global bifurcation in* B^H, *(if* $\lambda \in \mathbb{R}$, *if*

det $B(\lambda)^H$ *changes sign). The same result will happen for any of the sufficient conditions of the preceding sections when applied to* B^H, *for example
with special nonlinearities, and so on...*

If $\lambda \in \mathbb{R}$, $(\ker A)^H$ is one dimensional and $B(\lambda)^H$ changes sign, then
one may apply the Crandall-Rabinowitz theorem and obtain a smooth curve
in B^H. This is the equivariant branching lemma. More refined results will
be given below. In the next sections I shall take the reverse order of the
preceding sections, so that the role of the equivariant J−homomorphism
will be clearly seen.

3. Γ-Epi Maps

Problems of the form $F(\lambda, x) = 0$, where F is an equivariant map from $\wedge \times B$
into E (the action of Γ on $\wedge$ may be not trivial), may have "surfaces" of
solutions. Hence, it is natural to extend the notion of 0-epi maps to this
context. The results are taken from Ize et al. (1986).

Let $\tilde{E} = \wedge \times B$ and G be Γ−Banach spaces, where the actions are
isometries, let U be an open invariant subset of $\tilde{E}$ and S an arbitrary
invariant subset of $\tilde{E}$. As equivariant map $g : U \to G$ will be said to be
admissible on S and U if $g^{-1}(0) \cap S \subset V_0 \subset \bar{V}_0 \subset U$, where V_0 is an open,
invariant, bounded set.

Definition 3.1. a) $g : U \to G$ *admissible equivariant map on S and U is
said to be* $\underline{\Gamma - epi\ on\ S\ and\ U}$ *if* $g(x) = h(x)$ *has a solution x in $S \cap U$, for
any* $h : \tilde{E} \to G$, *compact equivariant and bounded support contained in U.*

b) *g is* $\underline{\Gamma - essential\ on S\ and\ U}$ *if for any open bounded and invariant
set V, with* $g^{-1}(0) \cap S \subset V \subset \bar{V} \subset U$, *then any continuous equivariant
extension* $\bar{g} : \bar{V} \to G$ *of* $g : \partial V \to G$, *with* $g - \bar{g}$ *compact on $\bar{V}$, has a zero
on* $S \cap V$.

As in Section 3, g is Γ−epi if and only if g is Γ−essential.
One has the following properties:

(P1). Existence. *If g is* Γ−*epi on S and U, then* $g^{-1}(0) \cap S \neq \phi$ *and
invariant.*

(P2). Localization. *If $g : U \to G$ is* Γ−*epi on S and U, then g is* Γ−*epi
on S and V for any open invariant V, with* $g^{-1}(0) \cap S \subset V_0 \subset \bar{V}_0 \subset V \subset U$.

(P3). Subgroups. *If $g : U \to G$ is equivariant and H is a closed subgroup
of Γ, then*

(a) *If g is H−epi on S and U, then g is* Γ−*epi on S and U.*

(b) *If S^H and U^H are Γ-invariant and g^H is Γ-epi on S^H and U^H, then g is Γ-epi on S and U.*

(c) *If g^H is zero-epi on S^H and U^H, then g is Γ-epi on S and U.*

The converses of (a), (b) and (c) are not true in general: for examples, see Ize et al. (1986).

(P4). Normalization. *If $0 \notin \partial U$, then $i : U \to E$ is Γ-epi on S and U if and only if $U_0^\Gamma \subset S$, where U_0^Γ is the connected component of 0 in U^Γ.*

(P5). Products. *If $G = G_1 \times G_2$, with action (γ_1, γ_2), then, if $g = (g_1, g_2)$ is Γ-epi on S and U, one has that g_2 is Γ-epi on $g_1^{-1}(0) \cap S$ and U.*

(P6). Homotopy. *If g is Γ-epi on S and U and $h : I \times U \to G$ is compact and equivariant, with $h(0, x) = 0$, and if $\{x \in S \cap U : g(x) = h(t, x)\} = A$, is such that $A \subset V_0 \subset \bar{V}_0 \subset U$, with V_0 open invariant and bounded, then $g(\cdot) - h(1, \cdot)$ is Γ-epi on S and U.*

(P7). Scaling. *If $\tilde{E} = E_1 \times E_2, G = G_1 \times G_2, S = E, g = (g_1, g_2)$ with $g_i(x_1, x_2) = L_i x_i - k_i(x_1, x_2)$ such that L_i is bounded linear and k_i is compact. Assume that $g^{-1}(0) \subset B_1^\epsilon \times B_2^\epsilon \cup A, g^{-1}(0) \cap (B_1 \times B_2 \backslash A)$ is closed, where $B_i^\epsilon = \{x_i : \|x_i\| < r_i - \epsilon\}, B_i \equiv B_i^0$ and A is a closed invariant set such that if $(x_1, x_2) \in A$, then (tx_1, x_2) and $(x_1, tx_2) \in A$ for all $t \geq 1$. Then g is Γ-epi on $E \backslash A$ if and only if g is Γ-epi on $B_1 \times B_2 \backslash A$.*

These properties are easy modifications of the ones for 0-epi map. One has also the following consequences:

Proposition 3.1. *Let g be Γ-epi on S and U. Then*

(a) *either $S \cap \partial V \neq \phi$ or $g(S \cap \bar{V}) \supset G^\Gamma$ for any open, bounded and invariant V such that $g^{-1}(0) \cap S \subset V \subset \bar{V} \subset U$. In particular if $G^\Gamma \neq \{0\}$ and g^Γ sends bounded, closed invariant subsets of $S \cap U$ into bounded subsets of G^Γ, then $S \cap \partial V \neq \phi$.*

(b) *either $S \cap U$ is unbounded or $(\overline{S \cap U}) \cap \partial U \neq \phi$ or there is V, as above, such that $g(S \cap \bar{V}) \supset G^\Gamma$.*

(c) *if g is bounded and proper on bounded and closed (in $\tilde{E}$) subsets of $S \cap U$, then there is an invariant set $\Sigma \subset S \cap U$ which is minimal, closed in U, such that: (i) g is Γ-epi on Σ (in particular $g^{-1}(0) \cap \Sigma \neq \phi$), (ii) if $\Sigma = \Sigma_1 \cup \Sigma_2$, where Σ_i are proper closed and invariant subsets with $\Sigma_1 \cap \Sigma_2 = \phi$, then $\Sigma_1 = \phi$ or $\Sigma_2 = \phi$, i.e. Σ/Γ is connected, (iii) Σ is minimal for any g_1, Γ-homotopic to g, (iv) if $G^\Gamma \neq \{0\}$, then Σ is either unbounded or $\bar{\Sigma} \cap \partial U \neq \phi$.*

The proof is given in the above reference and the arguments are similar to those used in Section III, taking care of the equivariance of the maps and the invariance of the sets.

Note that if $G^\Gamma = \{0\}$, then any map is $\Gamma-$epi provided $0 \in S \cap U$. Now, in order to get dimension results, one has to consider maps which have a nontrivial invariant part.

If $g : U \to G$ is invariant, then one obtains $\tilde{g} : U/\Gamma \to G$ by defining $\tilde{g}([x]) = g(\gamma x)$ and it is easy to see that g is 0-epi on S and U (i.e. with invariant maps) if and only if $\tilde{g}$ is 0-epi on $S/\Gamma \ U/\Gamma$. One has the following result:

Theorem 3.1. *Let g and Σ be as in (c) of Proposition 3.1. Assume that $G = G^\Gamma \oplus G_2$, with $G_2^\Gamma = \{0\}$ and $g = (g^\Gamma, g_2)$. Then there is an invariant minimal $\tilde{\Sigma} \subset g_2^{-1}(0) \cap \Sigma$ such that: (i) g^Γ is 0-epi invariant on $\tilde{\Sigma} \cap U$, in particular $\tilde{\Sigma}$ is either unbounded or $\tilde{\tilde{\Sigma}} \cap \partial U \neq \phi$ if $G^\Gamma \neq \{0\}$. (ii) If $\tilde{\Sigma}/\Gamma = \Sigma_1 \cup \Sigma_2$ with Σ_i closed and proper subsets of $\tilde{\Sigma}/\Gamma$, then $\dim(\Sigma_1 \cap \Sigma_2) \geq \dim G^\Gamma - 1$, in particular $\tilde{\Sigma}/\Gamma$ is connected and has local dimension at least $\dim G^\Gamma$. (iii) $\tilde{\Sigma}/\Gamma$ is minimal for any g_1 invariantly homotopic to g^Γ.*

The application to equivariant bifurcation will require the following hypotheses:

(1) $\dim \wedge < \infty, \wedge = \tilde{\wedge} \oplus \hat{\wedge}$, *invariant subspaces and* $\lambda_0 \in \wedge^\Gamma, \lambda_0 = \tilde{\lambda}_0 \oplus \hat{\lambda}_0$.

(2) $F : \wedge \times B \to E$, *with* $B = B^\Gamma \oplus \hat{B}, E = E^\Gamma \oplus \hat{E}$ *has the form* (F_0, F_1) *with* $F_0(\lambda, x_0, x_1) = L_0 x_0 - k_0(\lambda, x_0, x_1)$ *is invariant and* $F_1(\lambda, x_0, x_1) = L_1 x_1 - k_1(\lambda, x_0, x_1)$ *is equivariant and*

    (a) L_i *are bounded linear operators and proper on* $\bar{B}_i = \{x_i : \|x_i\| \leq r_i\}$

    (b) k_i *is compact.*

Note that $\hat{B}$ and $\hat{E}$ are invariant subspaces by the equivariant projection and note that one can allow relative condensing maps as in Section III.

(3) $k_0(\lambda, 0, 0) = 0$,

(4) *If* $F(\tilde{\lambda}, \hat{\lambda}_0, x_0, x_1) = 0$ *for* $x_1 \in \bar{B}_1, \|x_0\| \leq r_0 + \epsilon$ *and* $r_2 - \epsilon \leq \|\lambda - \lambda_0\| \leq r_2 + \epsilon$, *or* $x_1 \in \bar{B}_1, r_0 - \epsilon \leq \|x_0\| \leq r_0 + \epsilon$ *and* $\|\lambda - \lambda_0\| \leq r_2 + \epsilon$, *then* $x_1 = 0$. *Let* $\bar{B}_2 = \{\lambda : \|\lambda - \lambda_0\| \leq r_2\}$.

Theorem 3.2. *Assume* $(\hat{\lambda} - \hat{\lambda}_0, \|x_1\| - r_1/2, F(\tilde{\lambda}, \tilde{\lambda}_0, x_0, x_1))$ *is* $\Gamma-$*epi on* $B_2 \times B_0 \times B_1 \backslash \{x_1 = 0\}$. *Let* $S = \{(\lambda, x) : F(\lambda, x) = 0, x_1 \neq 0\}$. *Then S has a minimal closed (in S) invariant subset Σ such that the map* $(\hat{\lambda}^\Gamma - \hat{\lambda}_0, \|x_1\| - \phi(\|x_0\|, \|\lambda - \lambda_0\|))$ *is* $\Gamma-$*epi (invariant) on* Σ, *where* $\phi : \mathbb{R}^+ \times \mathbb{R}^+ \to [0, r_1/2]$ *is a continuous map such that* $\phi(s, t) = r_1/2$ *if* $0 \leq s \leq r_0$ *and* $0 \leq t \leq r_2, \phi(s, t) = 0$ *if* $s \geq r_0 + \epsilon$ *or if* $t \geq r_2 + \epsilon$. *The set Σ has the following properties:*

(a) $(\bar{\Sigma} \cap \{\hat{\lambda} = \hat{\lambda}_0\})/\Gamma$ *intersects* $B_2^\epsilon \times B_0^\epsilon \times \{0\}$ *and contains a closed connected subset, of local dimension at least one, which is either unbounded or intersects* $\wedge \times B^\Gamma \times \{0\}$ *outside* $B_2 \times B_0 \times \{0\}$.

(b) *If* $\Sigma/\Gamma = \Sigma_1 \cup \Sigma_2$, *with* Σ_i *proper bounded subsets of* Σ/Γ, *then* $\dim(\Sigma_1 \cap \Sigma_2) \geq \dim \hat{\wedge}^\Gamma$. *In particular* Σ/Γ *is connected and has local dimension at least* $\dim \hat{\wedge}^\Gamma + 1$.

The proof of this theorem follows easily from the preceding results and is given in Ize et al. (1986). The reason for singling out B^Γ, and E^H for H–equivariant maps, is the Hopf bifurcation problem where one may have stationary bifurcation. In Theorem 3.2. one looks for nonstationary solutions.

Note that if F satisfies (1), (2) and also the properties:

(5) $F_0 (\tilde{\lambda}, \hat{\lambda}_0, x_0, 0) \neq 0$ on $\bar{B}_2 \times \partial B_0$,

(6) $F_1(\tilde{\lambda}, \hat{\lambda}, x_0, x_1) \neq 0$ on $\{r_2 - \epsilon_2 \leq \|\lambda - \lambda_0\| \leq r_2 + \epsilon\} \times B_0 \times (B_1 \backslash \{0\})$ *then for* ϵ_1 *small enough, with* $2\epsilon_1 \leq r_1$, *there is an* ϵ *such that (4) is satisfied, with* r_1, *replaced by* $2\epsilon_1$. *The map* $(\hat{\lambda} - \hat{\lambda}_0, \|x_1\| - \epsilon_1, F(\lambda, x))$ *is then admissible on* $B_2 \times B_0 \times B_1$ *and on* $B_2 \times B_0 \times (B_1 \backslash \{0\})$ *and is* Γ-*deformable on these sets to* $(\hat{\lambda} - \hat{\lambda}_0, \|x_1\| - \epsilon, F_0(k, \hat{\lambda}_0, x_0, 0), F_1(\tilde{\lambda}, \hat{\lambda}_0, 0, x_1))$ *for any given* k *in* $\bar{B}_2$. *This follows from compactness arguments and simple homotopies.*

Note that hypotheses (5) and (6) are satisfied if $F(\tilde{\lambda}, \hat{\lambda}_0, x_0, x_1)$ verifies (1), (2) and

(5') $F_0(\tilde{\lambda}, \hat{\lambda}_0, x_0, 0) \neq 0$ on $B_2 \times \{0 < \|x_0\| \leq r_0'\}$ **or**

(5'') $F_0(\tilde{\lambda}, \hat{\lambda}_0, x_0, 0) \neq 0$ *for a sequence* $\epsilon_n \to 0, \|x_0\| = \epsilon_n, \lambda \in B_2$ *(i.e. there is no bifurcation of stationary solutions),* **and**

(6') $F_1(\tilde{\lambda}, \hat{\lambda}_0, x_0, x_1) = A_1(\tilde{\lambda})x_1 + k_1(\tilde{\lambda}, \hat{\lambda}_0, x_0, x_1)$ *with* $A_1(\tilde{\lambda})$ *invertible for* $\|\tilde{\lambda} - \hat{\lambda}_0\| = r_2, A_1(\tilde{\lambda})$ *continuous in the operator norm, and*

$$\|k_1(\tilde{\lambda}, \hat{\lambda}_0, x_0, x_1)\| \leq C\|x_1\|(\|\hat{\lambda} - \hat{\lambda}_0\|^\alpha + \|x_0\|^\alpha)$$

for some $\alpha > 0$, *and* x_0, x_1 *small.*

Now if $F(\lambda, x) = Ax - A(\lambda)x - g(\lambda, x)$, with $\wedge^\Gamma = \wedge, \lambda_0 = 0$, since A and $A(\lambda)$ are equivariant, then A and $A(\lambda)$ will decompose diagonally from $B^\Gamma \times \hat{B}$ into $E^\Gamma \times \hat{E}$, giving operators $A_0 - A_0(\lambda)$ and $A_1 - A_1(\lambda)$ which are Fredholm of index 0 if A is Fredholm of index 0 and λ is small, with bifurcating matrices $B_i(\lambda)$.

Proposition 3.2. (a) *If* F *satisfies* (1), (2), (5') *or* (5''), (6') *and* A_1 *is a Fredholm operator of index* $0, A_1(0) = 0$ *and* $A_1(\lambda)$ *is compact, then on*

$B_2 \times B_0 \times B_1$, *or on* $B_2 \times B_0 \times (B_1\backslash\{0\})$, *the map* $(\hat{\lambda}, \|x_1\| - \epsilon_1, F(\lambda, x))$ *is* Γ*-deformable to* $(\hat{\lambda}, \| y_1\| - \epsilon_1, F_0(k, 0, x_0, 0), B_1(\hat{\lambda}, 0)y_1, A_1z_1)$, *where* $x_1 = y_1 \oplus z_1, y_1$ *in* ker A_1.

(b) *If, in addition,* $F_0(\lambda, x_0, 0) = A_0(\lambda)x_0 - k_0(\lambda, x_0, 0)$ *is such that* $k_0(\lambda, x_0, 0) = o(\| x_0 \|)$ *and* $A_0(\hat{\lambda}_0)$ *is invertible for some* $\hat{\lambda}_0$, *then for* r_0 *small enough,* $(\hat{\lambda}, \|x_1\| - \epsilon_1, F(\lambda, x_0, x_1))$ *is* Γ*-epi on* $B_2 \times B_0 \times (B_1\backslash\{0\})$ *if and only if* $(\hat{\lambda}, \|y_1\| - \epsilon_1, x_0, B_1(\hat{\lambda}, 0)y_1, z_1)$ *is* Γ*-epi.*

Proof. See Ize et al. (1986) Properties 4.3 and 4.4.

Now, it may be difficult to prove that there is no bifurcation of stationary solutions. One may guarantee the existence of global branches, even if one is not sure of the nature of these solutions, via the following result:

(4') *If* $F(\tilde{\lambda}, \hat{\lambda}_0, x_0, x_1) = 0$ *for* $r_2 - \epsilon_2 \leq \|\lambda - \lambda_0\| \leq r_2 + \epsilon_2, (x_0, x_1) \in B_0 \times B_1$, *then* $(x_0, x_1) = 0$.

Theorem 3.3. *Assume* F *satisfies* (1), (2), (3) *and* (4'). *Let* S *be the set of solutions with* $x \neq 0$. *If* $(\hat{\lambda} - \hat{\lambda}_0, \|x_0\| + \|x_1\| - \epsilon_1, F(\lambda, x))$ *is* Γ*-epi on* $B_2 \times (B_0 \times B_1\backslash\{0, 0\})$, *then* S *has a minimal closed (in* S*) invariant subset* Σ *such that the map* $(\hat{\lambda}^{\Gamma} - \hat{\lambda}_0, \|x_0\| + \|x_1\| -\phi(\|\lambda - \lambda_0\|))$ *is* Γ*-epi (invariant) on* Σ, *where* ϕ *is a nonincreasing function, with* $\phi(x) = \epsilon_1$ *for* $0 \leq r \leq r_0, \phi(r) = 0$ *for* $r \geq r_0 + \epsilon$. *Moreover* Σ *has the properties* (a) *and* (b) *of Theorem 3.2.*

Note that (see Lemma 4.2. in the above reference), if F satisfies (1), (2), (3), (6') and

(7) $F_0(\tilde{\lambda}, \hat{\lambda}_0, x_0, 0) \neq 0$ *on* $(B_0\backslash\{0\})$ *and for* $r_2 - \epsilon_2 \leq \|\tilde{\lambda} - \tilde{\lambda}_0\| \leq r_2 + \epsilon_2$,
(8) $F_0(\tilde{\lambda}, \hat{\lambda}_0, r_0, 0)$ *is not 0-epi on* $\{\|x_0\| = \epsilon_1\}\&B_2 \times (B_0\backslash\{0\})$ *for all* ϵ_1, *with* $0 < \epsilon_1 < \bar{\epsilon}_1 < min(r_0, r_1)$, *for some* $\bar{\epsilon}_1$,

then $(\hat{\lambda} - \hat{\lambda}_0, \|x_0\| + \| x_1\| - \epsilon_1, F(\lambda, x))$ is admissible on $B_2 \times (B_0 \times B_1\backslash\{0, 0\})$ and is Γ-deformable to

$$(\hat{\lambda} - \hat{\lambda}_0, \|x_1\| - \epsilon_1, F_0(k, \hat{\lambda}_0, x_0, 0), B_1(\tilde{\lambda}, 0)y_1, A_1z_1)$$

for any k with $\|k - \tilde{\lambda}_0\| = r_2$. In particular, if $F_0(\tilde{\lambda}, \hat{\lambda}_0, x_0, 0)$ has the form given in Proposition 3.2. (b), then the first map is Γ-epi if and only if $(\hat{\lambda}, \|x_1\| - \epsilon_1, x_0, B_1(\tilde{\lambda}, 0)y_1, z_1)$ is Γ-epi and Theorem 3.3 may be applied.

It is clear that all these results point to a Γ-equivariant J-homomorphism. This will be made explicit below.

Γ-Degree

As in the nonequivariant case, a map will be Γ−epi on U if one is able to define a Γ−degree, i.e. for equivariant maps and invariant sets. Let then B and E be two Γ−spaces (with actions which are isometries). Let $\Omega \subset B$ be a bounded open invariant subset of B and $f : \bar{\Omega} \to E$ be an equivariant map, $f(\gamma u) = \tilde{\gamma} f(u)$, which is not 0 on $\partial\Omega$. As in Section III, $B = \mathbb{R}^M \times G, E = \mathbb{R}^N \times G$ and $f(u) = f(x_1, x_2) = (f_1(x_1, x_2), x_2 - f_2(x_1, x_2))$, with f_2 compact, $\mathbb{R}^M, \mathbb{R}^N$ and G are Γ−spaces.

In order to define an equivariant degree for f, with respect to Ω, $\deg_\Gamma(f; \Omega)$, one could try to use generic methods, as for the Leray-Schauder degree. This would be done in the following steps

1) Approximate f by $\tilde{f}$, also equivariant, such that 0 is a "regular equivariant value", by which I mean that $\tilde{f}^{-1}(0)$ is a finite number of isolated orbits with linearizations which are invertible when restricted to Poincaré sections. To ask that 0 is a regular value for $\tilde{f}$ (in the usual sense) is too strong. Even so, for $B \neq E$ (with the same action), the existence of such $\tilde{f}$ is unclear. Geba et al. (1992) have indicated that $\tilde{f}$ could be found, in the case where $B = \mathbb{R}^k \times E$, with trivial action of Γ on $\mathbb{R}^k$, such that f is Γ−homotopic on $\partial\Omega$ to $\tilde{f}$ and $\tilde{f}$ is "normal", in the sense that the set of zeros of $\tilde{f}$ with a given type (H) is away from the set of zeros with another type (K), provided $\dim W(H) = \dim W(K) = k$, and for which $\tilde{f}$ restricted to $\Omega_{(H)} = \{x \in \Omega$ with Γ_x conjugate to $H\}$ has the above property. This approximation was used by Dylawerski et al. (1991) for the construction of a S^1-degree when $k = 1$, and it is proved for the general case $\mathbb{R}^M \times G \to \mathbb{R}^N \times G$ and an abelian Γ in Ize–Vignoli (1993).

2) For such a generic, or normal, $\tilde{f}$, assuming it exists, one could define local equivariant indices, from the index of an element of a finite orbit, or from the index of a Poincaré section in case of an orbit of positive dimension. There are several difficulties: (a) the definition in case of a non-free action and, in particular, when the orbit space is not orientable, if Γ is not abelian, (b) the comparison between Poincaré indices for orbits which are far away: the Poincaré section is a local construction, hence, if one tries to continue these sections along a path joining two orbits (or even within the orbit), there is the possibility inverting the orientation of the section and get to an opposite index. In this case one could define a degree by considering only the parity (i.e. a mod 2-degree).

(c) How does one compare indices for orbits of different types? One may define the degree by considering the direct sum over the orbit types, but this may not be stable under perturbations, in particular for orbits of small dimension, i.e. with $\dim W(H) < k$.

(d) Will this degree be complete, in the sense that if f and g have the same degree on a ball, is it true that f and g are Γ−homotopic on the boundary of the ball?

The construction below will avoid all these difficulties. However, due to the "theorem on conservation of difficulty", one has to pay a price, and this will be the case in the computation of the degree.

The construction is parallel to the one given in Section II, Subsection 4: since Ω is bounded, then $\Omega \subset B_R$, a large ball with center at the origin, hence invariant. Let $\tilde{f} : B_R \to E$ be an equivariant extension with the usual compactness properties. Let N be a bounded, open, invariant, neighborhood of $\partial\Omega$ such that $\tilde{f}(u) \neq 0$ on $\bar{N}$. Let $\phi : B_R \to [0,1]$ be an invariant Urisohn function with $\phi(u) = 0$ in $\bar{\Omega}$ and $\phi(u) = 1$ outside $\Omega \cup N$. Let

$$F(t,u) = (2t + 2\phi(u) - 1, \tilde{f}(u)).$$

Then $\tilde{F}$ is an equivariant function such that $\tilde{F}(t,u) = 0$ only if $t = 1/2, u \in \Omega$ and $f(u) = 0$. Hence $\tilde{F} : S^B = \partial(I \times B_R) \to \mathbb{R} \times E\backslash\{0\} \cong S^E$, defines an element in $\Pi^{\Gamma}_{SB}(S^E)$, the homotopy group of equivariant maps between these two spheres. If $\dim B^{\Gamma} > 0$, then $\Pi^{\Gamma}_{SB}(S^E)$ in an abelian group (see Ize et al. (1989), Prop. A. 4).

Definition 4.1. *The Γ−equivariant degree of f with respect to Ω,* $\deg_{\Gamma}(f;\Omega)$, *is defined as* $[\tilde{F}]_{\Gamma}$ *in* $\Pi^{\Gamma}_{S_B}(S^E)$.

As before, it is easy to prove that this degree is independent of the construction and has the following properties.

(P1): Existence. *If $\deg_{\Gamma}(f;\Omega)$ is nontrivial, then there is a u in Ω, such that $f(u) = 0$.*

(P2): Homotopy invariance. *If $f(\tau,u) : [0,1] \times \bar{\Omega} \to E$ is Γ−equivariant and nonzero on $[0,1] \times \partial\Omega$, the $\deg_{\Gamma}(f(\tau,\cdot);\Omega)$ is constant.*

(P3): Excision. *If $f : \bar{\Omega} \to E$ is nonzero on $\bar{\Omega}\backslash\Omega_0, \Omega_0$ open and invariant subset of Ω, then $\deg_{\Gamma}(f;\Gamma) = \deg_{\Gamma}(f;\Omega_0)$.*

(P4): Suspension. *If $\tilde{f}$, the extension of f to B_R, is nonzero on $B_R\backslash\Omega$, then $\deg_{\Gamma}(f;\Omega) = \deg_{\Gamma}(\tilde{f};\Omega) = \Sigma[\tilde{f}]_{\Gamma}$, where Σ is the suspension by $2t - 1$.*

(P5): Additivity (up to one suspension). *If $\Omega = \Omega_1 \cup \Omega_2, \Omega_i$ open, invariant, such that $\Omega_1 \cap \Omega_2 = \phi$, then $\Sigma \deg_{\Gamma}(f;\Omega) = \Sigma \deg_{\Gamma}(f;\Omega_1) + \Sigma \deg_{\Gamma}(f;\Omega_2)$, where Σ is the suspension. (No suspension needed if Ω_2 is a ball).*

(P6): Hopf property. *If $\Omega = B_R$ and $\deg_{\Gamma}(f_1;\Omega) = \deg_{\Gamma}(f_2;\Omega)$ and Σ is one-to-one, then $f_1|_{\partial B}$ is Γ−homotopic to $f_2|_{\partial B}$.*

(P7): Universality property. *If* $\Delta_\Gamma(f;\Omega)$ *is any other degree with properties (P1)–(P3) and* Σ *is one to one, then, if* $\Delta_\Gamma(f;\Omega)$ *is nontrivial, this is also the case for* $\deg_\Gamma(f;\Omega)$.

Equivariant suspension. If V is a Γ–representation, then the <u>equivariant</u> suspension, Σ^V, by V of $f(u)$ is the map $(f(u),v)$.

Assume that G is written as $G_0 \times G_1$, where G_0 is a finite dimensional representation (so that G_1 is a Γ–space). Then, from the compactness properties of f_2, the map f is Γ–approximated by Γ–maps with finite range (in such a $\mathbb{R}^N \times G_0$). As in the construction of the Leray-Schauder degree, one needs to show that $\deg_\Gamma (f_0;\Omega \cap \mathbb{R} \times G_0) \cong \deg_\Gamma(\bar{f}_0;\Omega \cap \mathbb{R}^N \times \bar{G}_0)$ for two such approximations, where $\bar{G}_0 = G_0 \times V$ and $\bar{f}_0$ is the suspension by V. Hence one needs that Σ^V is one to one: see Ize et al. (1989) Proposition 3.1.

This has been proved for $\Gamma = S^1, M = 1, N = 0$ in Ize et al. (1992), and for Γ abelian, $B = \mathbb{R}^k \times E$, in Ize–Vignoli (1993), Theorem 9.1:

Proposition 4.1. *If* $\dim E^H \geq k+2 - \dim \Gamma/H$ *for all isotropy subgroups H of B with* $H < H_0$, *for* H_0 *isotropy subgroups of V, then* Σ^V *is one-to-one.* Σ^V *is also onto if all* H_0's *are isotropy subgroups of B. If* H_0 *is not an isotropy subgroup of B, then* Σ^V *will not be onto unless* $k = 0$ *and* $\Gamma/H_0 \simeq S^1$.

For the nonabelian case, one has the following result, if $B = \mathbb{R}^k \times E$, (see Namboodiri 1983): (Note that this statement differs from the one given in the above reference, but the proof asks for these hypotheses).

Σ^V *is onto if: 1)* $\dim E^H \geq k + 1$ *(for one-to-one* $\dim E^H \geq k + 2$) *for any isotropy subgroup H of B;* (2) $\dim E^{K'} - \dim E^H \geq k+1$ *(for one to one* $k + 2$) *for any H, isotropy subgroup for B and* K' *isotropy subgroup of* $B \times V$, *with* $(K') < (H_0), (K') < (H), H_0$ *isotropy subgroup for V but H is not a subgroup of a conjugate of* H_0.

This last result is weaker than Proposition 4.1 if Γ is abelian, and forces us to suspend by representations already contained in E_0, which might not be the case in applications. Hence the need to check, in each case, the suspension property, before taking the direct limit of the finite dimensional groups, although in principle, one does not need this approximation in order to define the Γ-degree

Now, if $\deg_\Gamma(f;\Omega)$ is nontrivial, then f is Γ–epi on Ω, since any compact perturbation of f, with support in Ω, coincides with f on $\partial\Omega$ and so gives the same degree. Thus, one may apply all the results of the preceding section. Furthermore, since the Γ–degree has the excision and additivity (up to one suspension) properties, one may compare local bifurcation indices. For instance, let $F(\lambda, x) = Ax - A(\lambda)x - g(\lambda, x)$, with $B = \mathbb{R}^M \times G$,

$E = \mathbb{R}^N \times G$, A of the form $Ax = (A_0x, x_2 - A_1x)$ with A_1 compact, $A(\lambda)$ compact, $A(\lambda) \to 0$ as $\lambda \to 0, g(\lambda, x) = o(\| x \|), \lambda \in \wedge = \tilde{\wedge} \times \hat{\wedge}$ is a Γ–space. Assume that the Γ–equivariant suspension theorem holds (Σ^V is one-to-one for all V) and suppose that, for some $\lambda_0 \in \wedge^\Gamma$, one has that $A - A(\tilde{\lambda}, \hat{\lambda}_0)$ is invertible for $\| \tilde{\lambda} - \tilde{\lambda}_0 \| \le \rho, \tilde{\lambda} \ne \tilde{\lambda}_0$. Let $\Sigma J^\Gamma [A - A(\tilde{\lambda}, \hat{\lambda}_0)] \equiv \deg_\Gamma(\| x \| - \epsilon, Ax - A(\tilde{\lambda}, \hat{\lambda}_0)x; B_\rho \times B_{2\epsilon})$, where $B_\rho = \{\tilde{\lambda} : \| \tilde{\lambda} - \tilde{\lambda}_0 \| < \rho\}, B_{2\epsilon} = \{x : \| x \| < 2\epsilon\}$.

As seen in Section II, Remark 1.5, if $\dim \tilde{\wedge} = 1$, then $J^\Gamma[A - A(\tilde{\lambda}, \hat{\lambda}_0)] = \Sigma[\deg_\Gamma(Ax - A(\tilde{\lambda} - \rho, \hat{\lambda}_0)x; B_{2\epsilon}) - \deg_\Gamma(Ax - A(\tilde{\lambda} + \rho, \hat{\lambda}_0)x; B_{2\epsilon})]$, since the change from $\tilde{\lambda}$ into $-\tilde{\lambda}$ gives the inverse in $\Pi^\Gamma_{S^B \times I}(S^{E \times I})$, as seen in Section II.

Theorem 4.1. *Under the above hypotheses, if $J^\Gamma[A - A(\tilde{\lambda}, \hat{\lambda}_0)]$ is nontrivial, then there is a branch Σ of solutions bifurcating from $(0, \lambda_0)$ with the following properties:*

(1) *$\bar{\Sigma} \cap \{\hat{\lambda} = \hat{\lambda}_0\}/\Gamma$ is either unbounded or returns to $x = 0$ outside B_ρ.*

(2) *The local dimension of Σ/Γ is at least $\dim \hat{\Lambda}^\Gamma + 1$.*

(3) *If the return points on Σ have a linearization, with the same invertibility properties assumed for λ_0, and Σ is bounded, then the sums of the local indices $J^\Gamma[A - A(\lambda)]_{\lambda_i}$ is trivial.*

(4) *If for some $H < \Gamma, F(\lambda, x)^H = 0$ on $B_\rho^H \times \{\hat{\lambda}_0\} \times B_{2\epsilon}^H$ implies $x^H = 0$, i.e. there is no local bifurcation in $(\wedge \times B)^H$, then (1), (2), (3) hold for a subset of Σ in the complement of B^H and the return points, if there are any, are in B^H.*

Proof. It is enough to note that $\deg_\Gamma(\tilde{\lambda} - \hat{\lambda}_0, \| x \| - \epsilon, F(\lambda, x); B_\rho \times \hat{\wedge} \times B_{2\epsilon})$ is nontrivial (since $\Sigma^{\hat{\wedge}}$ is one to one). Then one uses Theorem 3.3. The proof of (3) is as in Section II and (4) follows by complementing by $\| x_1 \| - \epsilon$ instead of $\| x \| - \epsilon$, where x_1 is in the complement of B^H and using Theorem 3.2.

Remark 4.1. As in Ize et al. (1992) section 4.5, one may replace the trivial solution $x = 0$ by any closed invariant set S_0 of solutions, such that (a) $(\lambda_0, x_0) \in S_0$, (b) if $F(\lambda, x) = 0$ for $\rho \le \| \lambda - \lambda_0 \| \le \rho + \eta, \hat{\lambda} = \hat{\lambda}_0$ and $\text{dist}(\lambda, x; S_0) \le 2\epsilon < \rho$, then $(\lambda, x) \in S_0$, c) if $\rho \le \text{dist}(x, \Gamma x_0) \le \rho + \eta$ and $\| \lambda - \lambda_0 \| \le \rho$ then $\text{dist}(\lambda, x; S_0) \ge 2\epsilon$. Then if $U_\epsilon = \{(\lambda, x) : \text{dist}(\lambda, x; S_0) < 2\epsilon, \| \lambda - \lambda_0 \| < \rho, \text{dist}(x, \Gamma x_0) < \rho\}, \tilde{U}_\epsilon = U_\epsilon \cap \{\hat{\lambda} = \hat{\lambda}_0\}, \tilde{S}_0 = S_0 \cap \{\hat{\lambda} = \hat{\lambda}_0\}$, let $\Psi_\epsilon(\lambda, x) = (\hat{\lambda} - \hat{\lambda}_0, \text{dist}(\lambda, x; S_0) - \epsilon, F(\lambda, x))$ and $\tilde{\Psi}_\epsilon(\tilde{\lambda}, x)$ the corresponding map with λ replaced by $\tilde{\lambda}$ (hence the term $\hat{\lambda} - \hat{\lambda}_0$ disappears). Then $\deg_\Gamma(\Psi_\epsilon; U_\epsilon)$ is well defined and $\deg_\Gamma(\Psi_\epsilon; U_\epsilon) = \Sigma^{\hat{\wedge}} \deg_\Gamma(\tilde{\Psi}_\epsilon; \tilde{U}_\epsilon)$.

Corollary 4.1. *If* $\deg_\Gamma(\Psi_\epsilon; U_\epsilon) \neq 0$, *let* C *be the connected component of* (λ_0, x_0) *in the closure of* $(S \backslash S_0) \cup (U_\epsilon \cap S_0)$ *Then* C *has a connected invariant subset* Σ *such that: (1) to (4) are valid with the set* $\{x = 0\}$ *replaced by* S_0. *Furthermore* $\bar{\Sigma} \cap S_0$ *has a closed subset* Σ_0, *with* Σ_0 / Γ *connected, which intersects* $\tilde{U}_\epsilon$ *and is either unbounded or meets* S_0 *again outside* $\tilde{U}_\epsilon$ *and* Σ_0 / Γ *has dimension at each point at least* $\dim \hat{\wedge}^\Gamma$. *If* $\dim \tilde{\wedge} = 1$ *and* $S_0 = \wedge \times \{0\}$, *then* Σ_0 *disconnects* S_0^Γ *in* $U_\epsilon \cap S_0^\Gamma$ *and for any pair* λ_1, λ_2 *in the components of* $\tilde{\lambda}_0 \pm \rho$ *in* $S_0^\Gamma \backslash \Sigma_0$, *then either* $\bar{\Sigma}$ *is unbounded above the segment joining* λ_1 *to* λ_2 *or* Σ *covers* λ_1 *or* λ_2.

The proof follows the above reference, for a general Γ instead of S^1.

Remark 4.1. Before studying the Γ−equivariant J-homomorphism, it might be interesting to state some results on $\Pi_{SB}^\Gamma(S^E)$, where, given the equivariant suspension theorem, I shall assume B and E finite dimensional.

Proposition 4.2. *If* $\dim B^H < \dim E^H$, *for all isotropy subgroups H for* B, *then* $\Pi_{SB}^\Gamma(S^E) = 0$.

If Γ is a finite group, this is proved in Adams (1982), Prop. 2.4. For a general Γ one may modify the proof given in Kosniowski (1974), Prop. 2.12, using equivariant obstruction theory. A direct proof is given, for S^1−maps in Ize et al. (1992) and, for abelian groups, in Ize–Vignoli (1993).

Proposition 4.3. *(T. tom Dieck, 1979, Theorem 8.4.1) If* $\dim B^H = \dim E^H$ *for all H and (1)* $\dim B^k \geq \dim B^H + 2$ *for any* K, H, *isotropy subgroups for B, with* $K < H$, *then* $\Pi_{SB}^\Gamma(S^E)$ *is nontrivial and is characterized by the set of all degrees of* $f^H : (S^B)^H \to (S^E)^H$ *with* $W(H)$ *finite. In fact, if* $K < H$ *are isotropy subgroups for B, then* $\deg(f^K)$ *is determined by* $\deg(f^H)$, *modulo* $|W(H)|$, *in the sense that fixing these* $\deg(f^H)$ *then all possible* $\deg(f^K)$, *for* $K < H$, *fill the whole residue class modulo* $|W(H)|$.

Under the above hypothesis and $B = E$, Komiya (1988) gives an explicit matricial formula for these congruences. If $B = E$ but (1) is not necessarily true, Hauschild (1977) and Rubinstein (1976), corrected by Dancer (1984), give a complete characterization of $\Pi_{SB}^\Gamma(S^E)$ in terms of the degrees of mappings associated to isotropy subgroups H, with $|W(H)| < \infty$ (of course if (1) holds, these mappings are given by f^H). In Ize–Vignoli (1993), one has the following

Proposition 4.4. *If* $\dim B^H = \dim E^H$, *for all H and Γ is abelian, then*

$$\Pi_{SB}^\Gamma(S^E) \cong \mathbb{Z} \times \mathbb{Z} \times ... \times \mathbb{Z}$$

with one copy of $\mathbb{Z}$ for each H such that $|\Gamma/H| < \infty$. Furthermore $[f]_\Gamma = (d_1\alpha_1, \ldots, d_r\alpha_r)$ on this product, with $\alpha_j > 0$ and $\alpha_j = 1$ if $B = E$. There are fixed integers $\beta_{ij} \neq 0, \beta_{ij} = 1$ if $B = E$, and $\epsilon_{ij} = 1$, if $H_i < H_j$, and 0 otherwise, such that, for all i :

$$\deg(f^{H_i}) = \Sigma_j \epsilon_{ij}\beta_{ij}\alpha_j d_j |\Gamma/H_j|.$$

By ordering the H_i's is decreasing order, these relations can be written in a lower triangular matricial way: $(\deg f^\Gamma, \ldots, \deg f^{H_r})^T = A(\alpha_1 d_1, \ldots, \alpha_r d_r)^T$ where $A_{ij} = \epsilon_{ij}\beta_{ij}|\Gamma/H_j|$ and the diagonal is $|\Gamma/H_i|$.

This says that, from the topological point of view, one could have restricted oneself to the study of f^H with the above congruences in order to study any equivariant map.

Note that in the case of S^1, H is S^1 or $\mathbb{Z}_k$. Hence $W(H)$ is S^1 for H a strict subgroup of S^1, that is $W(H)$ is finite only for $H = S^1$. Then, $\Pi^{S^1}_{SB}(S^E) \cong \mathbb{Z}$ and $[f]_\Gamma$ is given by $\deg(f^{S^1})$, the invariant part of f and $\deg_{S^1}(f;\Omega) = \deg(f^{S^1};\Omega^{S^1})$. In this case one may also compute $\deg(f;\Omega) = k \deg(f^{S^1};\Omega^{S^1})$, where k is an integer, independent of f, which depends only on the actions on B and $E, k = 1$ if $B = E$. See Ize et al. (1989) for a proof.

As a consequence of these results one has Borsuk-Ulam type theorems and a tie with equivariant category theory.

The next case is when one has one parameter and $\Gamma = S^1$, for example, for periodic solutions of autonomous differential equations where the parameter is the frequency, for periodic solutions with a fixed period and a first integral (the Lagrange multiplier is the parameter) or for critical points of S^1-invariant functionals (with a multiplier).

In this case $B = \mathbb{R} \times \mathbb{R}^l \times \mathbb{C}_1 \times \ldots \times \mathbb{C}_d$ with action on $\mathbb{C}_j$ given by $e^{im_j\phi}$ and $E = \mathbb{R}^l \times \mathbb{C}_1 \times \ldots \times \mathbb{C}_d$, with action on $\mathbb{C}_j$ given by $e^{in_j\phi}$. If $n_j = k_j m_j$ then $\dim B^H \leq \dim E^H + 1$ for all H's.

Propositon 4.5. *Under the above hypotheses, $\Pi^{S^1}_{SB}(S^E) \cong \Pi_{l+1}(S^l) \times \mathbb{Z} \times \ldots \times \mathbb{Z}$ with one $\mathbb{Z}$ for each isotropy subgroup H such that $\dim B^H = \dim E^H + 1$ and $\dim \Gamma/H = 1$. One has $[f]_{S^1} = (d_0, d_{H_1}, d_{H_2}, \ldots)$, where d_0 is the class of f^{S^1}, in $\mathbb{Z}_2$ if $l \geq 3$. Furthermore, given such a sequence and an open invariant set Ω, with $\Omega^H \neq \phi$ for any H with $d_H \neq 0$, then there is an equivariant function f such that $\deg_{S^1}(f;\Omega) = (d_0, d_{H_1}, \ldots)$.*

In infinite dimensions and f of the usual form, all but a finite number of the d_H are 0. If $k_j = 1$, Dylawerski et al. (1991) have defined a S^1−degree, using the approximation by normal maps, which coincides with the one defined above. Note that if one takes the space $\mathbb{C}^d$ with the standard action $e^{i\phi}$, one may define an equivariant map $\Theta : \mathbb{R}^{l+1} \times$

$\mathbb{C}^d \to B$, by $(\lambda, X_0, Z_1, \ldots, Z_d) \to (\lambda, X_0, Z_1^{m_1}, \ldots, Z_d^{m_d})$ and for a function $f(\lambda, X_0, z_1, \ldots, z_d) : B \to E$, one may define $\tilde{f}(\lambda, X_0, Z_1, \ldots, Z_d) : \mathbb{R}^{l+1} \times \mathbb{C}^d \to E$ by $\tilde{f}(\lambda, X_0, Z_1, \ldots, Z_d) = f(\lambda, X_0, Z_1^{m_1}, \ldots, Z_d^{m_d})$. Now $\mathbb{R}^{l+1} \times \mathbb{C}^d$ has only two isotropy subgroups: S^1 and $\{e\}$, thus, $\Pi^{S^1}_{S^{\mathbb{R}^{l+1}} \times \mathbb{C}^d}(S^E) \cong \mathbb{Z}_2 \times \mathbb{Z}$, $[\tilde{f}]_{S^1} = (\tilde{d}_0, \tilde{d})$. If $[f]_{S^1} = (d_0, d_{H_1}, \ldots)$, then, one has

$$\tilde{d}_0 = d_0, \tilde{d} = (\Sigma d_{H_j}/|H_j|)\Pi_j m_j = d_F \Pi_j$$

where the rational d_F is the Fuller degree of f. See the above reference, Theorem 3.4. and Ize et al. (1989).

The above result can be generalized to the case of an abelian group. In order to lighten the notation, I will take the case where $B = \mathbb{R}^k \times \tilde{B}$, such that dim $\tilde{B}^H = $ dim E^H for all isotropy subgroups of B. Some of results of Ize–Vignoli (1993) are the following:

Theorem 4.2. a) $\Pi^\Gamma_{S^B}(S^E) \cong A \times \mathbb{Z} \times \ldots \times \mathbb{Z}$, where A is a factor which depends on the isotropy subgroups H with dim $\Gamma/H < k$ and there is one $\mathbb{Z}$ for each H such that $\dim|\Gamma/H = k$. In particular if dim $\Gamma/H > k$, then H doesn't contribute to $[f]_\Gamma$. Furthermore if Ω is open invariant in B, then there is a map f such that $\deg_\Gamma(f; \Omega)$ has any given sequence of integers for the H's such that dim $\Gamma/H = k$ and $\Omega^H \neq \phi$. The group A has a specific decomposition in terms of decreasing dim Γ/H and the contributions to the $\mathbb{Z}$'s are given, in a matrix form, in terms of the $\Gamma-$degrees of the f^H's and there are explicit generators for these $\mathbb{Z}$'s.

b) If $k = 1$ and if, for any H such that dim $\Gamma/H = 0$ one has that dim $\tilde{B}^H - $ dim $\tilde{B}^K \geq 3$ if $\Gamma/H \simeq \mathbb{Z}_2$ or $\{e\}$, and at least 4 if $\Gamma/H \simeq \mathbb{Z}_p, p \geq 3$, for any $K > H$, then $A \simeq \Pi A_i$, with $A_i \cong \mathbb{Z}_{q_0} \times \ldots \times \mathbb{Z}_{q_m}$, if $\Gamma/H_i \cong \mathbb{Z}_{p_1} \times \ldots \times \mathbb{Z}_{p_m}$ with $q_0 = lcd(2, p_1, \ldots, p_m)$, $q_m = lcm(2, p_1, \ldots, p_m)$, $q_j = h_j/h_{j-1}$ where h_j is the largest common factor of all possible products of j of the p_i's. In particular if any two p_i, p_j are relatively prime and odd, then $A_i \cong \mathbb{Z}_{2|\Gamma/H_i|}$. In this case also one has specific generators for A_i and the same result for $\deg_\Gamma(f; \Omega)$.

For general Γ, the results of Geba et al. (1992) seem to indicate that $\Pi^\Gamma_{S^{B \times \mathbb{R}^k}}(S^B)$ has a factor of the form $\mathbb{Z} \times \ldots \times \mathbb{Z}$, with one $\mathbb{Z}$ for each H such that dim $W(H) = k$ and $W(H)$ has an orientation which is invariant under all left and right translations. It seems reasonable (from obstruction theory) to expect that if dim $W(H) > k$, then H will not contribute.

Note that most of the results, given in the preceding sections, which depend only on a degree being nonzero, can be extended without difficulties for the general Γ-degree.

V. THE EQUIVARIANT J-HOMOMORPHISM AND
SUFFICIENT CONDITIONS

As seen above, the conditions $\Sigma J^\Gamma[A - A(\lambda)] \neq 0$ and $\Sigma J^\Gamma[B(\lambda)] \neq 0$ are sufficient for equivariant global and local bifurcation. Here A is a Fredholm operator of index 0, equivariant with respect to $\Gamma, A(\lambda) = 0(\|\lambda\|)$, and has the right compactness properties. $A - A(\lambda)$ is invertible for $\lambda \neq 0$, λ small, and $B(\lambda)$ is the equivariant linear part of the bifurcation equation. Thus, $J^\Gamma[A - A(\lambda)]$ is the suspension of $J^\Gamma[B(\lambda)]$. Note that, since one is working on a sphere, one may consider directly $JB(\lambda)$ and not its suspension $\deg_\Gamma(\|x\| - \epsilon, B(\lambda)x)$.

For simplicity $\wedge \cong \mathbb{R}^k$ is invariant under Γ and, on $V = \ker A, B(\lambda)$ has the block diagonal form, $\mathrm{diag}(B_0, B_1^{\mathbb{R}}, \ldots, B_1^{\mathbb{R}}, \ldots)$, with B_0 on V^Γ. The basic set is $B_\rho \times B_{2\epsilon} = \{(\lambda, x) : \|\lambda\| < \rho, \|x\| < 2\epsilon\}$, with the maximum norm on V. The J-homomosphism is then the map: $\Pi_{k-1}(GL_\Gamma(V)) \to \Pi_{S^\wedge \times V}^\Gamma(S^V)$, given by

$$J^\Gamma[B(\lambda)] = [\|x\| - \epsilon, B(\lambda)x]$$

where $B(\lambda)$ is extended from S^{k-1} to B_ρ by $\|\lambda\|B(\lambda\rho/\|\lambda\|)$. Now $B(\lambda)$ can be written as the product of $(I, I, \ldots B_j(\lambda), \ldots, B_j(\lambda), I, \ldots)$ with $(B_0, B_1, \ldots I, \ldots, I, B_{j+1}, \ldots)$ and, since J^Γ is a morphism, one has

$$J^\Gamma(B(\lambda)) = \Sigma \Sigma^\Gamma J^\Gamma(B_j(\lambda), \ldots, B_j(\lambda)),$$

where Σ^Γ is the suspension by the other Γ-representations.

Remember that if $k = 1$, only the real representations are present and that $\mathrm{diag}(B_j^{\mathbb{R}}(\lambda), \ldots, B_j^{\mathbb{R}}(\lambda))$ is Γ-deformable to

$$\mathrm{diag}\left(\begin{pmatrix} a_j(\lambda) & 0 \\ 0 & I \end{pmatrix}, \ldots, \begin{pmatrix} a_j(\lambda) & 0 \\ 0 & I \end{pmatrix} \right),$$

where $a_j(\lambda) = \det B_j^{\mathbb{R}}(\lambda)$. By returning to $V_j^{\mathbb{R}}$ as absolutely irreducible representations with the original basis, one obtains $\Sigma^\Gamma a_j(\lambda)I$.

If $\lambda \epsilon \mathbb{R}^2$, then, besides the real representations, one has to consider the complex representations. There, $diag\, (B_j^{\mathbb{C}}(\lambda), \ldots, B_j^{\mathbb{C}}(\lambda))$ is Γ-deformable to

$$\mathrm{diag}\left(\begin{pmatrix} a_j(\lambda) & 0 \\ 0 & I \end{pmatrix}, \ldots, \begin{pmatrix} a_j(\lambda) & 0 \\ 0 & I \end{pmatrix} \right),$$

where $a_j(\lambda)$ is the complex determinant of $B_j^{\mathbb{C}}(\lambda)$. Then one obtains $\Sigma^\Gamma a_j(\lambda)I$.

There are few results on the equivariant Γ-homomorphism, and many of them are unreliable in the sense that they either assume tacitly a number of hypothesis or they are wrong. One of the problems is that it is not clear at all that the suspension is one-to-one. Another is the fact that the action is not free and so on... Thus, I shall present here only preliminary results, leaving for a further study a more complete analysis of the equivariant bifurcation.

Since $J^\Gamma[B(\lambda)] \neq 0$ implies local bifurcation, one should study under which circumstances $J^\Gamma[B(\lambda)] = 0$. It is then clear, that if this is the case, one has that $J^{\Gamma_0}[B(\lambda)] = 0$ for any $\Gamma_0 < \Gamma$. Furthermore if $H < \Gamma$, let $B^H = B|_{V^H}$. Since V^H is $N(H)$-invariant one has the following commutative diagram

$$
\begin{array}{ccc}
\Pi_{k-1}(GL_\Gamma(V)) & \xrightarrow{J^\Gamma} & \Pi^\Gamma_{S^{\wedge}\times V}(S^V) \\
\downarrow P_* & & \downarrow P_* \\
\Pi_{k-1}(GL_{N(H)}(V)) & \xrightarrow{J^{N(H)}} & \Pi^{N(H)}_{S^{\wedge}\times V^H}(S^{V^H})
\end{array}
$$

In particular, if $J^\Gamma[B(\lambda)] = 0$, then $J^{N(H)}[B(\lambda)^H] = 0$. Thus, $J^\Gamma[B_0(\lambda)] = 0$ for the invariant part, studied in Section II. Next, let H_0 be a maximal isotropy subgroup for $(V^\Gamma)^\perp$, which is a Γ-space. From the maximality of H_0, all points in $(V^\Gamma)^\perp \cap V^{H_0}$ have $\Gamma_x = H_0$, $W(H_0)$ acts freely on that space and $B(\lambda)$ has a diagonal form (see the comment in II) $B(\lambda) = \mathrm{diag}(B_0(\lambda), B^{H_0}(\lambda)^\perp)$. Since $J^{N(H_0)}[B(\lambda)^{H_0}] = \Sigma^{N(H_0)}J^{N(H_0)}[B_0(\lambda)] + \Sigma^{N(H_0)}J^{N(H_0)}[B^{H_0}(\lambda)^\perp]$ and the first term is 0 $(N(H_0) < \Gamma)$, one has that $\Sigma^{N(H_0)}J^{N(H_0)}[B^{H_0}(\lambda)^\perp] = 0$.

Note that, for the original decomposition of V, in irreducible V_j's, if $V_j \cap V^{H_0} \neq \{0\}$, one may choose the first elements of the basis of V_j as being in V^{H_0}. Then $B^{H_0}(\lambda)^\perp|_{V_j}$ will have the form $\mathrm{diag}(B_j(\lambda), \ldots, B_j(\lambda))$, where $B_j(\lambda)$ is repeated $\dim_K V_j \cap V^{H_0}$. Note that $V_j \cap V^{H_0}$ and $V_l \cap V^{H_0}$ may have equivalent $N(H_0)$-subrepresentations. If this is the case, and B_j and B_l have the same dimension, then the deformation $\begin{pmatrix} tB_j & -(1-t)B_jB_l \\ (1-t)I & tB_l \end{pmatrix}$, followed by a rotation and mimicked on all couples, is a valid $N(H)$-homotopy, including the case where $j = l$ and V_j has several equivalent $N(H)$-representations.

Next $(V^\Gamma)^\perp \cap (V^{H_0})^\perp$ is a $N(H_0)$-space. Let $H_1 < N(H_0)$, be a maximal isotropy subgroup. For any x in $V^{H_1} \cap (V^\Gamma)^\perp \cap (V^{H_0})^\perp$, one has $N(H_0)_x = \Gamma_x \cap N(H_0) = H_1$. Furthermore if $x \in V^{H_1} \cap V^{H_0}$, then either H_1 is not a subgroup of H_0 and x is fixed by H_0 and H_1, which, by the maximality of H_0, implies that $\Gamma_x = \Gamma$, or $H_1 < H_0$. Hence, on V^{H_1}, $B(\lambda)$ has the form $\mathrm{diag}(B_0(\lambda), B^{H_0}(\lambda)^\perp, B^{H_1}(\lambda)^\perp)$, with

$$
\Sigma^{N(H_0) \cap N(H_1)}J^{N(H_0)\cap N(H_1)}[B^{H_1}(\lambda)^\perp] = 0,
$$

(the term $B^{H_0}(\lambda)^{\perp}$ is not present if H_1 is not a subgroup of H_0). $N(H_0) \cap N(H_1)/H_1$ acts freely on $(V^{H_0})^{\perp} \cap V^{H_1}$.

Next $(V^{H_0})^{\perp} \cap (V^{H_1})^{\perp}$ is a $N(H_0) \cap N(H_1)$–space. Let H_2 be a maximal isotropy subgroup in $N(H_0) \cap N(H_1)$. Again for any x in $(V^{H_0})^{\perp} \cap (V^{H_1})^{\perp} \cap V^{H_2}, \Gamma_x \cap N(H_0) \cap N(H_1) = H_2$, while if x is in $(V^{H_0})^{\perp} \cap V^{H_1} \cap V^{H_2}$, then, if H_2 is not a subgroup of H_1, the maximality of H_1 would imply that $x = 0$, and if x is in $V^{H_0} \cap V^{H_1} \cap V^{H_2}$ and either $x \in V^{\Gamma}$ or $H_2 < H_1 < H_0$. Thus, $\Sigma^{\tilde{N}(H_2)} J^{\tilde{N}(H_2)}[B^{H_2}(\lambda)^{\perp}] = 0$, for $\tilde{N}(H_2) = N(H_0) \cap N(H_1) \cap N(H_2) > H_2$ and $\tilde{N}(H_2)/H_2$ acts freely on $(V^{H_0})^{\perp} \cap (V^{H_1})^{\perp} \cap V^{H_2}$.

Proposition 5.1. *If $J^{\Gamma}[B(\lambda)] = 0$, then for any sequence $H_0, H_1, H_2, \ldots$ as above, one has $\Sigma^{\tilde{N}(H_i)} J^{\tilde{N}(H_i)}[B^{H_i}(\lambda)^{\perp}] = 0$ and, on the corresponding space, $\tilde{N}(H_i)/H_i$ acts freely.*

For example, if $k = 1$, this implies that

$$[B^{H_i}(-\epsilon)^{\perp}x]_{\tilde{N}(H_i)} = [B^{H_i}(\epsilon)^{\perp}x]_{\tilde{N}(H_i)}.$$

This is always true if $\tilde{N}(H_i)/H_i$ has positive dimension, since $\tilde{N}(H_i)/H_i$ acts freely on $V^i \equiv \cap_{j<i}(V^{H_j})^{\perp} \cap V^{H_i}$ as S^1 or $N(S^1)$ in S^3 or S^3, giving to that space a complex or quaternionic structure, with $GL_{\tilde{N}(H_i)}(V^i)$ connected. If $\tilde{N}(H_i)/H_i$ is finite and acting freely on V^i, then, according to Proposition 4.3, these two equivariant maps are $\tilde{N}(H_i)/H_i$–homotopic if and only if they have the same degrees on isotropy subspaces, here only one, corresponding to V^i. From Remark 2.1, $\mathrm{Ind}(B^{H_i}(-\epsilon)^{\perp}x) = \mathrm{Sign} \det B^{H_i}(-\epsilon)^{\perp} = 1$, (from $1 + k|\tilde{N}(H_i)/H_i|$), and there is no change except if $\tilde{N}(H_i)/H_i \simeq Z_2$, acting as $\pm I$ on V^i, and this determinant changes sign. Note that, on the orthogonal complement of V^i in V^{H_i}, the corresponding H_j's contain strictly H_i and $\tilde{N}(H_i)/H_i$ acts as I there.

Now $B(\lambda)$ is Γ–homotopic to the Γ–suspension of

$$\mathrm{diag}(a_0(\lambda)I_0, a_1(\lambda)I_1, \ldots, a_n(\lambda)I_n),$$

where $a_j(\lambda)$ is the determinant on the real irreducible Γ–representation V_j of dimension equal to dim I_j and V_j, V_l are nonequivalent. Then, if $\tilde{N}(Hi)/Hi \cong Z_2$, the restriction $B^{H_i}(\lambda)$ on V^{H_i} is Z_2–homotopic to the Z_2–suspension of the 2×2 matrix

$$\begin{pmatrix} a_0(\lambda)\Pi a_j(\lambda)^{m_j} & 0 \\ 0 & \Pi a_j(\lambda)^{n_j} \end{pmatrix},$$

where $n_j = \dim (V_j \cap V^i)$ and $m_j = \dim V_j \cap (V^i)^{\perp} \cap V^{H_i}$ and the first term is invariant while the second is odd under the action of Z_2. Note that if H_i is the first subgroup in the lattice with $\tilde{N}(H_i)/H_i \simeq \mathbb{Z}_2$, then, from the above arguments and choosing $a_0(\lambda) = \lambda^2, a_j(\lambda) = \lambda$ for one j and λ^2 for the others, the term λ^{m_j} cannot change index, hence m_j must be even. Similarly, the second term will have the possibility of changing index only if some n_j is odd. If all n_j's are even, then one has to go on until one gets another $\mathbb{Z}_2$–action with n_j odd and $\Pi a_j(\lambda)^{n_j}$ changing sign. Since, for $\mathbb{Z}_2$, the suspension is one-to-one, one has the following recipe.

Proposition 5.2. a) *If* $\det B^\Gamma(\lambda)$ *changes sign, then there is bifurcation of invariant solutions.*

   b) *If for some* H_i *such that* $\tilde{N}(H_i)/H_i \simeq \mathbb{Z}_2$ *one has that* $\det B^{H_i}(\lambda)$ *changes sign, then there is bifurcation in* V^{H_i}.

Remark 5.1. The case of gradient maps is studied, using topological methods, in Bartsch (1992).

   A second instance where Proposition 5.1. can be applied is the case of an abelian group. Then $\tilde{N}(H_i) = \Gamma$ and, from the way they are constructed, H_i are the maximal isotropy subgroups on an irreducible Γ–representations, i.e. $\Gamma/H_i \cong \{e\}$ or $\mathbb{Z}_2$ for real representations and $\Gamma/H_i \cong \mathbb{Z}_p$ or S^1, $p \geq 3$, for complex representations. Hence, if $B(\lambda) = \mathrm{diag}(B_0(\lambda), B_1(\lambda), \ldots, B_r(\lambda), B_{1,p}(\lambda), \ldots, B_{s,p}(\lambda), B_1^{\mathbb{C}}(\lambda), \ldots, B_q^{\mathbb{C}}(\lambda))$, where $B_1(\lambda), \ldots, B_r(\lambda)$ are on nonequivalent $\mathbb{Z}_2$–real representations, $B_{j,p}(\lambda)$ on nonequivalent $\mathbb{Z}_p$–complex representations and $B_j^{\mathbb{C}}(\lambda)$ on nonequivalent S^1–representations, then

Proposition 5.3. *The relation* $J^\Gamma[B(\lambda)] = 0$ *implies that* $J[B_0(\lambda)] = 0, \Sigma_0 J^\Gamma[B_j(\lambda)] = 0, \Sigma^\Gamma J^\Gamma[B_j^{\mathbb{C}}(\lambda)] = 0$ *and* $\Sigma^\Gamma \Sigma_j J^\Gamma[B_{j,p}(\lambda)] = 0$ *for each* p.

Proof. In fact, $B^{H_i}(\lambda)^\perp$ reduces to these matrices on V^i since, if $\Gamma/H_i \cong \mathbb{Z}_2$, then there is no other isotropy group H, except Γ, such that $H_i < H < \Gamma$, while, if $\Gamma/H_i \cong S^1$ acts freely on V^i, there are no other harmonics in V^i but the fundamental. Hence, on V^{H_i}, S^1 acts as $e^{in_j\phi}$ and, if $n_j \neq 1$, one has a corresponding isotropy subgroup H which contains strictly H_i and for which one had $\Sigma^\Gamma J[B^H(\lambda)] = 0$. Finally, if $\Gamma/H_i \cong \mathbb{Z}_p$ acting as a $e^{2\pi i k m_j/p}$, for some $m_j, 0 \leq m_j < p$ and k between 0 and $p-1$, on the j-th coordinate in V^{H_i}. Hence, if H_j is the corresponding isotropy subgroup, $H_i < H_j$, then $\Gamma/H_j \simeq \mathbb{Z}_{n_j}$, where n_j divides p and $m_j/p = k_j/n_j$. Thus, if $n_j < p$, then $\Sigma^\Gamma J[B^{H_j}(\lambda)] = 0$ eliminates this term and, if $n_j = p$, one has that $H_j = H_i$ and the free action of Γ/H_i on V^i implies that m_j and p are relatively prime.   Q.E.D.

   Sharper results will be given in the next subsection. For the moment it is enough to note that, if $k = 2$, then $B_j^{\mathbb{C}}(\lambda)$ is S^1–homotopic to $\begin{pmatrix} a_j(\lambda) & 0 \\ 0 & I \end{pmatrix}$, with det $B_j^{\mathbb{C}}(\lambda) = a_j(\lambda)$ and that, see Ize et al. (1992), $J^{\Gamma}[B_j^{\mathbb{C}}(\lambda)] = 0$ if and only if $a_j(\lambda)$ has a zero winding number around 0, that is, if and only if $B_j^{\mathbb{C}}(\lambda)$ is deformable to I. In this case Σ^{Γ} is one to one. Similarly, Σ_0 is one-to-one if dim $V^i \geq 3$ and for $B_{j,p}(\lambda)$ one has the results of Theorem 4.2., i.e. $\Sigma_j J^{\Gamma}[B_{j,p}(\lambda)]$ belongs to $\mathbb{Z}_{2p}$, if p is odd, and to $\mathbb{Z}_2 \times \mathbb{Z}_p$, if p is even, and Σ^{Γ} is one-to-one provided each coordinate with $\Gamma/H \simeq \{e\}$ or $\mathbb{Z}_2$ is present at least three times and each with $\Gamma/H \cong \mathbb{Z}_p$, $p \geq 3$, at least twice. Furthermore if $\dim_{\mathbb{C}} V^i = d$, we define a standard action on $\mathbb{C}^d$ as $e^{2\pi i/p} X_j$ and equivariant maps ϕ, from $\mathbb{C}^d$ into V^i, and ψ, from V^i into $\mathbb{C}^d$, where $\mathbb{Z}_p$ acts on V_i, as $e^{2\pi i m_j/p}$, $j = 1, \ldots, d$:

$$\varphi(X_1, \ldots, X_d) = (X_1^{m_1}, \ldots, X_d^{m_d})$$
$$\psi(x_1, \ldots, x_d) = (x_1^{n_1}, \ldots, x_d^{n_d})$$

where $n_j \geq 1$ is such that $m_j n_j \equiv 1$, [mod p].

As in Ize (1985), one defines, for any equivariant map $f(x_1, \ldots, x_d)$ from V_i into itself, the equivariant maps $f(X_1^{m_1}, \ldots, X_d^{m_d}) = f \circ \varphi(X_1, \ldots, X_d)$ from $\mathbb{C}^d$ into V^i and $\psi \circ f \circ \varphi$ from $\mathbb{C}^d$ into itself. Here, $B_{j,p}(\lambda)$ is deformable to $a_{j,p}(\lambda) = \det B_{j,p}(\lambda)$ and one obtains the suspension of

$$\begin{pmatrix} a_{1,p}(\lambda) & \\ & a_{d,p}(\lambda) \end{pmatrix} \begin{pmatrix} x_1 \\ x_d \end{pmatrix}.$$

The corresponding map $\psi \circ f \circ \varphi$ will be

$$\begin{pmatrix} a_{1,p}(\lambda)^{n_1} & \\ & a_{d,p}(\lambda)^{n_d} \end{pmatrix} \begin{pmatrix} X_1^{n_1 m_1} \\ X_d^{n_d m_d} \end{pmatrix},$$

which is deformable in $\mathbb{C}^d$ to $\begin{pmatrix} \Pi a_{j,p}(\lambda)^{n_j} & \\ & I \end{pmatrix} \begin{pmatrix} X_1^{n_1 m_1} \\ X_d^{n_d m_d} \end{pmatrix}$. Using the techniques of Ize–Vignoli (1993), Theorem 8.3, it is then easy to see that $J^{\mathbb{Z}_p}[B(\lambda)] = \deg[\Pi a_{j,p}(\lambda)^{n_j}] \Pi n_j m_j$, [mod p]. Using the fact that the winding number of a product is the sum of the winding numbers and the periodicity theorem, one obtains:

Proposition 5.4. *If $k = 2$ and Γ is abelian, then the relation $J^{\Gamma}[B(\lambda)] = 0$ implies that $B_0(\lambda)$, $B_j(\lambda)$, $B_j^{\mathbb{C}}(\lambda)$, are deformable to I, provided* dim $V^{\Gamma} \geq 3$, dim $V_j^{\mathbb{R}} \geq 3$, *and* Σn_j Ind $(\det B_{j,p}(\lambda)) \equiv 0$, [mod p].

A more complete result will be given in Theorem 6.1.

VI. NECESSARY AND SUFFICIENT CONDITIONS FOR EQUIVARIANT BIFURCATION

Now I shall extend the results of Section II to the equivariant case, with some preliminary and simple consequences essentially for the abelian case and in the stable range. A more complete description will be published elsewhere.

As a first step it is easy to see that Theorem 2.1 of Section II extends to the equivariant case, i.e. let $B^k = \{\lambda : \|\lambda\| \le \rho\}$, $S^{d-1} = \{\eta : \|\eta\| = 1\}$, then

Proposition 6.1. *Let Γ act via isometries on $V \cong \mathbb{R}^d$, then $B(\lambda)\eta = S^{k-1} \times S^{d-1} \to V\backslash\{0\}$ extends to a $\Gamma-map$ $B(\lambda, \eta) : B^k \times S^{d-1} \to V\backslash\{0\}$ if and only if there is a $\Gamma-map$ $G(\lambda, x)$ continuous on $B^k \times V$ such that*

a) *$G(\lambda, x) = 0(\|x\|^2)$ for fixed λ and $G(\lambda, x) = o(\|x\|)$ uniformly in λ,*

b) *there is $\rho_1, 0 < \rho_1 \le \rho$, such that if $F(\lambda, x) = B(\lambda)x + G(\lambda, x) = 0$, for $\|\lambda\| \le \rho_1$, then $x = 0$ and there is $\epsilon_0 > 0$ such that if $F(\lambda, x) = 0$ for $\|x\| \le \epsilon_0, \|\lambda\| \le \rho$, then $x = 0$.*

In fact one may get a finer result by considering V^Γ and its orthogonal complement $V^\perp$, which is also a $\Gamma-$space. Let $x = X_0 \oplus Z$, $B(\lambda) = \text{diag}(B_0(\lambda), B^\perp(\lambda))$ and $F_0(\lambda, X) = F|_{V^\Gamma}$. Then, if $B(\lambda)\binom{\eta_0}{\eta}$ has a Γ-extension, this is also the case for $B_0(\lambda)\eta_0$ and there is $F_0(\lambda, X)$ with the above properties on V^Γ. Consider the problem of extending $(F_0(\lambda, X), B^\perp(\lambda)\eta)$ from $\partial(B^k \times B_0^{do}) \times S^{d-1} \to V\backslash\{0\}$ to a $\Gamma-$map $(B_0(\lambda, X, \eta), B^\perp(\lambda, X, \eta))$ from $B^k \times B_0^{do} \times S^{d-1} \to V\backslash\{0\}$, where $B_0^{do} = \{X_0 : \|X_0\| \le \epsilon_0\}$, $S^{d-1} = \{Z : \|Z\| = 1\}$.

Remark 6.1. In the case of the Hopf bifurcation problem $\nu\frac{dX}{dt} = L(\lambda)X + g(\lambda, X)$, the corresponding equations on the Fourier modes are $(i\nu n I - L(\lambda))X_n - g_n(\lambda, X) = 0$; hence $B_n(\lambda, \nu)$ is the Liapunov-Schmidt matrix corresponding to $in\nu I - L(\lambda)$ and, in particular for $n = 0$, the equation $F(\lambda, X_0)$ is independent of ν and $B_0(\lambda, \nu)$ cannot be invertible for $\|\lambda\| + |\nu| \le \rho$, unless $L(0)$ is invertible. Since the case of singular $L(0)$ should also be considered, then either there is a local bifurcation of stationary solutions, i.e. there is a sequence of ϵ_n going to 0, such that $L(\lambda)X_0 + g(\lambda, X_0)$ has solutions with $\|X_0\| = \epsilon_n$, or there is ϵ_0 such that this equation has no solutions, but $X_0 = 0$, for $\|X_0\| \le \epsilon_0, \|\lambda\| \le \rho$. In this last case, $\deg(L(\lambda)X_0 + g(\lambda, X_0); B^{do})$ is well defined and independent of λ.

Remark 6.2. If $F(\lambda, X, Z) = (B_0(\lambda)X + G_0(\lambda, X) + \tilde{G}(\lambda, X, Z), B^\perp(\lambda)Z + G(\lambda, X, Z))$, then, from the equivariance, $G(\lambda, X, 0) = 0$ and, from its definition, $\tilde{G}(\lambda, X, 0) = 0$. If these terms are smooth and of higher order, one expects $\|G(\lambda, X, Z)\|$, $\|\tilde{G}(\lambda, X, Z)\| \leq \tilde{C}(\|X_0\| + \|Z\|)\|Z\|$ with $\tilde{C}(0) = 0$.

Proposition 6.2. *Assume $B^\perp(\lambda)$ is invertible for $\|\lambda\| \neq 0$, $\|\lambda\| \leq \rho$, $B_0(\lambda)X + G_0(\lambda, X) = 0$, for $\|X\| \leq \epsilon_0$, $\|\lambda\| \leq \rho$, $X = 0$. Then $(B_0(\lambda)X + G_0(\lambda, X), B^\perp(\lambda)\eta)$ extends equivariantly from $\partial(B^k \times B_0^{do}) \times S^{d-1} \to V\backslash\{0\}$ to $(B_0(\lambda, X, \eta), B^\perp(\lambda, X, \eta) : B^k \times B_0^{do} \times S^{d-1} \to V\backslash\{0\}$ if and only if there are $G(\lambda, X, Z)$, $\tilde{G}(\lambda, X, Z)$, equivariant and invariant maps, and $\epsilon_1(G, \tilde{G}, \rho, \epsilon_0)$ with $\|G\|$, $\|\tilde{G}\| \leq C(\|Z\|)\|Z\|$ and $C(0) = 0$, such that the equation $F(\lambda, X, Z) \equiv (B_0(\lambda)X + G_0(\lambda, X) + \tilde{G}(\lambda, X, Z), B^\perp(\lambda)Z + G(\lambda, X, Z)) = (0, 0)$ implies, for $\|X\| \leq \epsilon_0$, $\|\lambda\| \leq \rho, \|Z\| \leq \epsilon_1$, that $X = 0, Z = 0$.*

Proof. **"If".** Since $B^\perp(\lambda)$ is invertible for $\|\lambda\| = \rho$, one may choose $\bar{\epsilon}_0 \leq \epsilon_0$, $\bar{\epsilon}_1 \leq \epsilon_1$, such that, on $S^{k-1} \times B_{\bar{\epsilon}_0}^{do} \times B_{\bar{\epsilon}_1}^d$, the only zero of $B^\perp(\lambda)Z + G(\lambda, X, Z)$ is $Z = 0$, thus the only zero of the couple is $X = 0, Z = 0$. Now, on $S_{\bar{\epsilon}_0}^{do-1}$, one has $\|B_0(\lambda)X + G_0(\lambda, X)\| \geq C$. Choose $\tilde{\epsilon}_1 \leq \bar{\epsilon}_1$ such that $\|\tilde{G}\| \leq C/2$ for $\|Z\| \leq \tilde{\epsilon}_1$. Then, on $\partial(B^{k-1} \times B_{\bar{\epsilon}_0}^{do}) \times S^{d-1}$, one may deform $F(\lambda, X, \tilde{\epsilon}_1\eta)$ to $(B_0(\lambda)X + G_0(\lambda, X), B^\perp(\lambda)\eta)$. Since $F(\lambda, X, \tilde{\epsilon}_1\eta)$ is nonzero on $B^{k-1} \times B_{\bar{\epsilon}_0}^{do} \times S^{d-1}$, the Borsuk equivariant extension theorem implies that the second map has a Γ−extension. By scaling X from $\bar{\epsilon}_0$ to ϵ_0, one will have the result.

"Only if". Let B and C be such that $B \leq \max(\|B_0\|, \|B^\perp\|) \leq C$ on $B^k \times B_0^{do} \times S^{d-1}$ and let A be such that $\|B_0(\lambda)\|, \|B^\perp(\lambda)\| \leq A\|\lambda\|$, $\|B_0(\lambda)X + G_0(\lambda, X)\| \leq A\|X\|$. Set $g(\|\lambda\|) = \min\|B^\perp(\tilde{\lambda})\eta\|$ on the set $\{\tilde{\lambda} : \|\lambda\| \leq \|\tilde{\lambda}\| \leq \rho\} \times S^{d-1}$. Then $g(0) = 0, g$ is continuous and nondecreasing. Define $g_0(\|X\|) = \min\|B_0(\lambda)Y + G_0(\lambda, Y)\| \|Y\|^{-1/2} \leq A\|X\|^{1/2}$ on the set $\{Y : \|X\| \leq \|Y\| \leq \epsilon_0\} \times B^{k-1}$.

Then $g_0(0) = 0, g_0$ is continuous and nondecreasing. Define $\varphi(\|\lambda\|, \|Z\|)$, dividing the quadrant $(\|\lambda\|, \|Z\|)$ in three regions, as in the proof of Theorem 2.1 in Section II. Define $\varphi_0(\|X\|, \|Z\|)$ as 1 if $\|Z\| \leq \|X\|^{1/3}/2$, (region III'), as 0 if $\|Z\| \geq \|X\|^{1/3}$ (region I') and a linear interpolation in $\|Z\|$ in the remaining region II'. Set

$$F^\perp(\lambda, X, Z) = (\|\lambda\|/(\varphi\|\lambda\| + (1 - \varphi)\rho)^{1/2}B^\perp(\lambda(\varphi\|\lambda\|$$
$$+ (1 - \varphi)\rho)/\|\lambda\|)Z + \|Z\|^2 B^\perp(\lambda\rho/(\|\lambda\|$$
$$+ \|Z\|^4), \epsilon_0 X/(\|X\| + \|Z\|^4), Z/\|Z\|)$$

$$F_0(\lambda, X, Z) = (\|X\|/(\varphi_0\|X\| + (1 - \varphi_0)\epsilon_0)^{1/2}(g_0(\|X\|)/g_0(\varphi_0\|x\|$$
$$+ (1 - \varphi_0)\epsilon_0))$$
$$[F_0(\lambda(\varphi_0 + (1 - \varphi_0)\rho/(\|\lambda\| + \|Z\|^4)), X(\varphi_0\|X_0\| + (1 - \varphi_0)\epsilon_0)/\|X\|]$$
$$+ \|Z\|^{3/2}g_0(\|Z\|^3)/CB_0(\lambda\rho/(\|\lambda\|$$
$$+ \|Z\|^4), \epsilon_0 X/(\|X\| + \|Z\|^4), Z/\|Z\|).$$

Here $F_0(\lambda, X) = B_0(\lambda)X + G_0(\lambda, X)$.

The following properties are easy to derive:

1) $\|F^\perp\| \le A(\rho\|\lambda\|)^{1/2}\|Z\| + C\|Z\|^2$

1') $\|F_0\| \le A(\epsilon_0\|X\|)^{1/2} + \|Z\|^{3/2}g_0(\|Z\|^3)$

2) $\|F^\perp - B^\perp(\lambda)Z\| \le \|Z\|(2A\rho^{1/2}h(\|Z\|)^{1/2} + C\|Z\|)$, where $h(\|Z\|)$ is the inverse of the increasing function $\|Z\| = (\|\lambda\|/\rho)^{1/2}g(\|\lambda\|)/(2C)$.

2') $\|F_0(\lambda, X, Z) - F_0(\lambda, X)\| \le \|Z\|^{3/2}(4\sqrt{2}A\epsilon_0^{1/2} + g_0(\|Z\|^3))$

3) $\|F^\perp\| \ge (\|\lambda\|/\rho)^{1/2}g(\|\lambda\|)\|Z\| - C\|Z\|^2$. Hence, $F^\perp \ne 0$ in regions II and III.

3') $\|F_0\| \ge \|X\|^{1/2}g_0(\|X\|) - \|Z\|^{3/2}g_0(\|Z\|^3)$, hence $F_0 \ne 0$ in regions II' and III'. Thus, any solution with $Z \ne 0$ has $\varphi = \varphi_0 = 0$. Furthermore, if at a zero, one has $\|B^\perp(\lambda\rho/(\|\lambda\| + \|Z\|^4), \epsilon_0 X/(\|X\| + \|Z\|^4), Z/\|Z\|)\| \ge B$, then $\|F^\perp\| \ge B\|Z\|^2 - A(\rho\|\lambda\|)^{1/2}\|Z\|$ and $F^\perp \ne 0$ if $\|Z\| \ge A(\rho\|\lambda\|)^{1/2}/B$, while if $\|B_0\| \ge B$ given the same arguments (recall the definition of B), then $\|F_0\| \ge B\|Z\|^{3/2}g_0(\|Z\|^3)/C - A(\epsilon_0\|X\|)^{1/2}g_0(\|X\|)/g_0(\epsilon_0)$ and $F_0 \ne 0$ if $\|X\| \le \|Z\|^{24/7}$ and $\|Z\| \le \min(1, Bg_0(\epsilon_0)/AC\epsilon_0^{1/2})^{4/3} \equiv \epsilon_1'$.

4) Let $\delta < \min(\rho, \epsilon_0)$ be such that $\|B^\perp(\lambda, Y, \eta) - B(\tilde\lambda)\eta\| \le g(\rho)/2$ if $\|\lambda - \tilde\lambda\| \le \delta$, $\|\tilde\lambda\| = \rho$, $\|Y\| \le \epsilon_0$, $\|\eta\| = 1$ and such that $\|B_0(\lambda, Y, \eta) - F_0(\lambda, X)\| \le g_0(\epsilon_0)\epsilon_0^{1/2}/2$ if $\|\lambda\| \le \rho$, $\|X - Y\| \le \delta$, $\|X\| = \epsilon_0$, $\|\eta\| = 1$. Then, in region I, one has $F^\perp(\lambda, X, \eta) = ((\|\lambda\|/\rho)^{1/2} + \|Z\|)B^\perp(\lambda\rho/\|\lambda\|)Z + \|Z\|^2[B^\perp(\lambda\rho/(\|\lambda\| + \|Z\|^4), \epsilon_0 X/(\|X\| + \|Z\|^4), Z/\|Z\|) - B^\perp(\lambda\rho/\|\lambda\|)\eta]$.

Hence, $\|F^\perp\| \ge ((\|\lambda\|/\rho)^{1/2} + \|Z\|)\, g(\rho)\|Z\| - \|Z\|^2 g(\rho)/2$ if $\|Z\|^4 \le \delta\|\lambda\|/(\rho-\delta)$ and $F^\perp \ne 0$ in this subregion of I. Similarly F_0 in the region I' is $((\|X\|/\epsilon_0)^{1/2}(g_0(\|X\|)/g_0(\epsilon_0)) + \|Z\|^{3/2}g_0(\|Z\|^3)/C)\, F_0(\lambda\rho/(\|\lambda\| + \|Z\|^4), \epsilon_0 X/\|X\|) + C^{-1}\|Z\|^{3/2}g_0(\|Z\|^3)[B_0(\lambda\rho/(\|\lambda\| + \|Z\|^4), \epsilon_0 X/(\|X\| + \|Z\|^4), Z/\|Z\|) - F_0(\lambda\rho(\|\lambda\| + \|Z\|^4), \epsilon_0 X/\|X\|)]$. Thus, if $\|Z\|^4 \le \delta\|X\|/(\epsilon_0 - \delta)$, one has that $\|F_0\| \ge ((\|X\|/\epsilon_0)^{1/2}(g_0(\|X\|)/g_0(\epsilon_0)) + \|Z\|^{3/2}g_0(\|Z\|^3)/(2C))\, g_0(\epsilon_0)\epsilon_0^{1/2}$.

Then if $(\rho_1, \bar\epsilon_1)$ is the intersection of the curves $\|Z\| = A(\rho\|\lambda\|)^{1/2}/B$ and $\|Z\|^4 = \|\lambda\|\delta/(\rho - \delta)$, i.e. $\rho_1 = (B/A)^4\rho^2\delta/(\rho - \delta)$ and $\bar\epsilon_1$ corresponds to the intersection of the curves $\|Z\| = \|X\|^{7/24}$ and $\|X\| = \|Z\|(\epsilon_0 - \delta)/\delta$, for $\|X\| = (\delta/(\epsilon_0 - \delta))^6$, then if $\|Z\| \le \epsilon_1 \equiv \min(\epsilon_1', \bar\epsilon_1, \tilde\epsilon_1)$, the conclusion of the the proposition holds for these numbers. Q.E.D.

Remark 6.3. If, as in Proposition 6.2, the map $(F_0(\lambda, X), B^\perp(\lambda)\eta)$, on $\partial(B^k \times B_0^{do}) \times S^{d-1}$ has an extension $(B_0(\lambda, X, \eta), B^\perp(\lambda, X, \eta))$ to $B^k \times B_0^{do} \times S^{d-1}$, then the map $(F_0(\lambda, X), B^\perp(\lambda)Z)$ has an extension from $S^{k-1} \times \partial(B_0^{do} \times B^d)$ to $B^k \times \partial(B_0^{do} \times B^d)$ by using $(B(\lambda, X, \eta), B^\perp(\lambda, X, \eta))$ on S^{d-1} and $(F_0(\lambda, X), B^\perp(\lambda)Z)$ on S_0^{do-1}. Hence Proposition 6.2. implies Proposition 6.1.

The converse is also true for the couple $(F_0(\lambda, X), B^\perp(\lambda)\eta)$. In fact, it is true for $(B_0(\lambda, X), B^\perp(\lambda)\eta)$, where $B_0(\lambda, X)$ is obtained in the following way: if $(F_0(\lambda, X), B^\perp(\lambda)Z)$ extends as $(B_0'(\lambda, X, Z), B^{\perp\prime}(\lambda, X, Z))$ from $S^{k-1} \times \partial(B_0^{do} \times B^d)$ to $B^k \times \partial(B_0^{do} \times B^d)$, then $B^{\perp\prime}(\lambda, X, 0) = 0$ and $B_0(\lambda, X) \equiv B_0'(\lambda, X, 0)$ is nonzero on $B^k \times S_0^{do-1}$ and extends $F_0(\lambda, X)$ from $S^{k-1} \times S_0^{do-1}$. Now, on $B^k \times S_0^{do-1} \times S^{d-1}$, one has that $(B_0'(\lambda, X, \eta), B^{\perp\prime}(\lambda, X, \eta))$ is Γ−homotopic, relative to S^{k-1}, to $(B_0(\lambda, X), B^\perp(\lambda)\eta)$, via $(B_0'(\lambda, X, t\eta), B^{\perp\prime}(\lambda, X, t\eta)/t)$ for $\epsilon \le t \le 1$, where ϵ is chosen so small that $B_0'(\lambda, X, \epsilon\eta)$ is nonzero (by the equivariance), and followed by a linear deformation to $(B_0, B^\perp(\lambda)\eta)$. Then, $(B_0(\lambda, X), B^\perp(\lambda)\eta)$ will have an extension from $\partial(B^k \times B_0^{do}) \times S^{d-1}$ to $B^k \times B_0^{do} \times S^{d-1}$, by using the Borsuk equivariant extension theorem (on $S^{k-1} \times B_0^{do} \times S^{d-1}$ the homotopy is constant and equal to $(F_0(\lambda, X), B^\perp(\lambda)\eta)$.

Now on $\partial(B^k \times B_0^{do}) \times S^{d-1}$, the homotopy $(B_0(t\lambda + (1 - t)\lambda_0, X), B^\perp(\lambda)\eta)$, with $\|\lambda_0\| = \rho$, is valid; hence $(B_0(\lambda_0, X), B^\perp(\lambda)\eta)$ has an extension and is also homotopic to $(F_0(\lambda, X), B^\perp(\lambda)\eta)$, via the some sort of deformation, on the above set.

Hence $(F_0(\lambda, X), B^\perp(\lambda)\eta)$ extends from $\partial(B^k \times B_0^{do}) \times S^{d-1}$ to $B^k \times B_0^{do} \times S^{d-1}$ if and only if $(F_0(\lambda, X), B^\perp(\lambda)Z)$ extends from $S^{k-1} \times \partial(B_0^{do} \times B^d)$ to $B^k \times B_0^{do} \times B^d$, assuming that $F_0(\lambda, X) \ne 0$ for $X \ne 0$.

Remark 6.4. In the same vein of ideas, if $(B_0(\lambda)X, B^\perp(\lambda)Z)$ extends from $S^{k-1} \times \partial(B_0^{do} \times B^d)$ to $B^k \times \partial(B_0^{do} \times B^d)$ via $(B_0(\lambda, X, Z), B^\perp(\lambda, X, Z))$, as in Proposition 6.1, then $J^\Gamma[B(\lambda)]$, the Γ-homotopy class of $(\|x\| - \epsilon, B(\lambda)x)$, on $\partial(B^k \times B_0^{do} \times B_{2\epsilon}^d)$, $\epsilon < \epsilon_0$, is zero, since this map is linearly deformable, on this sphere, to $(\|x\| - \epsilon, (\epsilon/\epsilon_0)B_0(\lambda, \epsilon_0 X/\epsilon, Z/\epsilon), \epsilon B^\perp(\lambda, \epsilon_0 X/\epsilon, Z/\epsilon))$, which is nonzero on the ball (B_0 and $B^\perp$ are extended equivariantly to the larger sets). Conversely, if $J^\Gamma[B(\lambda)] = 0$, hence $J[B_0(\lambda)] = 0$, then, if $k \le 2d_0 - 3$, from Theorem 2.3 in Section II, there is a $g_0(\lambda, X) = o(\|X\|)$, such that $F_0(\lambda, x) = B_0(\lambda)X + g_0(\lambda, X)$ is zero only at $X = 0$, if $\|\lambda\| \le \rho$. Then, on $\partial(B^k \times B_0^{do} \times B_{2\epsilon}^d)$, the maps $(\|x\| - \epsilon, B(\lambda)x)$ and $(\|Z\| - \epsilon, F_0(\lambda, X), B^\perp(\lambda)Z)$ are Γ−homotopic, hence, this last map extends to $B^k \times B_0^{do} \times B_{2\epsilon}^d$ via $(f(\lambda, X, Z), B_0(\lambda, X, Z), B^\perp(\lambda, X, Z))$. Set $I = [0, 2\epsilon]$ and let, on $\partial(I \times B^k \times B_0^{do})$ the invariant map $\tilde{F}_0(\lambda, X, r)$ defined as $(r - \epsilon, F_0(\lambda, X))$ if $r \ne 0$ and as $(f(\lambda, X, 0), B_0(\lambda, X, 0))$ if $r = 0$. Then, on $\partial(I \times B^k \times B_0^{do}) \times S^{d-1}$, the map $(\tilde{F}_0(\lambda, X, r), rB^\perp(\lambda)\eta)$ Γ−extends to $I \times B^k \times B_0^{do} \times S^{d-1}$, via $(f(\lambda, X, r\eta), B_0(\lambda, X, r\eta), B^\perp(\lambda, X, r\eta))$. The same happens to the map $(\tilde{F}_0(\lambda, X, r), B^\perp(\lambda)\eta)$ i.e. one is in the situation of Proposition 6.2, in the sense that $\tilde{F}_0(\lambda, X, r) \neq 0$ on $\partial (I \times B_0^{do}) \times B^k$. In particular taking λ_0, with $\|\lambda_0\| = \rho$ and $B_0(\lambda_0)$ invertible (hence deformable to I or to $diag\,(-1, 1, \ldots, 1)$), one may deform linearly λ to λ_0 in $\tilde{F}_0$ and $(r - \epsilon, B_0(\lambda_0)X, B^\perp(\lambda)\eta)$ as well as $(r - \epsilon, F_0(\lambda, X), B^\perp(\lambda)\eta)$, has an equivariant extension from $\partial(I \times B^K \times B_0^{do}) \times S^{d-1}$ to $I \times B^k \times B_0^{do} \times S^{d-1}$, i.e. $J^\Gamma[B(\lambda)] = 0$ corresponds to a suspension in Proposition 6.2. More precisely, if $(B_0(\lambda)X, B^\perp(\lambda)Z)$ extends from $S^{k-1} \times \partial(B_0^{do} \times B^d)$, that is if $(F_0(\lambda, X), B^\perp(\lambda)Z)$ has an extension (this is true, by the preceding remark, if and only if $(F_0(\lambda, X), B^\perp(\lambda)\eta)$ has an extension from $\partial(B^k \times B_0^{do}) \times S^{d-1}$), then $J^\Gamma(B(\lambda)) = 0$ and, if $k \leq 2d_0 - 3$ or $d_0 = 2$, or $d_0 = 4$ or 8 and $k \leq 2d_0 - 2$, one has that $(r - \epsilon, F_0(\lambda, X), B^\perp(\lambda)\eta)$ has an extension from $\partial(I \times B^k \times B_0^{do}) \times S^{d-1}$.

If $k < d_0$, then $B_0(\lambda)$ is deformable to $\begin{pmatrix} C(\lambda) & 0 \\ 0 & 1 \end{pmatrix}$. Assume that $J^\Gamma(C_0(\lambda)\tilde{X}, B^d(\lambda)Z) = 0$, where $X = \tilde{X} \oplus x_0$. Since $k \leq 2(d_0 - 1) - 3$ if $d_0 \geq 4$ and the above argument works for $d_0 = 3$ and, for $d_0 \leq 2$, then $B(\lambda)$ is deformable to I, and $(r - \epsilon, \tilde{F}_0(\lambda, \tilde{X}), B^\perp(\lambda)\eta)$ has an extension from $\partial(I \times B^k \times B_0^{do-1}) \times S^{d-1}$, where $\tilde{F}_0$ is the extension for $C_0(\lambda)\tilde{X}$. It is then clear that $(x_0, \tilde{F}_0(\lambda, \tilde{X}), B^\perp(\lambda)\eta)$ has an extension from $\partial(B^k \times B_0^{do-1}) \times S^{d-1}$ and that $(B_0(\lambda)X, B^\perp(\lambda)Z)$ has an extension from $S^{k-1} \times \partial(B_0^{do} \times B^d)$, using again the preceding remark. One now has the first part of the following result

Proposition 6.3. (a) *If $k < d_0$ and Σ_0, the suspension by a trivial variable, is one-to-one (i.e. if $J^\Gamma(B(\lambda)) = 0$ then $J^\Gamma(C_0(\lambda), B^\perp(\lambda)) = 0$ for some $C_0(\lambda)$), then there is no local linearized equivariant bifurcation if and only if $J^\Gamma(B(\lambda)) = 0$.*

(b) If $k < 2\,dim\,E^\Gamma - 2$ (with equality possible if $d_0 < dim\,E^\Gamma$ then there is no global linearized equivariant bifurcation if and only if $J^\Gamma(A_0 - A(\lambda)) = 0$.

Proof. (b). It is clear that the "only if" part of Theorem 3.2. in Section II works in the equivariant setting. Thus, if there is no linearized bifurcation, one has that $J^\Gamma(A_0 - A(\lambda)) = 0$. Conversely, if $J^\Gamma(A_0 - A(\lambda)) = 0$, then $J(A_0^\Gamma - A^\Gamma(\lambda)) = 0$ and there is no local linearized bifurcation, one thus has $F_0(\lambda, X) \neq 0$ for $X \neq 0$, as an extension of the invariant part (see Theorems 2.3 and 3.2. in Section II). Then $(r - \epsilon, F_0(\lambda, X), (A_0^\perp - A(\lambda)^\perp)\eta)$ has an extension from $\partial(I \times B^k \times B_0) \times \tilde{S}$, that is $(r - \epsilon, F_0(\lambda, X), (A_0^\perp - A(\lambda)^\perp)Z)$ has an extension on $\partial(I \times B^k) \times S$, as well as $(r - \epsilon, (A_0 - A(\lambda))x)$. But then, it is easy to see that Theorem 3.1 of Section II is valid in the equivariant context and that this fact is equivalent to the no global linearized equivariant bifurcation.

Q.E.D.

Let Γ be abelian. Hence $B(\lambda) = \mathrm{diag}(B_0(\lambda),\, B_1(\lambda), \ldots,\, B_{1,p}(\lambda), \ldots, B_1^{\mathbb{C}}(\lambda), \ldots)$, as in last subsection, where $B_j(\lambda)$ are real $d_j \times d_j$ matrices, with the action of Γ/H as $\{e\}$ or $\mathbb{Z}_2$, $B_{j,p}(\lambda)$ are complex $d_{jp} \times d_{jp}$ matrices with the action of Γ/H as $\mathbb{Z}_p$ and $B_j^{\mathbb{C}}(\lambda)$ are complex $d_j^{\mathbb{C}} \times d_j^{\mathbb{C}}$ matrices with action of Γ/H as S^1. As in Proposition 6.2, assume that $B_0(\lambda)X$ extends to $F_0(\lambda, X)$ which is 0 only at $X = 0$, $B_0(\lambda_0)$ is invertible for some λ_0 with $\|\lambda_0\| = \rho$, and $B_j, B_{j,p}, B_j^{\mathbb{C}}$ are invertible for $\lambda \neq 0$.

Recall that $P_*[C(\lambda)] = [C(\lambda)(0, \ldots, 0, 1)^T]$ in $\Pi_{k-1}(S^{d-1})$.

Theorem 6.1. *Let Γ be abelian and $B(\lambda)$ satisfy the above hypothesis. Assume $(F_0(\lambda, X),\ B^{\perp}(\lambda)\eta)$ has a $\Gamma-$equivariant nonzero extension from $\partial(B^k \times B_0^{do}) \times S^{d-1}$ to $B^k \times B_0^{do} \times S^{d-1}$. Then*

    a) *If $k < d_j$ and $k \neq 0 \bmod 4$, then $B_j(\lambda)$ is deformable to I.*

    b) *If $k < d_j$ and $k \equiv 0 \bmod 4$, then $B_j(\lambda)$ is deformable to*

$$
\begin{pmatrix} C_j(\lambda) & 0 \\ 0 & I \end{pmatrix},
$$

    with $C_j(\lambda)$ a $k \times k$ matrix, $J[B_j(\lambda)] = 0$ and $P_*[C_j(\lambda)]$ is even.

    c) *If $k = d_j$, then $\Sigma_0 J[B_j(\lambda)] = 0$ and $P_*[B_j(\lambda)] = 0$.*

    d) *If $k \leq 2d_{jp}$ or $2d_j^{\mathbb{C}}$ and k is odd, then $B_{j,p}$ or $B_j^{\mathbb{C}}$ is deformable to I.*

    e) *If $k \leq 2d_j^{\mathbb{C}}$, then $B_j^{\mathbb{C}}$ is deformable to I.*

    f) *If $k = 2s \leq 2d_{jp}$ for all j's and a given p, let $\Gamma/H \cong \mathbb{Z}_p$ act as $e^{2\pi i m_j/p}$, with m_j and p relatively prime and let n_j be such that*

$$
n_j m_j \equiv 1, \bmod p. \text{ Then } B_{jp}(\lambda) \text{ is deformable to } \begin{pmatrix} C_{jp}(\lambda) & 0 \\ 0 & I \end{pmatrix},
$$

    with $C_{jp}(\lambda)$ a $s \times s$ complex matrix and $P_*[\Pi C_{jp}^{n_j^s}(\lambda)]$ is a multiple of p. If $s = d_{jp}$ and there is only one B_{jp}, and there is no p' dividing p, except maybe one maximal, then B_{jp} is deformable to I.

    g) *Conversely, if any of the matrices is deformable to I, one has a $\Gamma-$extension which is 0 only if the corresponding variables are 0. In particular, if $k = 1$, and $\det B_j(\lambda)$ does not change sign, then there is a nonzero $\Gamma-$extension. If $k = 2$, $B_j(\lambda)$ are deformable to I, $\det B_j^{\mathbb{C}}(\lambda)$ have a zero winding number and, for all p's, one has $\Sigma_j n_j W(\det B_{j,p})$ is a multiple of p and, if p is odd, $\Sigma_j W(\det B_{jp})$ is even, where $W(\det B)$ is the winding number of $\det B$, then there is a $\Gamma-$extension, provided $d_0 \geq 3$ and $d_{jp} \geq 2$ for all j's.*

The proof of (e) was given in Ize (1985) using obstruction theory on a complex projective space. Here a direct argument will be given.

Proof. (a): If $\Gamma/H \simeq \mathbb{Z}_2$, then there is only one isotropy subgroup involved in V^H and, if $J^\Gamma[B^H(\lambda)] = 0$, one has $J[B^H(\lambda)] = \Sigma J[B_j(\lambda)] = 0$. If $k < d$ then Σ is an isomorphism and, from the Bott periodicity theorem, the conclusion follows. Furthermore (b), (c) and the part of (g) concerning the real matrices and the case $k = 1$ are similar, except for the statement on P_*.

(d) follows from the complex Bott periodicity, since $\Pi_{k-1}(GL(\mathbb{C}^d)) = 0$ if k is odd and $\mathbb{Z}$ if k is even with $B^{\mathbb{C}}(\lambda)$ deformable to $\begin{pmatrix} C^{\mathbb{C}}(\lambda) & 0 \\ 0 & I \end{pmatrix}$, and $C^{\mathbb{C}}(\lambda)$ a $s \times s$ matrix with $P_*[C^{\mathbb{C}}(\lambda)] = (s-1)!(-1)^{s-1}[C^{\mathbb{C}}(\lambda)]$.

For (e), the fact that the map Γ−extends implies that it S^1−extends, where $S^1 = \Gamma/H$. In this case S^1 acts trivially on the real and $\mathbb{Z}_p$ representations, since S^1 is connected, and acts as $e^{m_j \varphi}$, $j = 1, \ldots, l$, $m_1 < m_2 < \cdots$, on the rest of V^H, with $m_j = 1$ on $V_H = \{x : \Gamma_x = H\}$. (If $m_i < 0$, by taking conjugates, one gets $-m_j$). Order m_j in increasing numbers and note that if $m_j > 1$, then $\Gamma/H_j \simeq S^1$ and $H < H_j$; that is, using an induction argument, $B_j^{\mathbb{C}}(\lambda)$ is deformable to I. Hence one may assume that $(F_0(\lambda, X), B(\lambda)\eta_1, \eta_2, \ldots, \eta_l)$ has an S^1−extension $(G_0, \ldots, G_l)$ from $\partial(B^k \times B_0^{d_0}) \times S^{d-1}$ to $B_k \times B_0^{d_0} \times S^{d-1}$, where B_0 includes the real and $\mathbb{Z}_p$ representations and $S^{d-1} = \{\eta : \Sigma|\eta_j|^2 = 1\}$. The proof will be by induction on l.

If $l = 1$, $(F_0(\lambda, x), B(\lambda)(0, \ldots, 0, 1)^T) : \partial(B^k \times B_0^{d_0}) \to \mathbb{R}^{d_0} \times \mathbb{C}^d \backslash \{0\}$ has a nonzero extension to $B^k \times B_0^{d_0}$, that is, its class in $\Pi_{k+d_0-1}(S^{d_0+2d-1})$ is 0. By deforming $F_0(\lambda, X)$ to $F_0(\lambda_0, X)$ and to $B_0(\lambda_0)X$, with $B_0(\lambda_0)$ invertible, one obtains the d_0−suspension of $P_*[B(\lambda)]$. Since $k \leq 2d$, then $k < 4d + d_0 - 2$ and the suspension is an isomorphism, $P_*[B(\lambda)] = 0$. Hence if $k = 2d$, one has that $B(\lambda)$ is deformable to I, while if $k = 2s < 2d$, then $B(\lambda)$ being deformable to $\begin{pmatrix} C(\lambda) & 0 \\ 0 & I \end{pmatrix}$, with $C(\lambda)$ a $s \times s-$ matrix, $P_*[B(\lambda)] = 0$ independently of the extension process. Note that in the unstable case, $k > 2d$, if $\Sigma^{d_0} P_*[B(\lambda)] \neq 0$, one has bifurcation in all directions.

Let us write $\eta = (z_1, \ldots, z_s, \ldots, z_d)$ and assume that $B(\lambda)$ has the form $\begin{pmatrix} C(\lambda) & 0 \\ 0 & I \end{pmatrix}$.

Lemma 6.1. a) *Let* $(G_0, G)(\lambda, X, \eta) : \partial(B^k \times B^{d_0}) \times S^{2r-1} \to \mathbb{R}^{d_0} \times \mathbb{C}^d \backslash \{0\}$ *be* S^1−*equivariant, with the action* $e^{i\varphi}$ *on* $\mathbb{C}^r$ *and* $\mathbb{C}^d$. *Then, if* $k < 2(d+1-r)$, (G_0, G) *has a nonzero* S^1−*extension to* $B^k \times B^{d_0} \times S^{2r-1}$.

b) *If (G_0, G) and $(\tilde{G}_0, \tilde{G})$ are two equivariant extensions of (F_0, F) :
$\partial(B^k \times B^{do}) \times S^{2r-1} \to \mathbb{R}^{do} \times \mathbb{C}^d \backslash \{0\}$ to $B^k \times B^{do} \times S^{2r-1}$ and $k + 1 < 2(d+1-r)$, then (G_0, G) and $(\tilde{G}_0, \tilde{G})$ are S^1–homotopic on $B^k \times B^{do} \times S^{2r-1}$ via a homotopy which is (F_0, F) on $\partial(B^k \times B^{do}) \times S^{2r-1}$.*

Proof. (a). Let $\eta = (z_1, \ldots, z_r)^T$. For $\eta = (0, \ldots, 0, 1)^T$, we have that (G_0, G) is a map from S^{k+do-1} into $\mathbb{R}^{do \times 2d} \backslash \{0\}$, which extends, as above, if $k < 2d$. For $\eta = (0, \ldots, 0\ e^{i\psi})^T$, extends by defining $(G_0, G)(\lambda, X, 0, \ldots, 0, e^{i\psi}) = (G_0(\lambda, X, 0, \ldots, 0, 1), e^{i\psi}G(\lambda, X, 0, \ldots, 0, 1))$ giving an equivariant map. Then, for $\eta = (0, \ldots, 0, (1 - |z_r|^2)^{1/2}, z_r)^T$, one obtains a nonzero map from $\partial(B^k \times B^{do} \times B^2) = \partial(B^k \times B^{do}) \times B^2 \cup B^k \times B^{do} \times S^1$ by taking on the first piece the original map and, on the second piece, the extension just obtained. One has a continuous extension if $k < 2(d - 1)$. For $\eta = (0, \ldots, 0, e^{i\psi}(1 - |z_r|^2)^{1/2}, z_r)$ one defines $(G_0, G)(\lambda, X, \eta) = (G_0, e^{i\psi}G)(\lambda, X, 0, \ldots, 0, (1-|z_r|^2)^{1/2}, e^{-i\psi}z_r)$. It is easy to check that this map is well defined and equivariant.

One continues this process until $\eta = ((1-|z_2|^2 - \ldots)^{1/2}, z_2, \ldots, z_r)$, with a map on $\partial(B^k \times B^{do} \times B^{2(r-1)}) = \partial(B^k \times B^{do}) \times B^{2(r-1)} \cup B^k \times B^{do} \times S^{2r-3}$ into $\mathbb{R}^{do} \times \mathbb{C}^d \backslash \{0\}$, with the original map on the first piece and the previous extension on the second. This map is homotopically trivial if $k < 2(d+1-r)$ and one generates a S^1–map for η by defining $(G_0, G)(e^{i\psi}((1-|z_2|^2 - \ldots)^{1/2}, z_2, \ldots, z_r) = (G_0, e^{i\psi}G)((1 - |z_2|^2 \ldots)^{1/2}, e^{-i\psi}z_2, \ldots e^{-i\psi}z_r))$.

(b). On $\partial(I \times B^k \times B^{do}) \times S^{2r-1}$, define the S^1–map on (G_0, G) for $t = 0$, $(\tilde{G}_0, \tilde{G})$ for $t = 1$ and (F_0, F) on $I \times \partial(B^k \times B^{do}) \times S^{2r-1}$. Then one has a nonzero S^1–extension to $(I \times B^k \times B^{do}) \times S^{2r-1}$, giving the S^1–homotopy, if $k + 1 < 2(d + 1 - r)$.   Q.E.D.

The maps $(F_0(\lambda, x), 0, \ldots, 0, z_{s+1}, \ldots, z_d)$ and $(G_0, G)(\lambda, X, 0, \ldots, 0, z_{s+1}, \ldots, z_d)$ are two extensions of the first map from $\partial(B^k \times B^{do}) \times S^{2(d-s)-1}$ to $B^k \times B^{do} \times S^{2(d-s)-1}$; hence, from (b), they are S^1-homotopic extensions on the second set if $k + 1 < 2(d + 1 - r)$, with $r = d - s$ and $k = 2s$. Thus, the maps $(F_0(\lambda, X), C(\lambda)(0, \ldots, 0, (1 - \Sigma_{s+1}^d |z_j|^2)^{1/2})^T, z_{s+1}, \ldots, z_d)$ and $(G_0, G)(\lambda, X, 0, \ldots, 0, (1 - \Sigma_{s+1}^d |z_j|^2)^{1/2}, z_{s+1}, \ldots, z_d)$ coincide on $\partial(B^k \times B^{do}) \times B^{2(d-s)}$ and are S^1-homotopic on $B^k \times B^{do} \times S^{2(d-s)-1}$ with a fixed homotopy on the intersection of the two sets. Hence, the two maps are homotopic on $\partial(B^k \times B^{do} \times B^{2(d-s)})$. Since the second map has an extension to $B^k \times B^{do} \times B^{2(d-s)}$, given by the restriction of (G_0, G) to that set, then the first map also has an extension, that is, its class, as a map from $\partial(B^k \times B^{do} \times B^{2(d-s)})$ into $\mathbb{R}^{do} \times \mathbb{C}^d \backslash \{0\}$, is zero. This class is the degree of the map on $B^k \times B^{do} \times B^{2(d-s)}$, which is deformable to $(F_0(\lambda_0, X), C(\lambda)(0, \ldots, 0, 1)^T, z_{s+1}, \ldots, z_d)$ with degree equal to $P_*[C(\lambda)]$ $\deg(F_0(\lambda_0, X); B^{do}) = \pm P_*[C(\lambda)]$. Hence, $P_*[C(\lambda)] = 0$ and $[C(\lambda)] = 0$.

Assume that $l > 1$ and that the result is true for $l - 1$; then, if $(F_0(\lambda, X), B_j(\lambda)\eta_j)$ has the extension (G_0, G_j), this will be also the case on any V^K, with $K > H$, where there will less than l matrices and hence the corresponding B_j will be deformable to I. One is left with the case where $(F_0(\lambda, X), B_1(\lambda)\eta_1, \tilde{\eta})$, has the extension $(G_0, G_1, \tilde{G})$ and $B_1(\lambda)$ is deformable to $\begin{pmatrix} C(\lambda) & 0 \\ 0 & I \end{pmatrix}$ and $\eta_1 = (z_1, \ldots, z_s, \ldots z_{d_1})$. Assume

$$(*) \qquad G_1(\lambda, X, 0, \tilde{\eta}) = 0 \text{ for all } \tilde{\eta} \text{ with } \|\tilde{\eta}\| = 1.$$

This is true if $\Gamma_{\tilde{\eta}} > H$. To prove that $(*)$ holds in general requires an involved argument along the Extension Theorem 2.3. in Ize et al. (1992) noting that $V^H \cap \{\eta_1 = 0\}$ has dimension $k + d_0 + 2(d - d_1) \leq d_0 + 2d$. Another argument will be given later.

Lemma 6.2. (a) *Let $(G_0, G_1, \tilde{G})$ be defined on $S^{2(d-d_1+r)-1} \times \partial(B^k \times B_0^{d_o} \times \tilde{B})$ where $S^{2(d-d_1+r)-1} = \partial B = \{(\eta_1, \tilde{\eta}) : |\eta_1|^2 + |\tilde{\eta}|^2 = 1\}$, $\tilde{B} = B^{2(d-d_1)} = \{\tilde{\eta} : |\tilde{\eta}| < 1\}$, with action $e^{im_j\phi}$, with $m_0 = 0$, $m_1 = 1$, $m_j > 1$ for $j > 1$. Then, if $k < 2(d_1 - r + 1)$, one has an equivariant extension to $S^{2(d-r)-1} \times B^k \times B_0^{d_o}$.*

(b) If $k + 1 < 2(d_1 - r + 1)$, then any two extensions which coincide on $S^{2(d-r)-1} \times \partial(B^k \times B_0^{d_o} \times \tilde{B})$ are S^1-homotopic on $S^{2(d-r)-1} \times B^k \times B_0^{d_o}$.

Proof. (a) Let $\eta_1 = (z_1, \ldots, z_r)$. For $\eta_1 = (0, \ldots, 0, (1 - |\tilde{\eta}|^2)^{1/2})$, one has an element in $\Pi_{k+d_0+2(d-d_1)-1}(S^{d_o+2d-1})$ which is 0 if $k < 2d_1$ and one obtains an extension to $B^k \times B_0^{d_o} \times \tilde{B}$. For $\eta_1 = (0, \ldots, 0, e^{i\psi}(1 - |\tilde{\eta}|^2)^{1/2})$, let $G(\lambda, X, 0, \ldots, 0, e^{i\psi}(1 - |\tilde{\eta}|^2)^{1/2}, \tilde{\eta}) = \{e^{im_j\psi}G_j(\lambda, X, 0, \ldots, (1 - |\tilde{\eta}|^2)^{1/2}, e^{-im_i\psi}\tilde{\eta}_i)\}$ which is well defined and equivariant. For $\eta_1 = (0, \ldots, 0, (1 - |z_r|^2 - |\tilde{\eta}|^2)^{1/2}, z_r)$ one has a map on $\partial(B^k \times B_0^{d_o} \times B^{2(d-d_1)+2})$, which will be trivial if $k < 2(d_1 - 1)$. The extension for $(e^{i\psi}(1 - |z_r|^2 - |\tilde{\eta}|^2)^{1/2}, z_r)$ is via $e^{im_j\psi}G_j(\lambda, X, 0, \ldots, (1 - |z_r|^2 - |\tilde{\eta}|^2)^{1/2}, e^{-i\psi}z_r, e^{-im_i\psi}\tilde{\eta}_i)$. The last extension will be for $\eta_1 = ((1 - \Sigma_2^r|z_i|^2 - |\tilde{\eta}|)^{1/2}, z_2, \ldots, z_r)$, giving an element of $\Pi_{k+d_0+2(d-d_1+r-1)-1}(S^{d_o+2d-1}) = 0$, under the above hypotesis. The proof of (b) is similar. Q.E.D.

Remark 6.5. Theorem 2.2. In Ize et al. (1992) says that Lemma 6.2. in valid when $r = 0$ and when one has extensions to lower dimensional isotropy subspaces. Here, $(G_0, G_1, \tilde{G})$ and $(F_0, 0, \tilde{\eta})$ from $B^k \times B_0^{d_o} \times \partial B$ into $\mathbb{R}^{d_o} \times \mathbb{C}^d \backslash \{0\}$ are two extensions of the second map. Hence for $k = 2s$ and $r = d_1 - s$, they are S^1-homotopic. Since the first map extends to $B^k \times B^{d_o} \times S^{2d-1}$, this is true for the second and one may assume that $G_0 = F_0$, $\tilde{G} = \tilde{\eta}$, $G_1 = 0$ on $B^k \times B_0^{d_o} \times \partial\tilde{B}$. Note that here, when dealing with spheres, like $\partial \tilde{B}$, the dimension count on the domain is reduced by one.

The maps $(|\tilde{\eta}|G_0(\lambda, X, 0, \tilde{\eta}/|\tilde{\eta}|) + (1-|\tilde{\eta}|)F_0(\lambda, X), 0, \ldots 0, z_{s+1}, \ldots, z_{d_1}, |\tilde{\eta}|\tilde{G}(\lambda, X, 0, \tilde{\eta}/|\tilde{\eta}|)$ and $G(\lambda, X, 0, \ldots, 0, z_{s+1}, \ldots, z_{d_1}, \tilde{\eta})$ coincide on $\partial(B^k \times B_0^{do} \times \tilde{B})$ and extend to $S^{2(d-s)-1} \times B^k \times B_0^{do}$; hence they are S^1-homotopic on the last set, since $r = d_1 - s$. Then, on $\partial(B^k \times B_0^{do} \times B^{2(d-s)})$, the maps $(|\tilde{\eta}|G_0(\lambda, X, 0, \tilde{\eta}/|\tilde{\eta}|) + (1 - |\tilde{\eta}|)F_0(\lambda, X), C(\lambda) (0, \ldots, 0, (1 - \Sigma_{s+1}^{d_1}|z_j|^2 - |\tilde{\eta}|^2)^{1/2})^T, z_{s+1}, \ldots, z_{d_1}, |\tilde{\eta}|\tilde{G}(\lambda, X, 0, \tilde{\eta}/|\tilde{\eta}|))$ and $G(\lambda, X, 0, \ldots, (1-\Sigma_{s+1}^{d_1}|z_j|^2 - |\tilde{\eta}|^2)^{1/2}, z_{s+1}, \ldots z_{d_1}, \tilde{\eta})$ are homotopic and the second extends; hence the first has degree equal to 0. The square root may be deformed to 1 and λ (except in $C(\lambda)$) may be deformed to a λ_0 with $|\lambda_0| = \rho$, which is a valid deformation since for $|X| = \epsilon_0$, G_0 is F_0 and for $|\tilde{\eta}| = 1$, $(G_0, \tilde{G})$ is nonzero. The degree is then the degree of $(F_0(\lambda_0, X), C(\lambda)(0, \ldots, 0, 1)^T, z_{s+1}, z_{d_1}, \tilde{\eta})$, i.e. $\pm P_*[C(\lambda)] = 0$.

As noted above, (*) is difficult to prove, hence a different argument will be given: consider $\mathbb{C}^d = \{\xi_1, \ldots, \xi_l\}$, ξ_j in $\mathbb{C}^{d_j}$, with the standard action and $\tilde{\mathbb{C}}^d$ with the action $e^{iM\varphi}$, where $M = \Pi m_j$. Let $\phi : \mathbb{C}^d \to V$ be the equivariant map $\phi(\xi_1, \ldots \xi_l) = (\xi_1^{m_1}, \ldots, \xi_l^{m_l})$ and $\psi : V \to \tilde{\mathbb{C}}^d$ be the equivariant map $\psi(\eta_1, \ldots, \eta_l) = (\eta_1^{\tilde{m}_1}, \ldots, \eta_l^{\tilde{m}_l})$, with $\tilde{m}_j = m/m_j$.

It is clear that $(F_0(\lambda, X), \begin{pmatrix} C(\lambda) & 0 \\ 0 & I \end{pmatrix} \xi_1, \xi_2^{m_2}, \ldots, \xi_l^{m_l})$ has the equivariant extension $G(\lambda, X, \xi_1, \xi_2^{m_2}, \ldots, \xi_l^{m_l})$ from $\partial(B^k \times B_0^{do}) \times \tilde{S}^{2d-1}$ to $B^k \times B_0^{do} \times \tilde{S}^{2d-1}$, where $\tilde{S}^{2d-1} = \phi^{-1}(S^{2d-1})$. It is clear that Lemma 6.1 is valid in this context by changing the extension to points of the form $e^{i\psi}|\xi_r|$ to $e^{im_j\psi}G_j(\lambda, X, 0, \ldots, |\xi_r|, e^{-i\psi}\tilde{\xi})$. The first obstruction is for $r = d - s$ and it is the degree of $(F_0(\lambda_0, X), C(\lambda)(0, \ldots, 0, 1)^T, \xi_{s+1}, \ldots, \xi_r, \xi_2^{m_2}, \ldots, \xi_l^{m_l})$ which is $\pm \Pi_{j \geq 2} m_j^{d_j} P_*[C(\lambda)]$, hence the result.

Remark 6.6. If one does not use the induction hypothesis on l, then one has to consider the map

$$(F_0(\lambda, X), \begin{pmatrix} C_1(\lambda) & 0 \\ 0 & I \end{pmatrix} \eta_1, \ldots \begin{pmatrix} C_l(\lambda) & 0 \\ 0 & I \end{pmatrix} \eta_l).$$

The map, from $\partial(B^k \times B_0^{do}) \times \tilde{S}^{2d-1}$ into $\mathbb{R}^{do} \times \tilde{\mathbb{C}}^d \setminus \{0\}$, defined as $(F_0(\lambda, X), (C_1(\lambda)\xi_1^{m_1})^{\tilde{m}_j}, \ldots, (C_l(\lambda)\xi_l^{m_l})^{\tilde{m}_l}, \tilde{\xi}^M)$, where ξ_j are in $\mathbb{C}^s$, has an equivariant extension to $B^k \times B_0^{do} \times \tilde{S}^{2d-1}$. The same is true for the S^1-homotopic map

$$(F_0(\lambda, X), (C_1(\lambda)\xi_1^{m_1})^{\tilde{m}_1} + \xi_2^M, \ldots, C_{l-1}(\lambda)\xi_{l-1}^{m_{l-1}})^{\tilde{m}_{l-1}} + \xi_l^M,$$

$(C_l(\lambda)\xi_l^{M_l})^{\tilde{m}_l}, \tilde{\xi}^M)$. This map is nonzero on $\tilde{S}^{2d-1} \cap \{\xi_1 = 0\}$, hence the first obstruction for an extension will be the degree for $\xi_1^0 = (0, \ldots, 0,$

$(1 - \Sigma_{j\geq 2}|\xi_j|^2 - |\tilde{\xi}|^2)^{1/2})$ on $B^k \times B_0^{do} \times B^{2(d-s)}$. In order to compute this degree, one may use the fact that $P_*[C(\lambda)^{k^s}]$ and $(P_*[C(\lambda)])^k$, having the same degree, are homotopic. Hence $(C_1(\lambda)\xi_1^0)^{\tilde{m}_1}$ may be replaced by $C_1(\lambda)^{\tilde{m}_1^s}\xi_1^0$. This would correspond to the obstruction obtained for the map $(F_0(\lambda, X), C_1(\lambda)^{\tilde{m}_1^s}\xi_1^M, (C_2(\lambda)\xi_2^{m_2})^{\tilde{m}_2}, ..., \tilde{\xi}^M)$. One may rotate ξ_1 and ξ_2 in order to have a S^1–homotopy to $(F_0(\lambda, X), (C_2(\lambda)\xi_1^{m_2})^{\tilde{m}_2}, C_1(\lambda)^{\tilde{m}_1^s}\xi_2^M, ..., \tilde{\xi}^M)$. This map will have the same first obstruction as the map $(F_0(\lambda, X), C_2(\lambda)^{\tilde{m}_2^s}\xi_1^M, C_1(\lambda)^{\tilde{M}_1^s}\xi_2^M, ..., \tilde{\xi}^M)$ and, by rotation of the matrices, as the map $(F_0(\lambda, X), C_1(\lambda)^{\tilde{m}_1^s}C_2(\lambda)^{\tilde{m}_2^s}\xi_1^M, \xi_2^M, ..., \tilde{\xi}^M)$. Repeating this process, the obstruction will be the same for $(F_0(\lambda, X), \Pi_1^l C_j(\lambda)^{\tilde{m}_j^s}\xi_1^M, \xi_j^M, \tilde{\xi}^{\tilde{M}})$, which is $\pm M^{d-s}(s-1)!\Sigma_j \tilde{m}_j^s[C_j(\lambda)] = \pm M^d(s-1)!\Sigma_j[C_j(\lambda)]/m_j^s$. This number was called the generalized Fuller's degree in Ize (1985). The reader may consult this paper for the study of the unstable case, $k > 2d_j^{\mathbb{C}}$.

Continuation of the proof of Theorem 6.1: the $\mathbb{Z}_2$ case. It is easy to see that the equivalent of Lemma 6.1. takes the form:

> If $(G_0, G)(\lambda, X, \eta) : \partial(B^k \times B_0^{do}) \times S^{r-1} \to \mathbb{R}^{do} \times \mathbb{R}^d \backslash \{0\}$ is $\mathbb{Z}_2$–equivariant, with antipodal action on η and $\mathbb{R}^d$, then, if $k < d+1-r$, (G_0, G) has a nonzero $\mathbb{Z}_2$–extension to $B^k \times B_0^{do} \times S^{r-1}$ and any two $\mathbb{Z}_2$–extensions are $\mathbb{Z}_2$–homotopic on $B^k \times B_0^{do} \times S^{r-1}$ provided $k+1 < d+1-r$. If $k = d+1-r$, the obstruction is the degree of $(G_0, G)(\lambda, X, (1 - \Sigma_{j>1}x_j^2)^{1/2}, x_2, ..., x_r)$ with respect to $B^k \times B_0^{do} \times B^{r-1}$.

In fact it is enough to replace $e^{i\psi}$ by -1 in Lemma 6.1. Here, if $k \leq d$, then $B(\lambda)$ is homotopic to $\begin{pmatrix} C(\lambda) & 0 \\ 0 & I \end{pmatrix}$, with $C(\lambda)$ a $k \times k$ matrix (not unique).

The maps $(F_0(\lambda, X), 0, ..., 0, x_{k+1}, ..., x_d)$ and $(G_0, G)(\lambda, X, 0, ..., 0, x_{k+1}, ..., x_d)$ are two extensions of the first map from $\partial(B^k \times B_0^{do}) \times S^{d-k-1}$ to $B^k \times B_0^{do} \times S^{d-k-1}$. Hence they are not necessarily $\mathbb{Z}_2$–homotopic: the obstruction for this to be true is the degree of the map $F(t, \lambda, X, (1 - \Sigma_{k+2}^d x_j^2)^{1/2}, x_{k+2}, ..., x_d)$, with respect to $I \times B^k \times B_0^{do} \times B^{d-k-1}$, defined as, for example, a linear combination of the above two maps, with $x_{k+1} = (1 - \Sigma_{k+2}^d x_j^2)^{1/2}$.

Notice that the degree of $F(t, \lambda, X, -(1-\Sigma_{k+2}^d x_j^2)^{1/2}, x_{k+2}, ..., x_d) = (F_0, -\tilde{F})(t, \lambda, X, (...)^{1/2}, -x_{k+2}, ..., -x_d)$ is $(-1)^d(-1)^{d-k-1}$ times the first degree.

Now the maps $(F_0(\lambda, x), C(\lambda)(0, ... 0, (1-\Sigma_{k+1}^d x_j^2)^{1/2})^T, x_{k+1}, ..., x_d)$ and $(G_0, G)(\lambda, X, 0, ..., 0, (1-\Sigma_{k+1}^d x_j^2)^{1/2}, x_{k+1}, ..., x_d)$ coincide on $\partial(B^k \times B_0^{do}) \times B^{d-k}$ and are nonzero on $B^k \times B_0^{do} \times S^{d-k-1}$. The obstruction to the extension to $B^k \times B_0^{do} \times S^{d-1} \cap \{z_1 = ... = z_k = 0\}$ is their degree with respect to $B^k \times B_0^{do} \times B^{d-k}$. Obviously the degree for the second map is 0 and the degree for the first is $P_*[C(\lambda)]$. Note that if $k = d$, then there is no previous extension and $P_*[B(\lambda)] = 0$.

Denote the first map by $F_1(\lambda, X, \eta)$ and the second by $F_2(\lambda, X, \eta)$, with η in $\mathbb{R}^{d-k}$. Let $B_j^{d-k} = \{\eta : \|\eta\| < j\}$, for $j = 1$ or 2. On $B^k \times B_0^{do} \times B_2^{d-k}$ define a map $\tilde{F}$ given by $F_1(\lambda, X, \eta)$ if $\|\eta\| \leq 1$ and by $(\|\eta\| - 1)F_2(\lambda, X, \eta/\|\eta\|) + (2 - \|\eta\|)F_1(\lambda, X, \eta/\|\eta\|)$ if $1 \leq \|\eta\| \leq 2$. Note that on $\partial(B^k \times B_0^{do})$ the map is $(F_0(\lambda, X), 0, \eta/\|\eta\|)$ if $1 \leq |\eta| \leq 2$. It is easy to see that the degree of this map on $B^k \times B_0^{do} \times B_2^{d-k}$ is the degree of F_2 on $B^k \times B_0^{do} \times B_1^{d-k}$, that is 0, while, on $B^k \times B_0^{do} \times B_1^{d-k}$, this degree is $P_*[C(\lambda)]$. Hence $P_*[C(\lambda)] = -\deg(\tilde{F}; B^k \times B_0^{do} \times (B_2^{d-k} \backslash B_1^{d-k}))$. Now $\tilde{F}$ is a linear $\mathbb{Z}_2$-homotopy between F_1 and F_2 on $B^k \times B_0^{do} \times S^{d-k-1}$, which may be assumed to be nonzero if $x_{k+1} = 0$, from the equivalent of Lemma 6.2. Then the degree of $\tilde{F}$ may be computed on the hemispheres $x_{k+1} > 0$ and $x_{k+1} < 0$. Now, if (λ_0, X_0, η_0), with $x_{k+1} > 0$, is a zero, so is $(\lambda_0, X_0, -\eta_0)$. Hence the degree on the south hemisphere is $(-1)^d(-1)^{d-k}$ times the degree on the north hemisphere and the degree of $\tilde{F}$ is a multiple of $1 + (-1)^k$. Hence $P_*[C(\lambda)]$ is a multiple of $1 + (-1)^k$.

Note that this result is not surprising since in analyzing $\Pi_{k-1}(GL(\mathbb{R}^k))$ one may show that, for any $C(\lambda)$, $P_*[C(\lambda)]$ is 0 if k is odd and even if k is even, except if $k = 2, 4$ or 8, which are the only dimensions where there is a Hopf invariant equal to 1 : for these dimensions the above group has a generator β_0 with $P_*\beta_0 = 1$, $J\beta_0$ is the corresponding Hopf map, with $H(J\beta_0) = 1$.

End of the Proof of Theorem 6.1: the $\mathbb{Z}_p$ case. If a map has a $\Gamma-$ extension then it has a $\mathbb{Z}_p-$extension, where $\mathbb{Z}_p = \Gamma/H$. Also $V^H = \mathbb{R}^k \times \mathbb{R}^{do} \times \mathbb{R}^{d_1} \times \mathbb{C}^d$, where $\mathbb{Z}_p$ acts trivially on $\mathbb{R}^k \times \mathbb{R}^{do}$, as an antipodal map on $\mathbb{R}^{d_1}$, only if p is even, and as $e^{2\pi i m_j/p}$ on $\mathbb{C}^d$, $j = 1, \ldots l$; $m_1 = 1 < m_2 < \ldots < m_l$ and m_j and p may have common factors, corresponding to some $H_j > H$. Since one has a $\mathbb{Z}_2$ or trivial extension to $\mathbb{R}^k \times \mathbb{R}^{do} \times \mathbb{R}^{d_1}$, one may assume that the extension $(G_0, G_1, G)(\lambda, X, X_1, \eta)$ to $((G_0, G_1)(\lambda, X, X_1), B(\lambda)\eta)$ from $\partial(B^k \times B^{do} \times B^{d_1}) \times S^{2d-1}$ to $(B^k \times B^{do} \times B^{d_1}) \times S^{2d-1}$ satisfies that $(G_0, G_1)(\lambda, X, X_1)$ is zero only at $X = X_1 = 0$ and has degree, with respect to $B^{do} \times B^{d_1}$, equal to ± 1, since Proposition 6.2. and Remark 6.3 hold for this case. Of course, the same argument would work for $V^{H'}$ with $H' > H$ and so that $\Gamma/H' \cong \mathbb{Z}_{p'}$, with p' dividing p, instead of $p' = 2$.

Instead of following the extension process with the fundamental cell of Ize–Vignoli (1993), consider $\mathbb{C}^d$ with the standard action $e^{2\pi i/p}$ and the map $\phi : \mathbb{C}^d \to V^H$ given by $\phi(\xi_1, \ldots, \xi_l) = (\xi^{m_1}, \ldots, \xi_l^{m_l})$. Then $(G_0, G_1, G)(\lambda, X, X_1, \xi_1^{m_1}, \ldots, \xi_l^{m_l})$ will be a $\mathbb{Z}_p-$extension of $((G_0, G_1)(\lambda, X, X_1), B_j(\lambda)\xi_j^{m_j})$ from $\partial(B^k \times B^{do} \times B^{d_1}) \times \tilde{S}^{2d-1}$ to

$B^k \times B^{d_0} \times B^{d_1} \times \tilde{S}^{2d-1}$. Recall that here $k = 2s$.

Lemma 6.3. *Let (G_0, G_1, G) be a nonzero $\mathbb{Z}_p$−map from $\partial(B^k \times B^{d_0} \times B^{d_1}) \times S^{2r-1}$ into $\mathbb{R}^{d_0} \times \mathbb{R}^{d_1} \times \mathbb{C}^d \backslash \{0\}$. Then, if $k < 2(d - r + 1) - 1$, one has a $\mathbb{Z}_p$−extension to $B^k \times B^{d_0} \times B^{d_1} \times S^{2r-1}$ and two such extensions are $\mathbb{Z}_p$−homotopic if $k + 1 < 2(d - r + 1) - 1$. If $k = 2(d - r + 1)$, a first obstruction will be the degree of $(G_0, G_1, G)(\lambda, X, X_1, (1 - \Sigma_2^r |\xi_j|^2)^{1/2},$ $\xi_2, \ldots, \xi_r)$ with respect to $B^k \times B^{d_0} \times B^{d_1} \times B^{2(r-1)}$, and this degree is a multiple of p. If $k = 2(d - r)$, two extensions give rise to a map with degree a multiple of p, which is the only obstruction to the $\mathbb{Z}_p$−homotopy.*

Proof. If $\xi = (0, \ldots, 0, 1)^T$, one has an element of $\Pi_{k+d_0+d_1-1}(S^{d_0+d_1+2d-1})$, which is 0 if $k < 2d$. For $\xi = (0, \ldots, 0, e^{2\pi i n/p})$ one extends by taking $(G_0, G_1, G)(\lambda, X, X_1, \xi) = (G_0, (-1)^n G_1, e^{2\pi i n m_j/p} G) (\lambda, X, (-1)^n X_1,$ $0, \ldots, 0, 1)$. Then for $\xi = (0, \ldots, 0, e^{i\psi})$ with $0 < \psi < 2\pi/p$ one has an element in $\Pi_{k+d_0+d_1}(S^{d_0+d_1+2d-1})$ which is 0 if $k < 2d - 1$.

If this is the case one extends the map for $\xi = (0, \ldots, 0, e^{i\psi})$ with ψ in $[2\pi n/p, 2\pi(n + 1)/p]$ by taking $(G_0, G_1, G)(\lambda, X, X_1, \xi) = (G_0, (-1)^n G_1,$ $e^{2\pi i n m_j/p} G)(\lambda, X, (-1)^n X_1, 0, \ldots 0, e^{-2\pi i n/p} \xi)$.

One has an element of $\Pi_{k+d_0+d_1+1}(S^{d_0+d_1+2d-1})$ for $\xi = (0, \ldots, (1 - |\xi_r|^2)^{1/2}, \xi_r)$, and an extension for $\xi = (0, \ldots, e^{2\pi i n/p}(1 - |\xi_r|^2)^{1/2}, \xi_r)$ defined as $(G_0, (-1)^n G_1, e^{2\pi i n m_j/p} G) (\lambda, X, (-1)^n X_1, 0, \ldots 0, (1 - |\xi_r|^2)^{1/2},$ $e^{-2\pi i n/p} \xi_r)$ and so on... The last extension will be for $\xi = ((1 - \Sigma_2^r |\xi_j|^2)^{1/2},$ $\xi_2, \ldots, \xi_r)$ with an obstruction in $\Pi_{k+d_0+d_1+2(r-1)-1}(S^{d_0+d_1+2d-1})$ and for $\xi = (e^{i\psi}(\ldots)^{1/2}, \xi_2, \ldots, \xi_r)$ with an obstruction in the next group, which is 0 if $k < 2(d - r + 1) - 1$. If $k = 2(d - r + 1)$, then $B^k \times B^{d_0} \times B^{d_1} \times B^{2(r-1)}$ is divided in p sets with $Arg\, \xi_2$ in $[2\pi n/p, 2\pi(n + 1)/p]$, such that the map is non-zero on the boundary (from the extension process) and its degree on that sector is the degree of $(G_0, (-1)^n G_1, e^{2\pi i n m_j/p} G)(\lambda, X, (-1)^n X_1,$ $e^{-2\pi i n/p}(1 - \Sigma_2^r |\xi_j|^2)^{1/2}, e^{-2\pi i n/p} \xi_2, \ldots, e^{-2\pi i n/p} \xi_r)$ on the first sector, hence the same degree. The next obstruction will be in $\mathbb{Z}_2$. If $k = 2(d - r)$, one has the same situation, but in $\mathbb{Z}$, for the homotopy. Q.E.D.

Remark 6.7. It is clear that the same argument will work if the $\mathbb{Z}_2$−action on $\mathbb{R}^{d_1}$ is replaced by a $\mathbb{Z}_{p'}$−action on $\mathbb{R}^{d_1} \times \mathbb{C}^{d'}$ (hence p' must be even if p is even with antipodal action one $\mathbb{R}^{d_1}$ so that $\mathbb{R}^{d_1} \subset V^{H'}$. This is required by the mapping ϕ which is undefined on the real representations $\mathbb{R}^{d_1}$, in particular if d_1 is odd). Then $(-1)^n G_1$ has to be replaced by $e^{2\pi i m_j/p} G_1$ and $(-1)^n X_1$ by $e^{-2\pi i m_j/p} Z$, where $m_j/p = m'_j/p'$. Recall that in this case p' divides p.

Now, if all m_j's are 1 or if $B_j(\lambda)$ is deformable to I for $j \geq 2$, one has to consider the map $((G_0, G_1)(\lambda, X, X_1), C(\lambda)(0, \ldots, 0, (1 - \Sigma |z_j|^2)^{1/2})^T,$

$z_{s+1}, \ldots, z_{d_2}, \xi_2^{m_2}, \ldots, \xi_l^{m_l})$ on $\partial(B^k) \times B^{d_0} \times B^{d_1} \times B^{2(d-s)})$. By deforming λ in the first terms to λ_0 with $\|\lambda_0\| = \rho$, (G_0, G_1) can be taken as $(F_0(\lambda_0, X),$ $B_1(\lambda_0)X_1)$ with index ± 1 at $(X_0, X_1) = 0$ and the degree of this map is $\pm \Pi m_j^{d_j} P_*[(\lambda)]$ (here $r = d - s + 1, k = 2s$), while the degree of (G_0, G_1, G) is 0. As in the proof of the case $\mathbb{Z}_2$, one has that $\Pi m_j^{d_j} P_*[(\lambda)]$ is a multiple of p. Thus, if all m_j's are relatively prime to p, $P_*[(\lambda)]$ is a multiple of p. In general, the above relation will imply that $\Pi m_j^{d_j} P_*[C(\lambda)]$ is a multiple of p, where, in the product, there are only terms with m_j and p not relatively prime. By using Remark 6.7, one may also eliminate, for a minimal $H' >$ H, such that $\Gamma/H' \simeq \mathbb{Z}_{p'}$, with p' even if p is even, all m_j's such that $m_j/p = \tilde{m}_j/\tilde{p}_j$ with $\tilde{p}_j$ dividing p' and p' is the least common multiple of all such $\tilde{p}_j$'s. For each of the remaining m_j's, one has that l.c.m. $(p', p'_j) = p$ and m_j and p/p' have no common factor. Hence, one has that $P_*[C(\lambda)]$ is a multiple of p/p', for all such p', thus, also a multiple of the least common multiple of these p'.

Note that if $l = 1$ and $d_2 = s$, then one has directly that $P_*[B(\lambda)] = 0$; hence that $B(\lambda)$ is deformable to I, since this is the first extension to be made. Here $\mathbb{Z}_2$ may be replaced by $\mathbb{Z}_{p'}$, as in Remark 6.7.

In order to proceed with the proof, one needs an equivalent of Lemma 6.2.

Lemma 6.4. *Let* $(G_0, G_1, G_2, \tilde{G})$ *be a nonzero* $\mathbb{Z}_p$*–map defined on*

$$S^{2(d-d_2+r)-1} \times \partial(B^k \times B^{d_0} \times B^{d_1} \times \tilde{B}),$$

into $\mathbb{R}^{d_0} \times \mathbb{R}^{d_1} \times \mathbb{C}^d$, *where* $\tilde{B} = B^{2(d-d_2)} = \{\tilde{\eta} : |\tilde{\eta}| < 1\}$, *the sphere is the boundary of* $\{(\eta_2, \tilde{\eta}) : |\eta_2|^2 + |\tilde{\eta}|^2 < 1\}$, $\mathbb{Z}_p$ *acts as an antipodal map on* B^{d_1} *(if p is even), as* $e^{2\pi i m_j/p}$, *with* m_j *and* p *relatively prime, on* η_2 *in* $\mathbb{C}^r$, $0 \le r \le d_2$, *and as* $e^{2\pi i m_j/p}$ *on* $\tilde{B}$ *with no limitations on these* m_j's. *If* $r = 0$, *the map is required to have extensions on lower dimensional isotropy subspaces. Then, if* $k \le 2(d_2 - r)$, *the above map has a non-zero* $\mathbb{Z}_p$*–extension to* $S^{2(d-d_2+r)-1} \times B^k \times B^{d_0} \times B^{d_1}$. *If* $k+1 \le 2(d_2-r)$ *any two such extensions are* $\mathbb{Z}_p$*–homotopic on this last set, with a fixed homotopy on its boundary. If* $k = 2(d_2 - r)$, *two extensions give a map of degree which is a multiple of p and such that, if this degree is 0, the extensions are* $\mathbb{Z}_p$*–homotopic.*

Proof. The proof for the case $r \ge 1$ is along the lines of the preceding lemmas, since the action on η_2 is free. The argument for the homotopy is as in Lemma 6.3. The case $r = 0$ is more delicate and requires the extension Theorem 3.1 in Ize–Vignoli (1993): since one is working on the sphere, the conditions in this case is dim $V^H \le$ dim W^H, i.e. $k + d_0 + d_1 + 2(d - d_2) \le$ $d_0 + d_1 + 2d$.   Q.E.D.

Next, on $\partial(B^k \times B^{do} \times B^{d_1} \times B^{2(d-s)})$ the maps $((G_0, G_1)(\lambda, X, X_1),$ $C(\lambda)(0, \ldots, 0, (1 - \Sigma_{s+1}^{d_2}|z_j|^2 - |\tilde{\eta}|^2)^{1/2})^T, z_{s+1}, \ldots, z_{d_2}, \tilde{\eta})$ and $G(\lambda, X, X_1,$ $0, \ldots, 0, (\ldots)^{1/2}, z_{s+1}, \ldots, z_{d_2}, \tilde{\eta})$ coincide on $\partial(B^k \times B^{do} \times B^{d_1}) \times B^{2(d-s)}$ and differ on $B^k \times B^{do} \times B^{d_1} \times S^{2(d-s)-1}$ by a homotopy which has a degree which is a multiple of p. Here $r = d_2 - s$ and $k = 2s$. As in the proof of the $\mathbb{Z}_2$−case, this implies that the obstruction for the first map is a multiple of p and this obstruction is $P_*[C(\lambda)]$.

The next case is when $\tilde{B}(\lambda)\tilde{\eta}$ has the form $B_j(\lambda)\eta_j, j \geq 3$, with m_j and p not relatively prime (hence there are extensions to lower dimensional isotropy subspaces) and $m_2 = 1$ with $B_2(\lambda)$ of the form $\begin{pmatrix} C(\lambda) & 0 \\ 0 & I \end{pmatrix}$. From Lemma 6.4, with $r = 0$, one has a $\mathbb{Z}_p$−extension $(G_0, G_1, G_2, \tilde{G})$ for $|\tilde{\eta}| = 1$ and $\eta_2 = 0$ to $B^k \times B^{do} \times B^{d_1} \times \tilde{S}$, where G_2 is in $\mathbb{C}^s$. Then $F_1 \equiv$ $((1 - |\tilde{\eta}|)(G_0, G_1)(\lambda, X, X_1) + |\tilde{\eta}|(G_0, G_1, G_2)(\lambda, X, X_1, \tilde{\eta}/|\tilde{\eta}|), z_{s+1}, \ldots, z_{d_2},$ $|\tilde{\eta}|\tilde{G}(\lambda, X, X_1, \tilde{\eta}/|\tilde{\eta}|))$ is a nonzero $\mathbb{Z}_p$−extension of $((G_0, G_1)(\lambda, X, X_1), 0,$ $z_{s+1}, \ldots, z_{d_2}, B_j(\lambda)\eta_j)$ from $S^{2(d-s)-1} \times \partial(B^k \times B^{do} \times B^{d_1})$ to $S^{2(d-s)-1} \times$ $B^k \times B^{do} \times B^{d_1}$. Furthermore, the map $((1 - |\tilde{\eta}|) (G_0, G_1) (\lambda, X, X_1) +$ $|\tilde{\eta}|(G_0, G_1) (\lambda, X, X_1, \tilde{\eta}/|\tilde{\eta}|), C(\lambda)(0, \ldots, 0, (1 - \Sigma_{s+1}^{d_2}|z_j|^2 - |\tilde{\eta}|^2)^{1/2})^T +$ $|\tilde{\eta}|G_2(\lambda, X, X_1, \tilde{\eta}/|\tilde{\eta}|), z_{s+1}, \ldots, z_{d_2}, |\tilde{\eta}|\tilde{G}(\lambda, X, X_1, \tilde{\eta}/|\tilde{\eta}|))$ on $\partial(B^k \times B^{do} \times$ $B^{d_1} \times B^{2(d-s)})$, coincides with F_1 on $B^k \times B^{do} \times B^{d_1} \times S^{2(d-s)-1}$ and, on $\partial(B^k \times B^{do} \times B^{d_1}) \times B^{2(d-s)}$ is equal to $((G_0, G_1)(\lambda, X, X_1), C(\lambda)(0, \ldots,$ $0, (\ldots)^{1/2})^T, z_{s+1}, \ldots, z_{d_2}, B_j(\lambda)\eta_j))$. That is this map is a legitimate attempt for the extension to the next stage. The obstruction, the degree of this map, is, by Lemma 6.4, a multiple of p, and should be computed.

Assume for the moment that the extension F_1 is such that $G_2 = 0$. (This is in general not true since it would imply that there is no possible interaction between higher modes giving a nontrivial contribution to the fundamental mode. However it seems that, for the problem at hand with a linear leading term, this happens here. Another proof will be given below). Then one may compute the degree by deforming λ, everywhere except in $C(\lambda)$, to λ_0 with $\|\lambda_0\| = \rho$: on the boundary, with $S^{2(d-s)-1}$, the map F_1 is nonzero, on the other hand, on $\partial(B^{do} \times B^{d_1})$; then (G_0, G_1) is nonzero and, on $\partial B^k, C(\lambda)$ is invertible, hence one would need $|\tilde{\eta}| = 1$ with a nonzero F_1. It is then easy to conclude that the degree is $\pm P_*[C(\lambda)]$, a multiple of p.

Consider the case where $B_2(\lambda)$ is replaced by

$$\text{diag}\left(\begin{pmatrix} C_1(\lambda) & 0 \\ 0 & I \end{pmatrix}, \ldots, \begin{pmatrix} C_l(\lambda) & 0 \\ 0 & I \end{pmatrix}\right),$$

with actions $e^{2\pi i m_j/p}, m_1 = 1 < m_2 < \ldots < m_2$, with m_j and p relatively prime and $B_j(\lambda) = Id$ for the other terms. By going back to $(\xi_1, \ldots, \xi_l)$ with the standard action and the maps $\phi(\xi_1, \ldots, \xi_l) = (\xi_1^{m_1}, \ldots, \xi_l^{m_l})$, and

$\psi(\eta_1, \ldots, \eta_l) = (\eta_1^{n_1}, \ldots, \eta_l^{n_l})$, with $n_j m_j \equiv 1$, mod p, one obtains $((G_0, G_1)$ (λ, X, X_1), $(C_1(\lambda)\xi_1^{m_1})^{n_1}$, $\tilde{\xi}_1^{m_1 n_1}, \ldots, (C_l(\lambda)\xi_l^{m_l})^{n_l}$, $\tilde{\xi}_l^{m_l n_l}, \eta_j)$, where η_j corresponds to m_j not relatively prime to p. Replacing $(C_j(\lambda)\xi_j^{m_j})^{n_j}$ by $(C_j(\lambda)\xi_j^{m_j})^{n_j} + \xi_{j+1}^{m_{j+1} n_{j+1}}$ for $j = 1, \ldots, l-1$, one may repeat the arguments of Remark 6.6 and arrive at an obstruction which is the same, up to a multiple of p, as for the map $((G_0, G_1)(\lambda, X, X_1), C_1(\lambda)^{n_1^s} \ldots C_l(\lambda)^{n_l^s} \xi_1^{m_1 n_1}, \xi_2^{m_2 n_2}, \ldots \tilde{\xi}_l^{m_l n_l}, \eta_j)$; that is with a first obstruction $\pm \Pi(m_j n_j)^s P_*(\Pi C_j(\lambda)^{n_j^s})$, a multiple of p. This implies that $\Sigma n_j^s P_*(C_j(\lambda))$ is a multiple of p.

The last case is as above but with $B_j(\lambda)\eta_j$ and an extension on lower dimensional isotropy subspaces. Instead of proving that one has an extension with $G_2 = 0$, one may go back to Propositions 5.3 and 6.3. If $((G_0, G_1)$ $(\lambda, X, X_1), B(\lambda)\eta)$ has an extension, then $J^{\mathbb{Z}_p}[B(\lambda)] = 0$, hence $\Sigma^{\mathbb{Z}_p}(\text{diag}(B_{1,p}, ldots, B_{l,p})) = 0$. From Remark 6.4, this implies that $((G_0, G_1)$ $(\lambda, X, X_1), r - \epsilon, B_{1p}(\lambda)\eta_1, \ldots B_{lp}(\lambda)\eta_l, \tilde{\eta})$ has an extension. Here the suspension by $\tilde{\eta}$ provides an extension to $(\eta_1, \ldots, \eta_l) = 0$, which is of the form already studied above. The obstruction for this map is the suspension, due to the term $r - \epsilon$ of the preceding obstruction, but for Brouwer's degree, the suspension is an isomorphism.

It remains the case $k = 2$, which will be handled with a closer look at $J^{\mathbb{Z}_p}$. Let $\det B_{jp}(\lambda)$ have winding number r_j. Then $B_{jp}(\lambda)$ is deformable to $\begin{pmatrix} \lambda^{r_j} & 0 \\ 0 & I \end{pmatrix}$, where λ^{-1} means $\bar{\lambda}$. Then $J^{\mathbb{Z}_p}(B^{\mathbb{Z}_p}(\lambda)) = \Sigma_j \Sigma^{\mathbb{Z}_p} J(\lambda^{r_j})$ in the part of $\Pi_{S^{R^2} \times V}^{\mathbb{Z}_p}(S^V)$, with $V = E^{\mathbb{Z}_p}$, which is $\mathbb{Z}_p \times \mathbb{Z}_2$ if p is even and $\mathbb{Z}_{2p}$ if p is odd. In fact, if $\mathbb{Z}_p$ acts on z_j as $e^{2\pi i m_j / p}$ with $n_j m_j \equiv 1$, mod p, then the map η_j which is the suspension of $(\epsilon - |z_j|^2, \lambda z_j)$ generates this part together with another map $\tilde{\eta}$, with the relations $p(\eta_j + \tilde{\eta}) = 0, 2\tilde{\eta} = 0$ (see Ize–Vignoli: (1993), Theorem 8.3). Any map f which has the property that $f|_{z_j} \neq 0$, gives $[f] = r\eta_j + \gamma \tilde{\eta}$, with $\gamma = 0$ or 1 and r is deg$(f; B \cap \{z_j > 0\})$. Take $m_1 = 1$. Then $(z_1^{m_j n_j}, \lambda z_j)$ is $\mathbb{Z}_p$–deformable on $\partial B^2 \times \partial B$ to $(\lambda^{n_j} z_j^{n_j}, z_1^{n_j})$ and to $(\lambda^{n_j} z_1^{m_j n_j}, z_j)$, via $(\tau z_1^{m_j n_j} + (1-\tau)\lambda^{n_j} z_j^{n_j}, \tau \lambda z_j - (1-\tau) z_1^{m_j})$ and taking the second term to the n_j–power, and via a rotation of the form $(\tau z_j^{n_j} + (1 - \tau) z_1^{m_j n_j}, \tau z_1^{m_j} - (1 - \tau) z_2)$ for the second homotopy. Hence $(\epsilon^2 - |z_1|^2 - |z_j|^2, z_1^{m_j n_j}, \lambda z_j)$ is $\mathbb{Z}_p$–deformable on $\partial(B^2 \times B)$, to $(\epsilon^2 - |z_1|^2 - |z_j|^2, \lambda^{n_j} z^{m_j n_j}, z_j)$. Thus, $m_j n_j \eta_j + \tilde{\gamma}_j \tilde{\eta} = n_j \eta_1 + \gamma_1 \tilde{\eta}$ and $\eta_j = n_j \eta_1 + \gamma_j \tilde{\eta}$. If p is odd then, from $p(\eta_j + \tilde{\eta}) = 0 = p(\eta_1 + \tilde{\eta})$, one has that $\gamma_j = n_j - 1$. If p is even, then m_j and n_j are odd. Let $m_j n_j = 1 + kp$ and, on $\partial(|\lambda| \leq 1, |Z| \leq 2)$, consider the homotopic maps $(1 - |Z|^2, z_1', z_1(z_1^{kp} - ti2^{-kp}), \lambda z_j)$ and $(1 - |Z|^2, \lambda^{n_j} z_1', z_1(z_1^{kp} - ti2^{-kp}), z_j), 0 \leq t \leq 1$ decompose the set $\{|z_1| \leq 2\}$ in $B_1 = \{z_1 : |z_1| < 1/4\}$ and its complement B_2. The $\mathbb{Z}_p$–degree of the above maps is the sum of the degrees on B_1 and B_2. On B_1, one may deform z_1^{kp} to 0 and obtain η_j in the first case, $n_j \eta_1$ in the second. On B_2, one may use the homotopy $\tau(1 - |Z|^2) + (1 - \tau)(\epsilon^{2kp} - |z_1^{kp} - i2^{-kp}|^2 - |z_1'|^2 - |z_j|^2)$, where ϵ is so small that any disc, with center at a point with $z_1' = 0, z_j = 0, |z_1| = 1/2$, does not intersect B_2; hence, for $\tau = 0$, its degree is the same on the full set $\{|Z| < 2\}$. For $\epsilon < 1/2$, the map is nonzero if z_1 is real and nonnegative, hence its class is a multiple of $\tilde{\eta}$, computed on the boundary of the fundamental cell $\{0 \leq Arg\ z_1 \leq 2\pi/p\}$. There, one may deform z_1 to 1 and get $(z_1^{kp} - i2^{-kp})$ which has k zeros in the cell. For the first map one obtains $k\tilde{\eta}$ and for the second $kn_j\tilde{\eta}$. Thus, $\eta_j + k\tilde{\eta} = n_j\eta_1 + kn_j\tilde{\eta}$ and $\eta_j = n_j\eta_1 + k(n_j - 1)\tilde{\eta}$, and, since $m_j n_j = 1 + kp$, then $k(n_j - 1)$ and $n_j - 1$ have the same parity, $\eta_j = n_j\eta_1 + (n_j - 1)\tilde{\eta}$, witn $n_j - 1$ even if p is even.

Consequently, $J^{\mathbb{Z}_p}(B^{\mathbb{Z}_p}(\lambda)) = \Sigma_j r_j \eta_j = (\Sigma r_j n_j)\eta_1 + (\Sigma(n_j - 1)r_j)\tilde{\eta}$. Hence, if $\Sigma r_j \eta_j = 0$, one has that $\Sigma r_j n_j$ is a multiple of p. Furthermore, since $J(B^{\mathbb{Z}_p}(\lambda)) = 0$ i.e. non equivariant. Then Σr_j must be even, hence $0 = (\Sigma r_j n_j)(\eta_j + \tilde{\eta})$. Conversely, if $\Sigma n_j r_j$ is a multiple of p and Σr_j is even if p is odd, then $J^{\mathbb{Z}_p}(B^{\mathbb{Z}_p}(\lambda)) = 0$. Since $d_0 \geq 3$ and Σ_0 is an isomorphism, the last result of the theorem comes from Proposition 6.3.   Q.E.D.

Remark 6.8. If $m_{j'} = p - m_j$, then one may take $m_{j'} = -m_j$ and $n_{j'} = -n_j$. Then $(\lambda z_j + |Z|^2 \bar{z}_j', \lambda z_j' - |Z|^2 \bar{z}_j)$ has no zeros but $z_j = z_j' = 0$. In general $(\bar{\lambda}^{n_j} z_1^{m_j n_j} + |Z|^2 z_j^{n_j}, \lambda z_j - |Z|^2 z_1^{m_j})$ has no zeros but $z_1 = z_j = 0$, but this example is not linear dominant, except for $m_j n_j = 1$, i.e. the preceding case, near the origin.

Remark 6.9. Since $J^{\mathbb{Z}_p}(\lambda^{kp+n} z_1) = n\eta_1 + kp\tilde{\eta}$, it is easy to see that $J^{\mathbb{Z}_p}$ is onto $\mathbb{Z}_p$ if p is even and, if p is odd, onto $\mathbb{Z}_{2p}$. Hence under the conditions $d_0 \geq 3, d_j \geq 3, d_{jp} \geq 2$, the morphism J^Γ has, for $k = 2$, an image which is all of $\Pi^\Gamma_{S^{R^2} \times V}(S^V)$, except the $\mathbb{Z}_2$ parts for p even and kernel as discribed in (g) of Theorem 6.1.

Remark 6.10. There are many examples of applications of these results. Since the purpose of this paper is the topological content of bifurcation, I will only mention some of the most frequently used.

1) *Hopf bifurcation*: $\nu dX/dt - L(\lambda)X = g(\lambda, X)$, where $\lambda \in \mathbb{R}$ and $L(\lambda_0)$ is invertible for some λ_0. Assuming that $L(0)$ has eigenvalues $\pm i\nu_0 m_1, \ldots, \pm i\nu_0 m_s$ and that $L(\lambda)$ has its corresponding eigenvalues off the imaginary axis for $\lambda \neq 0$, then r_j, as the element in the S^1−degree of $J^{S^1}(\{in\nu I - L(\lambda))$ at $(\nu_0, 0)$, is the net crossing number of eigenvalues at $i\nu_0 m_j$.
See Ize et al. (1992) p. 132.

2) *Hopf bifurcation for systems with a first integral*, in particular for Hamiltonian systems, with $dX/dt = L(\lambda)X + g(\lambda, X)$ and the

first integral $V(X, \lambda)$ such that $\nabla V(X, \lambda) = H(\lambda)X + k(X, \lambda)$. If $L(\lambda)$ has the same hypothesis as in case (1) and ker $H(0) \cap \ker(im_j I - L(0)) = \{0\}$, then the element of the S^1−degree of $J^{S^1}(\dot{X} - L(\lambda)X - g(\lambda, X) - \nu \nabla V)$ at $(\lambda, \nu) = (0, 0)$ is the difference of the signature $\sigma_j^+(\rho) - \sigma_j^-(\rho)$, where $\sigma_j^+(\rho)$ is the sum of the signatures of $H(\rho)$ on the generalized eigenspaces of $L(\rho)$, with eigenvalues $i\beta$, with β close to m_j but larger than m_j, while $\sigma_j^-(\rho)$ corresponds to those smaller than m_j. See Dancer-Toland (1990) and Ize et al. (1992), p. 141 and 153 for the case of Hamiltonian systems.

   3) *Hopf bifurcation for equations with delays* : $dX/dt = \Sigma h_j(\lambda, X(t - r_j))$, which after the time scaling, $s = \nu t$, gives $\nu dX/ds = \Sigma h_j(\lambda, X((s - \nu r_j)/\nu))$, with 2π−periodic solutions. This problem is clearly S^1−equivariant, with the n−Fourier coefficient of h_j given by $e^{-inr_j \nu}$ times the Fourier coefficient of $h_j(\lambda, X(t/\nu))$. In particular, assuming $h_j(\lambda, 0) = 0$, for the linear terms, one obtains the indicial equation $i\nu n I = \Sigma k_j(\lambda) e^{-inr_j \nu}$, with $k_j(\lambda) = D_X h_j(\lambda, 0)$. If dim $X = 1$ and the equation is $x'(t) = -\lambda x(t) + \lambda f(x(t - r))$, with $f'(0) = -k < -1$, then the possible bifurcation points are for $\nu = (\nu_0 + 2m\pi)/nr$, where $\nu_0 = \text{Arcos}(-k^{-1})$, and $\lambda = (\nu_0 + 2m\pi)k/r(k^2 - 1)^{1/2}$. It is not difficult to prove that the S^1-index of $J(B_n(\lambda))$ is 1, for all n's. Thus, the bifurcating branch goes to ∞. If $r = 1$, then one may show that ν is different from $2m\pi$, see Mallet-Paret-Nussbaum (1992), and, if f is bounded, that $\|x\|_1 \leq K|\lambda|$. If furthermore $xf(x) < 0$, then one may show that the branch cannot go to negative λ's and that any periodic solution must have a zero. Thus, the global branch must go to ∞ in λ. In this case the branch corresponding to n is a scaling of the branch corresponding to 1. Many other situations can be analyzed in this setting.

   4) <u>Gradient</u> Γ −maps, as explained at the beginning of this section. For the case of S^1−gradient maps, see Dancer (1985) and Ize et al. (1992) p. 116.

   5) <u>Period doubling, symmetry breacking</u>, (for the S^1−case, see the above reference), *twisted periodic solutions* (see Fiedler (1988)) and so on...

References

Adams, J.F., Prerequisites (on equivariant stable homotopy theory) for Carlsson's lecture, *Lect. Notes in Math.* **1091** (1982), 483–532.

Alcaraz, D., Existence Theory for a model of Steady Vortex motion in Ideal Fluids, Oxford Ph.D. thesis, 1983.

Alexander, J.C., Bifurcation of zeros of parametrized functions, *J. Funct. Anal.* **29** (1978), 37–53.

Alexander, J.C., Calculating bifurcation invariants as elements of the homotopy of the general linear group. II, *J. Pure and Appl. Algebra*, **17** (1980), 117–125.

Alexander, J.C., A primer on connectivity, *Lect. Notes in Math.*, Springer-Verlag, **886** (1981), 445–483.

Alexander, J.C. and S.S. Antman, Global and local behavior of bifurcating multidimensional continua of solutions for multiparameter nonlinear eigenvalue problems, *Arch. for Rat. Mech. and Anal.* **76** (1981), 339–354.

Alexander, J.C. and S.S. Antman, Global behavior of solutions of nonlinear equations depending on infinite dimensional parameter, *Indiana Univ. Math. J.* **32** (1983), 39–62.

Alexander, J.C. and J. F.G. Auchmuty, Global branches of waves, *Manus. Math.* **27** (1979), 208–220.

Alexander, J.C. and P.M. Fitzpatrick, The homotopy of certain spaces of nonlinear operators and its relation to global bifurcation of the fixed points of parametrized condensing operators, *J. Funct. Anal.* **34** (1979), 87–106.

Alexander, J.C., I. Massabó and J. Pejsachowicz, On the connectivity properties of the solution set of infinitely parametrized families of vector fields, *Boll. Un. Mat. Ital.* **A. (6) 1** (1982), 309–312.

Alexander, J.C. and J.A. Yorke, The implicit function theorem and global methods of cohomology, *J. Funct. Anal.* **21** (1976), 330–339.

Alexander, J.C. and J.A. Yorke, Parametrized functions, bifurcation and vector fields on spheres. Anniversary Volume in Honnor of Mitropolsky, *Naukova Dumka*, **275** (1977), 15–17.

Alexander, J.C. and J.A. Yorke, Global bifurcation of periodic orbits, *Amer. J. Math.* **100** (1978), 263–292.

Alexander, J.C. and Yorke, J.A., Calculating bifurcation invariants as elements of the general linear group I, *J. Pure and Appl. Algebra*, **13** (1978), 1–8.

Alexander, J.C. and J.A. Yorke, On the continuability of periodic orbits of parametrized three-dimensional differential equations, *J. of Diff. Eq.* **49** (1983), 171–184.

Alligood, K.T., Homological indices and homotopy continuation, Ph.D. Thesis, Univ. of Maryland, 1979.

Alligood, K.T., Mallet-Paret, J. and J.A. Yorke, Families of periodic orbits: local continuability does not imply global continuability, *J. Diff. Geometry* **16** (1981), 483–492.

Alligood, K.T. and J.A. Yorke, Hopf bifurcation: the appearance of virtual periods in cases of resonance, *J. Diff. Eq.* **64** (1986), 375–394.

Amann, H. Fixed point equations and Nonlinear eigenvalue problems in ordered Banach spaces, *SIAM Reviews*, **18** (1976), 620–709.

Amann, H., Ambrosetti, A. and G. Mancini, Elliptic equations with non-invertible Fredholm linear part and bounded nonlinearities, *Math. Z.* **158** (1978), 179–194.

Antman, S.S., Buckled states of nonlinearly elastic plates, *Archiv. Rat. Mech. and Anal.* **67** (1978), 111–149.

Arnold, V.I., Lectures on bifurcation and versal systems, *Russ. Math. Surveys*, **27** (1972), 54–113.

Balonov, Z., Kushkuley, A. and P. Zabrejko, A degree theory for equivariant maps: geometric approach. To appear in *Top. Methods in Nonlinear Anal.*, 1993.

Bartsch. T., Verzweigung in Vektorraumbündels und äquivariante Verzweigung, Ph.D. Thesis, Univ. München, 1986.

Bartsch, T., Global bifurcation from a manifold of trivial solutions, *Univ. of Heidelberg Math. Inst.* **13** (1987).

Bartsch, T., A global index for bifurcation of fixed points, *J. Reine Math.* **391** (1988), 181–197.

Bartsch, T., The role of the J-homomorphism in multiparameter bifurcation theory, *Bull. Sei. Math.* **112** (1988), 177–184.

Bartsch. T. The global structure of the zero set of a family of semilinear Fredholm maps, *Nonlinear Anal. T.M.A.* **17** (1991), 313–332.

Bartsch, T., A simple proof of the degree formula for $\mathbb{Z}/p$–equivariant maps, Univ. of Heidelberg, preprint, to appear in *Math. Z.*, 1991.

Bartsch, T., Topological Methods for variational problems with symmetries, *Habilitationsschrift*, Heidelberg, 1992.

Bartsch, T. and M. Clapp, Bifurcation theory for symmetric potential operators and the equivariant cup-length, *Math. Z.* **204** (1990), 341–356.

Bazley, N.W., McLeod, J.B., Bifurcation from infinity and singular eigenvalue problems, *Proc. London Math. Soc.* **34** (1977), 231–244.

Benjamin, T.B., Applications of Leray-Schauder degree theory to problems in hydrodynamic stability, *Math. Proc. Cambridge Phil. Soc.* **79** (1976), 373–392.

Berestycki , H., On some nonlinear Sturm Liouville problems, *Jour. of Diff. Eq.* **26** (1977), 375–390.

M. Berger and M. Berger, *Perspectives in Nonlinearity. An Introduction to Nonlinear Analysis*, Benjamin Inc., 1968.

Berger, M.S., A Sturm Liouville theorem for nonlinear elliptic partial differential equations, *Ann. Scuola Norm. Sup. Pisa*, **3.20** (1966), 543–582.

Berger, M.S., A Bifurcation theory for nonlinear elliptic Partial Differential Equations and related Systems, J.B. Keller and S. Antman, Ed., Benjamin, 113–216, 1969.

Berger, M.S., Applications of global analysis to specific nonlinear eigenvalue problems, *Rocky Mountain J. of Math.* **3** (1973), 319–354.

Berger, M.S., *Nonlinearity and Functional Analysis*, Academic Press, 1977.

Berger, M.S. and Podolak, E., On nonlinear Fredholm operator equations, *Bull A.M.S.* **80** (1974), 861–864.

Berger, M.S. and D. Westreich, A convergent iteration scheme for bifurcation theory on Banach spaces, *J. Math. Anal. Appl.* **43** (1974), 136–144.

Böhme, R., Die Lösung der Verzweigungsgleichungen für nichtlineare Eigenwert-probleme, *Math. Z.* **127** (1972), 105–126.

Borisovich, Y.G., Topology and nonlinear functional analysis, *Russian Math. Surveys*, **37** (1979), 14–23.

Borisovich, Y.G., V.G. Zvyagin and Y.I. Sapronov, Nonlinear Fredholm maps and the Leray Schauder theory, *Russian Math. Surveys*, **32** (1977), 1–54.

Bredon, G.E., *Introduction to Compact Transformation Groups*, Academic Press. New York, 1980.

Bröcher, T. and T. tom Dieck, Representations of compact Lie groups, *Grad. Texts in Math.* **98**, Springer-Verlag, New York, 1985.

Browder, F.E., Nonlinear eigenvalue problems and group invariance, *Functional Analysis and Related Fields*, 1-58, Springer, 1970.

Browder, F.E., Nonlinear operators in Banach spaces, *Proc. Symp. P.M.* **18**, vol. 2, (1970), A.M.S.

Browder, F.E. and R.D. Nussbaum, The Topological degree for non compact nonlinear mappings in Banach spaces, *Bull. A.M.S.* **74** (1968), 671–676.

Buchner, M., Marsden, J. and S. Schecher, Applications of the blowing up construction and algebraic geometry to bifurcation problems, *J. Diff. Eq.* **48** (1983), 404–433.

Cantrell, R.S., A homogeneity condition guaranteeing bifurcation in multiparameter eigenvalues problems, *Nonlinear Anal. T.M.A.* **8** (1984), 159–169.

Cantrell, R.S., Multiparameter bifurcation problems and topological degree, *J. of Diff. Eq.* **52** (1984), 39–51.

Cantrell, R.S., A homogeneity condition guaranteeing bifurcation in multiparameter nonlinear eigenvalue problems, *Nonlinear Analysis, T.M.A.* **8** (1984), 159–169.

Cerami, G., Symmetry breaking for a class of semilinear elliptic problems, *Nonlinear Anal. T.M.A.* **10** (1986), 1–14.

Cesari, L., Functional Analysis, Nonlinear Differential Equations and the alternative method, *Lect. Notes in Pure and Applied Math.* **Vol. 19** (1976), Marcel Decker, 1–198.

Chang, K.C., Applications of homology theory to some problems in Differential Equations, *Proc. of Symp. in Pure Math.* **45** (1986), 253–262.

Chen, B., Jorge M.C. and A.A. Minzoni, Bifurcation of solutions for an inverse problem in potential theory, *Studies in Applied Math.* **86** (1992), 31–51.

Chiappinelli, R. and C.A. Stuart, Bifurcation when the linearized problem has no eigenvalues, *J. of Diff. Eq.* **30** (1978), 296–307.

Chossat, P. and G. Iooss, Primary and secondary bifurcation in the Couette-Taylor problem, *Japan J. of Appl. Math.* **2** (1985), 37–68.

Chow, S.N. and J.K. Hale, Methods of bifurcation theory, *Grundl. Math. Wiss.* **251**, Springer Verlag, 1982.

Chow, S.N. and R. Lauterbach, A bifurcation theorem for critical points of variational problems, *Nonlinear Anal. T.M.A.* **12** (1988), 51–61.

Chow, S.N. and J. Mallet-Paret, The Fuller index and global Hopf bifurcation, *J. Diff. Eq.* **29** (1978), 66–85.

Chow, S.N., J. Mallet-Paret and J.A. Yorke, Global Hopf bifurcation from a multiple eigenvalue, *Nonlinear Anal. TMA*, **2** (1978), 753–763.

Cicogna, G., Bifurcation from topology and symmetry arguments, *Bol. U.M.I.* **6** (1984), 131–138.

Conley, C., Isolated invariant sets and the Morse index, *CBMS regional Conf. Series in Math.* **38**, 1978.

Cosner, C., Bifurcation from higher eigenvalues in nonlinear elliptic equations: continua that meet infinity. Univ. of Miami, preprint, 1981.

Crandall, M.G. and P.H. Rabinowitz, Bifurcation from simple eigenvalues, *J. Func. Anal.* **8** (1971), 321–340.

Crandall, M.G. and P.H. Rabinowitz, The Hopf bifurcation theorem in infinite dimensions, *Arch. Rat. Mech. Anal.* **67** (1977), 53–72.

J. Cronin, 1964. Fixed points and topological degree in Nonlinear Analysis, *Mathematical Surveys*, **11** (1964), A.M.S. Providence.

Cronin, J., Eigenvalues of some nonlinear operators, *J. of Math. Anal. and Appl.* **38** (1972), 659–667.

Dancer, E.N., Bifurcation theory in real Banach space, *Proc. London Math. Soc.* **23** (1971), 699–734.

Dancer, E.N., Bifurcation theory for analytic operators, *Proc. London Math. Soc.* **26** (1973), 359–384.

Dancer, E.N., Global structure of the solutions of non-linear real analytic eigenvalue problems, *Proc. London Math. Soc.* **26** (1973), 745–765.

Dancer, E.N., On the structure of solutions of non-linear eigenvalue problems, *Indiana Univ. Math. J.* **23** (1974), 1069–1076.

Dancer, E.N., On the existence of bifurcating solutions in the presence of symmetries, *Proc. Royal Soc. Edinburg A*, **85** (1980), 321–336.

Dancer, E.N., Symmetries, degree, homotopy indices and asymptotically homogeneous problems, *Nonlinear Anal. T.M.A.* **6** (1982), 667–686.

Dancer, E.N., On the indices of fixed points of mappings in cones and applications, *J. of Math. Anal. and Appl.* **91** (1983), 131–151.

Dancer, E.N., Perturbation of zeros in the presence of symmetries, *J. Austral. Math. Soc.* **36** (1984), 106–125.

Dancer, E.N., A new degree for S^1-invariant gradient mappings and applications, *Annal. Inst. H. Poincaré, Anal. Non. Lin.* **2** (1985), 329–370.

Dancer, E.N. and J.F. Toland, Degree theory for orbits of prescribed period of flows with a first integral, *Proc. London Math. Soc.* **60** (1990), 549–580.

Dancer, E.N. and J.F. Toland, Equilibrium states in the degree theory of periodic orbits with a first integral, *Proc. London Math. Soc.* **61** (1991), 564–594.

Deimling, K., *Nonlinear Functional Analysis*, Springer Verlag, 1985.

Dellnitz, M., I. Melbourne and J.E. Marsden, Generic bifurcation of Hamiltonian vector fields with symmetry, *Nonlinearity* **15** (1992), 979–996.

Dugundji, J. and A. Granas, Fixed point theory I. Warzawa: PWN-Polish Scientific, 1982.

Dylawerski, G., Geba, K., Jodel, J. and W. Marzantowicz, An S^1-equivariant degree and the Fuller index, *Ann. Pol. Math.* **52** (1991), 243–280.

Eells, J., Fredholm structures, *Symp. Nonlinear Functional Anal.* **18** (1970), 62–85.

Elworthy, K.D. and A.J. Tromba, Degree theory on Banach manifolds. "Nonlinear Functional Analysis", *Proc. Symp. Pure Math.* Vol. **18/1** (1970), AMS, 86–94.

Erbe, L., Geba, K., Krawcewicz, W. and J. Wu, S^1-degree and global Hopf bifurcation theory of functional differential equations, *J. Diff. Eq.* **98** (1992), 277–298.

Esquinas, J. and J. López Gómez, Optimal multiplicity in local bifurcation theory. I. Generalized generic eigenvalues, *J. Diff. Eq.* **71** (1988), 71–92.

Esquinas, J., Optimal multiplicity in bifurcation theory. II. General Case, *J. Diff. Eq.* **75** (1988), 206–215.

Fadell, E.R. and P.H. Rabinowitz, Generalized cohomological index theories for Lie group actions with application to bifurcation questions for Hamiltonian systems, *Invent. Math.* **45** (1978), 139–174.

Fenske, C., An index for periodic orbits of functional differential equations, *Math. Ann.* **285** (1989), 381–392.

Field, M.J. and R.W. Richardson, Symmetry breaking and branching problems in equivariant bifurcation theory, I, *Arch. Mat. Mech. Anal.* **118** (1992), 297–348.

Fiedler, B., An index for global Hopf bifurcation in parabolic systems, *J. Reine u. Angew. Math.* **359** (1985), 1–36.

Fiedler, B., Global Hopf bifurcation of two-parameters flows, *Arch. Rat. Mech. Anal.* **94** (1986), 59–81.

Fiedler, B., Global Bifurcation of periodic solutions with symmetry, *Lect. Notes in Math.* **1309** (1988), Springer Verlag.

Fife, P.C., Branching Phenomena in fluid dynamics and chemical Reaction-diffusion theory, *CIME Lect. Notes* (1974).

Fitzpatrick, P.M., A generalized degree for uniform limits of A-proper mappings, *J. Math. Anal. Appl.* **35** (1970), 536–552.

Fitzpatrick, P.M., A-proper mappings and their uniform limits, *Bull. AMS* **78** (1972), 806–809.

Fitzpatrick, P.M., On the structure of the set of solutions of equations involving A-proper mappings, *Trans. AMS* **189** (1974), 107–131.

Fitzpatrick, P.M., Homotopy, linearization and bifurcation, *Nonlinear Anal. T.M.A.* **12** (1988), 171–184.

Fitzpatrick, P.M., The stability of parity and global bifurcation via Galerkin Approximation, *J. London Math. Soc.* **38** (1988), 153–165.

Fitzpatrick, P.M., Massabó, I. and Pejsachowicz, J., Complementing maps, continuation and global bifurcation, *Bull. A.M.S.* **9** (1983), 79–81.

Fitzpatrick, P.M., Massabó I. and Pejsachowicz, J., Global several parameters bifurcation and continuation theorems, a unified approach via complementing maps, *Math. Ann.* **263** (1983), 61–73.

Fitzpatrick, P.M., I. Massabó and J. Pejsachowicz, On the covering dimension of the set of solutions of some nonlinear equations, *Trans. A.M.S.* **296** (1986), 777–798.

Fitzpatrick, P.M. and J. Pejsachowicz, The fundamental group of the space of linear Fredholm operators and the global analysis of semilinear equations, *Contemp. Math.* **72** (1988), 47–87.

Fitzpatrick, P.M. and J. Pejsachowicz, A local bifurcation theorem for C^1-Fredholm maps, *Proc. Amer. Math. Soc.* **109** (1990), 995–1002.

Fitzpatrick, P.M. and J. Pejsachowicz, Parity and generalized multiplicity, *Trans. A.M.S.* **326** (1991), 281–305.

Fitzpatrick, P.M. and J. Pejsachowicz, Nonorientability of the index bundle and several parameter bifurcation, *J. of Funct. Anal.* **98** (1991), 42–58.

Fitzpatrick, P.M. and J. Pejsachowicz, The Leray-Schauder theory and fully non-linear elliptic boundary value problems, *Memoirs A.M.S.*, Vol. 101, No. **483** (1993).

Fitzpatrick, P.M., J. Pejsachowicz and P.J. Rabier, Topological degree for nonlinear Fredholm operators, *C.R. Acad. Sci. Paris*, **311** (1990), 711–716.

Fucik, S., Necas, J., and Soucek, V., Spectral Analysis of Nonlinear Operators, *Lect. Notes in Math.* **343** (1973), Springer-Verlag.

Fuller, F.B., An index of fixed point type for periodic orbits, *Am. J. Math.* **89** (1967), 133–148.

Furi, M., Martelli, M. and A. Vignoli, On the Solvability of nonlinear operator equations in normed spaces, *Ann. Mat. Pura Appl.* **124** (1980), 321–343.

Furi, M. and M.P. Pera, A continuation principle for forced oscillations on differentiable manifolds, *Pacific J. of Math.* **121** (1986), 321–338.

Gaines, R.E. and J.L. Mawhin, Coincidence degree and nonlinear differential equations, *Lect. Notes in Math.* **568** (1977), Springer Verlag.

Gavalas, G.R., *Nonlinear differential equations of chemically reacting systems*, Springer Verlag Tracts in Nat. Phil. **17**, 1968.

Geba, K. and A. Granas, Infinite dimensional cohomology theories, *J. Math. Pures et Appl.* **52** (1973), 147–270.

Geba, K., Krawcewicz, W. and J. Wu, An equivariant degree with applications to symmetric bifurcation problems, Preprint, 1993.

Geba, K., Marzantowicz, W., Global bifurcation of periodic orbits, *Topological Methods in Nonlinear Analysis*, **1** (1993), 67–93.

Geba, K., Massabó, I. and Vignoli, A., Generalized topological degree and bifurcation, *Nonlinear Functional Analysis and its Applications*, Reidel, (1986), 55–73.

Geba, K., Massabó, I, and Vignoli, A., On the Euler characteristic of equivariant vector fields, *Boll. UMI.* **4A** (1990), 243–251.

Golubitsky, M. and J.M. Gukenheimer (eds), Multiparameter bifurcation theory, *Cont. Math.* **56** (1986), AMS.

Golubitsky, M. and D.G. Schaeffer, Singularities and groups in bifurcation theory I, *Appl. Math. Sc.* **51** (1986), Springer Verlag.

Golubitsky, M., D.G. Schaeffer and I.N. Stewart, *Singularities and Groups in Bifurcation Theory II*, Springer Verlag, 1988.

Granas, A., The theory of compact vector fields and some of its applications to topology of functional spaces, *Rozprawy Mat.* **30** (1962), Warzawa.

Granas, A., The Leray-Schauder index and the fixed point theory for arbitrary ANRs, *Bull. Soc. Math. France*, **100** (1972), 209–228.

Guckenheimer, J., Holmes, P., *Nonlinear Oscillations, Dynamical Systems and Bifurcation of Vector Fields*, Springer Verlag, 1983.

Gurel, O., Rössler, O.E., Bifurcation theory and applications in scientific disciplines, *Annals New York Acad. Sci.* **316** (1979).

Hassard, B.J., Kazarinoff, N.D., Wan, Y-H., *Theory and Applications of Hopf Bifurcation*, Cambridge University Press, 1981.

Hauschild, H., Äquivariante homotopie I, *Arch. Math.* **29** (1977a), 158–167.

Hauschild, H., Zerspaltung äquivarianter Homotopiemengen, *Math. Ann.* **230** (1977b), 279–292.

Healey, T.J. and K. Kielhöfer, Symmetry and nodal properties in the global bifurcation analysis of quasi-linear elliptic equations, *Arch. Rat. Mech. Anal.* **113** (1991), 299–311.

Heinz, G., Lösungsverzweigung bei analytischen gleichungen mit Fredholm-operator vom Index null, *Math. Nachr.* **128** (1986), 243–254.

Hetzer, G., Stallbohm, V., Global behaviour of bifurcation branches and the essential spectrum, *Math. Nachr.* **86** (1978), 347–360.

Hoyle, S.C., Local solutions manifolds for nonlinear equations, *Nonlinear Anal. T.M.A.* **4** (1980), 285–295.

Hoyle, S.C., Hopf bifurcation for ordinary differential equations with a zero eigenvalue, *J. Math. Anal. Appl.* **74** (1980), 212–232.

Hernández, J., Bifurcación y soluciones positivas para algunos problemas de tipo unilateral. Tesis doctoral, Univ. Aut. de Madrid, 1977.

Iooss, G., *Bifurcation of Maps and Applications*, North Holland, 1979.

Iooss, G. and D.D. Joseph, *Elementary Stability and Bifurcation Theory*, Springer Verlag, 1980.

Ize, J., Bifurcation theory for Fredholm operators, *Memoirs A.M.S.* **174** (1976).

Ize, J., Periodic solutions for nonlinear parabolic equations, *Comm. in P.D.E.* **12** (1979), 1299–1387.

Ize, J. Introduction to bifurcation theory, Springer Verlag, *Lect. Notes in Math.* **957** (1982), 145–202.

Ize, J., Obstruction theory and multiparameter Hopf bifurcation, *Trans. A.M.S.* **289** (1985), 757–792.

Ize, J., Massabó, I., Pejsachowicz, J. and A. Vignoli, Structure and dimension of global branches of solutions to multiparameter nonlinear equations, *Trans. A.M.S.* **291** (1985), 383–435.

Ize, J., Massabó, I. and A. Vignoli, Global results on continuation and bifurcation for equivariant maps, *NATO-ASI*, **173** (1986), 75–111.

Ize, J., Necessary and sufficient conditions for multiparameter bifurcation, *Rocky Mountain J. of Math.* **18** (1988), 305–337.

Ize, J., Massabó, I. and A. Vignoli, Degree theory for equivariant maps I, *Trans. A.M.S.* **315** (1989), 433–510.

Ize, J., Massabó and A. Vignoli, Degree theory for equivariant maps: The general S^1-action, *Memoirs A.M.S.* **481** (1992).

Ize, J. and A. Vignoli, Equivariant Degree for abelian groups. Part I: Equivariant Homotopy groups. Preprint, 1993.

James, I.M., On the suspension sequence, *Annals of Math.* **65** (1957), 74–107.

Jorge, M.C. and A.A. Minzoni, Examples of bifurcation from a continuum of eigenvalues and from the continuous spectrum, *Quart. of Appl. Math.* **51** (1993), 37–42.

J.B. Keller and S. Antman, eds., *Bifurcation Theory and Nonlinear Eigenvalue Problems*, Benjamin, 1969.

Kielhöfer, H., Hopf bifurcation at multiple eigenvalues, *Arch. Rat. Mech. Anal.* **69** (1979), 53–83.

Kielhöfer, H., Multiple eigenvalue bifurcation for potential operators, *J. Reine Angew. Math.* **358** (1985), 104–124.

Kielhöfer, H., Interaction of periodic and stationary bifurcation from multiple eigenvalues, *Math Z.* **192** (1986), 159–166.

Kielhöfer, H., A bifurcation theorem for potential operators, *J. Funct. Anal.* **77** (1988), 1–8.

Kielhöfer, H., Hopf bifurcation from a differentiable viewpoint, *J. Diff. Eq.* **97** (1992), 189–232.

Kirchgässner, K. and P. Sorger, Branching analysis for the Taylor problem, *Quart. J. Mech. Appl. Math.* **22** (1969), 183–210.

Kirchgässner, K., Bifurcation in Nonlinear hydrodynamic stability, *SIAM Review*, **17** (1975), 652–683.

Komiya, K., Fixed point indices of equivariant maps and Moebius inversion, *Invent. Math.* **91** (1988), 129–135.

Kosniowski C., Equivariant cohomology and stable cohomotopy, *Math. Ann.* **210** (1974), 83–104.

Kötzner, P., Calculating homotopy classes and bifurcation, part I. Univ. of Augsburg, preprint, 1990.

Krasnosel'skii, M.A., *Positive Solutions of Operator Equations*, Noordhoff, 1964.

Krasnosel'skii, M.A., *Topological Methods in the Theory of Nonlinear Integral Equations*, Pergamon Press, MacMillan, New York, 1964.

Krasnosel'skii, M.A. and P.P. Zabreiko, Geometrical Methods in Nonlinear Analysis, *Grund Math. Wiss.* **263** (1984), Springer Verlag.

Küpper, T., Riemer, D., Necessary and sufficient conditions for bifurcation from the continuous spectrum, *Nonlinear Anal.* **3** (1979), 555–561.

Laloux, B. and J. Mawhin, Multiplicity, Leray-Schauder formula and bifurcation, *J. Diff. Eq.* **24** (1977), 309–322.

Landman, K.A. and S. Rosenblat, Bifurcation from a multiple eigenvalue and stability of solutions, *SIAM J. of Appl. Math.* **34** (1978), 743–759.

Leray, J. and Schauder, J., Topologie et equations fonctionnelles, *Ann. Sci. Ecole Normale Sup.* **51** (1934), 45–78.

Ljusternik, L. and L. Schnirelmann, *Methodes Topologiques dans les problèmes variationels.* Herman, Paris, 1934.

Lloyd, N.G., Degree theory, *Cambridge tracts in Math.* **73** (1978), Cambridge Univ. Press.

López Gómez, J., Hopf bifurcation at multiple eigenvalues with zero eigenvalue, *Proc. Roy. Soc. Edin.* **101** (1985), 335–352.

López Gómez, J., Multiparameter local bifurcation, *Nonlinear Anal. T.M.A.* **10** (1986), 1249–1259.

López Gómez, J., Multiparameter local bifurcation based on the linear part, *J. Math. Anal. and Appl.* **138** (1989), 358–370.

Ma. T, Topological degrees of set-valued compact fields in locally convex spaces, *Rozprawy Mat.* **42** (1972), Warszawa.

MacBain, J.A., Local and global bifurcation from normal eigenvalues II, *Pacific J. Math.* **74** (1978), 143–152.

Magnus, R.J., A generalization of multiplicity and the problem of bifurcation, *Proc. London Math. Soc.* **32** (1976), 251–278.

Magnus, R.J., On the local structure of the zero set of a Banach space valued mapping, *J. Funct. Anal.* **22** (1976), 58–72.

Mallet-Paret, J. and J.A. Yorke, Snakes: oriented families of periodic orbits, their sources, sinks and continuation, *J. Dif. Eq.* **43** (1982), 419–450.

Mallet-Paret, J. and R. Nussbaum, Boundary layer phenomena for differential-delay equations with state dependent time delays, *Arch. Rat. Math. Anal.* **120** (1992), 99–146.

Marino A, La biforcazione nel caso variazionale, *Conf. Sem. Mat. Univ. Bari*, **132** (1977).

Marsden, J.E., McCracken, M., *The Hopf Bifurcation and Its Applications*, Springer Verlag, 1976.

Marzantowicz, W., On the nonlinear elliptic equations with symmetries, *J. Math. Anal. Appl.* **81** (1981), 156–181.

Massabó, I. and J. Pejsachowicz, On the connectivity properties of the solutions set of parametrized families of compact vector fields, *J. Funct. Anal.* **59** (1984), 151–166.

Matsuoka, T., Equivariant function spaces and bifurcation points, *J. Math. Soc. Japan* **35** (1983), 43–52.

Mawhin, J., Topological degree methods in nonlinear boundary value problems, *CBMS* **40** (1977), AMS.

McLeod, J.B. and Turner, R.E.L., Bifurcation for Lipschitz operators with an application to elasticity, *Arch. R. Mech. and Anal.* **63** (1977), 1–45.

Milojevic, P.S., On the index and the covering dimension of the solution set of semilinear equations, *Proc. Symp. Pure Math. AMS* **45** (1986), 2, 183–205.

Morse, M., *The Calculus of Variations in the Large*, A.M.S., 1934.

Namboodiri, V., Equivariant vector fields on spheres, *Trans. A.M.S.* **278** (1983), 431–460.

Nirenberg, L., An application of generalized degree to a class of nonlinear problems, 3rd. Colloq. Anal. Funct., Liege, Centre Belge de Recherches Math., (1971), 57–73.

Nirenberg. L., Topics in nonlinear functional analysis, *Lect. Notes Courant Institute*, New York Univ., 1974.

Nirenberg. L., Variational and topological methods in nonlinear problems, *Bull. AMS* **4** (1981), 267–302.

Nirenberg, L., Comments on Nonlinear Problems, *Le Matimatiche* **36** (1981), 109–119.

Nishimura, T. Fukuda T. and K. Aoki, An algebraic formula for the topological types of one parameter bifurcation diagrams, *Arch. Rat. Mech.* **108** (1989), 247–266.

Nitkura, Y., Existence and bifurcation of solutions for Fredholm operators with nonlinear perturbations, *Nagoya Math. J.* **86** (1982), 249–271.

Nussbaum, R.D., Degree theory for local condensing maps, *J. Math. Anal. Appl.* **37** (1972), 741–766.

Nussbaum, R. D., A global bifurcation theorem with applications to functional differential equations, *J. Funct. Anal.* **19** (1975), 319–338.

Nussbaum, R.D., Some generalizations of the Borsuk-Ulam theorem, *Proc. London Math. Soc.* **35** (1977), 136–158.

Nussbaum, R.D., A Hopf bifurcation theorem for retarded functional differential equations, *Trans. A.M.S.* **238** (1978), 139–163.

Nussbaum, R.D., Differential-delay equations with two time lags, *Memoirs A.M.S.* **205** (1978).

Peitgen, H.O., Topologische Perturbationen bein globalen numerischen Studium nichtlinearer Eigenvert - und Verzweigungsprobleme, *Jber.d.Dt Math. Verein.* **84** (1982), 107–162.

Peitgen, H.O., Walter, H.O., (eds.): Functional differential equations and approximation of fixed points, *Lecture Notes in Mathematics*, **730** (1982), Springer Verlag.

Pejsachowicz, J., K-theoretic methods in bifurcation theory, *Contemporary Math.* **72** (1988), 193–205.

Petryshyn, W.V., Nonlinear equations involving noncompact operators. "Nonlinear Functional Analysis", *Proc. Symp. Pure Math.* **18/1** (1970), AMS, 206–223.

Petryshyn, W.V., On the approximation solvability of equations involving A-proper and pseudo-A-proper mappings, *Bull. AMS* **81** (1975), 223–312.

Petryshyn, W.V., Bifurcation and asymptotic bifurcation for equations involving A-proper mappings with applications to differential equations, *J. Diff. Eq.* **28** (1978), 124–154.

Petryshyn, W.V., Approximation-solvability of nonlinear functional and differential equation, M. Dekker, 1992.

Petryshyn, W.V., Fitzpatrick, P.M., A degree theory, fixed point theorems and mapping theorems for multivalued noncompact mappings, *Trans. AMS* **194** (1974), 1–25.

Pimbley, G., Eigenfunction branches of nonlinear operators, and their bifurcation, *Lectures Notes in Math.* **104** (1969), Springer Verlag.

Poincaré, H., Les figures equilibrium, *Acta Math.* **7** (1885), 259–302.

Rabier, P.J., Generalized Jordan chains and two bifurcation theorems of Krasnoselskii, *Nonlinear Anal. TMA.* **13**, **8** (1989), 903–934.

Rabier, P.J., Topological degree and the theorem of Borsuk for general covariant mappings with applications, *Nonlinear Anal. T.M.A.* **16** (1991), 393–420.

Rabinowitz, P.H., Some global results for nonlinear eigenvalue problems, *J. Functional Analysis* **7** (1971), 487–513.

Rabinowitz, P.H., Some aspects of nonlinear eigenvalue problems, *Rocky Mountain J. Math.* **3** (1973), 161–202.

Rabinowitz, P.H., On bifurcation from infinity, *J. Diff. Eq.* **14** (1973), 462–475.

Rabinowitz, P.H., Theorie du Degree Topologique et Applications (Lectures Notes), 1975.

Rabinowitz, P.H., A bifurcation theorem for potential operators, *J. of Funct. Anal.* **25** (1977), 412–424.

Romero Ruiz del Portal, F., Teoría del grado topológico generalizado y aplicaciones, Ph. Thesis, Madrid, 1990.

Rubinstein, R.L., On the equivariant homotopy of spheres, *Sissertationes Math.* **134** (1976), 1–48.

Sadovskii, B.N., Limit-compact and condensing operators, *Russian Math. Surveys* **27** (1972), 85–155.

Sather, D., Branching of solutions of nonlinear equations, *Rocky Mountain J. Math.* **3** (1973), 203–250.

Sattinger, D.H., Stability of bifurcating solutions by Leray-Schauder degree, *Arch. Rational Mech. Anal.* **43** (1971), 154–166.

Sattinger, D.H., Topics in Stability and Bifurcation Theory, *Lect. Notes in Math.* **309**, (1973), Springer-Verlag.

Sattinger, D.H., Group representation theory, bifurcation theory and Pattern formation, *J. of Funct. Anal.* **28** (1978), 58–101.

Sattinger, D.H., Group theoretic methods in bifurcation theory, *Lecture Notes in Math.* **762** (1979), Springer Verlag.

Sattinger, D.H., *Branching in the Presence of Symmetry*, Wiley, 1983.

Schaaf, R., Global behavior of solutions branches for some Neumann problems depending on one or several parameters, *J. für die Reine und Ang. Math.* **346** (1984), 1–31.

Schmidt, E., Zur theorie der linearen und nichtlinearen integralgleichungen, III, *Math. Ann.* **65** (1908), 370–399.

Schmitt, K., A study of eigenvalue and bifurcation problems for nonlinear elliptic partial differential equations via topological continuation methods, *Inst. Math. Pure et Appl. Louvain-la-Neuve*, 1982.

Schmitt, K. and Z.Q. Wang, On bifurcation from infinity for potential operators, *Diff. Int. Equations* **4** (1991), 933–943.

Sijbrand, J., Studies in nonlinear stability and bifurcation theory, Ph.D. Thesis, Utrecht, 1981.

Schwartz, J.T., *Nonlinear Functional Analysis*, Gordon and Breach, New York, 1969.

Stakgold, I., Branching solutions of nonlinear equations, *SIAM Rev.* **13** (1971), 289–332.

Steinlein, H., Borsuk's antipodal theorem and its generalizations and applications: a survey, *Topol. en Anal. Non Lineaire. Press de l'univ. de Montreal*, (1985), 166-235.

Stuart, C.A., Some bifurcation theory for k-set contractions, *Proc. London Math. Soc.* **27** (1973), 531–550.

Stuart, C.A., Bifurcation from the essential spectrum, *Lect. Notes in Math.* **1017** (1983), Springer, 575–596.

Stuart, C.A., Bifurcation from the continuous spectrum in $L^p(\mathbb{R})$, *Inter Sem. Numer. Math.* **79** (1987), Birkhäuser, 307–318.

Stuart, C.A., Guidance properties of nonlinear planar waveguides. Preprint, 1992.

Takens, F., Some remarks on the Böhme-Berger bifurcation theorem, *Math Z.* **125** (1972), 359–364.

Toland, J., Global bifurcation for k-set contractions without multiplicity assumptions, *Quart. J. Math.* **27** (1976), 199–216.

Toland, J. Global bifurcation for Neumann problems without eigenvalues, *J. Diff. Eq.* **44** (1982), 82–110.

tom Dieck, T., Transformation groups and representation theory, *Lect. Notes in Math.* **766** (1979), Springer Verlag.

Turner, R.E.L., Transversality in nonlinear eigenvalue problems. "Contributions to nonlinear functional analysis", Zarantonello, E. H. ed. Acad. Press, (1971), 37–68.

Vainberg, M.M., *Variational method and method of monotone operators in the theory of nonlinear equations*, Wiley, 1973.

Vainberg, M.M., Trenogin, V.A., *Theory of branching of solutions of nonlinear equations*, Noordhoff, Leyden, 1974.

Vanderbauwhede, A., Local bifurcation and symmetry, *Pitman Research Notes in Math.* **75** (1982).

Vignoli, A., L'Analisi Nonlineare nella teoria della biforcazione, *Enciclopedia delle Scienze Fiziche dell' Inst. dell'Enciclopedia Italiana*, Vol. 1, (1992), 134–144.

Wang, Z.Q., Symmetries and the calculation of degree, *Chin. Ann. of Math.* **10 B** (1989), 520–536.

Webb, J.R.L. and S.C. Welsh, Topological degree and global bifurcation, *Proc. Symp. Pure Math.* **45/II**. A.M.S., (1986), 527–531.

Welsh, S.C., Global results concerning bifurcation for Fredholm maps of index zero with a transversality condition, *Nonlinear Anal. T.M.A.* **12** (1988), 1137–1148.

Werner, B., Eigenvalue problems with the symmetry of a group and bifurcation, *NATO-ASI Series* **313** (1990), 71–88.

Westreich, D., Banach space bifurcation theory, *Trans. A.M.S.* **171** (1972), 135–156.

Westreich, D., Bifurcation at eigenvalues of odd multiplicity, *Proc. A.M.S.* **41** (1973), 609–614.

Westreich, D., Bifurcation at double characteristic values, *J. London Math. Soc.* (2), **15** (1977), 345–350.

Whitehead, G.W., On the homotopy groups of spheres and rotation groups, *Annals of Math.* **43** (1942), 634–640.

Whitehead, G.W., Elements of homotopy theory, *Graduate texts in Math.* **61** (1978), Springer Verlag.

Whyburn, G.T., *Topological Analysis*, 2nd. ed. Univ. Press, 1964.

Wolkowisky, J.H., A geometric theory of bifurcation, *Proc. Symp. Pure Maths.* **45/2** (1986), 553–564.

Zarantonello, E.H., *Contributions to Nonlinear Functional Analysis*, Acad. Press, 1971.

Zeidler, E., *Nonlinear Functional Analysis and its Applications*, Vol. III, Springer Verlag, 1984.

Zeidler, E., *Nonlinear Functional Analysis and its Applications*, Vol. IV. Springer Verlag, 1985.

Critical Point Theory and Applications to Differential Equations: A Survey

Paul H. Rabinowitz*

Mathematics Department
University of Wisconsin
Madison, WI 53706

Introduction

The purpose of this paper is to survey developments in the field of critical point theory and its applications to differential equations that have occurred during the past 20–25 years. This is too broad a theme for a single survey and we will focus on three particular areas. First we will examine contributions to the minimax approach to critical point theory. In particular the Mountain Pass Theorem, the Saddle Point Theorem, and variants thereupon will be discussed in Part 1. Then in Part 2, applications of critical point theory to the existence of periodic solutions of Hamiltonian systems of differential equations will be surveyed. Many of the different questions that have been studied will be described and a variety of results will be presented. Lastly, Part 3 deals with connecting orbits of Hamiltonian systems, mainly homoclinic and heteroclinic orbits. This last set of material represents the least developed subject matter treated here and therefore it can be described more completely than the earlier topics.

Even within Parts 1 and 2, there are important subtopics that will be omitted. In particular, except for a few remarks, there will be no discussion

---

* This research was sponsored in part by the National Science Foundation under Grant #MCS-8110556 and by the US Army under contract #DAAL03-87-k-0043. Any reproduction for the pruposes of the US Government is permitted.

of multiplicity results for critical points of functionals which are invariant under a group of symmetries in Part 1. The existence of periodic solutions of singular Hamiltonian systems will not be mentioned in Part 2. However both of these topics will be described in the survey of Ambrosetti in these proceedings. The variational approach to bifurcation theory also will not be covered.

There are many earlier surveys and several books that touch upon the topics treated here and which the interested reader can further consult. Indeed because of the size of the literature in the area, we have been able to mention only a small fraction of the relevant research so it is also worthwhile to check the bibliographies of the following: Ambrosetti [A1], Ambrosetti and Prodi [AP], Bahri [Ba1], Berestycki [B], Berger [Bg], Conley [CC], Chang [Ch1], DeFigueiredo [Df], Deimling [De], Ekeland [E1], Ghoussoub [Gh], Krasnoselski [K], Mawhin [Ma], Mawhin-Willem [MW], Nirenberg [N], Palais [P1], Rabinowitz [R1–2], Struwe [St1], and Zehnder [Z].

PART 1 – Critical Point Theory

§1. Introduction

Let E be a real Banach space and $I : E \to \mathbb{R}$ be continuously differentiable, i.e. $I \in C^1(E, \mathbb{R})$. Then u is a *critical point* of I if the Frechet derivative of I, $I'(u) = 0$, i.e.

$$I'(u)\varphi = 0 \tag{1.1}$$

for all $\varphi \in E$. The number $I(u)$ is then called a *critical value* of I. In applications to differential equations, finding a point u that satisfies (1.1) is equivalent to finding a weak solution of the differential equation. Hence critical point theory is a useful tool in the study of existence questions for certain classes of ordinary and partial differential equations.

During the 1960's, there was a great deal of work on Morse theory and minimax theory — sometimes called Ljusternik-Schnirelmann theory — in infinite dimensions by Palais [P1–2], Palais and Smale [PS], Schwartz [S1–2], Browder [Bw1–2], Berger [Bg], and many others. Subsequently Conley [CC] made important extensions of Morse theory introducing what is now called the Conley index. In this part of the survey, some of the progress in minimax results will be described. Minimax critical values of I simply are critical values which are characterized as a minimax:

$$c = \inf_{K \in \mathcal{S}} \max_{u \in K} I(u) \tag{1.2}$$

where $\mathcal{S}$ is some class of subsets of E which is chosen to take advantage of the topological nature of the level sets of I. Often the minimax has a related form

$$c = \inf_{g \in \Gamma} \max_{\theta \in A} I(g(\theta)) \qquad (1.3)$$

where A lies in some parameter space (e.g. $[0,1]$ or S^1, etc.) and Γ is a class of maps from A to E satisfying appropriate properties. The earliest results of this type mainly studied functionals which were bounded from below, and were invariant under some group of symmetries, and were defined on a Banach manifold which was invariant under the symmetries [LLS], [P2], [PS], [Bw1], [S1]. The focus here will be on functionals which are permitted to be highly indefinite, i.e. they may be unbounded from above or below, even modulo subspaces or submanifolds of finite dimension. In §2, the Mountain Pass Theorem and some variants will be discussed. The Saddle Point Theorem and some related results will be treated in §3. Then the notion of linking and generalized critical point theorems will be given in §4.

§2. The Mountain Pass Theorem

The Mountain Pass Theorem is one of the simplest and most geometrically appealing minimax results. It is also one of the most useful. There are many related versions of it and one due to Ambrosetti and the author [AR] will be presented here. Before doing so, some remarks are necessary. The formulation of all of the abstract critical point theorems that will be described later require some compactness hypotheses for the functional I. A condition of that type that has proved to be extremely useful was introduced by Palais and Smale in their work on Morse Theory in infinite dimensions. They called it condition (C) but it has come to be called the Palais-Smale condition. It will be denoted by (PS) here and says:

(PS)   Any sequence (u_m) for which $I(u_m)$ is bounded and $I'(u_m) \to 0$ as $m \to \infty$ possesses a convergent subsequence.

Actually one does not need the full global strength of (PS) but rather a local version:

$(PS)_c$ Any sequence (u_m) for which $I(u_m) \to c$ and $I'(u_m) \to 0$ as $m \to \infty$ possesses a convergent subsequence.

The number c, in any given situation, is usually determined by a minimax procedure. Weaker versions of $(PS)_c$ have been introduced. See, e.g. Cerami [Ce] or Schechter [Sc]. For simplicity, (PS) will be employed in what follows. In applications, one generally uses the boundedness of $I(u_m)$ and that $I'(u_m) \to 0$ as $m \to \infty$ to show that (u_m) is bounded. Then further hypotheses on the form of I yield the precompactness of (u_m). See, e.g. [A1], [R1], [St1] where (PS) is verified in many different settings.

Increasing attention is being paid to problems in which (PS) fails. There are a variety of such examples, e.g. limit exponent problems or equations on unbounded temporal or spatial domains. Experience with several such situations has shown that one can still get existence results in certain cases where one understands well enough how the (PS) condition breaks down. See, e.g. Bahri [Ba], Bahri-Coron [BC], Bahri-Rabinowitz [BaR], Coti Zelati-Ekeland-Séré [CZES], P. L. Lions [LPL], Sachs-Uhlenbeck [SU]. Some such cases will be mentioned in Part 3.

To continue, some notation is needed. Let $B_\rho(u)$ denote an open ball in E of radius ρ. For $I \in C^1(E, \mathbb{R})$ and $s, t \in \mathbb{R}$, let

$$I^s = \{u \in E \mid I(u) \leq s\}, \quad I_t = \{u \in E \mid I(u) \geq t\}$$

and $I_t^s = I^s \cap I_t$. Now the Mountain Pass Theorem [AR] can be stated:

Theorem 1.4. *Let E be a real Banach space and $I \in C^1(E, \mathbb{R})$. Suppose that $I(0) = 0$, I satisfies* (PS),

(I_1) *there are constants $\rho, \alpha > 0$ such that $I\big|_{\partial B_\rho(0)} \geq \alpha$*

and

(I_2) *there is an $e \in E \backslash B_\rho(0)$ such that $I(e) \leq 0$.*

Then I possesses a critical value $c \geq \alpha$. Moreover

$$c = \inf_{g \in \Gamma} \max_{\theta \in [0,1]} I(g(\theta)) \tag{1.5}$$

where

$$\Gamma = \{g \in C([0,1], E) \mid g(0) = 0, g(1) \in I^0 \backslash \{0\}\}. \tag{1.6}$$

The heuristic interpretation of Theorem 1.4 in the simplest setting, where 0 is a local minimum of I, is that if one stands at the origin, one is at the base of a valley surrounded by a mountain range. By (I_2), the range does not increase in height indefinitely in all directions. Hence there is hope of finding a mountain pass and indeed given (PS), this is the case. Simple examples show that without (PS), I need not possess another critical point. Finite dimensional analogues of Theorem 1.4 have been known for a long time, e.g. in the minimal surface literature although there it is generally assumed I has a pair of local minima and one concludes the existence of a third critical point of I. See also Birkhoff [Bi] for the earliest result we know of in this direction.

No proofs will be given in this survey. They can be found in the references cited. However there will be comments on the proofs. In particular, the usual proof of the Mountain Pass Theorem involves an indirect argument. There is a standard "deformation" type theorem that uses (PS) and arguments from the theory of ordinary differential equations to deform the set $I^{c+\epsilon}$ to $I^{c-\epsilon}$ for some small $\epsilon > 0$ provided that c is not a critical value of I. This allows one to choose $g \in \Gamma$ such that

$$\max_{\theta \in [0,1]} I(g(\theta)) \leq c + \epsilon \tag{1.7}$$

and then deform it to $h \in \Gamma$ such that

$$\max_{\theta \in [0,1]} I(h(\theta)) \leq c - \epsilon \tag{1.8}$$

contrary to (1.5). Another kind of proof using Ekeland's Theorem has been given. See, e.g. De Figueiredo [Df] or Mawhin-Willem [MW]. Variants of the above argument are used in the proofs of all minimax theorems, although of course in a manner suited to the minimax in question and this may involve a great deal of technical complexity.

We will illustrate the Mountain Pass Theorem and our later abstract results by simple applications to semilinear elliptic partial differential equations. Thus consider the boundary value problem

$$\begin{aligned} -\Delta u &= f(x, u) & x \in \Omega \\ u &= 0 & x \in \partial \Omega \end{aligned} \tag{1.9}$$

where $\Omega \subset \mathbb{R}^n$ is a bounded domain with a smooth boundary and $n \geq 3$. Assume f satisfies

(f_1) $f \in C^1(\mathbb{R}^n \times \mathbb{R}, \mathbb{R})$,

(f_2) (Sobolev growth condition) there are constants $a_1, a_2 \geq 0$ such that $|f(x, z)| \leq a_1 + a_2|z|^s$ where $1 < s < \frac{n+2}{n-2}$,

(f_3) $f(x, z) = o(|z|)$ as $z \to 0$,

(f_4) there are constants $\mu > 2$ and $r \geq 0$ such that for $|z| > r$,

$$0 < \mu F(x, z) \equiv \mu \int_0^z f(x, \xi)d\xi \leq zf(x, z).$$

Equation (1.9) under (f_1)–(f_4) is sometimes called a superlinear problem since

$$f(x, z)/|z| \to 0 \quad \text{as} \quad z \to 0 \quad \text{and} \quad \to \infty \quad \text{as} \ |z| \to \infty.$$

E.g. take $f(x, z) = |z|^{s-1}z$ with s as in (f_2).

It is not difficult to verify (see, e.g. [AR] or [R1]) that if $E = W_0^{1,2}(\mathbb{R}^n)$ under

$$\|u\|^2 = \left(\int_\Omega |\nabla u|^2 dx \right)^{1/2}$$

and

$$I(u) = \int_\Omega \left(\frac{1}{2} |\nabla u|^2 - F(x, u) \right) dx = \frac{1}{2} \|u\|^2 - \int_\Omega F(x, u) dx, \qquad (1.10)$$

then $I \in C^1(E, \mathbb{R})$ and critical points of I are weak solutions of (1.9). Hypotheses (f_1)–(f_2) and standard regularity results then show weak solutions of (1.9) are classical solutions. The functional I is easily seen to be highly indefinite and in particular unbounded from above or below.

To relate I to Theorem 1.4, note that $I(0) = 0$ and (f_1)–(f_3) imply that $I(u) = \frac{1}{2} \|u\|^2 + o(\|u\|^2)$ as $u \to 0$. Therefore I satisfies (I_1) of Theorem 1.4. Hypotheses (f_4) implies there are constants $a_3, a_4 > 0$ such that

$$F(x, z) \geq a_3 |z|^\mu - a_4 \qquad (1.11)$$

for all $z \in \mathbb{R}$ and since $\mu > 2$, (1.11) yields the existence of an $e \in E$ as required for (I_2). Indeed there is such an e along each say through 0. Finally using (f_1)–(f_4), (PS) can be verified and Theorem 1.4 yields a nonzero solution of (1.9). In fact using PDE techniques based on the Maximum Principle, it can be shown that (1.9) has a positive and a negative solution. See, e.g. [AR] or [R1] for more details.

There are many variations on and extensions of Theorem 1.4. Some will be mentioned in this section and in §4. First there are treatments of a "degenerate" mountain pass, i.e. $\alpha = 0$ in (I_1). See, e.g. [PSe] or [R1].

Theorem 1.12. *Let E be a real Banach space and $I \in C^1(E, \mathbb{R})$. Suppose $I(0) = 0$, I satisfies*

(I_1') *there is a $\rho > 0$ such that $I|_{\partial B_\rho(0)} \geq 0$,*

(I_2) *and (PS). Then I has a critical value $b \geq 0$. Moreover if $b = 0$, there is a critical point of I on $\partial B_\rho(C)$.*

In Theorem 1.12, $\partial B_\rho(0)$ can be replaced by any closed set, C, separating C and e. A slightly stronger result has been proved by Ghoussoub and Preiss [GP] who use Ekeland's Theorem and other arguments to show that if $c = 0$, there is a $u \in C$ such that $I(u) = 0$ and $I'(u) = 0$. This can also be proved by the arguments of [R1].

Another kind of extension of Theorem 1.4 involves weakening the assumption that I is C^2 (see, e.g. Chang [Ch2]), or weakening (PS) (see, e.g. Brezis-Nirenberg [BN1–2] and Brezis-Coron-Nirenberg [BCN]). Several people have also studied in what sense a mountain pass critical point is a saddle point. See e.g. Hofer [H1] and Pucci-Serrin [PSe1–2].

When E is a Hilbert space, $I'(u) = u - T(u)$ where T is compact, and u_0 is a mountain pass critical point, a natural question to ask is whether one can compute a Leray-Schauder index for this zero of I'. Such information has been obtained by Ambrosetti [A2] and by Hofer [H2]. They showed that under additional hypotheses the index is -1 or a related Morse index is 1. This result has proved useful in finding periodic solutions of prescribed minimal period for Hamiltonian systems. See Ekeland-Hofer [EH1–2]. Hofer [H3] has also used a mixture of variational, degree theoretic, and order theoretic ideas to get interesting multiplicity results for some applications.

In a related vein to [A2] and [H2], there has been a great deal of work both in the setting of Theorem 1.4 and for other situations to try to get information on Morse indices of isolated critical points. This has been carried out most extensively in the abstract setting by Ghoussoub and his collaborators (see, e.g. [Gh]) and in more concrete situations for differential equations. For the latter, the Morse index estimates have been used to get a priori bounds on solutions which lead in turn to existence or multiplicity results. See, e.g. Bahri-Berestycki [BB1], Bahri-Lions [BaL1–2], Coti Zelati-Ekeland-Lions [CZEL], and Tanaka [T1].

Another kind of generalization of Theorem 1.4 occurs when hypothesis (I_1) is weakened in the following way:

Theorem 1.13. *Suppose E is a real Banach space with the splitting $E = V \oplus X$ and V is finite dimensional. If I satisfies* (PS),

(I_3) *there are constants $\rho, \alpha > 0$ such that*

$$I\Big|_{\partial B_\rho(0) \cap X} \geq \alpha$$

and

(I_4) *there is an $e \in \partial B_1(0) \cap X$ and $R > \rho$ such that if*

$$Q = (\overline{B_R(0)} \cap V) \oplus \{re \mid 0 < r < R\},$$

then $I\Big|_{\partial Q} \leq 0$,

then I has a critical value $c \geq \alpha$ given by

$$c = \inf_{h \in \Gamma} \max_{u \in Q} I(h(u)) \tag{1.14}$$

where

$$\Gamma = \{h \in C(\bar{Q}, E) \mid h = \text{ id on } \partial Q\} \tag{1.15}$$

Part of the proof of Theorem 1.13 involves a topological degree argument showing that

$$h(\bar{Q}) \cap \partial B_\rho(0) \cap X \neq \phi \tag{1.16}$$

for all $h \in \Gamma$ and (1.16) immediately implies that $c \geq \alpha$. For a degenerate version of this theorem, see, e.g. Silva [Si1]. To see how Theorem 1.13 can be applied, consider the following modification of (2.6):

$$\begin{aligned} -\Delta u &= \lambda a(x)u + f(x, u), & x \in \Omega \\ u &= 0 & x \in \partial\Omega \end{aligned} \tag{1.17}$$

where f, Ω, and E are as earlier and the function $a(x) > 0$ in $\bar{\Omega}$ and lies in $C^1(\bar{\Omega})$. Then the associated functional is

$$I(u) = \int_\Omega \left[\frac{1}{2}|\nabla u|^2 - \frac{\lambda}{2}au^2 - F(x, u) \right] dx \tag{1.18}$$

and $I \in C^1(E, \mathbb{R})$. Dropping f in (1.17) yields the linear eigenvalue problem

$$\begin{aligned} -\Delta v &= \lambda a v, & x \in \Omega \\ v &= 0 & x \in \partial\Omega \end{aligned}. \tag{1.19}$$

This problem possesses a sequence of eigenvalues of finite multiplicity $0 < \lambda_1 < \lambda_2 \leq \ldots$ with $\lambda_k \to \infty$ as $k \to \infty$ and corresponding eigenfunctions v_k normalized via $\|v_k\| = 1$. If $\lambda \in (0, \lambda_1)$, for all $u \in E \backslash \{0\}$,

$$\int_\Omega (|\nabla u|^2 - \lambda u^2) dx > \left(1 - \frac{\lambda}{\lambda_1}\right) \|u\|^2. \tag{1.20}$$

It is then not difficult to verify that Theorem 1.4 can be applied again to obtain a solution of (1.17). However if $\lambda \geq \lambda_1$, (I_1) is no longer valid. Suppose $\lambda \in [\lambda_k, \lambda_{k+1})$ and set $V = \text{span}\{v_1, \ldots, v_k\}$. Then for $u \in X = V^\perp$

$$\int_\Omega (|\nabla u|^2 - \lambda u^2) dx \geq \left(1 - \frac{\lambda}{\lambda_{k+1}}\right) \|u\|^2. \tag{1.21}$$

Consequently (1.21) and an earlier argument show I satisfies (I_3). Suppose further that f satisfies

(f_5) $zf(x, z) \geq 0, \quad z \in \mathbb{R}, x \in \bar{\Omega}.$

Then it can be shown (see, e.g. [R1]) that for R sufficiently large, if Q is defined in Theorem 1.13, (I_4) holds. Finally (PS) is also satisfied and Theorem 2.10 applies here.

§3. The Saddle Point Theorem

In this section, the Saddle Point Theorem and a variant of it will be discussed.

Theorem 1.22 (Saddle Point Theorem [R1]). *Let E be a real Banach space with the splitting $E = V \oplus X$ where $V \neq \{0\}$ is finite dimensional. Suppose $I \in C^1(E, \mathbb{R})$, satisfies* (PS),

(I_5) *there is a bounded neighborhood D of 0 in V and a constant α such that $I\big|_{\partial D} \leq \alpha$, and*

(I_6) *there is a constant $\beta < \alpha$ such that $I\big|_X \geq \beta$.*

Then I possesses a critical value $c \geq \beta$ given by

$$c = \inf_{h \in \Gamma} \max_{u \in \bar{D}} I(h(u)) \tag{1.23}$$

where

$$\Gamma = \{h \in C(\bar{D}, E) \mid h = \text{ id on } \partial D\}. \tag{1.24}$$

Observe that here there is no known critical point to use as a starting point unlike the Mountain Pass Theorem where one generally has a local minimum of I at $u = 0$. A typical situation in which (I_5)–(I_6) can be satisfied occurs when I is concave on V, convex on X, and appropriately coercive along these subspaces. Indeed this special case was the source of the name "Saddle Point" Theorem. As in Theorem 1.13 a key ingredient of the proof here is that a degree theoretic agrument shows

$$h(\bar{D}) \cap X \neq \phi \tag{1.25}$$

for all $h \in \Gamma$. Thus (3.4) and (I_6) imply $c \geq \beta$.

Theorem 1.22 was motivated by an existence result for a partial differential equation due to Ahmad, Lazer, and Paul [ALP]. For other related results, see Amann [Am], and Castro-Lazer [CL]. We will briefly illustrate Theorem 1.22 by a version of the work of [ALP]. Consider again (1.17) where f now satisfies (f_1) and (f_2) with $s = 0$, i.e. $f(x, z)$ is bounded on $\bar{\Omega} \times \mathbb{R}$. Whenever 0 is not an eigenvalue of (1.19), an easy application of the Schauder Fixed Point Theorem gives a solution of (1.17) under these hypotheses. Thus suppose that $\lambda = \lambda_k < \lambda_{k+1}$ and further assume that f satisfies

(f_6) $F(x, z) \to \infty$ as $|z| \to \infty$ uniformly for $x \in \bar{\Omega}$.

Let $E = W_0^{1,2}(\Omega)$ and $V = \text{span}\{v_1, \ldots, v_k\}$ and $X = V^\perp$. Then for $u \in X$,

$$\int_\Omega (|\nabla u|^2 - \lambda_k a u^2) dx \geq \left(1 - \frac{\lambda_k}{\lambda_{k+1}}\right) \|u\|^2 \tag{1.26}$$

while since $F(x, z)$ grows at most linearly in z,

$$\left|\int_\Omega F(x, u) dx\right| \leq M_1 \|u\| \tag{1.27}$$

for some M_1 independent of u. Hence $I(u) \to \infty$ as $\|u\| \to \infty$ in X. Next if $u \in V$, write $u = u^0 + u^-$ where $u_0 \in \text{span}\{v_j \mid \lambda_j = \lambda_k\}$ and $u^- \in \text{span}\{v_j \mid \lambda_j < \lambda_k\}$. Then for such u,

$$I(u) \leq -M_2\|u^-\|^2 - \int_\Omega F(x, u^0 + u^-) dx \tag{1.28}$$

$$= -M_2\|u^-\|^2 - \int_\Omega F(x, u^0) dx$$

$$\quad - \int_\Omega (F(x, u^0 + u^-) - F(x, u^0)) dx$$

$$\leq -M_2\|u^-\|^2 + M_1\|u^-\| - \int_\Omega F(x, u^0) dx$$

which via (f_6) shows $I(u) \to -\infty$ as $\|u\| \to \infty$ in V. It follows that I satisfies (I_5) – (I_6) and another argument yields (PS). Therefore Theorem 1.22 applies and yields a critical point of I.

As was mentioned in §2, Morse index computations have also been made in the setting of the Saddle Point Theorem. See, e.g. Lazer-Solimini [LaS] and Ghoussoub [Gh].

Another somewhat different version of the Saddle Point Theorem has been given by Silva [Si1].

Theorem 1.29. *Let E be a real Banach space with the splitting $E = V \oplus X$ where V is finite dimensional. Suppose $I \in C^1(E, \mathbb{R})$, satisfies (PS),*

(I_7) *there is an $\alpha \in \mathbb{R}$ such that $I\big|_V \leq \alpha$*

and

(I_8) *there is a $\beta \in \mathbb{R}$ such that $I\big|_X \geq \beta$.*

Then I posseses a critical value $c \in [\beta, \alpha]$.

Note that (I_7)–(I_8) imply $\beta \leq I(0) \leq \alpha$. The critical value c can be given a minimax characterization but it requires a preliminary construction of a map $\eta \in C([0,1] \times E, E)$ such that for a given $r \geq 0$, $\eta(0, u) = u$ and $I(\eta(t, u)) \leq I(u)$ for $u \in E$ and $t \in [0,1]$, $\eta(1, u) \in I^{\beta - \epsilon}$ for some $\epsilon > 0$ and $u \in I^\alpha \backslash B_r(0)$. Moreover if P denotes the projector of E onto V, there exists $R > r$ such that $P\eta(t, u) \neq 0$ for $u \in \partial B_R(0) \cap V$ and $t \in [0,1]$. The map η is constructed using arguments based on the usual Deformation Theorem. Then if

$$\Gamma = \big\{ h \in C(\overline{B_R}(0) \cap V, E) \mid h(u) = \eta(1, u) \qquad (1.30)$$
$$\text{for } u \in \partial B_R(0) \cap V \big\},$$

$$c = \inf_{h \in \Gamma} \max_{u \in B_R(0) \cap V} I(h(u)) \qquad (1.31)$$

is a critical value of I. As in the earlier theorems, an important step in the proof of Theorem 1.29 is the intersection property:

$$h(B_R(0) \cap V) \cap X \neq \phi \qquad (1.32)$$

for all $h \in \Gamma$. We will turn to this point shortly in §4. Others have given variants of the Saddle Point Theorem. See, e.g. Liu [Li] and Schechter [Sc].

§4. Linking and a General Critical Point Theorem

An examination of the proofs of the abstract critical point theorems presented thus far shows they possess a common feature: the range of the elements of the associated set of maps, Γ, intersects a set on which by hypothesis I is bounded from below. E.g. in the Mountain Pass Theorem, for any $g \in \Gamma$, $g([0,1]) \cap \partial B_\rho(0) \neq \phi$ and hence $c \geq \alpha$. The analogous intersection in other settings was noted in (1.16), (1.25), and (1.32). This fact was recognized in particular in Benci-Rabinowitz [BR1] where an abstract critical point theorem was formulated which conceptually contains the earlier results and also generalizes them to allow splittings in which both V and X are infinite dimensional. The abstract hypothesis which leads to an intersection of sets is one involving linking. Thus a notion of linking will be introduced next and then a version of the result of [BR1] will be given.

Let E be a real Banach space with the splitting $E = E_1 \oplus E_2$. Let P_1, P_2 be the projectors of E onto E_1, E_2 associated with this splitting. Let

$$S = \big\{ \Phi \in C([0,1] \times E, E) \mid \Phi(0, u) = u \text{ and} \qquad (1.33)$$
$$P_2\Phi(t, u) = P_2 u - K(t, u) \text{ where } K \text{ is compact} \big\}.$$

The reason for the compactness assumption is that it permits the use of Leray-Schauder degree theory to verify an intersection property. Suppose

$S, Q \subset E$ with $Q \subset \tilde{E}$, a subspace of E. Then S and ∂Q are said to *link* if whenever $\Phi \in S$ and

$$\Phi(t, \partial Q) \cap S = \phi \text{ for all } t \in [0, 1], \tag{1.34}$$

then

$$\Phi(t, Q) \cap S \neq \phi \text{ for all } t \in [0, 1]. \tag{1.35}$$

We will give two examples of linking. First take the setting of the Mountain Pass Theorem. Set $E_1 = \{0\}$, $E_2 = E$, $Q = B_\rho(0)$, and $S = \{0\}$. If $0 \notin \Phi(t, \partial Q)$ for all $t \in [0, 1]$, then the Leray-Schauder degree of $\Phi(t, \cdot)$ with respect to the bounded, open set Q and the point 0, $d(\Phi(t, \cdot), Q, 0)$ is defined. Therefore for each $t \in [0, 1]$,

$$d(\Phi(t, \cdot), Q, 0) = d(\Phi(0, \cdot), Q, 0) = d(\mathrm{id}, Q, 0) = 1 \tag{1.36}$$

since Q is a neighborhood of 0. Hence for all $t \in [0, 1]$, there is a $q = q(t) \in Q$ such that $\Phi(t, q(t)) = 0$. Consequently S and ∂Q link.

Next take the setting of the Saddle Point Theorem. Let $S = X = E_1$ and $Q = D \subset V = E_2$. Suppose

$$\Phi(t, \partial D) \cap X = \phi \tag{1.37}$$

for $t \in [0, 1]$ and set $\Psi(t, u) = P_2\Phi(t, u) = u - K(t, u)$ for $u \in E_2$. Then (1.37) implies $\Psi(t, \cdot) \neq 0$ on ∂D and $d(\Psi(t, \cdot), D, 0)$ is defined. Moreover

$$d(\Psi(t, \cdot), D, 0) = d(\mathrm{id}, D,)) = 1 \tag{1.38}$$

since D is a neighborhood of 0. Consequently there is a $q(t) \in D$ such that $\Phi(t, q(t)) \in X$ for each $t \in [0, 1]$ and S and ∂D link.

Now a general critical point theorem which in essence contains the earlier such results will be stated:

Theorem 1.39. *Let E be a real Hilbert space with the splitting $E = X \oplus X^\perp$ and corresponding projectors P_X and $P_{X^\perp}$. Suppose that $I \in C^1(E, \mathbb{R})$, satisfies (PS), and*

(I_9) $I(u) = \frac{1}{2}(Lu, u) + \psi(u)$ *where $Lu = L_X Pu + L_{X^\perp} Pu$ and L_X, $L_{X^\perp}$ are bounded, self-adjoint operators on X and $X^\perp$ respectively,*

(I_{10}) $\psi'(u)$ *is compact,*

and

(I_{11}) *there is a subspace $\tilde{E} \subset E$ and sets $S \subset E$, $Q \subset \tilde{E}$ and constants $\alpha > \omega$ such that*

(i) $S \subset X$ and $I\big|_S \geq \alpha$,

(ii) Q is bounded and $I\big|_{\partial Q} \leq \omega$,

(iii) S and ∂Q link.

Then I possesses a critical value $c \geq \alpha$.

Once again c has a minimax characterization. Let

$$\Gamma = \{h \in C([0,1] \times E, E) \mid h \text{ satisfies } (\Gamma_1) - (\Gamma_3)\} \tag{1.40}$$

where

(Γ_1) $h(0,u) = u$,

(Γ_2) $h(t,u) = u$ if $u \in \partial Q$,

(Γ_3) *there is a $\theta \in C(([0,1] \times E, \mathbb{R})$ and $K \in C([0,1] \times E, E)$ with K compact such that*

$$h(t,u) = e^{\theta(t,u)L}u + K(t,u).$$

Then

$$c = \inf_{h \in \Gamma} \max_{u \in Q} I(h(1,u)) \tag{1.41}$$

is a critical value of I. As in the earlier cases, part of the proof involves showing

$$h(1,Q) \cap S \neq \phi. \tag{1.42}$$

For other versions of Theorem 1.39, see e.g. [BR1], Hofer [H4], Ni [Ni] and Silva [Si1]–[Si2]. Another notion of linking — "local linking" — was given by Li and Liu [LL]. In his thesis [Si1] Silva introduced a more general version of linking he called "linking of deformation type" and used it to obtain generalizations of many earlier results, especially to degenerate situations. Unfortunately much of his work has not yet been published. However some of his results and extensions thereof subsequently have been discovered independently by others. For other work in this direction see also Schechter-Tintarov [ST] and Liu-Willem [LW]. A recent theorem which contains Theorem 1.39 has been obtained by Costa and Magalhães [CM].

We conclude this section with two final remarks. First one of the reasons for the importance of Theorem 1.39 is that it allows us to treat functionals, I, which are more highly indefinite than the semilinear elliptic applications treated thus far. In particular X and $X^\perp$ may be infinite dimensional. Such situations occur naturally when dealing with general Hamiltonian systems for which X and $X^\perp$ correspond roughly to subspaces on which the associated action integral is respectively positive and negative definite. This will be illustrated in Part 2. Finally if in addition to the hypothese made earlier, I happens to be invariant under a group of symmetries, e.g. $\mathbb{Z}_2$, $\mathbb{Z}^n$, S^1, etc., the level sets of I have a rich structure and one can generally get multiple critical points of I for this setting. Indeed such results go back to the origins of the minimax theory of Ljusternik and Schnirelmann [LS] and are among the most fascinating parts of the subject. They will be dealt with in the survey of Ambrosetti in this volume and we will refer to some of them in the applications that follow in Part 2.

Part 2. Periodic Solutions of Hamiltonian Systems

§1. Introduction

Hamiltonian mechanics is a fertile source of applications of critical point theory. While connections between Hamiltonian mechanics and the calculus of variations have long been known, it is only relatively recently that variational methods have emerged as an important existence tool for certain kinds of orbits. This late arrival is partially due to the fact that some of the necessary technical tools from critical point theory were only discovered in the 1970's–1980's. It is also partially due to the point of view of most people who work in dynamical systems. They tend to think in terms of the initial value problem rather than the global approaches of critical point theory. Thus it is not surprising that much of the recent progress that has been achieved is due to the efforts of researchers coming from the direction of critical point theory.

The first area of application of the global methods of the calculus of variations to Hamiltonian systems was to the existence of periodic solutions for Hamiltonians which are smooth or at least C^1 functions. There has been a great deal of progress in this area and the subject has matured sufficiently so that there are now two books by Ekeland [E1] and Mawhin-Willem [MW] that deal primarily with such topics. In the past few years some of the ideas and techniques that have been developed in the course of this research have begun to be applied in four other directions:

(a) symplectic geometry,

(b) singular Hamiltonian systems such as arise in celestial mechanics,

(c) connecting orbits of Hamiltonian systems,

(d) periodic solutions on Lorenz manifolds.

By (a), we refer to the work on topics like symplectic capacities and the Weinstein conjecture [W1] initiated by Ekeland and Hofer [EH3], Viterbo [V], and others. See also Hofer and Viterbo [HV] and Hofer and Zehnder [HZ1]. Topic (b) will be treated extensively in this volume by Ambrosetti.

The study of (d) by variational methods has been carried out by Benci, Fortunato, and their students and collaborators. See, e.g. [BFG]. By (c), we mean global orbits such as homoclinics and heteroclinics.

Topic (c) will be discussed in Part 3. Here in Part 2, the wealth of material on periodic solutions of Hamiltonian systems will be surveyed. Since there are so many results, we will focus on general Hamiltonian systems and note only in passing that analogous and often sharper results can be obtained for second order Hamiltonian systems. Section 2 will set up the technical variational framework for treating these problems. Periodic solutions of prescribed energy will be discussed in §3 and the existence of solutions of prescribed period will be treated in §4.

Another interesting set of applications of variational methods has been to infinite dimensional Hamiltonian systems representing semilinear wave equations in one space dimension. See, e.g. Brezis [Br] for a survey of work in this direction.

§2. The Technical Framework

In this section, the technical setup for studying periodic solutions of Hamiltonian systems will be described. Let $p, q \in \mathbb{R}^n$ and $H \in C^1(\mathbb{R}^{2n}, \mathbb{R})$. Then Hamilton's system of ordinary differential equations is

$$\dot{p} = -\frac{\partial H}{\partial q}(p, q) \equiv -H_q(p, q) \tag{2.1}$$

$$\dot{q} = \frac{\partial H}{\partial p}(p, q) \equiv H_q(p, q)$$

It can be written more simply as

$$\dot{z} = \mathcal{J} H_z(z), \qquad \mathcal{J} = \begin{pmatrix} 0 & -\mathrm{id} \\ \mathrm{id} & 0 \end{pmatrix} \tag{HS}$$

where $z = (p, q)$ and here id denotes the $n \times n$ identity matrix. If $z(t)$ is a solution of (HS), then

$$\frac{d}{dt} H(z(t)) = H_z(z) \cdot \mathcal{J} H_z(z) \equiv 0. \tag{2.2}$$

Therefore $H(z(t)) \equiv$ constant, i.e. "energy" is conserved along a solution. The special cases of $H(z) = K(p) + V(q)$ or $K(p, q) + V(q)$, where K represents the kinetic energy and V the potential energy, arise often in mechanics. Indeed several results will be presented for the simplest such case of $K(p) = \frac{1}{2}|p|^2$. Then (HS) reduces to the second-order Hamiltonian system:

$$\ddot{q} + V_q(q) = 0. \tag{HS2}$$

Forced Hamiltonian systems will also be studied, i.e. H or V depend explicitly on t in a time periodic fashion and (HS), (HS2) become

$$\dot{z} = J H_z(t, z), \qquad \text{(FHS)}$$
$$\ddot{q} + V_q(t, q) = 0, \qquad \text{(FHS2)}$$

respectively.

To formulate (HS) as a variational problem, some preliminaries are needed. Let $L^2(S^1, \mathbb{R}^{2n})$ denote the set of $2n$-tuples of 2π periodic functions which are square integrable. Any such function, z, has a Fourier expansion:

$$z = \sum_{j \in \mathbb{Z}} a_j e^{ijt}$$

where $a_j \in \mathbb{C}^{2n}$, $a_{-j} = \bar{a}_j$, and

$$\sum_{j \in \mathbb{Z}} |a_j|^2 < \infty.$$

Define $E = W^{\frac{1}{2}, 2}(S^1, \mathbb{R}^{2n})$, the subspace of $L^2(S^1, \mathbb{R}^{2n})$ of functions which possess "half" a derivative under the norm

$$\|z\|^2 = \sum_{j \in \mathbb{Z}} (1 + |j|) |a_j|^2.$$

For smooth $z = (p, q) \in E$, where p and q are n-tuples, let $p \cdot q$ denote the $\mathbb{R}^n$ inner product of p and q and define $A(z)$, the so-called action integral, by

$$A(z) = \int_0^{2\pi} p(t) \cdot \dot{q}(t) dt. \qquad (2.3)$$

Then

$$|A(z)| \leq \text{const. } \|z\|^2. \qquad (2.4)$$

Hence A extends to all of E as a continuous quadratic form.

Let E^0, E^+, E^- denote respectively the subspaces of E on which A is null, positive definite and negative definite. It is not difficult to find a basis for these spaces – see e.g. [R1] – for which E^0, E^+, E^- are mutually orthogonal with respect to the $L^2(S^1, \mathbb{R}^{2n})$ inner product as well as with respect to the bilinear form

$$B[z, \varsigma] = \int_0^{2\pi} (p \cdot \dot{\psi} + \dot{q} \cdot \varphi) dt$$

associated with A. Here $z = (p, q)$ and $\zeta = (\varphi, \psi)$. Thus if $z \in E$, $z = z^0 + z^+ + z^- \in E^0 \oplus E^+ \oplus E^- = E$. Moreover

$$(|z^0|^2 + A(z^+) - A(z^-))^{1/2} \tag{2.5}$$

can be taken as an equivalent norm in E and in this norm,

$$A(z) = \|z^+\|^2 - \|z^-\|^2, \tag{2.6}$$

thus displaying the indefinite nature of the action integral.

Suppose further that $H(z)$ satisfies a power growth condition:

$$|H(z)| \le a_1 + a_2 |z|^s, \qquad 1 \le s < \infty. \tag{H_1}$$

Then it is not difficult to show that

$$\mathcal{H}(z) = \int_0^{2\pi} H(z(t))dt \in C^1(E, \mathbb{R}) \tag{2.7}$$

as is

$$I(z) = A(z) - \mathcal{H}(z). \tag{2.8}$$

Moreover critical points of I on E are classical 2π-periodic solutions of (HS). See, e.g. [R1].

There is a similar variational formulation of (FHS) where the underlying space depends on the period, T, of H. Equations (HS2) and (FHS2) can be treated more simply. E.g. for (HS2), the corresponding functional is

$$I(q) = \int_0^{2\pi} \left[\frac{1}{2}|\dot{q}|^2 - V(q(t)) \right] dt \tag{2.9}$$

on $E = W^{1,2}(S^1, \mathbb{R}^n)$.

In the following two sections many of the problems concerning periodic solutions of (HS), (FHS) etc. or equivalently critical points of I, will be discussed.

§3. Periodic Solutions of Prescribed Energy

A large number of problems have been studied for (HS), (FHS), and their second-order analogues. They can be divided into two main classes: prescribed energy problems and prescribed period problems. This section is devoted to the first topic. For prescribed energy problems, one fixes the energy, e.g. $H(z) = 1$ and seeks a periodic solution of (HS) on $H^{-1}(1)$. This is a natural question due to the fact that solutions of (HS) preserve energy. The multiplicity of periodic solutions on a given energy surface has been investigated as well.

The earliest results for the fixed energy case go back to the work of Seifert in the early 1950's [Sf]. They are based on geodesic arguments from geometry. The subject then essentially lay dormant until 1978. Then related results were obtained independently from two rather different points of view. In particular in [R3], it was shown that

Theorem 2.10. *Suppose $H \in C^1(\mathbb{R}^{2n}, \mathbb{R})$ and $H^{-1}(1)$ is a compact manifold which bounds a starshaped region. Then $H^{-1}(1)$ contains a periodic orbit.*

Simultaneous with Theorem 2.10, Weinstein [W2] proved the same result under the stronger assumption that $H^{-1}(1)$ bounds a convex region. He used extensions of Seifert's approach together with a new trick. The proof in [R3] employed a finite dimensional approximation argument with appropriate estimates enabling passage to a limit. The finite dimensional problem was solved using an S^1 version of a theorem of Ljusternik-Schnirelmann. Subsequently more direct approaches were found, two of which will be briefly indicated next.

Note that the period of a periodic solution of (HS) on $H^{-1}(1)$ is a priori unknown. Thus making a change of time scale, one seeks a 2π periodic solution of

$$\dot{z} = \lambda \mathcal{J} H_z(z) \tag{2.11}$$

for some λ, where $\lambda = T/2\pi$, T being the unknown period. The most direct approach to this problem is to find a solution of (2.11) as a critical point of $A(z)$ on the submanifold of E given by $\mathcal{H}(z) = 1$. At a critical point of this constrained variational problem, (2.11) holds where λ is the associated Lagrange multiplier. Moreover, since for a solution of (2.11), $H(z(t)) \equiv$ constant,

$$\mathcal{H}(z) = 1 = H(z(t)), \tag{2.12}$$

i.e. the critical point lies on $H^{-1}(1)$. There are technical difficulties in making this approach work, in particular in showing that $\mathcal{H}^{-1}(1)$ is a C^1 manifold and in (PS) being satisfied.

The second approach to the fixed energy problem reduces it to a fixed period one. This will be illustrated for the special case of Theorem 2.10 in which possibly after a translation, $H^{-1}(1)$ bounds a starshaped neighborhood of 0. Then for all $z \in \mathbb{R}^{2n} \backslash \{0\}$, there is a unique $\alpha(z) \in \mathbb{R}^+$ and $w(z) \in H^{-1}(1)$ such that $z = \alpha(z)w(z)$. Define $\bar{H}(z) = \alpha(z)^4$. Then $\bar{H} \in C^1(\mathbb{R}^{2n} \backslash \{0\}, \mathbb{R})$, is homogeneous of degree 4, and $\bar{H}^{-1}(1) = H^{-1}(1)$. Suppose one has a nontrivial 2π periodic solution of

$$\dot{z} = \mathcal{J}\bar{H}_z(z). \tag{2.13}$$

Then $\bar{H}(z(t)) \equiv \beta > 0$ and $\zeta(t) = \gamma z(t)$ satisfies

$$\dot{\zeta} = \gamma^{-2} J \bar{H}_z(\zeta) \tag{2.14}$$

and

$$\bar{H}(\zeta) = \gamma^4 \beta. \tag{2.15}$$

Choosing γ so that $\gamma^4 \beta = 1$, then ζ lies on $H^{-1}(1)$ and is a periodic solution of (HS). In §3, we will describe fixed period results that yield the 2π periodic nontrivial solution of (2.13) that is needed for this analysis to succeed.

Returning to the general question of periodics of prescribed energy, Weinstein [W2] observed that the cases treated by the papers [W1], [R3] possessed a common geometrical feature, namely $H^{-1}(1)$ is a compact hypersurface of contact type with $H^1(S, \mathbb{R}) = 0$. That led him to conjecture that $H^{-1}(1)$ should contain a periodic orbit whenever this structure was present. Weinstein's conjecture was settled by Viterbo [V2]. Hofer and Zehnder [HZ2] then simplified his proof and extended the result somewhat to show:

Theorem 2.16. *Suppose $H \in C^1(\mathbb{R}^{2n}, \mathbb{R})$ and $H^{-1}(1)$ is a manifold bounding a compact region. Then there is a sequence $0 \neq \epsilon_m \to 0$ as $m \to \infty$ such that $H^{-1}(1 + \epsilon_m)$ contains a periodic orbit, z_m.*

Hofer and Zehnder used a variant of the fixed period argument given above together with a version of Theorem 1.39 to prove Theorem 2.16. As a corollary of Theorem 2.16, one gets the existence of a periodic solution of (HS) on $H^{-1}(1)$ whenever one can find m independent upper and (positive) lower bounds for the period T_m of z_m. Such estimates follow from upper and lower bounds on the action integral and have been obtained for a broad class of Hamiltonians in [BR2] and [BHR]. The next corollary gives an important special case.

Corollary 2.17. *If $H(p, q) = \frac{1}{2}|p|^2 + V(q)$, where $V \in C^1(\mathbb{R}^n, \mathbb{R})$, and $V^{-1}(1)$ is compact, and $V'(q) \neq 0$ on $V^{-1}(1)$, then (HS) has a periodic solution on $H^{-1}(1)$.*

Some small extensions of Theorem 2.16 were made in [R4] and Struwe [St2]. However the basic open question remains: if $H^{-1}(1)$ is a compact hypersurface in $\mathbb{R}^{2n}$, e.g. diffeomorphic to S^{2n-1}, does there exist a periodic solution of (HS) on it.

Next we turn to the multiplicity of periodic solutions of (HS) on $H^{-1}(1)$. This is a difficult question and not much has been proved. The main breakthrough here was made by Ekeland and Lasry [EL] who found:

Theorem 2.18. *Suppose* $H \in C^1(\mathbb{R}^{2n}, \mathbb{R})$ *and* $H^{-1}(1)$ *is a manifold which bounds a strictly convex neighborhood of* 0. *Let*

$$R = \max_{z \in H^{-1}(1)} |z| \quad \text{and} \quad r = \min_{z \in H^{-1}(1)} |z|.$$

If

$$1 < \frac{R}{r} < \sqrt{2}, \tag{2.19}$$

then $H^{-1}(1)$ *contains* n *geometrically distinct periodic solutions.*

Some of the key ingredients of the proof of Theorem 2.18 are (i) the S^1 symmetry of the functionals (2.3) and (2.7), i.e. A and $\mathcal{H}$ are invariant under the maps $z(t) \rightarrow z(t + \theta)$ for all $\theta \in [0, 2\pi]$; (ii) use of an appropriate symmetric critical point theorem, and (iii) a comparison argument exploiting the "pinching condition" (2.19).

Variants of Theorem 2.18 have been given several authors. E.g. Berestycki, Lasry, Mancini, and Ruf [BLMR] proved a related theorem when $H^{-1}(1)$ is nested between a pair of ellipsoids. Their result contains in particular a bifurcation theorem of Weinstein [W3] as a limiting case. Ambrosetti and Mancini [AM1] have an analogue of [BLMR] when $H^{-1}(1)$ is nested between starshaped regions. In another direction various researchers such as van Groesen [VG1] and Girardi [Gr] have weakened (2.19) assuming further symmetries for H like $H(z) = H(-z)$. See also Bartsch and Clapp [BCl] who treat multiplicity questions for a more general class of symmetries.

For convex Hamiltonians, Ekeland and Lassoued [ELs], Szulkin [Sz], and Ekeland-Hofer [EH4] have obtained versions of the following result:

Theorem 2.20. *If* $n \geq 2$, $H \in C^1(\mathbb{R}^{2n}, \mathbb{R})$, *and* $H^{-1}(1)$ *is a manifold which bounds a compact convex region, then* $H^{-1}(1)$ *contains at two distinct periodic solutions.*

Aside from Theorem 2.20, it remains an open question as to whether multiplicity results can be obtained for the fixed energy problem for (HS) without a pinching condition.

We conclude this section with some results for "brake orbits". For simplicity we will just describe them for (HS2) although they arise more generally when H is even in p. Suppose $V^{-1}(1)$ is a compact manifold bounding a neighborhood D of 0 in $\mathbb{R}^n$. A brake orbit of (HS2) is a solution of the equation such that $q(0) \in V^{-1}(1)$, and for some $T > 0$, $q(t) \in D$, $0 < t < T$, and $q(T) \in V^{-1}(1)$.

Since

$$\frac{1}{2}|\dot{q}(t)|^2 + V(q(t)) = 1, \tag{2.21}$$

$\dot{q}(0) = 0 = \dot{q}(T)$. Therefore extending q as an even function about 0 and T yields a $2T$ periodic solution of (HS). The earliest results on the existence of brake orbits are due to Seifert [Sf]. His work was extended by Weinstein [W1]. See also Bolotin [B1]. These authors used arguments from geometry. Subsequently others have used minimax methods to get existence and multiplicity results for brake orbits in the spirit of Theorem 2.16 and 2.18. See, e.g. van Groesen [VG1], Ambrosetti-Benci-Long [ABL], [R4]. Offin [O] has also treated a situation where D may be unbounded.

§4. Periodic Solutions of Prescribed Period

A considerable amount of research has been devoted to prescribed period problems. Before going into details, we will give an overview of this topic. An initial classification of problems is into autonomous equations such as (HS) and (HS2) and periodically forced equations like (FHS) and (FHS2). Existence and multiplicity of solutions have been studied for both classes. For autonomous problems, one can seek solutions, z, for which the prescribed period, T, is the *minimal period*, i.e. $z(t)$ has no period smaller than T. For forced problems, one can ask for *subharmonic* solutions, i.e. solutions whose minimal period is an integer multiple of the period of forcing. Within each class, results have been obtained for three types of growth behavior of H: (i) superquadratic, i.e. $H(z)/|z|^2 \to \infty$ as $|z| \to \infty$, (ii) subquadratic, i.e. $H(z)/|z|^2 \to 0$ as $|z| \to \infty$, and (iii) H is asymptotically quadratic as $|z| \to \infty$.

To describe the situation more fully, we begin with the autonomous case. Perhaps the first variational results here were for superquadratic problems [R3] where the following was proved:

Theorem 2.22. *Suppose* $H \in C^1(\mathbb{R}^{2n}, \mathbb{R})$ *and satisfies*

(H_1) $H(z) \geq 0$ *for* $z \in \mathbb{R}^{2n}$,

(H_2) $H(z) = o(|z|^2)$ *as* $z \to 0$.

and

(H_3)   *there are constants* $\mu > 2$ *and* $r \geq 0$ *such that*

$$0 < \mu H(z) \leq z \cdot H_z(z) \qquad \text{for} \quad |z| > r.$$

Then for each $T > 0$, *(HS) possesses a nonconstant* T *periodic solution.*

The original proof of Theorem 2.22 employed a finite dimensional approximation argument together with careful estimates to get a nontrivial limit. Now the result can be proved much more directly as an application of Theorem 1.39. See, e.g. [BR1] or [R1]. In fact using an S^1 symmetric version of Theorem 1.39 and comparison arguments, a much stronger version of Theorem 2.22 can be proved:

Theorem 2.23 [R5]. *Let $H \in C^1(\mathbb{R}^{2n}, \mathbb{R})$ and satisfy (H_3). Then for any $T, R > 0$, (HS) has a T-periodic solution, z, with $\|z\|_{L^\infty} \geq R$.*

Thus merely under (H_3), (HS) has infinitely many distinct T periodic solutions. Using Theorem 2.22 or 2.23, one can prove some of the fixed energy results of §3, such as Theorem 2.10, via the trick that was indicated earlier. There are also analogues of Theorems 2.22 (and 2.23) for (HS2).

A natural question to ask in the setting of Theorem 2.22 is whether there is a periodic solution having minimal period T. Some initial progress was made by Ambrosetti and Mancini [AM2] and by Girardi and Matseu [GM1–2]. Their results were improved by Ekeland and Hofer [EH1] who showed:

Theorem 2.24. *Suppose $H \in C^2(\mathbb{R}^{2n}, \mathbb{R})$, satisfies (H_1)–(H_3), and H is strictly convex. Then for each $T > 0$, (HS) possesses a T periodic solution with minimal period T.*

The proof of Theorem 2.24 involves the so-called dual variational transformation to a problem to which the Mountain Pass Theorem can be applied, and a combination of index results of Ekeland [E] and results of Hofer [H3]–[H4] mentioned earlier on the behavior of a functional near a critical point of mountain pass type. Subsequently there have been improved versions of Theorem 2.24 replacing the convexity assumption by milder ones. See, e.g. Girardi-Matseu [GM3]. There has also been work on the second order case. See, e.g. [GM3], [Lo1–2]. One cannot expect to get solutions having an arbitrary minimal period in the setting of Theorem 2.23. See, e.g. [R1] or [MW] for counterexamples.

There are also versions of Theorem 2.22 for (FHS) and (FHS2) with T being the period of forcing. However for (FHS), further hypotheses are required for H to compensate for the loss of energy conservation. On the other hand, as was mentioned earlier, one can also attempt to find subharmonic solutions for the forced problems. The following result illustrates this:

Theorem 2.25. *Suppose $H \in C^1(\mathbb{R} \times \mathbb{R}^{2n}, \mathbb{R})$ with $H(t, z)$ being T periodic in t. If further $H(t, z)$ satisfies (H_1)–(H_3) as a function of z and*

(H_4) *there are constants $\alpha, \beta > 0$ such that for $|z| > \beta$, $|H_z(t, z)| \leq \alpha z \cdot H_z(t, z)$,* *then for each* $k \in \mathbb{N}$, (FHS) *possesses a* kT *periodic solution* $z_k(t)$. *Moreover infinitely many of these functions are distinct.*

Improvements of this theorem have been obtained by various authors. E.g. Ekeland and Hofer [EH2] treated a strictly convex Hamiltonian for which they find geometrically distinct solutions for each $k \in \mathbb{N}$. Other results in this direction can be found in [E] and [MW].

For other results on subharmonics such as global analogues of the classical Birkhoff-Lewis Theorem, see e.g. Benci-Fortunato [BF] and Felmer [F1]. Also for second order equations where certain changes in sign were allowed see Felmer-Silva [FS], Han [Ha], and Lassoued [La].

Observe that in Theorem 2.22 and 2.25, $z \equiv 0$ is a solution of the corresponding Hamitonian system. Another question of interest here is what happens if the equation is perturbed so that $z = 0$ is no longer a solution. E.g. suppose H satisfies (H_2)–(H_3) and we consider

$$\dot{z} = \mathcal{J}(H_z(z) + f(t)) \tag{2.26}$$

where f is T periodic. The perturbation $f(t)$ destroys both the S^1 symmetry and autonomous nature of the problem and these features played an important role in Theorems 2.22 and 2.23. The first results for (2.26) were obtained by Bahri and Berestycki [BB2] who proved.

Theorem 2.27. *Suppose* $H \in C^2(\mathbb{R}^n, \mathbb{R})$ *and* $f \in C(\mathbb{R}, \mathbb{R}^n)$ *with* f *being* T *periodic in* t. *If* H *satisfies* (H_3) *and*

(H_5) $|H(t)| \leq a_1 + a_2|z|^s$ *where* $s < 2\mu$,

then (2.26) *possesses an unbounded sequence of solutions.*

The proof of Theorem 2.27 involves a nice blend of analytical and topological arguments including minimax and Morse-theoretic elements and comparison arguments. Subsequently another proof was found by Long [Lo3–4] who allowed H to be C^1 and weakened (H_5). Related results were obtained for (HS2) under milder hypothesis. See Bahri-Berestycki [BB3] and Long [Lo5].

Thus far we have been discussing superquadratic Hamiltonian, i.e. some version of (H_3) is satisfied. Another class of problems for which many interesting results have been obtained involve subquadratic Hamiltonians, i.e. $H(z)/|z|^2 \to 0$ as $|z| \to \infty$, and a few such results will be described next. When H is also convex, the dual variational method can be used to transform (HS) to a new problem which can sometimes be treated successfully by minimization arguments. Roughly this means writing (HS) as

$$-\mathcal{J}\dot{z} = H_z(z), \tag{2.28}$$

and using the convexity of H or monotonicity of H_z to invert H_z and convert (2.28) to

$$z = H_z^{-1}(-J\dot{z}). \tag{2.29}$$

Then taking $\dot{z}$ as a new independent variable (i.e. making a Legendre transform), (2.29) can be reformulated as a variational problem. This approach was introduced by Clarke [Cl1] and used successfully by Clarke and Ekeland [CE] to treat the following subquadradic situation:

Theorem 2.30. *Suppose $H \in C^1(\mathbb{R}^{2n}, \mathbb{R})$, H is convex with $H(0) = 0$, $H_z(0) = 0$, and*

(H$_6$) $H(z)/|z|^2 \to \infty$ as $|z| \to 0$,

(H$_7$) $H(z)/|z|^2 \to 0$ as $|z| \to \infty$.

Then for all $T > 0$, (HS) possesses a T periodic solution with minimal period T.

Theorem 2.30 is the first variational result we know of establishing the existence of solutions having a prescribed minimal period. There are analogues for (HS2), see, e.g. [E], [MW]. Several generalizations have been given. However if one drops (H$_6$), it cannot be expected that there are solutions of minimal period T for each $T > 0$. See e.g. [MW] for a counterexample. On the other hand, several authors have shown without (H$_6$) that there are nonconstant periodic solutions for large T. See, e.g. [BR1], Brezis-Coron [BC] and Mawhin-Willem [MW]. The following such result is a generalization of work of Brezis-Coron by Mawhin-Willem [MW].

Theorem 2.31. *If $H \in C^1(\mathbb{R}^{2n}, \mathbb{R})$ is convex, satisfies (H$_7$), and*

(H$_8$) $H(z) \to \infty$ as $|z| \to \infty$,

then there is a $T_0 > 0$ such that for each $T > T_0$, (HS) possesses a periodic solution $z(t)$ with minimal period T. Moreover $\min_{t \in [0,T]} |z(t)| \to \infty$ as $T \to \infty$.

There are also version of the above results for (FHS) and (FHS2). See, e.g. Ekeland [E], and Mawhin-Willem [MW]. Here is a fairly general one due to Willem [Wi].

Theorem 2.32. *Suppose $H \in C^1(\mathbb{R} \times \mathbb{R}^{2n}, \mathbb{R})$ where H is T periodic in t, convex in z, and satisfies (H$_7$)–(H$_8$) with respect to z, the latter uniformly in t. Then for each $k \in \mathbb{N}$ there exists a kT periodic solution, z_k, of (FHS). Moreover as $k \to \infty$, $\|z_k\|_{L^\infty} \to \infty$ and the minimal period T_k of z_k approaches infinity.*

If H is not convex, one can use versions of Theorem 4.7 to get existence results for subquadratic problems. Here are two such results for (FHS) due to Silva [Si1–2]:

Theorem 2.33. *Suppose* $H \in C^1(\mathbb{R} \times \mathbb{R}^{2n}, \mathbb{R})$ *with* H *being* T-periodic in t,

(H_9) *there is an* $M > 0$ *such that*

$$|H_z(t, z)| \leq M \quad \text{for} \quad t \in [0, T], \ z \in \mathbb{R}^{2n},$$

and

(H_{10}) $H(t, z) \to \infty$ *(or to* $-\infty$*) as* $|z| \to \infty$, *uniformly for* $t \in [0, T]$.

Then for each $k \in \mathbb{N}$, *(FHS) possesses a* kT *periodic solution,* z_k, *with* $\|z_k\|_{L^\infty} \to \infty$ *as* $k \to \infty$.

Theorem 2.34. *Suppose* $H \in C^1(\mathbb{R} \times \mathbb{R}^{2n}, \mathbb{R}) \not\equiv 0$ *is* T *periodic in* t *and satisfies*

(H_{11}) $H(t, z) \to 0$ *as* $|z| \to \infty$ *uniformly for* $t \in [0, T]$,

(H_{12}) $H_z(t, z) \to 0$ *as* $|z| \to \infty$ *uniformly for* $t \in [0, T]$,

and

(H_{13}) $H(t, z) \leq 0$ *(or* ≥ 0*) for* $t \in [0, T]$, $z \in \mathbb{R}^{2n}$.

Then there is a sequence $(k_j) \subset \mathbb{N}$, $k_j \to \infty$ *as* $j \to \infty$ *and distinct* $k_j T$ *periodic solutions,* z_j, *of (FHS).*

Giannoni [Gi1] earlier proved versions of Theorems 2.33 and 2.34 for (FHS2) using a variant of Theorem 1.22.

Another interesting and important class of subquadratic problems arises when

(H_{14}) H is T periodic in t and T_i periodic in q_i, $1 \leq i \leq 2n$.

When (H_{14}) holds, if z is a solution of (FHS), so is $z + (k_1 T_1, \ldots, k_{2n} T_{2n})$ for each $k = (k_1, \ldots, k_{2n}) \in \mathbb{Z}^{2n}$. Thus solutions occur in $\mathbb{Z}^{2n}$ equivalent families. A major result for this class of equations and which settled a conjecture of Arnold is due to Conley and Zehnder [CZ]:

Theorem 2.35. *Suppose* $H \in C^2(\mathbb{R} \times \mathbb{R}^{2n}, \mathbb{R})$ *and* (H_{14}) *is satisfied. Then (FHS) possesses at least* $n + 1$ *distinct families of solutions.*

If all of the solutions of (FHS) correspond to nondegenerate critical points of the associated functional, Conley and Zehnder showed (FHS)

has at least 2^n distinct families of solutions [CZ]. Some authors have obtained generalizations of Theorem 2.35 which replace H by $Q + H$, Q being quadratic in z. See, e.g. Chang [Ch3] and Mawhin-Willem [MW]. There are also related results for (FHS2), see, e.g. Chang [Ch2], Felmer [F3–4], Fournier-Willem [FW], Mawhin-Willem [MW], and Rabinowitz [R6].

To conclude this section, one final class of problems that is intermediate to the super- and subquadratic results mentioned this far will be discussed. Thus consider (FHS) where

(H_{15}) $H_z(t, z) = A_\infty z + o(|z|)$ as $|z| \to \infty$ uniformly in t.

and the matrix A_∞ is independent of t. Condition (H_{15}) is satisfied if H is appropriately asymptotically quadratic at infinity. Various authors have given existence and multiplicity results for periodic solutions of (HS), (FHS), etc. in such settings. The first results we know of are due to Clark [C] for (HS2) with H even. Amann and Zehnder [AZ1] initiated this question for (FHS). They obtained several results including:

Theorem 2.36. *If* $H \in C^2(\mathbb{R} \times \mathbb{R}^{2n}, \mathbb{R})$, *is* T *periodic in* t, *satisfies* (H_{15}),

(H_{16}) $H_{zz}(t, z)$ *is uniformly bounded,*

and

(H_{17}) $\sigma(\mathcal{J}A_\infty) \cap \frac{2\pi i}{T} \mathbb{Z} = \phi$,

(FHS) *possesses at least one* T-*periodic solution.*

Condition (H_{17}) is a nonresonance condition at ∞. If H is also asymptotically quadratic at $z = 0$, i.e.

(H_{18}) $H_z(t, z) = A_0 z + o(|z|)$ as $|z| \to 0$ uniformly in t,

then the solution given by Theorem 2.36 may be $z(t) \equiv 0$. However if additional hypotheses, such as a nonresonance condition at 0 and the nonvanishing of a symplectic invariant are satisfied, one still gets a nontrivial solution of (FHS). See [AZ1] for more details. Amann and Zehnder also obtained results for the autonomous case [AZ2]. Several authors have obtained generalizations and variations on these results. See, e.g. Benci [Be], Costa-Willem [CW], and Long [Lo6].

Part 3. Connecting Orbits

§1. Introduction

In this final section, global variational approaches to connecting orbits of Hamiltonian systems will be discussed. The term "connecting orbit" does not have a standard meaning. Here it refers to an orbit defined on an infinite or semi-infinite time interval with well defined, at least in an asymptotic sense, initial and terminal states. For most of the results presented here, these asymptotic states will be prescribed. Research by variational methods on these problems only began in the past few years so it has not yet seen the great amount of activity that has occurred for critical point theory proper and for periodic solutions of Hamiltonian systems. Therefore a more complete survey can be given of the work done here than in Parts 1 and 2. Most of the results here are in the setting of (HS2) and treat homoclinic or heteroclinic orbits with the limit states being equilibria.

Essentially two questions have been studied for connecting orbits, but in different contexts. First is the existence of one (or more) basic homoclinic or heteroclinic solution. Secondly there are results which show roughly that near formal sums or concatenations of these basic trajectories, there are actual solutions which are referred to as *multibump* solutions.

Before describing these results, a few observations are in order. The mathematical problems encountered here are in general more complex than those one deals with for periodic solutions. One major reason is that there is less compactness available for connecting orbit problems since one is dealing with spaces of functions defined on all of $\mathbb{R}$ rather than a bounded interval. Consequently it is generally not possible to verify the (PS) condition here and the treatment of the problem becomes more difficult. On the other hand, dealing with $\mathbb{R}$ provides a lot of room and makes possible the multibump phenomena. Analyzing how Palais-Smale sequences behave plays a crucial part in the analysis. Fortunately insights gained in previous work in other contexts proved to be helpful here. We refer to work by Sachs and Uhlenbeck [SU] on the Yang-Mills equations, Bahri and Coron [BC] on limit exponent problems for semilinear elliptic partial differential equations, Bahri [Ba] on critical points at infinity, and P. L. Lions [LPL] on concentration compactness. Another difficulty faced here, especially when treating heteroclinic solutions is the lack of an obvious candidate for a function space in which to pose the variational problem. Moreover one may have to modify the naturally occurring functional since contributions from asymptotic states may make it infinite.

These caveats aside, in §2, the existence of multibump homoclinic solutions for certain superquadratic Hamiltonian systems will be discussed and in §3 a variety of results for heteroclinic solutions will be presented.

§2. Homoclinic Solutions

Recall that $z(t)$ is a homoclinic solution of (FHS) or more precisely $z(t)$ is homoclinic to $\xi \in \mathbb{R}^{2n}$ if $H_z(t, \xi) = 0$ for all t and $z(t) \to \xi$ as $|t| \to \infty$.

In this section several results on homoclinic solutions of (FHS) will be described.

We begin with work of Coti Zelati, Ekeland, and Séré [CZES]. They studied (FHS) for $H(t, z) = Az \cdot z + R(t, z)$ where A and R satisfy $R \in C^2(\mathbb{R} \times \mathbb{R}^{2n}, \mathbb{R})$ and is T periodic in t,

(A) A is a hyperbolic matrix, i.e. the eigenvalues of A have a nonzero real part,

(R_1) R is a strictly convex,

(R_2) there are constants $a_1, a_2 > 0$ and $\mu > 2$ such that

$$a_1 |z|^\mu \leq R(t, z) \leq a_2 |z|^\mu, \qquad z \in \mathbb{R}^{2n}$$

and R is superquadratic as $|z| \to \infty$, i.e. R satisfies (H_3) for all $z \in \mathbb{R}^{2n}$. Note that by (H_3) and the form of H, $z = 0$ is an equilibrium solution of (FHS). Moreover since H is T-periodic in t, if $z(t)$ is homoclinic to 0, so is $z(t + jT)$ for all $j \in \mathbb{Z}$. This family of translates geometrically represent the same homoclinic solution of (FHS). In [CZES], it was proved:

Theorem 3.1. *Suppose $H(t, z) = Az \cdot z + R(t, z)$ with $R \in C^2(\mathbb{R} \times \mathbb{R}^{2n}, \mathbb{R})$ and R is T periodic in t. If (A), (R_1)–(R_2) and (H_3) for R holds, then (FHS) possesses at least 2 geometrical distinct solutions which are homoclinic to 0.*

The associated variational problem will not be formulated here. Instead it will be shown how to do this for a related simpler problem for (FHS2) later in this section. The proof of Theorem 3.1 involves using the convexity of R to make the dual variational transformation and applying a mountain pass argument to the new problem. This is not a straightforward process since (PS) is not satisfied. However results of P. L. Lions [LPL] are used to help study the breakdown of (PS) and get an existence result. An additional indirect variational argument gives a second solution. Hofer and Wysocki [HW] and Tanaka [T2] have proved generalizations of Theorem 3.1 dropping (R_1) and using rather different variational arguments.

In an important sequel to [CZES], Séré [Se1] proved:

Theorem 3.2. *Under the hypotheses of Theorem 3.1, (FHS) possesses infinitely many solutions which are geometrically distinct and homoclinic to 0.*

In fact Séré proved a stronger result. As in [CZES], the dual variational principle is used to transform the variational problem associated with (FHS) to a new one with a corresponding functional, I. Although I does not satisfy

(PS), it otherwise satisfies the qualitative hypotheses of the Mountain Pass Theorem and therefore a minimax value c as in (1.5) can be defined. Let $\mathcal{K}$ denote the set of nontrivial critical points of I. Séré showed if $\mathcal{K}$ modulo the translational symmetry noted earlier, i.e. $\mathcal{K}/\mathbb{Z}$ is finite, c is a critical value of I. He then introduced a new functional, $\hat{I}$, closely related to I, that he called a split functional and a class of sets with respect to which he obtained a minimax value for $\hat{I}$. This value turned out to be $2c$. Finally assuming that $\mathcal{K}/\mathbb{Z}$ is finite, he used a deformation argument to contradict the minimax characterization of $2c$. Thus $\mathcal{K}/\mathbb{Z}$ cannot be finite. He further noted that modifications in his argument show that if

(*) there is an $\alpha > 0$ such that $\mathcal{K}^{c+\alpha}/\mathbb{Z}$ is finite,

then $\mathcal{K}^{2c+\alpha}_{2c-\alpha}/\mathbb{Z}$ is infinite.

Motivated by [CZES] and [Se1], Coti Zelati and Rabinowitz [CZR1] studied homoclinics for (HS2) under related hypotheses. This variational problem is technically simpler to formulate than [CZES] so it will be briefly indicated how this is done. Consider

$$\ddot{q} - L(t)q + V_q(t, q) = 0 \qquad (3.3)$$

where L and V satisfy

(L) $L(t)$ is a continuous T periodic matrix which is positive definite symmetric for all $t \in [0, T]$,

(V_1) $V \in C^2(\mathbb{R} \times \mathbb{R}^n, \mathbb{R})$ and is T periodic in t,

(V_2) $V_{qq}(t, 0) = 0$,

(V_3)   there is a $\mu > 2$ such that

$$0 < \mu V(t, q) \leq q \cdot V_q(t, q) \quad \text{for all } t \in \mathbb{R} \text{ and } q \neq 0.$$

A solution of (3.3) is homoclinic to 0 if $q(t)$ and $\dot{q}(t) \to 0$ as $|t| \to \infty$. For simplicity, let $T = 1$ and $L(t) = $ id. Set $E = W^{1,2}(\mathbb{R}, \mathbb{R}^n)$. Then $q \in E$ implies $q(t) \to 0$ as $|t| \to \infty$. Let

$$I(q) = \int_{\mathbb{R}} \left[\frac{1}{2}(|\dot{q}|^2 + |q|^2) - V(t, q) \right] dt. \qquad (3.4)$$

Then $I \in C^1(E, \mathbb{R})$ and critical points of I are classical solutions of (3.3). Moreover if $q \in E$, (3.3) and (V_1)–(V_3) imply $\dot{q} \in E$ and therefore $\dot{q}(t) \to 0$ as $|t| \to \infty$. Consequently critical points of I on E are homoclinic solutions of (3.3).

The existence of one such solution of (3.3) was shown in [R7] under slightly milder hypotheses than those mentioned above using an approximation argument which determined the homoclinic as a limit of subharmonics.

Assuming (V_1)–(V_3), it readily follows that I satisfies the qualitative hypotheses of the Mountain Pass Theorem but once again (PS) fails. In fact let c denote the mountain pass minimax value associated with I as given by (1.5). For $j \in \mathbb{Z}$ and $q \in E$, define

$$\tau_j q(t) = q(t - j). \tag{3.5}$$

Then for all $j \in \mathbb{Z}$ and $q \in E$,

$$I(\tau_j q) = I(q) \tag{3.6}$$

so I possesses a $\mathbb{Z}$ symmetry. Suppose (q_j) is a (PS) sequence for I corresponding to c, i.e.

$$I(q_j) \to c, \quad I'(q_j) \to 0, \ j \to \infty. \tag{3.7}$$

Then for $m(j)$ sufficiently large and $m(j) \to \infty$ as $j \to \infty$,

$$I(q_j + \tau_{m(j)} q_j) \to 2c, \quad I'(q_j + \tau_{m(j)} q_j) \to 0, \ j \to \infty. \tag{3.8}$$

However $q_j + \tau_{m(j)} q_j$ will not possess a convergent subsequence. If, e.g. c were a critical value of I and q a corresponding critical point, we could take $q_j \equiv q$ and $m(j) = j$. A similar argument shows (PS) fails at kc for all $k \in \mathbb{N}$, $k \geq 2$. The numbers kc, however play an important role in the existence of multiple homoclinics for (3.3). Indeed

Theorem 3.9 [CZR1]. *Suppose* (L), (V_1)–(V_3) *are satisfied and*

($*$) *there is an $\alpha > 0$ such that $\mathcal{K}^{c+\alpha}/\mathbb{Z}$ is finite.*

Then c is a critical value of I and for each $j \in \mathbb{N}$, $j \geq 2$, $\mathcal{K}^{jc+2}_{jc-\alpha}/\mathbb{Z}$ is infinite.

Actually a stronger result is valid here. To state it precisely is rather technical. Therefore sacrificing accuracy for clarity, it can be described as follows: For any small $r > 0$ and $k \geq 2$, there is an $\ell = \ell(r, k)$ such that whenever $v_1, \ldots, v_k \in \mathcal{K}$ with $I(v_i) = c$, $1 \leq i \leq k$, if $|\ell_i - \ell_j| \geq \ell$ for $i \neq j$,

$$B_r \left(\sum_1^k \tau_{\ell_i} v_i \right) \cap \mathcal{K} \neq \phi. \tag{3.10}$$

In other words, near any sum of sufficiently large translates of elements of $\mathcal{K} \cap I^{-1}(c)$, there is an actual solution of (3.3). Since the numbers $|\ell_i - \ell_j|$ are large, this is a 'k-bump' solution. Hence (3.3) has infinitely many k-bump homoclinic solutions if $(*)$ holds. On the other hand if $(*)$ fails, there are still infinitely many homoclinic solutions of (3.3).

In a recent preprint, Séré [Se2] has obtained stronger results in the setting of Theorem 3.2. He shows that ℓ can be chosen independently of k and (3.10) holds (although B_r is taken in a different norm). This enables him to pass to a limit to find ∞-bump solutions of (FHS). He also weakens $(*)$.

Note that if (FHS) or (3.3) are autonomous, $(*)$ fails since if $z(t)$ is a solution, so is $z(t + \theta)$ for all $\theta \in \mathbb{R}$. Moreover simple examples show the conclusion of Theorem 3.9 is false. Thus some sort of condition like $(*)$ is needed.

There is a classical result in the theory of dynamical systems (see, e.g. [KS]), which says that if for equations like (FHS) or (3.3), one has a solution homoclinic e.g. to 0 and if the stable and unstable manifolds for the associated Poincaré (or time T) map intersect transversally at the corresponding homoclinic point, there is a rich and complicated structure of homoclinic, heteroclinic, and periodic orbits nearby. In the global approaches taken in [Se1]–[Se2] and [CZR1], condition $(*)$ and its analogues seem to be playing the role of the transversal intersection hypothesis. Moreover the homoclinics obtained variationally roughly correspond to a subset of those given by the symbolic dynamics in [KS]. In the one case we know of where a direct comparison of variational and classical results has proved possible, Bessi [Bs] studied an interesting model problem in which his analogue of $(*)$ proves to be weaker than the usual Melnikov type assumption. In some work in progress on a sequel to [CZR1], the authors have found periodic solutions of (3.3) with arbitrarily large periods and which are near (in an appropriate sense) the homoclinic solutions for some $\Sigma \tau_{\ell_i} v_i$. These multi-bump periodic solutions again appear to correspond to another subset of those solutions given by the symbolic dynamics of [KS]. A more precise understanding of the variational and classical approaches remains to be made.

To give an idea of how variational methods give rise to multibump solutions, a brief sketch will be given of the proof of the Theorem 3.9. First an analogue of the set Γ of (1.6) is required. For $\theta = (\theta_1, \ldots, \theta_j) \in [0, 1]^j$, set $0_i = (\theta_1, \ldots, \theta_{i-1}, 0, \theta_{i+1}, \ldots, \theta_j)$ and $1_i = (\theta_i, \ldots, \theta_{j-1}, 1, \theta_{i+1}, \ldots, \theta_j)$. Define for $j \geq 2$,

$$\Gamma_j = \{ G = g_1 + \cdots + g_j \mid g_i \text{ satisfies } (g_1) - (g_3), 1 \leq i \leq j \} \qquad (3.11)$$

where

(g$_1$) $g_i \in C([0,1]^j, E)$, $1 \leq i \leq j$,

(g$_2$) $g_i(0_i) = 0$, $g_i(1_i) \in I^0 \backslash \{0\}$, $1 \leq i \leq j$,

(g$_3$) there are numbers $p_1 < p_2 < \cdots < p_{j-1} \in \mathbb{R}$ which are independent of θ such that supp $g_1(\theta) \subset (-\infty, p_1)$, supp $g_2(\theta) \subset (p_1, p_2), \ldots$, supp $g_j(\theta) \subset (p_{j-1}, \infty)$.

In (g$_3$), for $q \in E$, supp q denotes the support of q as a mapping of $\mathbb{R}$ to $\mathbb{R}^n$. A version of these sets for $j = 2$ was already used by Séré [Se1]. Define

$$b_j = \inf_{G \in \Gamma_j} \max_{\theta \in [0,1]^j} I(G(\theta)). \tag{3.12}$$

Then it is not difficult to show that

Proposition 3.13. $b_j = jc$.

Now the idea of the proof of (3.10) for e.g. $j = 2$ is as follows. For any $\epsilon > 0$, a $g \in \Gamma$ can be chosen such that g has compact support and

$$\max_{\sigma \in [0,1]} I(g(\sigma)) \leq c + \frac{\epsilon}{2}. \tag{3.14}$$

Then for all ℓ large, if $G(\theta) = g(\theta_1) + \tau_\ell g(\theta_2)$, it follows that $G \in \Gamma'_2$ and

$$I(G(\theta)) \leq 2c + \epsilon. \tag{3.15}$$

Now if $\mathcal{K}^{2c+\alpha}_{2c-\alpha}/\mathbb{Z}$ were finite, for any r sufficiently small, using the known behavior of (PS) sequences for (3.4), one can deform $G(\theta)$ to $\hat{G}(\theta)$ satisfying

$$I(\hat{G}(\theta)) \leq 2c - \epsilon \tag{3.16}$$

and

$$\|\hat{G}(\theta) - G(\theta)\| \leq r \tag{3.17}$$

provided that $\epsilon(r)$ is sufficiently small compared to r. If $\hat{G}$ were in Γ_2, (3.16) would contradict Proposition 3.13. Unfortunately $\hat{G}$ need not belong to Γ_2 but by (3.17), it is near Γ_2. Indeed (3.17) can be used to help modify $\hat{G}$ to construct $H \in \Gamma_2$ such that

$$\max_{\theta \in [0,1]^2} I(H(\theta)) \leq 2c - \frac{\epsilon}{2}. \tag{3.18}$$

Now the desired contradiction obtains and completes the sketch of the proof of Theorem 3.9.

In research subsequent to [CZR1], the authors have obtained related results for a family of semilinear elliptic partial differential equations on $\mathbb{R}^n$ [CZR2]. Li [Li] has exploited some of the ideas from [Se1] and [CZR1]–[CZR2] and introduced new ones to study some limit exponent problems for semilinear elliptic partial differential equations and related problems in geometry. Alama and Li [AL] have extended the results of [CZR1] and [CZR2] weakening the periodicity hypotheses to an assumption of asymptotic periodicity. They also obtained related results for time periodic solutions of nonlinear Schrödinger equations. For some related results to [Se1] and [CZR1] on manifolds, see, e.g. Giannoni [Gi2] and Giannoni-Rabinowitz [GR].

We conclude this section with two recent results. First Séré [Se3] has studied homoclinics for the fixed energy problem for (HS). In particular he found:

Theorem 3.19. *Assume* $H \in C^2(\mathbb{R}^{2n}, \mathbb{R})$, $H^{-1}(1)$ *is compact and has exactly one critical point of* H *at* $z = 0$ *with* $H_{zz}(0)$ *nondegenerate and* $JH_{zz}(0)$ *hyperbolic. If also* $H^{-1}(1)\backslash\{0\}$ *is of contact type, then* (HS) *possesses a solution on* $H^{-1}(1)$ *which is homoclinic to* 0.

Secondly, as was noted earlier, Theorem 3.9 excludes the autonomous case and in fact, without some extra hypotheses, there are counterexamples to the result. Recently Ambrosetti and Coti Zelati [ACZ] have studied the autonomous case of (3.3) with $L = $ id and under somewhat stronger hypotheses than (V_1)–(V_3). They further assume

(V_4) $V_q(q) \cdot q < V_{qq}(q) \cdot q \cdot q$ for all $q \in \mathbb{R}^n\backslash\{0\}$

as well as a pinching condition:

(V_5) there are functions φ_1, φ_2 homogeneous of degree $\alpha \geq \mu$ such that

$$\varphi_1(q) \leq V(q) \leq \varphi_2(q)$$

for all $q \in \mathbb{R}^n$.
   Set

(3.20) $a = \min_{|x|=1} \varphi_1(x), \qquad b = \max_{|x|=1} \varphi_2(x)\,.$

Then Ambrosetti and Coti Zelati show:

Theorem 3.21. *If* V *satisfies* (V_1)–(V_5) *and*

$$\frac{b}{a} < 2^{\frac{\alpha-2}{2}}\,,$$ (3.22)

then (3.3) has at least two geometrically distinct solutions which are homoclinic to 0.

§3. Heteroclinic Solutions

In this section, the existence of heteroclinic solutions of (HS2) and (FHS2) will be discussed. We begin with [R8] which treats (HS2) for potentials periodic in the components of V. Thus these are systems of multiple pendulum type.

Suppose that V satisfies

(V_6) $V(q)$ is T_i periodic in q_i, $1 \le i \le n$.

For simplicity, let $T_i = 1$, $1 \le i \le n$. Set

$$\mathcal{M} = \{\xi \in \mathbb{R}^n \mid V(\xi) = \max_{\mathbb{R}^n} V\}. \tag{3.23}$$

By (V_6), $\mathcal{M}$ is the union of lattices since $x \in \mathcal{M}$ implies $x + j \in \mathcal{M}$ for all $j \in \mathbb{Z}^n$. Suppose further that

($\mathcal{M}$) $\mathcal{M}$ consists of isolated points.

Then it was show in [R8] that

Theorem 3.24. *If $V \in C^1(\mathbb{R}^n, \mathbb{R})$ and satisfies (V_6) and ($\mathcal{M}$), for each $\xi \in \mathcal{M}$, there is an $\eta \in \mathcal{M}\backslash\{\xi\}$ and a heteroclinic solution q of (HS2) such that $q(t) \to \xi$ as $t \to -\infty$ and $q(t) \to \eta$ as $t \to \infty$.*

Since (HS2) is time reversible, an immediate consequence of Theorem 3.24 is:

Corollary 3.25. *Under the hypotheses of Theorem 3.24, for each $\xi \in \mathcal{M}$, there is an $\eta \in \mathcal{M}\backslash\{\xi\}$ and a heteroclinic solution, q of (HS2) such that $q(t) \to \xi$ as $t \to \infty$ and $q(t) \to \eta$ as $t \to -\infty$.*

Unlike most of the results presented thus far, Theorem 3.24 can be proved by a minimization argument. Taking, without loss of generality, $V(0) = 0$, set

$$I(q) = \int_{\mathbb{R}} \left(\frac{1}{2}|\dot{q}|^2 - V(q)\right) dt. \tag{3.26}$$

Then the heteroclinic solution given by Theorem 3.24 can be obtained as the infimum of I over the class of $W^{1,2}_{\text{loc}}(\mathbb{R}, \mathbb{R}^n)$ functions which tend to ξ as $t \to -\infty$ and to $\mathcal{M}\backslash\{\xi\}$ as $t \to \infty$. The original proof in [R8] involved an additional approximation argument but this simpler minimization succeeds. For a related result on manifolds, see Benci-Giannoni [BG].

By taking an appropriate choice of origin in $\mathbb{R}^n$, it can be assumed that $0 \in \mathcal{M}$ and therefore $\mathcal{M} \supset \mathbb{Z}^n$. In [R8], variant of the above minimization argument also proved:

Theorem 3.27. *If $V \in C^1(\mathbb{R}^n, \mathbb{R})$, satisfies (V_6), and $\mathcal{M} = \mathbb{Z}^n$, then for each $\xi \in \mathbb{Z}^n$, there are at least $2n$ distinct heteroclinic orbits of (HS2) which emanate from ξ and $2n$ which terminate at ξ.*

A stronger multiplicity results under the hypotheses of Theorem 3.24 was proved in [R9]. Another result was obtained in this setting for heteroclinic chains. By a *heteroclinic chain* of solutions of (HS2) we mean a finite number of points $p_0 \, p_1, \ldots, p_k$ and corresponding heteroclinic solutions Q_i of (HS2), $1 \leq i \leq k$ such that Q_i joins p_{i-1} and p_i.

Theorem 3.28 [R10]. *Under the hypotheses of Theorem 3.24, for each $\xi, \eta \in \mathcal{M}$, there is a heteroclinic chain of solutions of (HS2) such that $p_0 = \xi$ and $p_n = \eta$.*

The heteroclinic chain is obtained by taking the infimum of I in (3.26) over the class of $W_{loc}^{1,2}$ functions joining ξ and η.

The system (HS2) under (V_6) can be viewed as a simple model of the multiple pendulum equations. Results related to the theorem just stated for (HS2) have been obtained for the actual multiple pendulum problem and more general systems by Felmer [F5].

Theorems 3.24, 3.27, and 3.28 all involve orbits joining maxima of the potential V. Other results in this spirit have been obtained by Rabinowitz and Tanaka [RT] and some of them will be mentioned next.

Theorem 3.29. *Let Ω be a bounded neighborhood of 0 in $\mathbb{R}^n$, $V \in C^1(\bar{\Omega}, \mathbb{R})$, and*

(V_7) $V(x) < V(0)$ *for* $x \in \bar{\Omega} \backslash \{0\}$.

Then there is a solution, $q(t)$ of (HS2) such that $q(0) \in \partial\Omega$, $q(t) \in \Omega$ for all $t > 0$, and $q(t) \to 0$, $\dot{q}(t) \to 0$ as $t \to \infty$.

As above, Theorem 3.29 is proved by minimizing

$$I(q) = \int_0^\infty \left(\frac{1}{2} |\dot{q}|^2 - V(q) \right) dt \qquad (3.30)$$

over the class of $W_{loc}^{1,2}(\mathbb{R}, \mathbb{R}^n)$ curves in Ω which start on $\partial\Omega$ and end at 0. A related result is

Theorem 3.31. *Let $V \in C^1(\mathbb{R}^n, \mathbb{R})$ and assume*

(V_8) $V(0) = 0$ *and* 0 *is a strict local maximum of* V.

Let $\mathcal{V} = \{x \in \mathbb{R}^n \mid V(x) < 0\} \cup \{0\}$ *and let* Ω *be the component of* $\mathcal{V}$ *containing* 0. *If*

(V_9) Ω *is bounded and* $V'(x) \neq 0$ *for* $x \in \partial\Omega$,

there is a solution q of (HS2) *such that $q(0) \in \partial\Omega$, $\dot{q}(0) = 0$, $q(t) \in \Omega$ for all $t > 0$, and $q(t)$, $\dot{q}(t) \to 0$ as $t \to \infty$.*

Using the time reversibility of (HS2) and extending q to $\mathbb{R}$ via $q(-t) = q(t)$ then yields

Corollary 3.32. *Under the hypotheses of Theorem* 3.31, *there is a solution, q, of* (HS2) *homoclinic to* 0 *and such that $q(0) \in \partial\Omega$, $\dot{q}(0) = 0$.*

This last result was proved independently by Ambrosetti and Bertotti [AB] who found the homoclinic as the limit as $T \to \infty$ of solutions satisfying a related boundary value problem on $[-T, T]$. The solutions of the boundary value problem were obtained via the Mountain Pass Theorem.

Suppose that $\partial\Omega$ contains ℓ components in the setting of Theorem 3.31. Then one can give a lower bound in terms of ℓ on the number of homoclinics as in Corollary 3.32 plus the number of periodic brake orbits of (HS2) that move back and forth between components of $\partial\Omega$. A related result was obtained by Bolotin and Kozlov [Kz] when no homoclinics are present but by rather different arguments.

When $\Omega = \mathbb{R}^n$, the following variant of Theorem 3.29 obtains:

Theorem 3.33. *If $V \in C^1(\mathbb{R}^n, \mathbb{R})$, satisfies (V_7) with $\Omega = \mathbb{R}^n$ and*

(V_{10}) $\overline{\lim}_{|x| \to \infty} V(x) < V(0)$,

then for each $\xi \in \mathbb{R}^n$, there is a solution q of (HS2) *such that $q(0) = \xi$ and $q(t)$, $\dot{q}(t) \to 0$ as $t \to \infty$.*

Bolotin and Kozlov [Kz] have obtained a more general result than Theorem 3.33 using arguments from Riemannian geometry. It is also possible, in the setting of Theorem 3.33, to show there exists a solution of (HS2) which joins 0 and infinity (possibly in finite time) [RT].

Next a result will be described for (FHS2) which involves both homoclinic and heteroclinic solutions. Suppose $V \in C^2(\mathbb{R} \times \mathbb{R}^n, \mathbb{R})$, is T periodic in t, and further satisfies

(V_{11}) there is a $\xi \in \mathbb{R}^n \setminus \{0\}$ such that for all $t \in \mathbb{R}$,

(i) $V(t, 0) = 0 = V(t, \xi)$,

(ii) $V_q(t, 0) = 0 = V_q(t, \xi)$,

(iii) $V_{qq}(t, 0)$ and $V_{qq}(t, \xi)$ are negative definite,

and

(V_{12}) there is a $V_0 > 0$ such that $\overline{\lim\limits_{|q| \to \infty}} \, V(t, q) < -V_0$ for all $t \in \mathbb{R}$

By (V_{11}), 0 and ξ are equilibrium solutions of (FHS2). Setting

$$I(q) = \int_{\mathbb{R}} \left[\frac{1}{2}|\dot{q}|^2 - V(t, q) \right] dt \qquad (3.34)$$

and

$$c(0, \xi) = \inf_{q \in \Gamma(0, \xi)} I(q), \qquad (3.35)$$

$$c(\xi, 0) = \inf_{q \in \Gamma(\xi.0)} I(q), \qquad (3.36)$$

where

$$\Gamma(0, \xi) = \{ q \in W^{1,2}_{\mathrm{loc}}(\mathbb{R}, \mathbb{R}^n) \mid q(t) \to 0 \qquad (3.37)$$
$$\text{as } t \to -\infty \text{ and } q(t) \to \xi \text{ as } t \to \infty \}$$

and

$$\Gamma(\xi, 0) = \{ q \in W^{1,2}_{\mathrm{loc}}(\mathbb{R}, \mathbb{R}^n) \mid q(t) \to \xi \qquad (3.38)$$
$$\text{as } t \to -\infty \text{ and } q(t) \to 0 \text{ as } t \to \infty \},$$

then slight modifications of the arguments of [R8] show there exist $v \in \Gamma(0, \xi)$ and $w \in \Gamma(\xi, 0)$ such that $I(v) = c(0, \xi)$ and $I(w) = c(\xi, 0)$. Now one can use v and w as building blocks to construct multibump solutions of (FHS2) [R11]. To do so an analogue of $(*)$ from §2 of Part 3 is required:

$(*)$ v and w are isolated minima (in $\| \cdot \|_{W^{1,2}(\mathbb{R},\mathbb{R}^n)}$) of I in $\Gamma(0, \xi)$, $\Gamma(\xi, 0)$ respectively.

Here $(*)$ means if v and v_1 minimize I in $\Gamma(0, \xi)$, there is a $\nu > 0$ and independent of v and v_1 such that $\|v - v_1\|_{W^{1,2}(\mathbb{R},\mathbb{R}^n)} \geq \nu$. Note that although neither v nor v_1 lies in $E \equiv W^{1,2}(\mathbb{R}, \mathbb{R}^n)$, it is not difficult to show that $v - v_1 \in E$.

Using v and w, a construction somewhat in the spirit of [CZR1] can be given to obtain multibump solutions for (FHS2). A precise statement of a theorem here requires too much preparation so we will content ourselves with an informal description of the result in the simplest case. For $\ell \in \mathbb{N}$, consider the heteroclinic chain formed by v and $\tau_\ell w$, where τ_ℓ was defined in (3.5). This chain does not lie in E but by modifying v near $t = \infty$ and w near $t = -\infty$, for any $\rho > 0$ and $\ell = \ell(\rho)$ large enough, an approximation $\mathcal{P} = \mathcal{P}(\rho, \ell) \in E$ can be found which is L^∞ close to the heteroclinic chain and $|I(\mathcal{P}) - c(0, \xi) - c(\xi, 0)| = o(1)$ as $\rho \to 0$. Then it can be shown there is an $r_0 > 0$ such that for $0 < r < r_0$, there exists an actual solution of (FHS2) within r of $\mathcal{P}$ in E for all but finitely many ℓ provided that $0 < \rho \leq \rho_0(r)$. This gives infinitely many homoclinic solutions of (FHS2) of "2-chain" type. Similarly there are homoclinic solutions near the heteroclinic formed by w and $\tau_\ell v$ for all large ℓ. Likewise starting with the heteroclinic chain formed by $v, \tau_\ell w$, and $\tau_{\ell_2} v$ yields heteroclinic solutions of 3-chain type. More generally for each $j \in \mathbb{N}$, there are infinitely many homoclinics to 0 or to ξ of $2j$-chain type and heteroclinics to 0 and ξ and to ξ and 0 of $(2j + 1)$-chain type.

Turning to another question, observe that all of the results for homoclinics or heteroclinics that have been cited thus far in this and the previous section have had asymptotic states as $|t| \to \infty$ that are equilibrium solutions of the associated Hamiltonian system. What can one say when there is a different type of asymptotic behavior for the solution, e.g. periodic limit states. A difficulty in attempting to find homoclinics or heteroclinics in such a setting is that the functional that arises naturally may be infinite when evaluated at any candidate for a solution that possesses the requisite asymptotic behavior. Thus one may have to modify the problem in some appropriate way. Indeed such an approach was taken by Bolotin who stated and sketched some results in [Bo2]. In fact [Bo2] is the earliest variational work we know of for homoclinic or heteroclinic orbits of Hamiltonian systems. Bolotin treated the existence of a homoclinic solution asymptotic to a periodic solution in a more geometrical setting than shall be described here. A complete proof has recently appeared in the framework of solutions homoclinic to invariant tori of symplectic diffeomorphisms [Bo3]. Bolotin uses an approximation argument, minimizing a compactified and modified functional. He then has to analyze the behavior of his approximate solution as a small parameter goes to 0. For this, he requires results of Mather on minimal invariant sets [M].

Our survey concludes with a simple recent result which was motivated by Bolotin's work and which is a partial extension of Theorem 3.24. Thus consider

$$\ddot{q} + W_q(tq) = f(t) \qquad (3.39)$$

where $W \in C^2(\mathbb{R} \times \mathbb{R}^n, \mathbb{R})$ and satisfies

(V_{13}) W is T periodic in t and T_i periodic in q_i, $1 \leq i \leq n$.

Concerning f, suppose $f \in C(\mathbb{R}, \mathbb{R}^n)$ and

(f_1) f is T periodic in t,

(f_2) $[f] \equiv \displaystyle\int_0^T f(t)dt = 0,$

and

(V_{14}) $V(t,q) = W(t,q) - f(t) \cdot q$ is time reversible: $V(t,q) = V(-t,q)$.

For simplicity suppose $T = 1 = T_1$, $1 \le i \le n$. It follows from e.g. [R6] that

$$I_1(q) = \int_0^1 \left(\frac{1}{2}|\dot{q}|^2 - V(t,q) \right) dt \tag{3.40}$$

possesses a minimum in the class E_1 of one periodic functions in $W^{1,2}([0,1],$ $\mathbb{R}^n) \equiv E$. Moreover it is not difficult to show using (V_{14})that

$$c_1 = \inf_{q \in E_1} I_1(q) = \inf_{q \in E} I_1(q) \tag{3.41}$$

and if

$$M = \{q \in E \mid I_1(q) = c_1\} \tag{3.42}$$

then $M \subset E_1$. By (V_{13}), if $q \in M$, so is $q + j$ for all $j \in \mathbb{Z}^n$. Now we have

Theorem 3.43 [R12]. *Let* $W \in C^2(\mathbb{R} \times \mathbb{R}^m, \mathbb{R})$ *and* $f \in C(\mathbb{R}, \mathbb{R}^n)$ *satisfy* (V_{13})–(V_{14}), *and* (f_1)–(f_2) *and assume*

(M_1) M *consists of isolated points.*

Then for each $\bar{q} \in M$, *there is a solution,* Q, *of* (3.39) *and* $v \in M \backslash \{\bar{q}\}$ *such that* $Q(t) \to \bar{q}(t)$ *uniformly as* $t \to -\infty$ *and* $Q(t) \to v(t)$ *uniformly as* $t \to \infty$.

The heteroclinic orbit Q also has a variational characterization which will be sketched next. Set

$$L(q) = \frac{1}{2}|\dot{q}|^2 - V(t,q). \tag{3.44}$$

For $p \in \mathbb{Z}$ and $q \in W_{\text{loc}}^{1,2}(\mathbb{R}, \mathbb{R}^n)$, define

$$a_p(q) = \int_p^{p+1} (L(q) - L(\bar{q}))dt. \tag{3.45}$$

By (3.41), $a_p(q) \ge 0$. For $q \in W_{\text{loc}}^{1,2}(\mathbb{R}, \mathbb{R}^n)$, define

$$J(q) = \sum_{p \in \mathbb{Z}} a_p(q). \tag{3.46}$$

At first glance, (3.46) may seem to be a cumbersome way of writing

$$\int_{\mathbb{R}} (L(q) - L(\bar{q}))dt. \qquad (3.47)$$

However the sum (3.46) may be finite even if the integral (3.47) is not, i.e. cancellations may occur in each term of (3.46) that do not take place for (3.47) in the sense of conditional convergence.

Now set

$$\Gamma(\bar{q}) = \{q \in W^{1,2}_{\text{loc}}(\mathbb{R}, \mathbb{R}^n) \mid (3.49)\ (i)\text{--}(ii)\ \text{hold}\} \qquad (3.48)$$

where

(i) $q(t) \to \bar{q}(t)$ uniformly as $t \to -\infty$

(ii) $q(t) \to \mathcal{M}\backslash\{\bar{q}(t)\}$ uniformly as $t \to \infty$. $\qquad (3.49)$

Then Q of Theorem 3.43 can be obtained as the solution of the variational problem

$$\inf_{q \in \Gamma(\bar{q})} J(q). \qquad (3.50)$$

The details and related results can be found in [R12].

References

[A1] Ambrosetti, A., Critical Points and Nonlinear Variational Problems, Mémoire 49, Soc. Math. de France, 1992.

[A2] Ambrosetti, A., *Elliptic equations with jumping nonlinearities*, J. Math. Phys. Sci. (Madras), 1984, 1–10.

[AB] Ambrosetti, A. and M. L. Bertotti, *Homoclinics for second order conservative systems*, Proc. Conf. in honor of L. Nirenberg, to appear.

[ABL] Ambrosetti, A., V. Benci, and Y. Long, *A note on the existence of multiple brake orbits*, to appear in Nonlinear Analysis: TMA.

[ACZ] Ambrosetti, A. and V. Coti Zelati, *Multiple homoclinic orbits for a class of conservative systems*, preprint.

[AL] Alama, S. and Y. Y. Li, *On "multibump" bound states for certain semilinear elliptic equations*, preprint.

[ALP] Ahmad, S. A. C. Lazer and J. L. Paul, *Elementary critical point theory and perturbations of elliptic boundary value problems at resonance*, Indiana Univ. Math. J. **25**, 1976, 933–944.

[Am] Amann, H., *Saddle points and multiple solutions of differential equations*, Math. Z. **196**, 1979, 127–166.

[AM1] Ambrosetti, A. and G. Mancini, *On a theorem by Ekeland and Lasry concerning the number of periodic Hamiltonian trajectories*, J. Differential Equations **43**, 1981, 1–6.

[AM2] Ambrosetti, A. and G. Mancini, *Solutions of minimal period for a class of convex Hamiltonian systems*, Math. Ann. **255**, 1981, 405–421.

[AP] Ambrosetti, A. and G. Prodi, A Primer of Nonlinear Analysis, Cambridge Studies in Advanced Math. **34**, 1993.

[AR] Ambrosetti, A. and P. H. Rabinowitz, *Dual variational methods in critical point theory and applications*, J. Funct. Anal. **14**, 1973, 349–381.

[AZ1] Amann, H. and E. Zehnder, *Nontrivial solutions for a class of nonresonance problems and applications to nonlinear differential equations*, Ann. Scuola Norm. Sup. Pisa Cl. Sci. (4) **7**, 1980, 539–603.

[AZ2] Amann, H. and E. Zehnder, *Periodic solutions of asymptotically linear Hamiltonian systems*, Manus. Math. **32**, 1980, 149–189.

[Ba] Bahri, A., Critical Points at Infinity in some Variational Problems, Wiley, 1989.

[B] Berestycki, H., *Solutions periodiques de systemes hamiltoniens*, Sem. Bourbaki 1982/83, expose 603, Astérisque, 1983, 105–128.

[BaL1] Bahri, A. and P. L. Lions, *Remarks on the variational theory of critical points and applications*, C. R. Acad. Sc. Paris, Sci. I. Math. **301**, 1985, 145–148.

[BaL2] Bahri, A. and P. L. Lions, *Solutions of superlinear elliptic equations and their Morse indices*, Comm. Pure Appl. Math. **45**, 1992, 1205–1215.

[BaR] Bahri, A. and P. Rabinowitz, *Periodic solutions of Hamiltonian systems of 3-body type*, Ann. IHP–Analyse Nonlin. **8**, 1991, 561–649.

[BB1] Bahri, A. and H. Berestycki, *A perturbation method in critical point theory and applications*, Trans. Amer. Math. Soc. **267**, 1981, 1–32.

[BB2] Bahri, A. and H. Berestycki, *Forced vibrations of superquadratic Hamiltonian systems*, Acta. Math. **152**, 1984, 143–197.

[BB3] Bahri, A. and H. Berestycki, *Existence of forced oscillations for some nonlinear differential equations*, Comm. Pure Appl. Math. **37**, 1984, 403–442.

[BC] Bahri, A. and J–M. Coron, *On a nonlinear elliptic equation involving the critical Sobolev exponent: the effect of the topology of the domain*, Comm. Pure Appl. Math. **41**, 1981, 253–294.

[BCl] Bartsch, T. and M. Clapp, *The compact category and multiple periodic solutions of Hamiltonian systems on symmetric starshaped energy surfaces*, Math. Ann. **293**, 1992, 523–542.

[BCN] Brezis, H., J. M. Coron, and L. Nirenbergy, *Free vibrations of a semi-linear wave equation and a theorem of Rabinowitz*, Comm. Pure Appl. Math. **33**, 1980, 667–689.

[Be] Benci, V., *A geometrical index for the group S^1 and some applications to the research of periodic solutions of O.D.E.'s*, Comm. Pure Appl. Math. **34**, 1981, 393–432.

[BF] Benci, V. and D. Fortunato, *'Birkhoff–Lewis' type result for a class of Hamiltonian systems*, Manus. Math. **59**, 1987, 441–456.

[BFG] Benci, V., D. Fortunato, and F. Giannoni, On the existence of periodic trajectories in static Lorentz manifolds with nonsmooth boundary, Nonlinear Analysis: A tribute in honor of G. Prodi, Quad. del. Sc. Norm. Sup. Pisa, A. Ambrosetti and A. Marino ed., 1991.

[Bg] Berger, M. S., *Nonlinearity and functional analysis*, Academic Press, New York, 1978.

[BG] Benci, V. and F. Giannoni, *Homoclinic orbits on compact manifolds*, J. Math. Anal. and Appl. **157**, 1991, 568–576.

[BHR] Benci, V., H. Hofer, and P. H. Rabinowitz, *A remark on a priori bounds and existence for periodic solutions of Hamiltonian systems*, Periodic Solutions of Hamiltonian Systems and Related Topics, P. H. Rabinowitz et. al. eds., D. Reidel, 1987, 85–88.

[Bi] Birkhoff, G. D., *Dynamical systems with two degrees of freedom*, Trans. Amer. Math. Soc. **18**, 1917, 199–300.

[BLMR] Berestycki, H., J. M. Lasry, G. Mancini and B. Ruf, *Existence of multiple periodic orbits on star-shaped Hamiltonian surfaces*, Comm. Pure Appl. Math. **38**, 1985, 253–290.

[BN1] Brezis, H. and L. Nirenberg, *Positive solutions of nonlinear elliptic equations involving critical Sobolev exponents*, Comm. Pure Appl. Math. **36**, 1982, 437–477.

[BN2] Brezis, H. and L. Nirenberg, *Remarks on finding critical points*, Comm. Pure Appl. Math. **44**, 1991, 939–963.

[Bo1] Bolotin, S., *Libration motions of natural dynamical systems*, Vestnik Moskov. Univ. Ser. I Matem. Mekh. **6**, 1978, 72–77.

[Bo2] Bolotin, S., *Existence of homoclinic motions*, Vestnik Moskov. Univ. Ser I, Matem. Mekh **6**, 1983, 98–103.

[Bo3] Bolotin, S., *Homoclinic orbits to invariant tori of symplectic diffeomorphisms and Hamiltonian systems*, preprint.

[Br] Brezis, H., *Periodic solutions of nonlinear vibrating strings and duality principles*, Bull. Amer. Math. Soc. **8**, 1983, 409–426.

[BrC] Brezis, H. and J. M. Coron, *Periodic solutions of nonlinear wave equations and Hamiltonian systems*, Amer. J. Math. **103**, 1981, 559–570.

[BR1] Benci, V. and P. H. Rabinowitz, *Critical point theorems for indefinite functionals*, Invent. Math. **52**, 1979, 241–273.

[BR2] Benci, V. and P. H. Rabinowitz, *A priori bounds for periodic solutions of Hamiltonian systems*, Ergodic Th. and Dynam. Sys. **8**, 1988, 27–31.

[Bs] Bessi, H., preprint.

[Bw1] Browder, F. E., *Infinite dimensional manifolds and nonlinear eigenvalue problems*, Ann. of Math. (2) **82**, 1965, 459–477.

[Bw2] Browder, F. E., *Nonlinear eigenvalues and group invariance*, Functional Analysis and Related Fields (F. E. Browder, ed.), Springer-Verlag, Berlin and New York, 1970, 1–58.

[C] Clark, D., *On periodic solutions of autonomous Hamiltonian systems of ordinary differential equations*, Proc. AMS **39**, 1973, 579–584.

[CC] Conley, C. C., *Isolated invariant sets and the Morse index*, CBMS Regional Conf. Ser. in Math., no. 38, Amer. Math. Soc., Providence, R. I., 1978.

[Ce] Cerami, G., *Un criterio di esistenza per i punti critici su varieta illimitate*, Rend. Acad. Sci. Let. Ist. Lombardo **112**, 1978, 332-336.

[CE] Clarke, F. and I. Ekeland, *Hamiltonian trajectories having prescribed minimal period*, Comm. Pure Appl. Math. **33**, 1980, 103–116.

[Ch1] Chang, K. C., *Infinite Dimensional Morse Theory and Multiple Solution Problems*, Birkhaüser, 1993.

[Ch2] Chang, K. C., *Variational methods for nondifferentiable functionals and their applications to partial differential equations*, J. Math. Anal. Appl. **80**, 1981, 102–129.

[Ch3] Chang, K. C., *On the periodic nonlinearity and the multiplicity of solutions*, Nonlinear Analysis TMA **13**, 1989, 527–538.

[Ch4] Chang, K. C., *Applications of homology theory to some problems in differential equations and analysis*, Nonlinear Analysis and Applications, F. E. Browder ed., Proc. Symp. Pure Math. **V45** AMS, 1986, 253–262.

[Cl1] Clarke, F., *A classical variational principle for periodic Hamiltonian trajectories*, Proc. A.M.S. **76**, 1979, 186–189.

[Cl2] Clarke, F., *Periodic solutions of Hamiltonian inclusions*, J. Diff. Eq. **40**, 1981, 1–6.

[CL] Castro, A. and A. Lazer, *Applications of a maximum principle*, Rev. Columbiana, Mat. **10**, 1976, 141–149.

[CM] Costa, D. G. and C. A. Magalhães, *A unified approach to a class of strongly indefinite functionals*, preliminary version.

[Co1] Coffman, C. V., *A minimum-maximum principle for a class of nonlinear integral equations*, J. Analyse Math. **22**, 1969, 391–419.

[CW] Costa, D. and M. Willem, *Multiple critical points of invariant functionals and applications*, J. Nonlin. Analysis **9**, 1986, 843–852.

[CZ] Conley, C. and E. Zehnder, *The Birkhoff-Lewis fixed point theorem and a conjecture of V. Arnold*, Inv. Math. **73**, 1983, 33-49.

[CZEL] Coti Zelati, V., I. Ekeland, and P. L. Lions, to appear.

[CZES] Coti Zelati, V., I. Ekeland, and E. Séré, *A variational approach to homoclinic orbits in Hamiltonian systems*, Math. Ann. **288**, 1990, 133–160.

[CZR1] Coti Zelati, V. and P. H. Rabinowitz, *Homoclinic orbits for second order Hamiltonian systems possessing superquadratic potentials*, J. Amer. Math. Soc. **4**, 1992, 693–727.

[CZR2] Coti Zelati, V. and P. H. Rabinowitz, *Homoclinic type solutions for a semilinear elliptic PDE on $\mathbb{R}^n$*, Comm. Pure Appl. Math., **45**, 1992, 1217–1269.

[De] Deimling, K., *Nonlinear Functional Analysis*, Springer, 1985.

[Df] De Figueiredo, D. G., Lectures on the Ekeland Variational Principle with Applications and Detours, Tata Inst. of Fund. Research, 1989.

[E1] Ekeland, I., Convexity Methods in Hamiltonian Mechanics, Springer, 1990.

[EH1] Ekeland, I. and H. Hofer, *Periodic solutions with prescribed minimal period for convex autonomous Hamiltonian systems*, Invent. Math. **81**, 1985, 155-188.

[EH2] Ekeland, I. and H. Hofer, *Subharmonics for convex nonautonomous Hamiltonian systems*, Comm. Pure Appl. Math. **40**, 1987, 1–36.

[EH3] Ekeland, I. and Hofer, *Symplectic topology and Hamiltonian dynamics*, Math. Z. **200**, 1989, 355–378.

[EH4] Ekeland, I. and H. Hofer, *Convex Hamiltonian energy surfaces and their closed trajectories*, Comm. Math. Phys. **113**, 1987, 419–467.

[EL] Ekeland, I. and J.-M. Lasry, *On the number of periodic trajectories for a Hamiltonian flow on a convex energy surface*, Ann. of Math. (2) **112**, 1980, 283–319.

[ELs] Ekeland, I. and L. Lassoued, *Multiplicité des trajectories fermées d'un système Hamiltonian sur une hypersurface d'énergies convexe*, Ann. IHP – Analyse nonlin. **4**, 1987, 1–29.

[F1] Felmer, P. L., *Subharmonic near an equilibrium point for Hamiltonian systems*, Manu. Math. **66**, 1990, 359–396.

[F2] Felmer, P. L., *Periodic solutions of 'superquadratic' Hamiltonian systems*, (to appear) J. Diff. Eq.

[F3] Felmer, P. L., *Periodic solutions of spatially periodic Hamiltonian systems*, J. Diff. Eq. **98**, 1992, 143–168.

[F4] Felmer, P. L., *Multiple periodic solutions for Lagrangian systems in T^n*, Nonlinear Analysis, TMA **15**, 1990, 815–831.

[F5] Felmer, P. L., *Heteroclinic orbits for spatially periodic Hamiltonian systems*, Analyse nonlin. IHP **8**, 1991, 477–497.

[FM] Fonda, A. and J. Mawhin, *Multiple periodic solutions of conservative systems with periodic nonlinearity*, preprint.

[FS] Felmer, P. L. and E. Silva, *Subharmonics near an equilibrium for some second order Hamiltonian systems*, (to appear) Proc. Royal Soc. Edinburgh.

[FW] Fournier, G. and M. Willem, *Multiple solutions of the forced double pendulum problem*, preprint.

[Gh] Ghoussoub, N., Duality and Perturbation Methods in Critical Point Theory, preprint.

[Gi1] Giannoni, F., *Periodic solutions of dynamical systems by a Saddle Point Theorem of Rabinowitz*, Nonlinear Analysis TMA **13**, 1989, 707–719.

[Gi2] Giannoni, F., *On the existence of homoclinic orbits on Riemannian manifolds*, preprint.

[GM1] Girardi, M. and M. Matseu, *Some results on solutions of minimal period to superquadratic Hamiltonian equations*, Nonlinear Analysis TMA **7**, 1983, 475–482.

[GM2] Girardi, M. and M. Matseu, *Solutions of minimal period for a class of nonconvex Hamiltonian systems and applications to the fixed energy problem*, Nonlinear Analysis TMA **10**, 1986, 371–382.

[GM3] Girardi, M. and M. Matseu, *Periodic solutions of convex Hamiltonian systems with a quadratic growth at the origin and superquadratic at infinity*, Ann. Mat. Pura. Appl. **147**, 1987, 21–72.

[GP] Ghoussoub, N. and D. Preiss, *A general mountain pass principle for locating and classifying critical points*, Ann. IHP – Analyse nonlin. **6**, 1989, 321–330.

[Gr] Girardi, M., *Multiple orbits for Hamiltonian systems on starshaped surfaces with symmetries*, Ann. IHP – Analyse nonlin. **1**, 1984, 285–294.

[GR] Giannoni, F. and P. H. Rabinowitz, *On the multiplicity of homoclinic orbits on Riemannian manifolds for a class of second order Hamiltonian systems*, (to appear) Nonlin. Diff. Eq. and Appl.

[Ha] Han, Z., *Periodic solutions of a class of dynamical systems of second order*, J. Diff. Eq. **90**, 1991, 408–417.

[H1] Hofer, H., *A geometric description of the neighborhood of a critical point given by the Mountain Pass Theorem*, J. London Math. Soc. (2) **31**, 1985, 566–570.

[H2] Hofer, H., *A note on the topological degree at a critical point of mountain pass type*, Proc. AMS **49**, 1984, 309–315.

[H3] Hofer, H., *Variational and topological methods in partially ordered spaces*, Math. Ann. **261**, 1982, 493–514.

[H4] Hofer, H., *On strongly indefinite functionals with applications*, Trans. Amer. Math. Soc. **275**, 1983, 185–214.

[HV] Hofer, H. and C. Viterbo, *The Weinstein conjecture in the presence of holomorphic spheres*, preprint.

[HW] Hofer, H. and K. Wysocki, *First order elliptic systems and the existence of homoclinic orbits in Hamiltonian systems*, preprint.

[HZ1] Hofer, H. and E. Zehnder, *A new capacity for symplectic manifolds*, Analysis, et cetera (P. Rabinowitz and E. Zehnder eds.) Academic Press, 1990, 405–428.

[HZ2] Hofer, H. and E. Zehnder, *Periodic solutions on hypersurfaces and a result by C. Viterbo*, Inv. Math. **90**, 1987, 1–9.

[K] Krasnoselski, M. A., Topological methods in the theory of nonlinear integral equations, Macmillan, New York, 1964.

[KS] Kirchgraber, U. and D. Stoffer, *Chaotic behavior in simple dynamical systems*, SIAM Review **32**, 1990, 424–452.

[Kz] Kozlov, V. V., *Calculus of variations in the large and classical mechanics*, Russ. Math. Surv. **40**, 1985, 37–71.

[La] Lassoued, L., *Periodic solutions of a second order superquadratic system with a change in sign in the potential*, J. Diff. Eq. **93**, 1991, 1–18.

[LaS] Lazer, A. and S. Solimini, *Nontrivial solutions of operator equations and Morse indices of critical points of Minimax type*, Nonlinear Analysis TMA **12**, 1988, 761–775.

[Li] Liu, J. Q., *A generalized saddle point theorem*, J. Diff. Eq., 1989.

[LL] Li, S. and J. Liu, *Some existence theorems on multiple critical points and their applications*, Kexue Tonghao **17**, 1984,

[Lo1] Long, Y., *The minimal period problem of periodic solutions for autonomous superquadratic second order Hamiltonian systems*, (to appear) J. Diff. Eq.

[Lo2] Long, Y., *The minimal period problem for even autonomous superquadratic second order Hamiltonian systems*, preprint, 1992.

[Lo3] Long, Y., *Periodic solutions of perturbed superquadratic Hamiltonian systems*, Ann. Sc. Norm. Sup. Pisa, Ser 4, **17**, 1990, 35–77.

[Lo4] Long, Y., *Periodic solutions of superquadratic Hamiltonian systems with bounded forcing terms*, Math. Z. **203**, 1990, 453–467.

[Lo5] Long, Y., *Multiple solutions of perturbed superquadratic second order Hamiltonian systems*, Trans. AMS **311**, 1989, 749–780.

[Lo6] Long, Y., *A Maslov-type index theory and asymptotically linear Hamiltonian systems*, Dynamical Systems and Related Topics, K. Shirawa, ed., World Scientific, 1991, 333-341.

[LPL] Lions, P. L., *The concentration-compactness principle in the calculus of variations*, Ann. IHP – Analyse nonlin. **1**, 1984, 101–145.

[LS] Ljusternik, L. and L. Schnirelmann, Methodes topologique dans les problémes variationnels, Hermann and Cie, Paris, 1934.

[LW] Li, S. and M. Willem, *Applications of local linking to critical point theory*, preprint.

[LY] Li, Y. Y., *On $-\Delta u = K(x)u^5$ in $\mathbb{R}^3$*, (to appear) Comm. Pure Appl. Math.

[LZ] Long, Y. and E. Zehnder, *Morse Theory for forced oscillations of asymptotically linear Hamiltonian systems*, Stochastic processes, Physics, and Geometry, S. Albevario et al. ed., World Scientific, 1990, 528–563.

[M] Mather, J., *Action minimizing invariant measures for positive definite Lagrangian systems*, Math. Z. **207**, 1991, 169–207.

[Ma] Mawhin, J., *Periodic solutions of ordinary differential equations: the Poincaré' heritage*, Differential Topology-Geometry and Related Fields, (Russian ed.), Teubner, 1985, 287–307.

[MW] Mawhin, J. and M. Willem, Critical Point Theory and Hamiltonian Systems, Springer, 1989.

[N] Nirenberg, L., *Variational and topological methods in nonlinear problems*, Bull. Amer. Math. Soc. (N.S.) **4**, 1981, 267–302.

[Ni] Ni, W. M., *Some minimax principles and their applications in nonlinear elliptic equations*, J. Analyse Math. **37**, 1980, 248–275.

[O] Offin, D., *A class of periodic orbits in classical mechanics*, J. Diff. Eq. **66**, 1987, 90–117.

[P1] Palais, R. S., *Critical point theory and the minimax principle*, Proc. Sympos. Pure Math., vol. 15, Amer. Math. Soc., Providence, R. I., 1970, 185–212.

[P2] Palais, R. S., *Lusternik-Schnirelmann theory on Banach manifolds*, Topology **5**, 1966, 115–132.

[PS] Palais, R. S. and S. Smale, *A generalized Morse Theory*, Bull. AMS **70**, 1964, 165–171.

[PSe1] Pucci, P. and J. Serrin, *Extensions of the mountain pass theorem*, J. Funct. Analysis **59**, 1984, 185–210.

[PSe2] Pucci, P. and J. Serrin, *The structure of the critical set in the mountain pass theorem*, Trans. Amer. Math. Soc. **299**, 1987, 115–132.

[R1] Rabinowitz, P. H., Minimax Methods in Critical Point Theory with Applications to Differential Equations, CBMS Reg. Conf. Ser. in Math. **65**, 1986.

[R2] Rabinowitz, P. H., *Periodic solutions of Hamiltonian systems: A survey*, SIAM J. Math. Anal **13**, 1982, 343–352.

[R3] Rabinowitz, P. H., *Periodic solutions of Hamiltonian systems*, Comm. Pure Appl. Math. **31**, 1978, 157–184.

[R4] Rabinowitz, P. H., *On a theorem of Hofer and Zehnder, Periodic Solutions of Hamiltonian systems and Related Topics*, P. H. Rabinowitz et. al. eds., D. Reidel, 1987, 245–253.

[R5] Rabinowitz, P. H., *Periodic solutions of large norm of Hamiltonian systems*, J. Differential Equations **50**, 1983, 33–48.

[R6] Rabinowitz, P. H., *On a class of functionals invariant under a $\mathbb{Z}^n$ action*, Trans. AMS **310**, 1988, 303–311.

[R7] Rabinowitz, P. H., *Homoclinic orbits for a class of Hamiltonian systems*, Proc. Roy. Soc. Edin. **114A**, 1990, 33–38.

[R8] Rabinowitz, P. H., *Periodic and heteroclinic orbits for a periodic Hamiltonian system*, Ann. IHP-Analyse nonlin. **6**, 1989, 331–346.

[R9] Rabinowitz, P. H., *Some recent results on heteroclinic and other connecting orbits of Hamiltonian systems*, (to appear) Proc. Conf. on Variational Methods in Hamiltonian Systems and Elliptic Equations.

[R10] Rabinowitz, P. H., *A variational approach to heteroclinic orbits for a class of Hamiltonian systems*, Frontiers in Pure and Applied Mathematics, R. Dautray ed., 1991, 267–278.

[R11] Rabinowitz, P. H., *Homoclinic and heteroclinic orbits for a class of Hamiltonian systems*, Calculus of Var. **1**, 1993, 1–36.

[R12] Rabinowitz, P. H., *Heteroclinics for a reversible Hamiltonian system*, preprint, 1993.

[RT] Rabinowitz, P. H. and K. Tanaka, *Some results on connecting orbits for a class of Hamiltonian systems*, Math. Z. **206**, 1991, 473–499.

[S1] Schwartz, J. T., *Generalizing the Lusternik-Schnirelmann theory of critical points*, Comm. Pure Appl. Math. **17**, 1964, 307–315.

[S2]   Schwartz, J. T., Nonlinear functional analysis, Gordon & Breach, New York, 1969.

[Sc]   Schechter, M., *A bounded mountain pass lemma without the (PS) condition and applications*, Trans. Amer. Math. Soc. **331**, 1992, 681–704.

[Se1] Séré, E., *Existence of infinitely many homoclinic orbits in Hamiltonian systems*, Math. Z. **209**, 1992, 27–42.

[Se2] Séré, E., *Looking for the Bernoulli shift*, preprint.

[Se3] Séré, E., *Homoclinic orbits on compact hypersurfaces in $\mathbb{R}^n$ of restricted contact type*, preprint.

[Sf]   Seifert, H., *Periodische Bewegungen mechanischer systeme*, Math. Z. **51**, 1948, 197–216.

[Si1] Silva, E., *Critical point theorems and applications to differential equations*, thesis, University of Wisconsin–Madison, 1988.

[Si2] Silva, E., *Linking theorems and applications to nonlinear elliptic equations at resonance*, Nonlinear Analysis: TMA **16**, 1991, 455–477.

[St1] Struwe, M., Variational Methods and their Applications to Partial Differential Equations and Hamiltonian Systems, Springer, 1990.

[St2] Struwe, M., *Existence of periodic solutions of Hamiltonian systems on almost every energy surface*, preprint.

[ST]   Schechter, M. and C. Tintarov, *Pairs of critical points produced by linking subsets with applications to semilinear elliptic problems*, Bull. Soc. Math. Belg. **44**, 1992, 249–261.

[SU]   Sacks, P. and K. Uhlenbeck, *On the existence of minimal immersions of 2-spheres*, Ann. of Math. **113**, 1981, 1–24.

[Sz1] Szulkin, A., *Morse theory and existence of periodic solutions of convex Hamiltonian systems*, Bull. SMF, 1988.

[T1]   Tanaka, K., *Morse indices at critical points related to the symmetric mountain pass theorem and applications*, Comm. PDE **14**, 1989, 99–128.

[T2]   Tanaka, K., *Homoclinic orbits in a first order Hamiltonian system: Convergence of subharmonic orbits*, Ann. IHP – Analyse nonlin. (to appear).

[V]   Viterbo, C., *A proof of the Weinstein conjecture in $\mathbb{R}^{2n}$*, Ann. IHP – Analyse nonlin. **4**, 1987, 337–356.

[VG1] van Groesen, E. W. C., *Existence of multiple normal mode trajectories on convex energy surfaces of even classical Hamiltonian systems*, J. Differential Equations **57**, 1985, 70–89.

[W1]   Weinstein, A., *On the hypotheses of Rabinowitz' periodic orbit theorems*, J. Diff. Eq. **33**, 1979, 353-358.

[W2] Weinstein, A., *Periodic orbits for convex Hamiltonian systems*, Ann. of Math. (2) **108**, 1978, 507–518.

[W3] Weinstein, A., *Normal modes for nonlinear Hamiltonian systems*, Inv. Math. **20**, 1973, 45–57.

[Wi] Willem, M., *Subharmonic oscillations of convex Hamiltonian systems*, J. Nonlinear Anal. **9**, 1985, 1303–1311.

[Z] Zehnder, E., *Periodische Losungen von Hamiltonishe Systemen*, Jahresber. Deutsch. Math. – Verein **89**, N1, 1987, 33–59.

Symplectic Topology: An Introduction

Claude Viterbo

Département de Mathématique
Bâtiment 425
Université de Paris-Sud
F-91405 Orsay Cedex

The aim of this survey is to review progress in symplectic topology during the last 25 years, that is, since 1968. Our task is made a little easier by the fact that symplectic topology was only born around 1983.

We shall first try to give a concrete illustration of one of the basic results in symplectic topology, called Gromov's squeezing theorem. We shall view it as a classical uncertainty principle.

1. The Classical Uncertainty Principle, Symplectic Rigidity

Let us consider the oldest (?) mechanical system, made of a certain number of planets undergoing interaction forces deriving from a scalar potential. The motion is determined by the knowledge of the position and momentum of each planet at a given time (and of course by the equations). Assume these quantities to be only approximately known at time 0 with an error of $\Delta q_i^0, \Delta p_i^0$ for the position and momenta of planet i. We want to know how accurately the position and momenta of a given planet, e.g. the earth, may be foretold from the information we have. Before Gromov's result, we could only use Liouville's theorem on volume preservation by Hamiltonian flow. It implies that if $\Delta q_i^T, \Delta p_i^T$ denote the unaccuracy on the position and momenta at time T, we have

$$\prod_{i=1}^{n} \Delta q_i^T \cdot \Delta p_i^T \geq \prod_{i=1}^{n} \Delta q_i^0 \cdot \Delta p_i^0.$$

That there must be other restrictions is physically obvious from the following argument: Assume we are only interested in the future position and momentum of the earth, but we also measure the position and momenta of some distant star (Sirius) at time 0. Then, as we are not interested in the position and momentum of Sirius at time T, we may allow $\Delta q^T_{\text{Sirius}}$ and $\Delta p^T_{\text{Sirius}}$ to be arbitrarily large, thus the above inequality will be satisfied with $\Delta q^T_{\text{Earth}}$ and $\Delta p^T_{\text{Earth}}$ arbitrarily small.

As a consequence of Gromov's theorem, we may prove our intuition to be correct. Indeed, the Liouville inequality may be improved to:

$$\Delta q_j(T) \cdot \Delta p_j(T) \geq \inf_i \Delta q_i(0) \cdot \Delta p_i(0)$$

Recall that the Heisenberg uncertainty principle states that if a particle's position, q, is known with accuracy Δq, while its momentum, p, is known with accuracy Δp, we must have

$$\Delta q \cdot \Delta p \geq \hbar/2,$$

where $\hbar$ is a physical constant. Now it is usually claimed that this statement has no counterpart in classical mechanics, but this is not exactly so. What is indeed true, is that there is no universal constant that restricts our knowledge. On the other hand, once we measure a certain phenomenon with a given precision, our forecast can never improve this precision as measured by the quantity $\Delta q \cdot \Delta p$.

In particular there is no hope of improving our forecast concerning a given particle by increasing beyond any limit the number of planets or stars taken into account in our computations unless they are measured with more precision than the best measure on a nearby object. We wish to point out that the above fact is by no means obvious. First of all, it is false for non Hamiltonian systems, as may be seen for a dissipative system in a potential well: eventually we know that our particle will rest at the bottom of the well (thus q is known and $p = 0$).

The above result deals only with conservative motions. After the invention of modern mechanics by Galilei and Newton, a unified formalism was invented, now called Hamiltonian mechanics. The idea is to describe the motion using equations involving the position and speed of a particle, in the energy. Assuming all masses to be one, a conservative motion is described by the Newtonian equation

$$\ddot{x} = -\nabla V(x)$$

while its Hamiltonian form is

$$\dot{x} = \frac{\partial H}{\partial p} \quad \dot{p} = -\frac{\partial H}{\partial x}$$

where $H(x,p) = 1/2p^2 + V(x)$ is the energy, that is classically the sum of the kinetic and potential energy. For a general C^2 function H, that is also allowed to depend on time, the second equation describes a flow in $\mathbb{R}^{2n}$ called Hamiltonian flow of H. It has two important properties:

1) the flow preserves the symplectic form $\sum_{i=1}^{n} dp_i \wedge dq_i$
2) if H does not depend on time, then H is constant on the trajectories (conservation of energy)

The first and apparently mysterious property, implies in particular that the flow will preserve the volume form (also known as the Liouville measure) ω^n. As a result, we get Poincaré recurrence theorem: if the flow stays in a bounded region (in particular if H is time independent, we need only that the level sets of H be compact and that H be bounded from below) a trajectory will eventually always return in the neighborhood of the starting point. It is well known that this apparent paradox, predicting that, coffee and milk mixed in a cup will after some time separate again (to the despair of capuccino lovers), is solved because the time needed for this is greater than several billion years.

It is not hard to see that, in $\mathbb{R}^{2n}$, the property of preserving the symplectic form, in fact characterizes those maps which can be written as time one of a Hamiltonian flow. Such maps are called symplectic, and since they are volume preserving they must be diffeomorphisms. In a general symplectic manifold symplectic diffeomorphisms and up to some finite dimensional deformation that may be seen as a cohomology space.

Thus a slight generalization of the uncertainty principle stated above would read

Theorem: *There exists a symplectic map Ψ sending*

$$\prod_{i=1}^{n} [q_i^0 - \Delta q_i^0, q_i^0 + \Delta q_i^0] \times [p_i^0 - \Delta p_i^0, p_i^0 + \Delta p_i^0]$$

into

$$\prod_{i=1}^{n} [q_i^T - \Delta q_i^T, q_i^T + \Delta q_i^T] \times [p_i^T - \Delta p_i^T, p_i^T + \Delta p_i^T]$$

only if

$$\inf_i \Delta q_i^T \Delta p_i^T \geq \inf_i \Delta q_i^0 \Delta p_i^0.$$

Similarly, if $B^2(r)$ is the ball of radius r in the q_1, p_1 plane, and $B^{2n}(r')$ the ball of radius r' in $\mathbb{R}^{2n}$, there is a symplectic map sending $B^{2n}(r')$ into $B^2(r) \times \mathbb{R}^{2n-2}$ only if $r' \leq r$.

An apparently unrelated question is: how really different are symplectic maps from ordinary volume preserving maps? A mathematical way to rephrase this question is as follows. Let Ψ be a volume preserving diffeomorphism. Under which assumption is there a sequence Ψ_n of symplectic diffeomorphisms, converging for the $\mathbf{C^0}$ topology to Ψ? It turns out that the rather surprising answer is that Ψ must be itself symplectic.

It is interesting to compare this situation with the Riemannian one, and volume geometry.

Let Ψ be an isometry. This a priori means that $\langle d\Psi(x)\xi, d\Psi(x)\eta \rangle = \langle \xi, \eta \rangle$ for any pair of vectors ξ, η. This looks like a first order differential condition, i.e. if Ψ_n is a sequence satisfying the condition, it will be satisfied by its limit Ψ provided it is a C^1 limit.

But we know that a Riemannian metric determines a unique distance on U, and that a map is an isometry if and only if it preserves this distance. Now, to preserve a distance is a continuous condition, i.e. if Ψ_n is a sequence such that $d(\Psi_n(x), \Psi_n(y)) = d(x, y)$ for all x, y in U, and if Ψ is simply the C^0-limit of the Ψ_n, then $d(\Psi(x), \Psi(y)) = d(x, y)$, thus Ψ is an isometry. In short we proved that the isometry group is closed for the C^0 topology. Note, however, that one may prove that this group is always a finite dimensional, Lie group; thus our result is not so striking (there are not so many invariant metrics on such a group: see the appendix).

We now turn to an example where the group is infinite dimensional. Let Ω be the volume form, and $\mathrm{Diff}_\Omega(U)$ the set of volume preserving diffeomorphisms. This is a very large group, since for example in $\mathbb{R}^3$ the flow of any divergence free vector field will be in the group. Again to be in this group is a priori a differential equation: $\det(\frac{\partial f_i}{\partial x_j}) = 1$.

But again, we may notice that preserving the volume is equivalent to preserving the measure, that is $\mu(f(A)) = \mu(A)$ for all subsets A of U. Now since the Lebesgue measure is continuous for the Hausdorff topology,* that is, $d(V, W) = \sup_{x \in V} d(x, W) + \sup_{y \in W} d(x, V)$. Thus if f_n converges C^0 to f, $f_n(A)$ converges to $f(A)$ for the Hausdorff topology, hence since $\mu(f_n(A)) = \mu(A)$, we have $\mu(f(A)) = \mu(A)$, whenever A is bounded with C^1 boundary, and this proves that f is volume preserving.

Let us point out some common features of these examples. A subgroup G of $\mathrm{Diff}(U)$ is defined, using a differential relation. While G is obviously closed for the C^1 topology, it turns out that it is also closed for the C^0 topology. This phenomenon is called rigidity (Gromov). The proof is based on the construction of a function defined on the set of subsets of U, invariant

---

* This is true only if we restrict ourselves to sets with sufficiently regular boundary, C^1 regularity is good enough. It is easy to check that it is sufficent to preserve the measure of these sets to preserve the measure of any set.

by G and continuous for the Hausdorff topology (indeed in the riemannian case to preserve the distance is equivalent to preserving the diameter of sets).

In other words we have

Theorem 2: *The subgroup* Diff_ω *of symplectic diffeomorphisms is close for the C^0 topology in the group of volume preserving diffeomorphisms.*

In the next section, we shall try to explain two approaches to the proof of Theorems 1 and 2, both through the construction of symplectic invariants.

2. Construction of Symplectic Invariants

a. Elliptic methods point of view, the width

This is the original approach due to Gromov, and is based on the relationship between symplectic geometry and complex algebraic geometry. The relationship is clear in that any algebraic manifold has a natural symplectic form; this is a two-form that may be written in local coordinates as $\sum dp_i \wedge dq_i$. However, a symplectic manifold does not have to be algebraic. Results of Thurston, McDuff and Gompf, show that many restrictions on the topology of a Kähler manifold do not hold for symplectic manifolds. However, symplectic manifolds always have "almost complex" structure. In the case of $\mathbb{R}^{2n}$, this consists in giving for each point in $\mathbb{R}^{2n}$, a matrix $J(z)$, such that $J(z)^2 = -Id$. A J-complex plane through z is a plane generated by $\xi, J(z)\xi$, or else, invariant by $J(z)$. The simplest example occurs when $J(z) = i$ where i is a square root of -1 and we identified $\mathbb{R}^{2n}$ with $\mathbb{C}^n$. In fact if near each point we may find local coordinates such that we are reduced to this special case, the almost complex structure is said to be integrable, and one ususally talks about a complex structure.

Now, given an almost complex structure, we may talk about J- holomorphic curves: these are surfaces in $\mathbb{R}^{2n}$ such that at each point z on the surface, its tangent space is a complex J-plane. It is important to point out that for a holomorphic curve, there is always a local parametrization $u(x, y)$ of our surface, called the isothermal parametrization, which is J-holomorphic if and only if u satisfies the Cauchy-Riemann equation $\frac{\partial}{\partial x}u + J(u(x, y))\frac{\partial}{\partial y}u = 0$. This is an elliptic equation, a crucial feature.

Now we have to bring in the symplectic structure. We shall say that an almost complex strucure on $\mathbb{R}^{2n}$ is admissible, if $\omega(\xi, J(x)\xi) > 0$ for each x, ξ. As a result, the integral of ω over any piece of holomorphic curve is strictly positive. This is the geometric property that yields a priori estimates that are needed in the analysis.

It is then easy to define a symplectic invariant, called "width" by Gromov, as follows: we set

$$w(U) = \sup_{J} \inf_{\Sigma} \int_{U \cap \Sigma} \omega.$$

Here J varies in the set of admissible almost compex structures, while Σ varies in the set of J holomorphic surfaces, closed (i.e. without boundary) in U, containing some given point in U. This is clearly a symplectic invariant, since a symplectic diffeomorphism preserves the set of admissible almost complex structure. Moreover w is clearly monotone, that is, if $U \subset V$, $w(U) \leq w(V)$ (because $\int_{\Sigma \cap (V-U)} \omega \geq 0$). So the second part of the theorem above will follow if we can prove that

$$w(B^{2n}(r)) = w(B^2(r') \times \mathbb{R}^{2n-2}) = \pi r^2.$$

Here, lies the main difficulty, and is the point we shall try to explain, the proof of the first part of the theorem, being similar in all aspects.

We first deal with the inequality $w(B^{2n}(r)) \geq \pi r^2$. For this we choose J to be the standard complex structure, and the common point to all surfaces to be the origin. Thus we have to prove that, for any holomorphic curve, the area intercepted by the radius r ball is greater than πr^2. Now a holomorphic curve for the standard complex structure is minimal, and it is known that for minimal surfaces through the origin, the quantity $\frac{1}{r^2} \text{area}(B(r) \cap \Sigma)$ is increasing with r. Since the limit, as r goes to 0 is π (this is obtained, replacing Σ by its tangent plane), the required inequality follows.

Now, the inequality $w(B^2(r)) \times \mathbb{R}^{2n-2}) \leq \pi r^2$. For this, it is sufficient to prove that for any admissible almost complex structure J, there is a closed J-holomorphic curve through the origin, such that its projection on $B^2(r)$ is onto. The simplest way to see this, is to compactify the situation. This is done by embedding $B^2(r)) \times \mathbb{R}^{2n-2}$ into $S^2(r+\epsilon) \times \mathbb{C}P^{n-1}$, where $S^2(r+\epsilon)$ is the sphere of area $\pi(r+\epsilon)^2$. One then has to prove that the set of J-holomorphic spheres homotopic to $S^2(r+\epsilon) \times \{a\}$ through a given point is non empty. This is proved by the continuation method, since we deal with an elliptic equation, and since for J-holomorphic maps, $\int_{S^2} |Du|^2 = \int_{S^2}[|\partial u|^2 + |\bar{\partial} u|^2] = \int_{S^2}[|\partial u|^2 - |\bar{\partial} u|^2] = \int_{S^2} \omega = \pi(r+\epsilon)^2$ (since this last quantity is homotopy invariant), this gives an a priori estimate, that, joined with ellipticity enables one to use the continuation method.

This concludes our computations of w, since the inequality $w(B^{2n}(r)) \leq w(B^2(r)) \times \mathbb{R}^{2n-2})$ obviously follows from the monotonicity of the width.

b. The approach through closed characteristics

We shall start with a Riemannian model. Assume we want to find the Riemannian version of the width. This would be a real number attached, let's say to each subset of $\mathbb{R}^n$, invariant by isometry, and monotone. Now while Riemannian invariants are very easy to construct, since (contrary to the symplectic case) the group $O(n)$ of isometries of $\mathbb{R}^n$ is finite dimensional, hence the set of invariants that we may identify to the quotient of the set of open sets by the action of $O(n)$ is very large. Finding the monotone ones is harder, even though more exciting and useful. Let us try to construct some.

Here the (Riemannian) width is defined as follows: let $\mathbb{R}P^{n-1}$ be the set of hyperplanes in $\mathbb{R}^n$, and for v in $\mathbb{R}P^{n-1}$, $b(v)$ the smallest distance of two affine hyperplanes, both parallel to v, such that U is contained between them. Then $w(U) = \inf_v b(v)$ is isometry invariant and clearly monotone. The same is true of $d(U) = \sup_v b(v)$, but this is nothing other than the diameter of U, a well-known invariant. Now, at least if U is convex, we may consider billiard trajectories on U. These are closed broken lines, contained in U, broken only on the boundary ∂U of U, and the angles of the two segments issuing from a point on the boundary satisfy the law of reflection (see Figure 2). Now there is a very simple way to obtain billiard trajectories. Indeed, consider any set of k points $(x_1, \ldots, x_k)$ on U, and consider the function $\sum_{i=1}^{k} |x_i - x_{i+1}|$ (we set $x_{k+1} = x_1$). Then $(x_1, \ldots, x_k)$ is a critical point of this function if and only if the segments $[x_i, x_{i+1}]$ make a billiard trajectory. Since the function $\sum_{i=1}^{k} |x_i - x_{i+1}|$ is bounded from above, it achieves a maximum on U. It is easy to prove that the maximum is achieved for some $(x_1, \ldots, x_k)$, with all the x_i on the boundary ∂U. Moreover, set

$$d_k(U) = \sup_{(x_1, \ldots, x_k) \in U^k} \sum_{i=1}^{k} |x_i - x_{i+1}|$$

Then d_k is a monotone riemannian invariant. Moreover, we may extend the definition of d_k, from the set of convex sets to the set of all subsets of $\mathbb{R}^n$, by defining $d_k(U) = d_k(\hat{U})$ where $\hat{U}$ is the convex hull of U. Since $U \subset V$ implies $\hat{U} \subset \hat{V}$ we see that d_k is still a monotone invariant. Now, while it is clear that the maximum of the function $\sum_{i=1}^{k} |x_i - x_{i+1}|$ on U^k is monotone, it is clear that the length of other billiard trajectories, with no variational characterization, will not yield monotone invariants (however, they are obviously Riemannian invariants).

In the symplectic case, the situation is very similar, the role of billiard trajectories being played by periodic orbits for a Hamiltonian system having ∂U as an energy level.

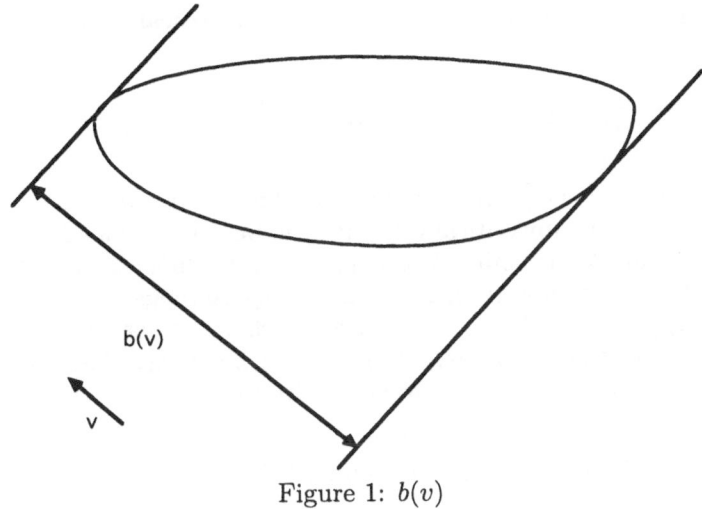

Figure 1: $b(v)$

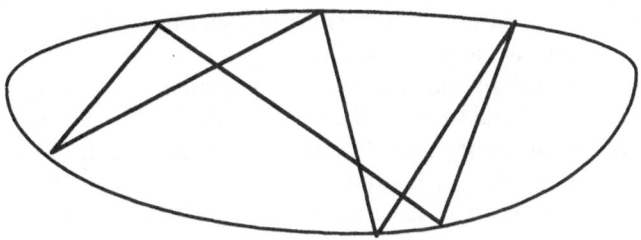

Figure 2: Billiard trajectory

To see further why periodic orbits or fixed points are related to rigidity properties, we may consider the Conley-Zehnder Theorem. This tells us that at any time one map of a Hamiltonian flow on the T^{2n} torus has at least $2n + 1$ fixed points. On the other hand, for $n > 1$, there are volume preserving maps of the torus with no fixed points. Because a C^0 limit of maps with a fixed point necessarily has one, such a volume preserving map cannot be approximated by the time one maps of Hamiltonian flows. This of course does not show why only symplectic maps may be approximated, but shows that some maps cannot, and how this phenomenon is related to the existence of periodic orbits.

Using a method similar to the construction of measures through Radon measures (i.e. considering the measure as an element in the dual of $C_0^0(U)$) we shall associate a real number to any compact supported function on $\mathbb{R}^{2n}$. This is done as follows. Indeed, let H be a time independent Hamiltonian; then the periodic orbits are obtained as critical points of the action functional:

$$A_H(z) = \int_0^1 [-pdq + H]dt \ , \ \ z \in C^1(S^1, \mathbb{R}^{2n}).$$

Now the number $c(H)$ will be a critical value of this functional. However, we may not choose any critical value, it must be one "obtained by minmax". Let us explain the meaning of this in the finite dimensional setting. Let $\mathcal{F}$ be a family of subsets in X, invariant by isotopies, and set $c_{\mathcal{F}}(f) = \inf_{F \in \mathcal{F}} \sup_{x \in F} f(x)$. The author's favorite choice is obtained by taking $\mathcal{F}$ to be the set of chains representing a given homology class. The crucial properties of $c_{\mathcal{F}}(f)$, both easy to check, are that:

   (i) (monotonicity) $f \le g$ then $c_{\mathcal{F}}(f) \le c_{\mathcal{F}}(g)$;
   (ii) (continuity) $|f - g| \le \epsilon$ then $|c_{\mathcal{F}}(f) - c_{\mathcal{F}}(g)| \le \epsilon$.

Applying this, suitably adapted for the infinite dimensional setting, to the action functional, we get a real number $c(H)$, which will be the action of some one-periodic orbit of the Hamiltonian vector field of H. We have to show that such a number is invariant by symplectic transformation, that is $c(H \circ \psi) = c(H)$ for any symplectic map ϕ. Now on $\mathbb{R}^{2n}$ we may always asume that ψ is the time one map of a Hamiltonian isotopy ψ_t. We remark also that the Hamiltonian flow of $H \circ \psi_t$ is $\psi_t^{-1} \circ \phi_s \circ \psi_t$; thus it has the same periodic orbits as ϕ_s and they have the same action. Because the set of critical values of $A_{H \circ \phi_t}$, for fixed t, has measure zero (Sard's theorem) and does not depend on t, the number $c(H \circ \psi_t)$ is a continuous function of t, with values in a set of measure zero. It is then necessarily constant so that $c(H) = c(H \circ \psi)$. Now for U a subset in $\mathbb{R}^{2n}$, we set

$$c(U) = \sup_{\text{supp}(H) \subset U} c(H).$$

It is clear once again that $c(U)$ is symplectically invariant (this immediately follows from $c(H \circ \psi) = c(H)$) and monotone. Once more the difficulty lies in proving that the invariant is non trivial, and in our special case that $c(B^{2n}(r)) = c(B^2(r)) \times \mathbb{R}^{2n-2}) = \pi r^2$. Now the πr^2 comes from the fact that this is the area of a closed characteristic on the boundary of $B^{2n}(r)$. Let us explain this point. Note that since H is time independent, and $c(H)$ is monotone, given a sequence of Hamiltonians H_n such that $c(H_n)$ converges to $c(B^{2n}(r))$, we may find a sequence K_n such that $K_n(q,p) = K_n(|q|^2 + |p|^2)$ is radially symmetric, the support of K_n is also in $B^{2n}(r)$, $H_n \le K_n$. Thus since $c(H_n) \le c(K_n)$ the limit of $c(K_n)$ is also $c(B^{2n}(r))$. Now the periodic orbits of K_n must be great circles, or constants. For k_n bell-shaped, the constant will give negative critical values of A_{K_n}. Since

$c(K_n)$ is a positive critical value of A_{K_n}, and since the other periodic orbits have action $k \cdot \pi r^2. - \epsilon_n \geq \pi r^2 - \epsilon_n$, we see that the limit, as n goes to infinty, is at least πr^2. This proves the inequality $c(B^{2n}(r)) \geq \pi r^2$. The reverse inequality requires a more detailed study of the variational problem.

Finally, still another question should be adressed. Until now we always assumed that the potential was known with absolute precision, an obviously unrealistic hypothesis. Now, assuming the Hamiltonian to be only approximately known, in the C^0 topology, is it possible to conclude something about the flows? Obviously, since the flow depends on the first derivative of H, it is not reasonable to expect too much. However, consider a sequence H_n of Hamiltonians, having C^0 limit H (that we assume to be C^1), and denote by ψ_n and ψ their associated flows. Then, if the ψ_n converge C^0 to anything at all, the limit must be ψ. We shall not prove this result; it is based on the methods explained in the following section.

3. Generating Functions

As should now be clear, one of the goals of symplectic topology is to find properties of symplectic objects of topological character. Besides the symplectic maps, other objects are of interest—in particular, Lagrange submanifolds. These are submanifolds in a symplectic manifold such that the symplectic form vanishes on them, and with dimension half the dimension of the ambient manifold. They appear in many instances. For example, they may be thought of as generalizations of symplectic maps. Indeed, if ψ is a symplectic diffeomorphism of (M, ω), its graph, that is $\{(z, \psi(z)) \mid z \in M\}$ is a Lagrange sumbanifold in $(M \times M, \omega \times -\omega)$.

Also if H is an integrable system, the space is decomposed as a union of invariant Lagrange submanifolds (tori) and the K.A.M. theory tells us that perturbations still have many invariant Lagrange submanifolds.

As a last example, solutions of Hamilton-Jacobi equations, $H(q, Du) = 0$ $u : \mathbb{R}^n \to \mathbb{R}$ may be associated with certain Lagrange submanifolds in $\mathbb{R}^{2n}$.

In this section we shall give a description of a subclass of these manifolds. For our purposes, it will also be easier to assume that the Lagrange submanifold L lives in T^*N where N is a compact manifold. The simplest such manifold is then obtained as the graph of the differential of a function $f : N \to \mathbb{R}$: $L_f = \{(x, df(x)) \mid x \in N\}$. Moreover, if, for instance, $N = T^n$, then any Lagrange submanifold in T^*T^n that is a graph, is of the type $p + L_f$ where p is a constant vector. In other words, up to a translation in p, L coincides with L_f. Indeed $p(q)dq$ will be a closed form, and such forms differ from an exact form by a constant.

Now let ψ_t be a hamiltonian isotopy, and Γ_t be its graph in $(T^*T^n \times$

$T^*T^n, dp \wedge dq - dP \wedge dQ)$. There is a symplectic covering of this manifold by T^*T^{2n} given by

$$(x, y, \xi, \eta) \rightarrow (x + \xi/2, y + \eta/2, x - \xi, y - \eta).$$

Since we cannot hope to write L as the graph of df, we shall consider a more general description of L. Indeed let $S : N \times \mathbb{R}^k \rightarrow \mathbb{R}$ be a C^2 function, and let L_S be the Lagrange submanifold defined by

$$L_S = \{(x, \frac{\partial S}{\partial x}(x, v)) \mid \frac{\partial S}{\partial \xi}(x, v) = 0\}.$$

That this is an exact Lagrange submanifold is easy to see since $pdq = dS$ on L. It is useful to consider intersection points of L with the zero section. These are in one-to-one correspondence with critical points of S. While for compact N such points always exist when $L = L_f$, they need not exist for L_S, since a function on $N \times \mathbb{R}^k$ does not need in general to have a critical point. However, it is well known that some assumptions at infinity will guarantee the existence of such critical points. We shall mainly be concerned with prescribing that S behaves like a nondegenerate quadratic form at infinity.

Given a Lagrange submanifold in T^*N two natural questions arise.

Is it possible to find a generating function of L that is quadratic at infinity?

If the answer to the above question is yes, is it possible to describe all generating functions quadratic at infinity for L?

The first question does not have a complete answer. However a sufficient condition for the existence of such a generating function is that L may be written as $\phi_1(0_N)$ where ϕ_t is a Hamiltonian isotopy. A heuristic explanation of this may be easily given. Indeed, the simplest generating function (that is not properly speaking one) is the action functional. To see this, one should consider the space of all paths in T^*N, $\mathcal{P} = \{(q, p) : [0, 1] \rightarrow T^*N \mid p(0) = 0\}$ and $\pi : \mathcal{P} \rightarrow N$ given by $\pi(q, p) = q(1)$. We denote by ξ the variable describing $\pi^{-1}(q)$.

Let $A(q, p) = \int_0^1 [p\dot{q} - H(t, q, p)]dt$; then

$$DA_H(q, p)(\delta q, \delta p) = \int_0^1 \left[\left(\dot{q} - \frac{\partial H}{\partial p} \right) \delta p - \left(\dot{p} - \frac{\partial H}{\partial q} \right) \delta q \right] dt + p(1)\delta q(1).$$

Thus $\frac{\partial A_H}{\partial \xi} = 0$ if and only if (q, p) satisfies Hamilton's equations, in other words, if and only if $(q(t), p(t)) = \phi_t(q(0), 0)$, where ϕ_t is the Hamiltonian flow. Unfortunately, A_H is neither defined on a finite dimensional manifold, nor is it quadratic at infinity. In practice however, it is possible by considering finite dimensional reductions of A_H to obtain a genuine generating function, quadratic at infinity, for $\phi_1(0_N)$.

Note also that generating functions are relevant to the problem of Lagrange intersection. Indeed, there is a one-to-one correspondence between critical points of a generating function, and intersection points of L with the zero section, since if $\frac{\partial S}{\partial \xi}(x, v) = 0$ and $\frac{\partial S}{\partial x}(x, v) = 0$, the point $(x, 0)$ is in $L \cap 0_N$.

The second question is slightly more complicated. It is first important to notice that, given a generating function quadratic at infinity, it is easy to construct new ones using the following procedure:

$\tilde{S}(x, v, w) = S(x, v) + Q(w)$ where Q is a non degenerate quadratic form;

$\tilde{S}(x, v) = S(x, z(x, v))$ where the map $(x, v) \rightarrow (x, z(x, v))$ is a fiber preserving diffeomorphism such that $S(x, z(x, v))$ is still quadratic at infinity (there is no need to require that $(x, v) \rightarrow (x, z(x, v))$ is linear at infinity).

Then under the assumption that $L = \phi_1(0_N)$ we have that, starting from two generating functions quadratic at infinity, we may choose a sequence of operations (i) and (ii) in order to end up with the same function. Two such generating functions are said to be stably equivalent*.

This has many pleasant consequences. In particular, given a generating function, we may look for its critical levels. We shall only consider minmax critical values. Clearly provided, the set $\mathcal{F}$ is invariant by diffeomorphisms, we have that $c_{\mathcal{F}}(S)$ is unchanged through the second operation. The first operation is slightly more tricky, since we must associate to $\mathcal{F}$ a family in $N \times \mathbb{R}^k$ a family $\tilde{\mathcal{F}}$ in $N \times \mathbb{R}^k \times \mathbb{R}^q$. Now if X is a subset in $N \times \mathbb{R}^k$, we may consider $X \times \mathbb{R}^{q-}$ ($q-$ is the index of the quadratic form Q). Then, provided we also allow compact supported deformations of such sets, this describes the family $\tilde{\mathcal{F}}$. However it is unfortunately not true that $c_{\mathcal{F}}(S) = c_{\tilde{\mathcal{F}}}(\tilde{S})$, but only $c_{\mathcal{F}}(S) \geq c_{\tilde{\mathcal{F}}}(\tilde{S})$. It is not even true that as q increases, the sequence must stabilize. In case $X = S^p$ is a sphere, $\mathcal{F}$ is an element u in $\pi_p(N)$; this is connected with the fact that the image of u in the "stable limit" $\lim_{r \to \infty} \pi_{r+p}(\Sigma^r N)$ may be zero.

However if we take $\mathcal{F}$ to be the set of X representing some nonzero cohomology class, the equality $c_{\mathcal{F}}(S) = c_{\tilde{\mathcal{F}}}(\tilde{S})$ holds because of the Künneth formula. In fact for each cohomology class α in $H^*(N)$, there is a critical

---

* This is not totally correct, because we deliberately overlooked the fact that there is a third operation on generating functions, namely addition of a constant that does not modify L. In general however there are normalizations which remove this undeterminacy.

level $c(\alpha, S)$ such that $c(\alpha, S) = c(\alpha, \tilde{S})$. As a result, we may denote this level by $c(\alpha, L)$.

In particular, as N is a compact manifold, there are two distinguished classes in $H^*(N)$, the zero dimensional one, which we denote by 1, and the top dimensional one, denoted by μ_N. In de Rham cohomology, the first one is represented by the constant function with value one, the second by a volume form with integral one. Then we may consider $c(1, L)$ and $c(\mu_N, L)$. These yield a "norm" on the set of Lagrange sumanifolds Hamiltonianly isotopic to the zero section by taking $N(L) = c(\mu, L) - c(1, L)$. In other words, we have $N(L) = 0$ if and only if L coincides with the zero section. This norm is quite useful since many problems in symplectic topology may be expressed using Lagrange submanifolds. The simplest and most frequent case occurs in $\mathbb{R}^{2n}$ with L the graph of a compact supported Hamiltonian diffeomorphism: we then set $L_t = \{(z, \phi_t(z)) \mid z \in \mathbb{R}^{2n}\}$. Using the symplectomorphism $\mathbb{R}^{2n} \times \mathbb{R}^{2n} \to T^*\mathbb{R}^{2n}$ given by $(q, p, Q, P) \to (\frac{q+p}{2}, \frac{Q+P}{2}, p - q, Q - P)$, we send the diagonal into the zero section. Thus the image of L_0 is the zero section, and the image of L_t, denoted by Γ_t, is isotopic to the zero section. Therefore it has a generating function quadratic at infinity. Moreover, since Γ_t coincides with the zero section outside a compact set, we may compactify the situation to get $\bar{\Gamma}_t$ a compact Lagrange submanifold of T^*S^{2n}. We will thus be able to associate to ϕ two real numbers, $c_+(\phi) = c(\mu, \bar{\Gamma})$ and $c_-(\phi) = c(1, \bar{\Gamma})$. It is not hard to prove many useful properties of these numbers. In particular they are invariant by conjugation, that is $c_+(\psi\phi\psi^{-1}) = c_+(\phi)$, and the same holds for c_-. Thus, as in the previous sections, we will get symplectic invariants. Once these are constructed, most properties follow easily using the nontrivial inequality

$$c_+(\phi\psi) \leq c_+(\phi) + c_+(\psi)$$

and the fact that $c_+(\phi)$ is a critical value of the generating function, the critical points of which are in one-to-one correspondence with fixed points of ϕ, in view of the fact that the zero section is the image of the diagonal, and the fixed points are but the points of intersection of the graph and the diagonal. Thus we see that there are two fixed points x_+ and x_- corresponding to $c_+(\phi)$ and $c_-(\phi)$.

The main idea of this approach is that, no matter which variational principle we associate to a fixed point or an intersection problem, the invariants obtained from it, in other terms the critical levels obtained by minmax, must be independent of the choice of the variational principle. As an example, let $T(q, \dot{q})$ be a Lagrangian, convex in $\dot{q}$. We consider the energy $E(q) = \int_0^1 T(q, \dot{q})dt$. It is defined on the set $\mathcal{C} = \{q \in C^1([0, 1], \mathbb{T}^n) \mid \dot{q}(0) = 0\}$. The projection map $\pi : \mathcal{C} \to \mathbb{T}^n$ is given by $q \to q(1)$. The

Lagrange submanifold thus generated is given by

$$L = \left\{ \left(q(1), \frac{\partial T}{\partial \dot{q}}(q(1), \dot{q}(1))\right) \mid \dot{q}(0) = 0 \ \frac{d}{dt}\frac{\partial T}{\partial \dot{q}} - \frac{\partial T}{\partial q}(q(t), \dot{q}(t)) = 0 \right\}.$$

Then if ϕ_t is the Hamiltonian flow associated to the Legendre transform of T; $H(q, p) = \sup_{\dot{q} \in \mathbb{R}^n} [(p, \dot{q}) - T(q, \dot{q})]$. Now we have:

$$c(1, L) = \inf_C E$$

$$c(\mu, L) = \sup_x \inf_{q \in C \mid q(1) = x} E(q).$$

4. Historical Remarks

The oldest result that has something to do with symplectic topology may be considered to be the Poincaré–Birkhoff theorem. This asserts that a diffeomorphism of the annulus that rotates the inner and outer boundary circles in opposite directions has at least two fixed points. Such maps occur naturally as Poincaré maps for a Hamiltonian flow that has a periodic orbit bounding a disk transverse to it. The first return map is an area preserving map from the disk into itself. It has a fixed point by Brouwer's theorem (no need to be area preserving) and if the infinitesimal rotation near this fixed point differs from the inverse of the period of the periodic orbit, by blowing up the fixed point, we get an area preserving map of the annulus that rotates the inner and outer circle by different amounts, and by iteration we may reduce our problem to the previous one so that we get two more fixed points of the Poincaré map, thus two periodic orbits. And by iteration, we get infinitely many periodic orbits.

This statement, conjectured by Poincaré, proved by Birkhoff was generalized in several directions. First to a topological result: one need not have an area preserving diffeomorphism; it is enough to assume that the image of any open set U is not contained in V with $V \subset U$ and $U - V$ has nonempty interior. It is also easy to see that the above statement is implied by the following result where μ is the Lebesgue measure on the plane.

Let f be an area preserving map of the torus T^2 so that it lifts to $\tilde{f}$ so that $\int_{[0,1]^2} \tilde{f}(x) - x d\mu = 0$. Then f has at least three fixed points.

The "rotationless" condition $\int_{[0,1]^2} \tilde{f}(x) - x d\mu = 0$ is necessary in order to exclude pure rotations (that have no fixed points!). The result on the annulus follows by gluing two annuli along their boundaries to form a torus, and taking for f the same map on both sides. Then among the three fixed points, two of them must be in the same annulus, hence the Poincaré–Birkhoff theorem.

On the other hand, in the Russian translation of Poincaré's works, Arnold conjectured that this statement has a symplectic generalization in higher dimensions. Note that the obvious statement for volume preserving maps is definitely false in higher dimensions. Many years later, the Arnold conjecture for the $2n$-torus was (with slightly stronger assumptions) proved by Conley and Zehnder. It marked the starting point of genuine symplectic topology. Their result may be stated as follows: *Let f_t be a Hamiltonian flow on T^{2n}, endowed with the standard symplectic structure $\sum_{i=1}^{n} dp_i \wedge dq_i$. Then f_1 has at least $n + 1$ fixed points.*

Here the Hamiltonian flow is the standard one, that is given by the equations

$$\dot{q} = \frac{\partial H}{\partial p}(t, q, p) \quad \dot{p} = -\frac{\partial H}{\partial q}(t, q, p).$$

It is not hard to see that in two dimensions, it implies the "rotationless" condition.

Quite independently and a few months later, Gromov explored the fruitful approach of pseudo-holomorphic curves. Yet another point of view was explored by Y. Eliashberg, more closely related to contact geometry, a "projective version" of symplectic geometry. This had also been studied by D. Bennequin, who discovered the first exotic contact structures (in dimension three).

But already from the Conley-Zehnder theorem, we may, as M. Herman noticed, deduce some kind of rigidity statement. Indeed, as we said, there are volume preserving maps of the torus that have no fixed points. Such a map can never be approximated by the time one map of a Hamiltonian flow because the C^0 limit of maps with fixed points must have a fixed point. Because the condition of being Hamiltonian is more restrictive than being merely symplectic, this statement is not exactly what we wanted, but is quite close to it. It is thus clear (a posteriori!) how fixed points rigidity and variational methods were related to each other.

Later, merging Gromov and Conley-Zehnder's approaches, came Floer homology. Its basic principle is that in the infinite dimensional loop space, the action functional

$$A(z) = \int_{S^1} [(J\dot{z}, z) - H(t, z)] dt$$

has a weak L^2 gradient which trajectories are given by the equation

$$\frac{\partial z_s}{\partial s} = J\dot{z}_s - \nabla H(t, z_s).$$

In other words, the map $u(s, t) = z_s(t)$ satisfies the equation

$$\bar{\partial} u = -\nabla H(u).$$

This is a perturbation of the Cauchy-Riemann equation, but Gromov's analysis still holds. Floer used this approach in his fundamental contribution towards the solution of the Arnold conjecture.

Appendix: Rigidity for Finite Dimensional Lie Groups

Let H be a finite dimensional Lie group, and d_1, d_0 be two distances such that

(i) d_1, d_0 are invariant: $d_i(hx, hy) = d_i(x, y)$
(ii) $d_0 \le d_1$
(iii) $d_1(\exp(\epsilon x, e) \le C\epsilon||x||$ for $|| \cdot ||$ a norm on T_eH.

Then the two metrics are equivalent on bounded sets.

Given d as above, consider the sequence $u_n(x) = nd(\exp(\frac{x}{n}), e)$. Let $\nu(x)$ be the upper limit of this sequence. We claim that in fact $\nu(x)$ is the limit of the sequence. Indeed, if $n = kp + r$ we have $\exp(\frac{x}{(kp+r)}) = \exp(\frac{x}{kp}) \exp(\frac{xr}{kp \cdot (kp+r)}) \exp(\frac{r}{kp(kp+r)} a(x, kp, r))$ This follows from the Campbell-Hausdorff formula:

$$\exp(\epsilon a) \exp(\epsilon b) = \exp\left(\epsilon(a + b) + \frac{\epsilon^2}{2}[a, b] + \epsilon^3 \ldots\right).$$

Thus, assuming that $d(\exp(\frac{x}{N}), e) \le C/N$ for x in a bounded subset of T_eH, we get

$$(kp + r) \exp\left(\frac{x}{(kp + r)}\right) \ge kpd\left(\exp\left(\frac{x}{kp}\right), e\right) - rd\left(\exp\left(\frac{x}{kp}\right), e\right)$$
$$- (kp + r)d\left(\exp\left(\frac{r}{kp(kp + r)} a(x, kp, r)\right), e\right).$$

All the terms have limit zero with k, but the first one. Since the triangle inequality implies $pd(\exp(\frac{x}{p}), e) \le kpd(\exp(\frac{x}{kp}), e)$ the limit of the right hand side is greater than $\nu(x)$, hence equal to it. Now it is easy to check that $\nu(x + y) \le \nu(x) + \nu(y)$. It is now clear that ν is a norm on T_eH. Since H is finite dimensional all norms must be equivalent. Now we overlooked the fact that ν could be infinity. But this is prevented by (ii), or that it could be zero, but using the triangle inequality, we have: $d(\exp(\epsilon x), e) \le nd((\exp(\epsilon \frac{x}{n}), e)$. Thus $\nu(x) = 0$ would imply $d(\exp(\epsilon x), e) = 0$ for ϵ small enough. This is impossible.

The assumption $d(\exp(\frac{x}{N}), e) \le C/N$ will be satisfied by d_0 provided it is satisfied by d_1 because of (ii). Thus if ν_1 and ν_0 are the norms on T_eH obtained as above, since they are equivalent (T_eH is finite dimensional), and the convergence of $u_n(x)$ to $\nu(x)$ is uniform on compact sets (apply Dini's theorem to the decreasing sequence $u_{n^k}(x)$ with continuous limit $\nu(x)$) we have that $nd_1(exp\frac{x}{n}, e) \leq C \cdot nd_0(exp\frac{x}{n}, e)$ for x in a compact set. But this implies that the metrics d_0 and d_1 are equivalent on bounded subsets, hence they are both complete if and only if one of them is complete.

As a result if $\|\cdot\|_1$ and $\|\cdot\|_0$ are the C^1 and C^0 metrics on the group G of diffeomorphisms. Then if d_0 and d_1 are the induced metrics on the finite dimensional subgroup of isometries of the manifold, we have that $d_0 \leq d_1$. Thus the metrics are equivalent. Since H is closed for $\|\cdot\|_1$ in G it is complete, hence it is complete for d_0. Now, if h_n is a sequence in H having a limit for d_0 in G, it will be a Cauchy sequence for $\|\cdot\|_0$ hence for d_0. The completeness of H for d_0 implies that h_n has a limit in H for d_0, hence for d_1 or else for $\|\cdot\|_1$. This concludes the proof of the rigidity in the case of finite dimensional symmetry groups.

We thus proved:

Theorem: *Let H be finite dimensional subgroup of* Diff(M). *If the group is closed in* Diff(M) *for the C^k topology for some k, it is closed for the C^0 topology.*

References

[B] Bennequin, D., Problèmes elliptiques surfaces de Riemann et structures symplectiques, Séminaire Bourbaki 1985-1986, exposé 657, *Astérisque*, vol. 138.

[Bi] Birkhoff, G., *Dynamical Systems*, AMS Publications 1924.

[C] Chaperon, M., Quelques questions de géométrie symplectique, Séminaire Bourbaki 1982-1983, *Astérisque*, vol. 105–106, pp. 231–249.

[Co-Z] Conley, C. and Zehnder, E., Morse type index theory for flows and periodic solutions for Hamiltonian equations, *Comm. Pure Appl. Math.*, vol. 37 (1984), pp. 207–253.

[E–H1] Ekeland, I. and Hofer, H., Symplectic topology and Hamiltonian dynamics, *Math. Zeitschrift*, vol. 200 (1989), pp. 355–378.

[E-H2] Ekeland, I., Hofer, H., Symplectic topology and Hamiltonian dynamics II, *Math. Zeitschrift*, vol. 203 (1990), pp. 355–378.

[G1] Gromov, M., Pseudo holomorphic curves on almost complex manifolds, *Inventiones Math.*, vol. 82 (1985), pp. 307–347.

[G2] Gromov, M., Soft and Hard Symplectic Geometry, in *Proceedings of the International Congress of Mathematicians 1986*, vol.1 (1987), pp. 81–98.

[V1] Viterbo, C., Capacités symplectiques et applications, Séminaire Bourbaki, juin 89, exposé 714, *Astérisque*, vol. 177–178 (1990).

[V2] Viterbo, C., Symplectic topology as the geometry of generating functions, *Mathematische Annalen*, vol. 692 (1992), pp. 685–710.

Progress in Nonlinear Differential Equations and Their Applications

Editor
Haim Brezis
Département de Mathématiques
Université P. et M. Curie
4, Place Jussieu
75252 Paris Cedex 05
France
and
Department of Mathematics
Rutgers University
New Brunswick, NJ 08903
U.S.A.

Progress in Nonlinear Differential Equations and Their Applications is a book series that lies at the interface of pure and applied mathematics. Many differential equations are motivated by problems arising in such diversified fields as Mechanics, Physics, Differential Geometry, Engineering, Control Theory, Biology, and Economics. This series is open to both the theoretical and applied aspects, hopefully stimulating a fruitful interaction between the two sides. It will publish monographs, polished notes arising from lectures and seminars, graduate level texts, and proceedings of focused and refereed conferences.

We encourage preparation of manuscripts in some form of TeX for delivery in camera-ready copy, which leads to rapid publication, or in electronic form for interfacing with laser printers or typesetters.

Proposals should be sent directly to the editor or to: Birkhäuser Boston, 675 Massachusetts Avenue, Cambridge, MA 02139